ULL
a302
D1180864

ULL
WITHDRAWN

ECOLOGY OF TROPICAL AND SUBTROPICAL VEGETATION

ECOLOGY OF TROPICAL AND SUBTROPICAL VEGETATION

HEINRICH WALTER
Professor in the University of Stuttgart–Hohenheim

Translated by
D. MUELLER-DOMBOIS
Associate Professor of Botany
University of Hawaii, Honolulu

Edited by
J. H. BURNETT
Sibthorpian Professor of Rural Economy
University of Oxford

VAN NOSTRAND REINHOLD COMPANY
NEW YORK CINCINNATI TORONTO LONDON MELBOURNE

OLIVER & BOYD
Tweeddale Court, Edinburgh, EH1 1YL
A Division of Longman Group Limited

English edition first published 1971

Van Nostrand Reinhold Company Regional Offices:
Cincinnati New York Chicago Millbrae Dallas

Van Nostrand Reinhold Company International Offices:
London Toronto Melbourne

Library of Congress Catalog Number 70-177407

Published by Van Nostrand Reinhold Company
450 West 33rd Street, New York, N.Y. 10001

Printed in Great Britain by
T. and A. Constable Ltd, Edinburgh

This book is dedicated in memory
of the great plant geographer
A. F. W. SCHIMPER
and to our friends the
world over with a warm
appreciation for all their help

Contents

Preface to the First Edition xi

Preface to the Second Edition xv

Editor's Preface to the English Edition xvii

I Introduction
1 General considerations concerning vegetation and plant communities 1
2 Succession, climax and vegetation zones 10
3 Competition 14
4 The production of dry matter and the leaf-area index 21
5 Growth capacity and competitive ability 24
6 Life forms and competitive ability 28
7 Energy cycles and production of dry matter in ecosystems 33
8 The elucidation and significance of the ecological water relationships of plants 34
9 Zonal climatic types 44
10 The representation of climatic types by climatic diagrams 51
11 Soil zones of the world 57
12 Principal features of a three-dimensional classification of the earth's vegetation 62
References 68

II The Continuously Wet Tropical Rain Forest
1 General features 72
2 The climate of the region of the humid tropics 75
3 Microclimate in the tropical rain forest 82
4 Soil relations in the tropical rain forest 89
5 Structure of the tropical rain forest 95
6 Ecology of the trees in the tropical rain forest 102
7 Other ecological types of plant of the rain forest 117
(*a*) Shrub layer 117
(*b*) Herb layer 118
(*c*) Lianas 123
(*d*) Hemi-epiphytes 126
(*e*) Epiphytes 128
(*f*) Saprophytes and parasites 138
8 Conclusion 139
References 139

III Other Vegetation Types of the Humid Tropics
1 Swamp and aquatic vegetation 144
2 Mangroves 150
3 Strand vegetation 166
4 Vegetation of dry habitats 169
5 The landscape of the cultivated tropics 171
References 174

IV Cooler Rain Forests of Higher Altitudes in Tropical Mountains
1 General features of the altitudinal vegetation belts in the tropics 177
2 Evergreen subtropical rain forests 179
3 Tropical montane rain forests of Java 182
4 Tropical alpine vegetation of the Andes 183
5 Altitudinal vegetation belts on Kilimanjaro 194
References 205

V Tropical Semi-evergreen and Deciduous Forests
1 The ecological significance of annual leaf-fall 207
2 Semi-evergreen and rain-green tropical forests 209
3 Tropical dry-woodland and thorn-scrub thickets 218
4 Tropical parkland areas 227
5 Tropical grassland 230
References 236

VI Natural Savannahs as a Transition to the Arid Zone
1 The savannah concept 238
2 Grass and trees as competitors 241
3 The competition-equilibrium in the savannah 248
4 Invasive scrub—a threat to farm management in the savannah zone 252
5 Edaphically determined vegetation in the savannah-grassland zone 256
6 Vegetation zones in relation to decreasing amounts of rainfall in the subtropical region 259
References 264

VII General Features of the Vegetation of Subtropical Arid Regions
1 The concept of arid regions 266
2 Sources of water for plants in arid areas: decreasing vegetation density and unequal water distribution in the soil 270
3 The importance of soil texture to the water relations of arid regions 278
4 The principle of relative constancy of habitat with changing ecological niche (or biotope) 281
5 Water relations and the main ecological plant types of the non-saline soils of arid regions 285
6 Halophytes and the salt factor 293
References 296

VIII The Sonoran Desert
1 Climate and habitat conditions 299
2 Classification of the vegetation 304
3 Ecological plant types of the Sonoran desert 310
(*a*) Ephemerals 310
(*b*) Poikilohydrous pteridophytes 314
(*c*) General features of the perennial species 314
(*d*) Succulents 316
(*e*) Characteristic shrubs 327
(*f*) Dwarf-shrubs with soft, hairy leaves (malacophyllous plants) 333
4 Conclusion 335
References 336

IX The Namib Fog-desert
1 Precipitation 338
2 The ecological significance of fog 342

3 The vegetation of the Namib and its ecology 345
(*a*) Plant communities of level plains 346
(*b*) Plants of rocks 353
(*c*) Vegetation of erosion channels and dry-valleys 361
(*d*) The ecology of *Welwitschia mirabilis* 369
References 373

X THE CHILEAN-PERUVIAN COASTAL DESERT AND ITS FOG-OASES
1 Fog as a source of water 375
2 The Loma vegetation 381
References 385

XI THE KARROO
1 General features 387
2 Vegetation of the Upper Karroo 391
3 Ecological investigations 393
(*a*) The behaviour of succulents and halophytes 393
(*b*) Experiments with poikilohydrous ferns 398
(*c*) Water relations of Karroo dwarf-shrubs 399
(*d*) The response of sclerophyllous woody plants to lack of water 399
References 401

XII THE ARID REGIONS OF CENTRAL AUSTRALIA
1 A comparison of Australia with Southern Africa 402
2 Rainfall patterns in Central Australia 404
3 The vegetation 409
4 Ecological investigations 415
(*a*) Saltbush semi-desert 415
(*b*) Spinifex grassland (Porcupine grasses) and Mulga (*Acacia aneura* scrub) in Central Australia 422
(*c*) Moist habitats in Central Australia 427
References 431

XIII THE SAHARA
1 Introduction 433
2 Soil and vegetation relations 437
(*a*) The stone desert or hamada 437
(*b*) The pebble or gravel deserts (serir or reg) 438
(*c*) The sand desert (erg or areg) 439
(*d*) The dry-valleys (wadis or oueds) 441
(*e*) Pans, dayas, sebchas or schotts 442
(*f*) Oases 444
3 On the biology of Saharan plants 447
4 Ecological investigations in the Algerian Sahara and in Libya 449
5 Relationships of the vegetation in the Tibesti mountains 457
References 459

XIV THE EGYPTIAN-ARABIAN DESERT INCLUDING SINAI AND NEGEV
1 Habitats and plant communities 461
(*a*) Vegetation of the stone-desert 464
(*b*) Vegetation of the wadis 466
(*c*) Vegetation of the pebble desert 467

2 Ecological researches near Cairo 468
(*a*) Investigations near the Cairo-Suez road 469
(*b*) Investigations in the Wadi Hoff near Helwân 478
3 The vegetation along the Mediterranean coast of Egypt 483
4 Rainfall in Egypt in the past 491
5 Ecological situations on the Sinai peninsula 493
6 Ecological investigations in the Negev desert 497
7 Transitional areas to the Mediterranean sclerophyllous zone 508
References 516

INDEX 521

The laboratory of the ecologist
is God's nature
And his area of work is the
entire world

Preface to the first edition

WHEN I was asked in the spring of 1957 to revise the fourth edition of Schimper's *Pflanzengeographie* I was not prepared to undertake this task.

Schimper's book in its original form is a classical work. Any attempt to introduce more recent viewpoints into the same old frame would have necessarily led to an imbalanced presentation. Moreover, the more general ecological questions which Schimper treated in his first part had already been dealt with in my *Phytologie* Vol. III, 1 (second edition 1960) entitled *Standortslehre*. For this reason I considered it better to start with an entirely new orientation and to change the title correspondingly. But, I have made much use of the very valuable illustrations of the third edition of Schimper-Faber.[1] Only good photographs convey a correct idea of vegetation. The outline of my whole series is planned as follows (all in German):

Vol. I. The tropical and subtropical zones.
Vol. II. The temperate and arctic zones.

A continuation is anticipated in the form of vegetation monographs covering continents individually or other large contiguous geographic areas, such as the following:

1. The Eurosibiric area including the Mediterranean region.
2. The North American region.
3. The South and East Asian region including Indonesia.
4. The African region south of the Sahara.
5. The South American region.
6. The Australian region including New Zealand.
7. The oceans.

It might appear too ambitious for one person to attempt such a treatment of the earth's vegetation. However, my purpose is not to produce detailed handbooks. Rather an attempt is made to summarise our current knowledge of the salient features of the earth's plant cover in a reasonably concise form. In this way I hope to point out the existing gaps in our knowledge.

Volumes I and II will present, primarily, general ecological phenomena by stressing those common to many vegetation zones. For this, functional relations will have to be emphasised, and only those regions can be covered for which relevant experimental results are available.

[1] The author lost all his photographs from the years 1927-1941 during the Second World War.

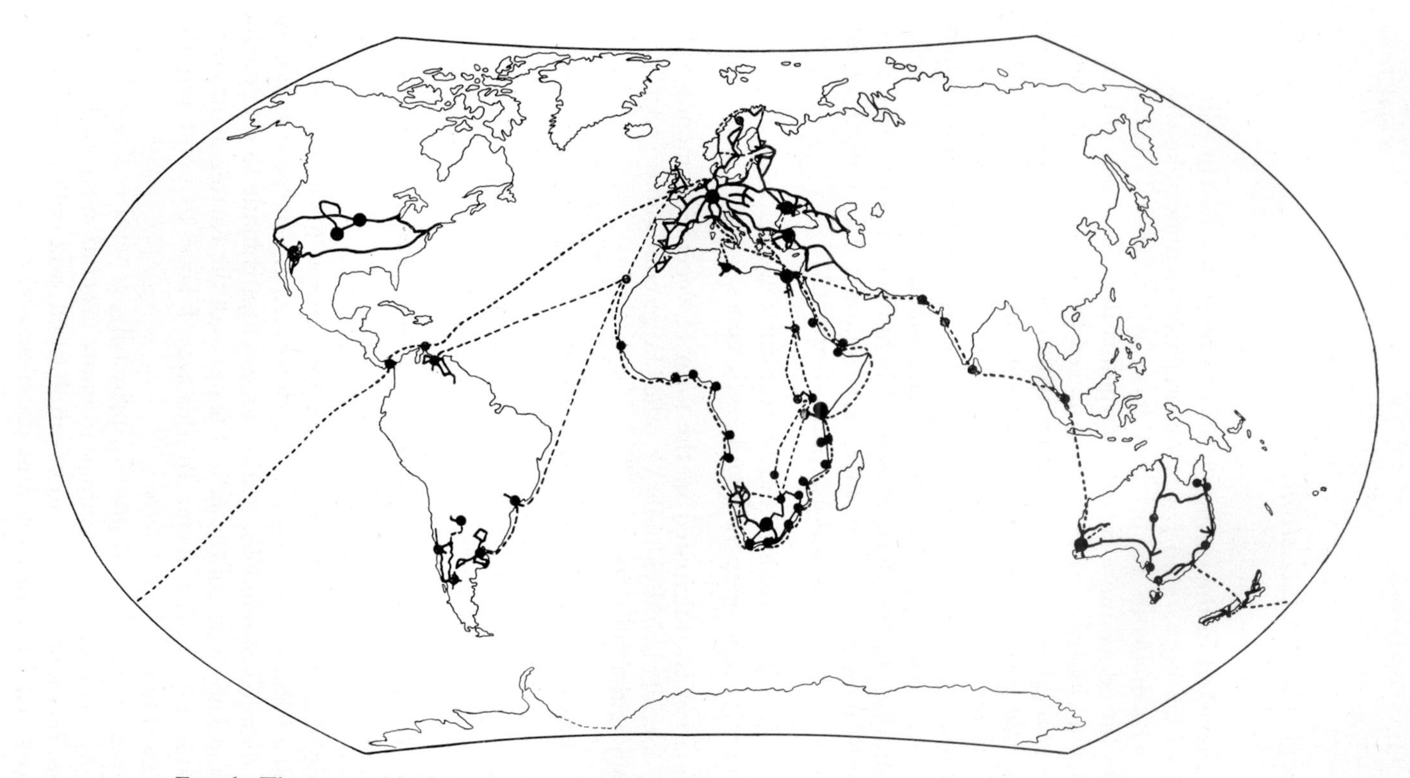

FIG. 1. The areas visited by the Author up to 1968. ———by land; · · · · · · ·by sea or air. Large dots—ecological investigations at botanical institutes—or experimental stations.

However, my last large expedition in 1958-59 to Australia and New Zealand has convinced me of the need also for analysing similarly the floristic relations. The flora, which can be understood only in a historical context, determines which elements in the vegetation are available for development of plant communities. In comparison with the Old World we will find in equivalent habitats in Australia or New Zealand not only floristically but also physiognomically quite different plant communities. For example, the deciduous tree life forms, the frost resistant needle trees and the desert succulents are entirely lacking.

These floristic differences are noticeable in each vegetation zone. Yet, there are many floristic similarities among the different vegetation zones of a given continent having the same floristic history. It is intended to illustrate these in the vegetation monographs. A compact review will be given for each major geographic area in 100-150 pages. Each will discuss floristic relations and vegetation patterns, and will be accompanied by many vegetation maps.

Whether I shall see the whole work to completion is not for me to decide. However, a beginning will be made. Each part will be complete within itself. For the monographs I hope for the support of other investigators who are familiar with particular regions and are prepared to write about them independently.

Although I was fortunate enough to see all the vegetation zones and floristic regions of the earth during my many expeditions (Fig. 1), the life-span of one person is too short to investigate all of them thoroughly. Today's modern transport has made the earth a small place. For this reason in particular the need becomes increasingly urgent to develop a comprehensive synthetic approach simultaneously with ever more detailed analytical studies.

Die Vegetation der Erde is the direct continuation of my *Einführung in die Phytologie* (Introduction to Phytology), Eugen Ulmer Verlag, Stuttgart. Therefore, I have had to refer repeatedly in this text to discussions given in the introductory *Phytologie*. The form of this continuation is an extrapolation of a world-wide scale.

Some of the greatest problems I faced were questions of nomenclature. Plant names have often been changed in the interim and it would be impossible to determine all synonyms and the most recent and correct names for the entire plant kingdom. It is therefore possible that a species is mentioned in the text under different names or cited by an older name. Unfortunately I had to accept this or I would have had to forgo the task of writing this book. Name corrections or other critical comments will always be gratefully accepted.

Finally I wish to express my deep-felt gratitude to everyone who has given me the opportunity during the last 30 years for research far beyond the borders of Europe. These are the universities and ministries which granted the necessary leaves of absence, the Rockefeller Foundation which financed the first expedition and the Deutsche Forschungsgemeinschaft and other organisations which approved travel funds. Government departments and scientific institutions in North America, Africa, Turkey, Persia and Egypt as well as in Australia and New Zealand have supported our work in most generous ways. The same is true for my many friends the world over, especially those in South-West Africa, whose hospitality I will never forget and whom it is not possible for us to mention here. This work, however, could not have been done without the professional assistance of my wife, Dr Erna Walter, who accompanied me on all the expeditions regardless of strain and discomfort. Thanks are due also to all my young co-workers who assisted me in the evaluation and analysis of my results.

The publisher has already supported this work through the publication of the *Klimadiagramm-Weltatlas*, which formed the basis for the zonation of vegetation; he has met all my wishes in the best way possible.

H. Walter

Stuttgart-Hohenheim, 1961.

Preface to the second edition

THE first edition went out of print in only one year. It was very favourably accepted. In spite of this I considered it necessary to incorporate a few changes, extensions and complementary passages. Therefore, most chapters have been revised, and the introductory section on competition has been enlarged with further detail.

Also, this second edition is not a description of vegetation in the normal sense, it gives instead a comprehensive account of ecological, or more precisely, eco-physiological results. Experimental ecology began in the first half of this century with investigations of the water economy of the plant. It focused particularly on arid-zone research. Therefore, many published results deal with arid environments. Also, ecological analyses and syntheses are easiest to carry out in extreme environments. This explains the greater elaboration of arid regions in the present volume. Moreover, the desert regions are ecologically so different that they could only be treated individually.

Tropical areas with more or less long rainy seasons in the summer have so far received the smallest share of attention in eco-physiological research. For this reason my discussion of these areas had to be rather brief. Many papers are written about the humid tropics, but only few give experimental results. Of particular importance here is the light factor and thus photosynthesis. But so far we still have no simple method for field measurements. Also, the questions of dry-matter production and competition have as yet received little attention in the tropics.

A few readers commented about the lack of a detailed bibliography. It was not my intention to cite more references than can be read and evaluated by an author. References that support our discussions are cited at the respective places. Further literature citations can be found in these references.

Work on Volume II has already begun and I hope that it will be completed.[1] Simultaneously with this second edition appears *Die Vegetation von Nordamerika und Mittelamerika*, which is the first of the previously announced vegetation monographs emphasising the more descriptive aspects. My appreciation goes to Professor R. Knapp, Giessen, who kindly accepted the writing of this monograph.[2]

To all my co-workers again my deep-felt thanks.

H. WALTER

Stuttgart-Hohenheim, 1963.

[1] Walter, H. 1968. *Die Vegetation der Erde in öko-physiologischer Betrachtung*. Band II, Die gemässigten und arktischen Zonen. Fischer, Jena and Stuttgart.

[2] The next monographs have been published: Hueck, K. 1966. *Die Wälder Südamerikas*; Knapp, R. 1971. *Die Vegetation Afrikas*. Fischer, Stuttgart.

Editor's preface to the English edition

THE plant sciences developed in the North Temperate zone and are still, in essence, orientated towards temperate vegetation and the problems, physiological, ecological and biological, of temperate plants. This is especially notable in respect of physiology and ecology although there is an increasing corpus of tropical descriptive ecology. However, in addition, the majority of the world's plant scientists live in the temperate zones, or have been trained in the traditions of temperate zone botany and, as yet, only a minority of these have either visited or worked in the tropics.

The demands of the world's population for food, especially in tropical and subtropical areas, and the apparent need both to exploit the natural resources and develop industrialisation in such regions puts them at severe risk as areas of intense biological interest. One hopes that such areas will not suffer a polluted biological fate as readily and as devastatingly as those of the temperate zones.

So it is particularly appropriate that as wide a range as possible of plant scientists, pure or applied, should be able to turn to source books at this time. For many years the successive needs of German-speaking biologists has been met in full measure by Professor Walter's *Die Vegetation der Erde*: this English-language edition should supply the need for English-speaking scientists. Moreover, the work is especially appropriate for the times, not only is it authoritative—Professor Walter has personally visited most of the regions himself (*vide* Fig. 1)—but it is written in an eco-physiological context which is of particular value to both pure and applied plant sciences. Incidentally, it also provides a useful account and demonstration, for English-speaking readers, of Professor Walter's views on water relations and their applications in plant ecology—an area in which there has hitherto been some misunderstanding. But the great virtue of the work is that it presents on a broad, embracing scale the vegetation of the tropics and sub-tropics in language understandable to, and adaptable by, ecologists, physiologists, agronomists and indeed, the intelligent non-scientist.

I shall be very surprised if its publication does not materially quicken the interests of plant scientists in tropical vegetation.

J. H. BURNETT

Note on the Translation

It is desirable to relate this work to the German editions. A magnificent basic translation was carried out by Professor Mueller-Dombois and my role has been to ensure that the English is as idiomatic and appropriate as possible. The 2nd German edition was the basic text used but Professor Walter has modified the order and inserted a number of short new passages. This book, therefore, is a little closer to the 3rd German edition which is about to be published. It differs from the older German editions in one other respect, namely the plates. Owing to the unfortunate loss of Professor Walter's originals in the Second World War and the non-availability of some of the original blocks the number of plates is reduced. Although unfortunate it does not affect the basic arguments of the book which are its essence.

I. *Introduction*

1. General considerations concerning vegetation and plant communities

THE object of research on vegetation is not the individual plant but the plant community. A plant community comprises several plant species that may vary greatly in numbers. The plant cover of a geographical area is composed of several plant communities which together form the vegetation of this area. This statement immediately poses the question of the criteria which constitute a plant community. There is hardly a topic in the literature on vegetation that has received so much discussion as this question. I have no intention, at this point, of reviewing the different philosophies of plant communities or of discussing theoretical questions but intend to concern myself with the actual plant cover of the earth's surface. I shall, therefore, restrict this discussion to my own viewpoint, which has determined the treatment of vegetation presented in this book. There are essentially two opposing views concerning the nature of plant communities in the literature (compare, for example, Ellenberg, 1956; Poore, 1962).

1. One group of authors holds the view that plant communities are units with sociological characteristics. Therefore, they use the terms 'plant sociology' or 'phytocoenology' to designate the science dealing with plant communities. Rather diverse viewpoints have been presented within this group. While one group considers the plant community to be something like an organism, others consider it to be a more complex unit composed of several smaller or layered communities, the synusiae.

Some authors emphasise discontinuity between plant communities while others emphasise continuity and transition. The main objectives of phytosociologists are to recognise and describe plant communities in detail and to classify them into a system similar to that used by the plant taxonomist. However, no absolute criteria have been found for the classification of plant communities into a taxonomic system and discussions on this topic have not always been rewarding. Authors that might be named as representatives of this first group are: Clements, Weaver, Tansley, Sukachev, Alechin, Du Rietz, Gams, Lippmaa, Braun-Blanquet, Tüxen, Scharfetter, Schennikow, etc. (see Cain and de Oliviera, Castro, 1959; Whittaker, 1962; Ellenberg, 1956).

2. The representatives of the other viewpoint consider the individual plant as the only concrete unit in nature. Plants grow in communities which, however, cannot be clearly defined as units. According to this

view the plant cover is regarded as a continuum, which means that it consists of continuously changing combinations of plants which can be delimited only with great difficulty or not at all. Exponents of this viewpoint are Ramenski, Negri, Gleason, Cain, Masen, Curtis, Brown, Whittaker, Poore, etc. Mathematical analysis also gives support to this viewpoint (Goodall, 1961, 1963).

I have adopted a viewpoint that lies between these two main trends and believe that it corresponds most closely to the reality of nature. The recurrence of similar groups of plants or species combinations in similar habitats can hardly be denied. This implies the existence of certain biotic communities in nature, which can be recognised and studied.

They can be easily delimited in a restricted area with differing habitats. However, if one traces a particular plant community over a long distance with an accompanying change in macroclimate, one observes that one plant species after the other disappears from the community while new ones appear until, finally, one is dealing with an entirely different combination of species. If one were to compare only the two plant groupings at the ends of such a long-distance transect one might easily describe them as two different plant communities. However, if one considers the gradual changes along this transect, it becomes difficult to draw a definite boundary. Such gradual changes are rare in central and southern Europe, where much phytosociological work has been done. Strictly, natural biotic communities are non-existent in this part of Europe and the influence of man has resulted in clearly delimited areas of forest, grassland, and weed communities. Biotic communities that resemble those under natural conditions, such as bogs, are present only as widely scattered remnant communities which, therefore, lack continuity as well.

However, areas that are only thinly populated, such as south-west Africa, offer excellent examples for observing a gradual transition (Walter and Walter, 1953). This is brought about by the appearance and disappearance of certain associated species. A gradual change of macroclimate is certainly not associated with a sudden change in plant communities; the transition is usually continuous. For this reason one cannot deny the independence of a single species even within its community and the basic unit of vegetation is, therefore, not the plant community but the individual species, as was also pointed out clearly by Ellenberg (1956).

If climate varies over relatively short distances, as occurs for example on mountain slopes and if the plant communities are each determined by a dominant species, the transition from one community to the next can be quite abrupt. We observe this in changes with altitude in our mountainous areas where, for example, beech forest is replaced by spruce forest as elevation increases. In spite of the fact that the climate changes more or less gradually along the slope, the change from one community to the next is relatively abrupt. However, this is only due to the fact that these communities are each determined by a dominant life form. The community change is associated therefore, with the replacement of one species by another. The separation of altitudinal communities is rather difficult in tropical forests or in mountain forests of the southern hemisphere, both of which are made up of many dominant tree species. Here the changes between communities are more gradual. Fig. 2 illustrates the differences in continuity among communities by a sketch. Situation A indicates an idealised change in

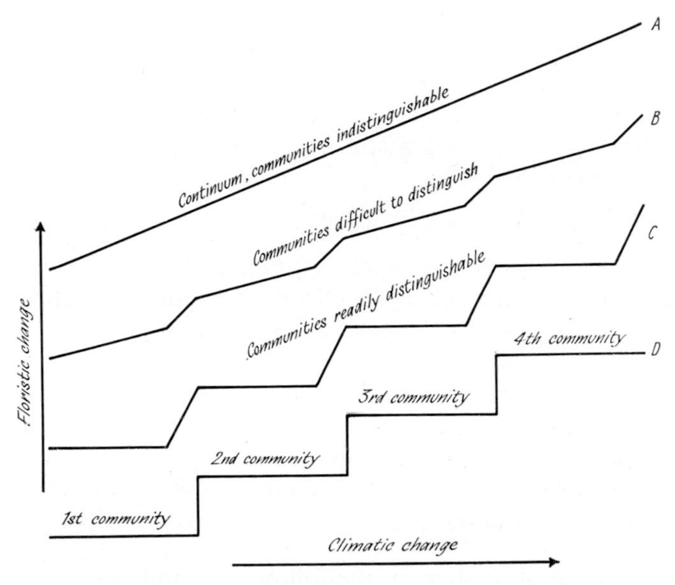

FIG. 2. Scheme showing floristic changes of plant communities associated with uni-directional changes in climate. Further explanation in text.

plant cover in the form of a continuum that is correlated with a gradual change in climate; differentiation of distinct plant communities would thus be impossible. Situation D illustrates the other extreme. Here, the plant communities are sharply separated and the plant cover changes in steps without transition zones. This extreme never occurs in nature. Actual conditions in nature resemble more closely the situations shown as C and B. In situation C, plant communities are distinguished with relative ease, because they are separated by only short transition zones. These are made relatively short in the figure, but they could be even broader than the zones represented by the plant communities themselves. Even under such conditions it is often assumed that the community remains floristically fairly constant within its distributional range in spite of some deviation in macroclimate. This assumption also does not appear to match the reality of nature, except perhaps in those habitats in which the influence of soil, or man, on the aspect of the vegetation is stronger than that of climate (halophytic and dune communities, weed and meadow communities, etc.). However, as a rule, any variation in macroclimate is reflected floristically in the species combination of the community. Situation B, therefore, appears to resemble most closely the conditions found in nature. The distinctness of the communities may be quite obvious or very difficult to detect, depending upon the reaction of the plant community to changes in macroclimate and on the relative width of the transition zones.

In spite of my emphasis on the individuality of the species in a plant community, I have to admit that the species combination of an undisturbed community is not an arbitrary one but depends on rather definite principles. These are stated in the following definition:

A plant community is understood to be a more or less stable combination of naturally occurring species, which are in an ecological equilibrium with one another and their environment.[1]

This equilibrium is brought about:

1. By the fact that certain species compete with one another, their competitive ability being determined by their physiologic-morphological constitution as well as by the particular environment in which they grow.
2. By the fact that some species depend on others which have created the ecological conditions that ensure the existence of a particular community structure.
3. By the occurrence of complementary species which find their own niches in time, or space, within the same habitat without necessarily competing with one another.

This is illustrated further by the following examples.

In the hardwood areas of central Europe we find that the important hardwood species, such as beech (*Fagus sylvatica*), oak (*Quercus robur* and *Q. petraea*) and *Carpinus betulus*, constantly compete with one another. The shade-tolerant beech is the most successful competitor in habitats that are physiologically favourable for its growth. These occur in the montane region, particularly on lime-rich soils, where beech is by far the most common and dominant species. However, oak replaces beech where the climate becomes warmer and drier, or in the Atlantic area, where moisture increases with decreasing nutrient content of the soil. The same is true of alluvial bottomlands which are flooded for longer periods. Oak is often associated with *Carpinus*, whose wide distribution in central Europe is mostly due to man. It regenerates easily as coppice and was, therefore, favoured by earlier forest management practices. *Carpinus* is more shade-tolerant than oak and replaces beech toward the east where the climate becomes more continental, limiting the range of beech. Therefore, either one or the other hardwood

[1] Pioneer communities are excluded from this definition.

species may compete more successfully, depending on the particular climatic or edaphic conditions of the area, or all three species may be represented equally as result of equal competition pressure.

Similar competitive relations also occur among the plants of the undergrowth. However, here the light factor plays an important role in addition to the others and the light relations are dependent on the tree canopy. Thus, while the plants of the undergrowth are competing amongst each other, they also depend as a group on the trees. The canopy produces the microclimate favourable for optimum growth of many undergrowth plants. These species are protected by shade from the competitive pressure of heliophytes. Not only does the tree layer change the light intensity near the forest floor but, by its roots, it also influences the absorption of water and mineral nutrients from the soil (cf. pp. 26-7). At the same time, the tree layer is also the dominant humus former because of its large annual litter contribution. This humus provides appropriate conditions for the growth of humus-inhabiting organisms, particularly if these are saprophytes as, for example, *Neottia nidus-avis* in beech forests. On the other hand litter production may inhibit development of a moss layer.

In even closer interdependency, of course, are parasites in relation to their hosts, e.g. *Lathraea squamaria*, the *Melampyrum* species, and the mistletoes (*Viscum*), etc.

The upper layer is not always the independent, and the lower the dependent layer. For example, *Sphagnum* cover may become the determining factor in raised bogs. Here, only those species able to grow on the nutrient-poor substrate at a rate sufficiently fast not to be overgrown by the peat moss, can maintain themselves. The growth rate of *Sphagnum* is so rapid in the moist-oceanic climate that pine (*Pinus sylvestris*) is unable to compete with it. However, in the continental east *Sphagnum* development is more inhibited so that raised bogs become forested and there we find pine-bog forests with an undergrowth of *Sphagnum*.

Not all the plants of a community, however, are competing with one another. Some may occupy different niches of phenological development by completing their life cycle at different periods. These are called complementary species. For example, in the European beech forest *Ranunculus ficaria* flowers in the spring when the ground is still moist and before the trees come into foliage. At nearly the same time the growth of *Corydalis* and *Anemone nemorosa* begins. Later, when *Ranunculus ficaria* has almost disappeared and the others are declining, *Mercurialis perennis* and *Asperula odorata* begin their full development and, still later, *Milium effusum* or *Sanicula europaea*.

Similarly well-developed seasonal aspects are shown in our fresh, nutrient-rich meadows. Here we find the sequence of *Bellis perennis*, *Taraxacum officinale*, *Cardamine pratensis*, *Ranunculus acer*, *Lychnis flos-cuculi*, *Tragopogon pratensis*, *Chrysanthemum leucanthemum*, *Anthriscus silvestris*, *Crepis biennis*, etc. Of course, the changing aspect is mostly due to the different flowering periods. But, in most plants, this also coincides with the period of maximum vigour and dominance. While the early species are fruiting, the flowering of the late ones has just started. Rosette-plants must complete their development before that of the tall grasses. Similar striking, seasonal aspects are found in the east European steppe areas, in the North American prairies and in the subtropical savannahs. These are but a few examples.

In all these cases the species were complementary in a chronological sense. However, there are situations in which species are complementary in space. This is well shown in the stratification of vegetation layers above the ground. In temperate forests we find an upper and lower tree layer, a shrub layer, an upper, middle, and lower herb layer and finally a ground layer with mosses that are only a little higher than the forest floor. In the virgin tropical forest the entire vertical space may be utilised by foliage material, while only the lower layers may be missing due to lack of light near the forest floor.

Even in grassland, for example in our meadows, a stratification can be recognised into upper grasses, lower grasses with equal-sized herbs and a ground layer of many rosette-plants and some mosses.

However, complementation in space is not restricted to stratification above ground. The same is true also for root systems in the soil, although their stratification is concealed from direct observation. The above-ground stratification is best expressed in forests. In non-forest communities the below-ground stratification may be better developed than that above ground. For example, such below-ground stratification is well shown in the *Koeleria glauca-Jurinaea* community on sand-dunes of the upper Rhine lowland. Here, rooting of mosses and lichens is restricted to the soil surface, while the many annuals, which develop in early spring, root between 5 and 15 cm depth. Most perennials root at 20 to 60 cm depth and only the deep-rooters among these, which are the autumn-flowering plants, penetrate deeper than 150 cm. The same applies to the steppe in central Anatolia and to the North American prairie (cf. Walter, 1960, pp. 273ff).

Complementation among plants leads to communities in which, depending on local site factors, the growing season for plants is completely utilised. At the same time this brings about the full utilisation of the space available above and below ground.

All natural plant communities that are in equilibrium with their local site factors increase their dry-matter production to the maximum possible under these conditions. Such communities can be thought of as being saturated with both plant species and individual plants. This applies even to the open-rock-fissure communities and desert formations.

This situation explains why adventitious species have such difficulties in penetrating into undisturbed communities.

Thus, the plant community represents a system that is closely interrelated with its environment from which it cannot be dissociated.

Therefore, the study of vegetation is a part of ecology, provided that it is not understood entirely as a descriptive science. Only the clarification of causal ecological relationships can lead to a deeper understanding of the vegetation cover. An ecological equilibrium, however, is not static but dynamic. Within each community there is a circulation of matter and energy; plants come and go. The point of dynamic equilibrium is not constant because the environmental factors vary from year to year. Drought years alternate with wet years and the seasons may sometimes be above, or below, the average temperature. Thus, such environmental variations favour certain species over others. At the same time there is a tendency towards a norm in the response of a plant community. Certain catastrophic years may produce gaps in a community and the re-attainment of an equilibrium may take a long time. One can observe such stages very easily in virgin forests, for example, along the tracks of tropical cyclones.

On the other hand, a continuous, unidirectional change of environmental factors also causes changes in the entire community. If the unidirectional change lasts for a long time, e.g. the continuous lowering of the ground-water table caused by the rise of a land surface, or the constant increase in depth of fine soil due to weathering, a continuous, unidirectional change in species combinations will likewise occur. This is called succession (see next Section).

It is man, above all, who interferes with the equilibria of plant communities, causing numerous secondary successions. Vegetation always has a tendency to re-establish its equilibrium, but this is practically never attained in highly populated countries. If man's intervention is sufficiently regular it becomes part of the environment and a corresponding anthropogenic equilibrium may become established. Therefore, more-or-less stable, secondary plant communities can arise. Nearly all central European plant communities are maintained through such an anthropogenic equilibrium, for example, our forests, our grasslands and arable land, the heather moors and sheep-grazing lands, the ruderal communities and the aquatic plant communities in eutrophic waters, etc. The different degrees of community modification result only from differing intensities of human intervention (cf. Ellenberg, 1963).

In less intensively cultivated land areas of other continents the natural plant cover is also frequently more or less modified by man as a consequence of logging and grazing, through the burning of grasslands and by shifting agriculture in forest areas.

Areas with truly natural vegetation are rapidly decreasing.

The importance of establishing reserve-areas for the study of problems relating to improved management practices cannot be overstressed. Such areas should not be selected only in beautiful landscapes that are attractive to tourists. Unfortunately, experience has shown that the public and government in those countries still possessing a natural vegetation cover only become aware of this need when it is already too late.

We may conclude from the foregoing discussion concerning the criteria of the plant community that it shows by its composition and structure a very close interrelationship with the habitat, where habitat, in the ecological sense, is understood to be the totality of the physical and chemical factors which constitute the plant environment (cf. Walter, 1960). The claim of the phytosociologists is absolutely correct that the plant community itself, by its species composition and frequency relations, represents the best indicator of its local environment. This is especially true with regard to the reflection of average environmental conditions over long periods of time, for direct measurements of local site factors will only give momentary values. Long-period measurements are usually impractical. However, the plant cover can only be employed as an indicator of habitat factors when the plant community habitat relations are well known, and this requires an accurate ecological analysis.

From the fact that plant communities reflect any change in habitat, we may also conclude that while definite boundaries between communities are correlated with abrupt changes in habitat, community boundaries are more difficult to determine where the habitat changes are gradual.

But even under those conditions, where habitat changes are abrupt, an orderly classification of the vegetation cover is possible only if we define a certain, limited number of types. We recognise as types those plant communities which recur repeatedly in a region under similar habitat conditions. These types are called plant associations. They are characterised by the list of species composing the association and by the vertical structure (i.e. layering) and frequency relations among the species. The methods employed are presented in detail by Ellenberg (1956, pp. 16-66).

There are no objective criteria for classification, which is always somewhat subjective. Successful classification requires an exact knowledge of the entire area, much experience and a sixth sense of skilful discrimination. The relative size or degree of inclusiveness of the types recognised is a matter of expediency, which depends on the purposes for which the classification is intended. Braun-Blanquet recognised at the outset, with great skill, relatively broadly conceived types in the Alps, in the Mediterranean area and in central Europe, which explains the success of his approach.

At the same time Braun-Blanquet developed a method of vegetation analysis for the assessment of plant communities. It relies on visual estimates of density relations, is not too time consuming and is of sufficient accuracy for many practical purposes.

No criticism of significance can be raised against this method, the world-wide adoption of which would be desirable. Most of the criticism of the Braun-Blanquet approach was directed against a certain dogmatism of the school, such as viewing the types as fixed units of a higher order, or against the development of a strictly hierarchical system following the principles developed for plant taxonomy.

From an ecological viewpoint the dominant species are the most important for the structure of the community, since the whole biological community would be destroyed if they were removed. However, for purposes of characterisation, one can use, in addition, ecologically more specialised plants, whose presence is often correlated with very special habitat conditions and thus show a narrow ecological amplitude. These were called 'Leitpflanzen' (indicator species) by Gradmann, Braun-Blanquet called them 'Charakterarten' (characteristic species) and Tüxen referred to them recently as 'Kennarten' (diagnostic species). Such species have little significance in respect of the ecological facies and structure of the community if they are not also dominant species. However, they form valuable indicators even though of more local importance.

The major reason European phytosociologists have attributed so much importance to indicator species of the ground vegetation appears to arise from the fact that the tree layer has been modified so much by forest management in central Europe. In contrast to this, the herb and shrub layers have been much disturbed in other countries where forest grazing has been practised. In such situations and under natural conditions it appears more appropriate to use the dominant tree layer as the primary criterion for classifying the vegetation, while plants of lower layers may be used for characterising further subdivisions.

If we are confronted with the task of having to assess the vegetation relations of a given region, we must always begin by classifying from above, i.e. we first distinguish the larger units which may then be further subdivided upon closer study. Where to start depends entirely on the kind of objective, that is, on the area chosen. This is the entire world in my case. Therefore, I have begun, initially, with very broad relationships. In other cases it may be a continent such as North America, or a smaller area, such as the Apennine Peninsula, or just a mountain area, such as the Black Forest. In special cases we may have to deal with very small areas, such as a forest stand, a bog or a meadow, but we should always begin with the general then proceed to the particular. It will only be possible to obtain a profound insight into the vegetation of a region in this way. How much subdividing one does depends upon the questions asked or the purposes pursued. A rather finely graded classification is desirable for application to forestry and range management. The delimitation of a definite lowest unit cannot be made.

The study of vegetation is not only that part of plant science in the geobotanical area but, at the same time, it borders on geography itself (cf. Schmithüsen, 1959). Thus if one wishes to establish a satisfactory assessment of the entire plant cover of a certain region, geographical aspects cannot be ignored in an investigation of vegetation. Plant lists of typical plant communities alone—as published in recent phyto-sociological papers—are as unsatisfactory for this purpose as are floras, in which all species of a region are merely set out.

2. Succession, climax and vegetation zones

As mentioned before, the plant community is in dynamic equilibrium with its environment. However, this dynamic equilibrium may constantly shift in a certain direction, thereby introducing some new vegetational elements, while others disappear. Eventually, an entirely new plant community may result. This change of communities with time is called succession. One distinguishes between autogenic and allogenic causes of succession. The first refers to causes originating within the plant community itself, the second to changes caused through alterations in its environment.

The concept of succession was initially described by Cowles in North America and then used by Clements as the basis for his vegetation classification. He assumed that pioneer communities in aquatic habitats, on bare rock, on sand, or saline soils would start a series of communities that, in the course of time, would undergo many successional changes into the highest form of development, the climax community. The climax community was thought to be controlled entirely by climate; that is, its species composition was believed to exhibit the vegetation characteristic for the climate of the zone. Clements considered all other plant communities of the same climatic zone as temporary stages of different successional series. He referred to hydroseres, if the successional series was believed to have started from aquatic habitats, to xeroseres, if the series was believed to have originated from rock, or to haloseres, if they were believed to have begun from saline soils.

According to Clements these primary successions were predominantly autogenic. Tansley and Braun-Blanquet accepted Clements' viewpoint of vegetation dynamics, although Braun-Blanquet (1951) presented a modified interpretation in the second edition of his *Pflanzensoziologie*.

However, this viewpoint is not supported by historical facts. The invasion of the earth's surface by plants began *c*. 500 million years ago. The changes that have occurred since this initial invasion were related to the development of new plant forms (pteridophytes, gymnosperms, angiosperms) and to changes in climate. In the northern hemisphere, invasion of large areas of newly exposed substrata only occurred after the retreat of the large glaciers during the Quarternary Period. But these changes cannot be considered as autogenic primary succession, because the glaciated areas were invaded by plant species from refugia and the invasion was related to changes in climate (cf. Walter, 1954b).

Some stages in primary succession can still be observed in certain places, for example, where remnants of older stages are preserved below more recent ones as in the silting up of lakes. Older stages can also be detected in dunes or salt marshes on the basis of old maps (cf. Chapman, 1959). The invasion of landslide scars in mountainous areas can also be observed over long periods of time. Yet these processes are of limited, local importance and proceed only at a very slow rate. Usually, such changes are not autogenic, but allogenic successions. The rapid silting up of lakes usually occurs only locally, namely, at river inlets where sedimentation is fast. Likewise, dense vegetation can only become established in coastal dune areas at some distance from the shoreline, where sand-shifting and wind intensity have decreased. Invasion of landslide areas is possible only where soil downwash from above has ceased. All these cases are characterised by a change in habitat and the vegetation is merely adapting itself to the changed environmental conditions.

The concept of succession is based mainly on the assumption that one may draw conclusions about a change in time from observations of spatial variations. But this is not valid unless there is direct proof of a temporal sequence at the same locality.

Near lake shores one can usually observe a zonation in the vegetation, beginning with hydrophytic and shoreline communities and usually ending with various forest communities. One can only be certain that one is dealing with a temporal sequence, however, if one can observe lacustrine deposits at the base of the soil profiles. Such a zonation is usually, therefore, merely an ecological series of plant communities which have formed along a water table gradient. Such community zones can be very stable.

Factual support for successions following a xero- or halosere is extremely difficult to obtain, because evidence of earlier vegetation is only very rarely visible. Therefore, in all such situations one should only talk of ecological series, which are related to an environmental gradient, such as increasing depth of soil in rock outcrop areas, decreasing salinity of the soil, etc.

A much more important role is played by secondary successions. These can be observed wherever the plant cover is suddenly destroyed, in part or completely, and where reinvasion takes place on the more or less denuded soil, for example, after destruction of a forest by fire, clear felling, windfall, landslides, flooding, etc. The same is true where a habitat is modified by man and where particular factors suddenly cease to operate, for example, development after swamp drainage, the endykement of a salt marsh, a field left fallow or an abandoned hay meadow. Secondary successions usually proceed rapidly and changes can be observed from year to year. In this connection it is not important whether the cause of disturbance is natural or due to human influence. Today, the latter case is usually true, even in sparsely populated areas (for example, by burning of grassland, etc.).

The concept of the climax is, in its extreme form, as vague as the succession concept, that is, the assumption that the vegetation is moulded into its final state by climate alone. The vegetation depends primarily on its total environment. The controlling environmental factors (temperature, moisture, light, chemical and mechanical parameters) are conditioned by climate, and also by topography and soil. In extreme topographical or soil conditions the influence of macroclimate becomes secondary. The relation of vegetation to different macroclimates can only be compared on relatively level areas and on soils that are neither poorly nor excessively drained, in which the water available for plant growth is a direct function of regional precipitation. Moreover, the soils should not be extreme in nutrient content. Under these conditions we can find in almost all climatic zones a corresponding type of vegetation, which I call *zonal vegetation.* By zonal vegetation I understand rather broadly conceived vegetation units, which occupy the larger part of the earth's surface in anthropogenically undisturbed areas. The gradual change of zonal vegetation in relation to climate is readily seen from an aeroplane when crossing a continent, disregarding the patterns caused by man. One also notices that the vegetation patterns on extreme soils (dune areas, salt marshes, swamps or rock-outcrop areas), which may dominate locally, do not correspond to that of the zonal vegetation. In such extreme soil conditions the vegetation is more strongly influenced by edaphic than by climatic factors and I refer to such vegetation as *azonal.* However, azonal vegetation is not entirely independent of the zonal climate. For example, dune communities are floristically quite different in the temperate and tropical zones, although generally quite similar in physiognomy. This is true also for halophytic and aquatic communities.

One can also observe that segments of zonal vegetation may often occur outside their zone proper in neighbouring areas, where they occur in local habitats that are particularly favourable for their formation. For example, forest commonly extends deep into the steppe zone in moist river valleys, or steppe vegetation may occur as islands in a forest zone especially on dry and warm habitats. Here too we are dealing with mature plant communities which, however, do not correspond to the zonal vegetation of the region. I refer to such vegetation as *extrazonal.* The prerequisite for this phenomenon is that the change in macroclimate is compensated for by the habitat of the extrazonal community. This I call the principle of relative habitat constancy, which will be referred to again in Chap. VII, 4 (cf. Walter, 1954b, p. 41).

For example, the temperature relationships are very similar in boreal forest over the vast level terrain of northern Europe and wherever it occurs either north of its zone, on warm southern exposures, or south of it on cool northern exposures, or in deep canyons. We can observe that the plant communities of the eastern European steppes are likewise found extrazonally in central Europe where they are restricted to sunny slopes with shallow, excessively drained, calcareous soils. The higher precipitation in central Europe is compensated for in such habitats by increased evaporation and decreased water storage capacity of the soil. Therefore, the water relations of such habitats are similar to those of the dry steppe climate.

The concept of zonal vegetation corresponds somewhat to that of climax but in a much more restricted way and quite differently from the vague concept of primary succession in its extreme form. All attempts to save the climax theory by substituting for it a polyclimax theory, or by introducing such concepts as climax groups or climax population are unsatisfactory as they are all based on the hypothesis of primary succession. Nevertheless, a dynamic viewpoint in research on vegetation is satisfactory, provided that it is not based on speculation and remains in the realm of factual evidence.

The determination of the zonal vegetation is extremely difficult in areas in which the vegetation has been changed by human influence over hundreds or thousands of years. If there are no remnants of natural vegetation left, one must attempt to obtain historical evidence. In such cases one is limited to a very general characterisation of zonal vegetation only and one has to forgo more exact information. For example, little more can be said about the zonal vegetation of central Europe than that it was a mixed hardwood forest with a large admixture of beech (*Fagus sylvatica*). The composition was probably changed in favour of oak (*Quercus robur* and *Q. petraea*) in the area dominated by the Atlantic climate and in favour of *Carpinus betulus* in areas with very low precipitation.

The concept of zonal vegetation should be used only for the classification of vegetation of whole continents; that is, on a broad geographic scale. The influence of macroclimate on vegetation is more important on a broad geographic scale, where edaphic influences may be neglected. However, the influences of soil and topography should not be overlooked in more local vegetation studies.

Climate changes much more rapidly with altitude than with latitude. Therefore, its importance as a factor in the classification of vegetation extends over much shorter distances in mountainous than in level terrain. Plant communities that are associated with altitudinal changes in climate and which correspond, in a general sense, to the zonal vegetation of level areas, are referred to as *altitudinal belts*. These too should only be determined from vegetation occurring on relatively level topographic positions. Slope and aspect not only change the temperature but also the water relations and vegetation in such sites is analogous to the extrazonal vegetation of level terrain. It is not always easy to find level topographic positions in rough mountainous terrain.

In such cases it is advisable to examine sites on northern and southern exposures in order to define the altitudinal belt, which may correspond to the average characteristics of the vegetation at the two contrasting sites. On the other hand, it may be better to determine the altitudinal belts separately for each main exposure. This is obligatory where a mountain ridge separates two macroclimates, for the vegetation on the opposite slopes is then not comparable at all (Walter, 1954a).

3. Competition

We have noted earlier that competition is a very important factor in the development of a plant community. This had already been emphasised by Clements, Weaver and Hanson (1929). In spite of this, competition has not received sufficient attention as a factor in ecological research.

Competition can be observed wherever two plants grow together closely in a limited space when each tries to utilise the available light, soil-water and nutrients as much as possible to its own advantage. The factors mediating competition are of a purely physico-chemical nature (Clements *et al.*, 1929; Weaver and Clements, 1938; Schmid, 1944; Boysen-Jensen, 1949; Ellenberg, 1954). It has not yet been proved whether excretion of metabolites plays an additional role in competition. Symbiotic relations among heterotrophic micro-organisms are probably related to factors based on exchange of metabolites. However, autotrophic plants are not dependent upon absorption of metabolites from their surroundings. Therefore, allelopathic influences could only be produced by certain growth inhibitors, which would need to be effective in very small concentrations. So far it has only been shown that such effects are possible in laboratory experiments and with plants that never occur together in nature. In these cases they were caused by root excretions or by substances which diffuse, or can be extracted from, dead parts of plants (Klapp, 1955; Grümmer, 1955; Rademacher, 1959). Such direct effects would appear quite possible when one considers the close contact that exists among the root systems of different plants in a community. Martin (1957; see also Ahlgren and Aamodt, 1939; Börner, 1960; Kolb, 1961) was able to demonstrate that excretions from intact roots are insignificant, while those from damaged roots may become significant. However, such substances—in this case it was scopoletin—decompose rapidly in the soil. Ahlgren and Aamodt found that different pasture plants, when grown in mixtures, produced a smaller amount of dry matter than when grown in pure populations. These investigators believed the effect to be caused by reciprocal growth-inhibiting root excretions. However, Donald (1946) did not support their conclusions. In spite of the fact that he used the same species in 27 different combinations, he did not find a reciprocal growth inhibition. Certain substances are released into the soil during litter decomposition. It is also possible that there is an accumulation of certain ethereal oils in the air layer near the soil surface of a plant community during hot, calm days. However, there is no clear evidence of the importance of these substances as selective factors in a plant community, and thus we have no answer as yet to this problem.

For example, absence of natural regeneration of *Araucaria cunninghamii* in forests of South Queensland with *Backhousia* in the understorey, was attributed to toxic effects of dehydroangustione, which forms the major proportion of oils in *Backhousia* leaves. This substance inhibited germination of *Araucaria* seeds in laboratory experiments when its concentration was increased above 1%. However, field experiments have shown that an increase in *Backhousia* leaves in the litter did not have any inhibiting effect. On the contrary, it even had a promoting effect on *Araucaria* germination, despite the presence of a 1-2% concentration of dehydroangustione (Cannon *et al.*, 1962).

Competition for light, soil-water and nutrients increases with the number of individuals per area and with their size. Small seedlings may be quite dense without interfering with one another, but competition becomes increasingly effective with continued growth. The larger individuals and the faster growing ones will expand their leaves to take full advantage of light. They will produce more organic matter than the smaller, shaded ones. Because of this, they get ahead more and more. In addition, their root systems will be better nourished and penetrate deeper at a faster rate so that their water and nutrient supply will be more favourable. The suppressed individuals lag behind and, as soon as their nutrient and water supply becomes insufficient, they perish often without having reached the flowering or fruiting stage.

Competition for light is only between individuals of the same layer. This applies, for example, to tree seedlings in their early stages when they compete with the herbs, which may suppress the tree seedlings completely. However, if the latter are able to survive and show yearly height increments, they finally overcome this shading effect. The trees gain dominance and the herbs then have to adapt themselves to the limited light beneath the tree canopy or they disappear altogether. Thus the lower layers are always the dependent ones with regard to light. Should the tree and shrub layers be interrupted, however, the herb layer may participate in the competition for soil water and nutrients. For example, the grasses form the dominant layer in natural savannah areas, and the woody plants are forced to utilise water left over after the rainy season (cf. Chap. VI, 2). Peat mosses (*Sphagnum* spp.), which form peaty ground layers in forests, can even bring about the death of the trees if they inhibit respiration of the roots through waterlogging the soil.

It is necessary to distinguish between two kinds of competition.

1. Between individuals of the same species—intraspecific competition.

2. Between individuals of different species—interspecific competition.

The results are very different. Intraspecific competition is particularly important in mono-cultures in agriculture and forestry. If the number of crop plants growing in a field is very high, the individuals are weakened by competition and the result is a low yield. If seeding is scattered or plantings widely spaced, competition is practically zero and each plant has an opportunity for maximum development, but the over-all crop yield may be below the maximum. In addition, there is a considerable danger from weeds. Thus, seeding and planting standards have been developed empirically for field crops of certain climatic zones to obtain the optimum yields. Similarly, the forester prefers, initially, a close stand which favours the growth of knot-free wood and later periodic thinning of the stand to avoid lowering the yield through competition.

Pure stands composed of one species even occur naturally; for example, on burned-over land in the boreal forest zone, where *Pinus* spp. re-seed quickly after fire; on saline mud flats at the sea coast, where thousands of individuals of *Salicornia herbacea* often appear; in deserts where, after rain, the seeds of one species may be washed together to germinate in dense groups; or in steep moist, forested glens in central Europe, where one often finds hordes of *Impatiens noli-tangere*. However, these naturally occurring pure stands are never as uniform as artificial ones and selection always takes place soon; the more vigorous individuals grow up, they flower and fruit, and the feeble ones die out. Therefore, intraspecific competition is favourable for the survival of a species, for an originally dense population even protects a species from competition with other species.

While establishing shelter belts in the east European Steppe, the following observations were made. If acorns of oak were planted singly, the young tree seedlings were suffocated by grass and succumbed to competition. If, on the other hand, acorns were planted in clumps, the young seedlings established themselves in dense groups. The outer ones were overgrown by the neighbouring grasses, but the one in the centre was protected by the outer ones and had sufficient time to overtop the grass vegetation and to develop a strong root system. In time it became a tree (Orlenko, 1955).

Interspecific competition has quite different consequences. Here there is a fight for survival of the species. The species that succeeds best also fruits abundantly and ensures its effective reproduction. On the other hand, a suppressed species disappears completely, or remains sterile, or fruits only very sparingly so that it recedes from year to year. However, a particular competition-equilibrium will become established with time in the more permanent plant communities. There are usually several equally aggressive plant species, which establish a mixture of characteristic proportions in their respective layers, for example, several tree species in a hardwood mixed forest, or hardwood and softwood species together, or several tall grasses together with tall herbs in grassland communities or on the steppe.

One can recognise the constant competitive relationship among these species by the fact that even the slightest change in habitat factors results in a change in their frequencies and by the fact that an individual plant never attains as vigorous a growth as if it is grown in isolation, for example, in a flower bed. Thus, plants in a community suffer somewhat from undernourishment. This shows, yet again, that the carrying capacity of the habitat is utilised to the full by a plant community and that there is no room for foreign invaders. It thus explains the stability of the undisturbed plant community.

In connection with competition, the site factors are of decisive importance, in so far as these are correlated with their intensity and their combined effectiveness, in favouring one or the other species. Dry periods inhibit the development of moisture-loving species more than that of drought-resistant ones; frost periods cause dieback of frost-sensitive, evergreen species and this usually results in their defeat, through competition, by cold-resistant species. A change in site factors will rarely cause the disappearance of a species from a natural plant community by direct influence, but rather by indirect influences which weaken the competitive strength of one species relative to another.

Therefore, the same environmental factor, for example, a late frost, may have quite different consequences for any given species depending on the kind of competing species. In one case a species may be weakened by an environmental factor and then fully subordinated by a powerful competitor, in another case, while it may still be injured, it may still have a chance of retaining its position in the community. This shows how careful one should be in inferring direct correlations between environmental factors and the presence of species. Such relationships are almost always of an indirect nature.

This becomes very clear from a ground-water experiment done at Hohenheim (Ellenberg, 1956, 1963; Walter, 1960). *Alopecurus pratensis*, *Arrhenatherum elatius* and *Bromus erectus* are three important grasses in central Europe, of which the first dominates in wet, the second in well-drained, and the third in dry meadows. However, they all showed their optimum growth at intermediate depths of the water table after being sown as rows of pure seeds on sloping soils in tanks having a gradient of water-table depths from 0-150 cm (Fig. 3). Yet, when grown in mixed seed-rows a different ecological behaviour was observed (Fig. 4). The strongest competitor, *Arrhenatherum*, developed best in the optimal habitat, with an intermediate depth of water-table, and pushed *Alopecurus* to the moist side and *Bromus* to the dry side, particularly on soils with low nitrogen levels (Ellenberg, 1954). In time there was also an influx of weeds from seeds arriving at random. But, in

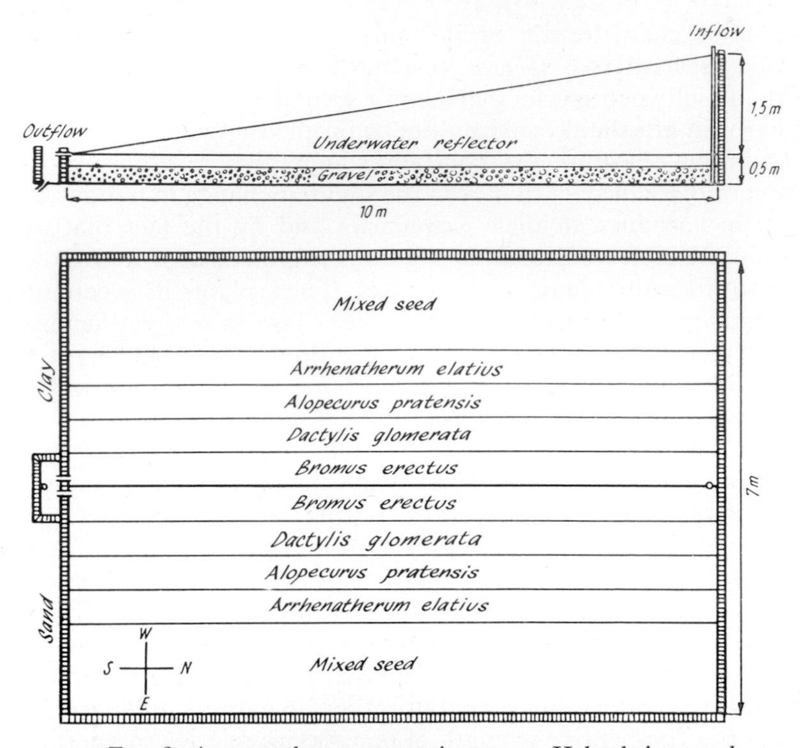

FIG. 3. A ground-water experiment at Hohenheim: each tank has a constant height of water-table. Above: side view; below: view from above with experimental layout 1953. Further explanation in text.

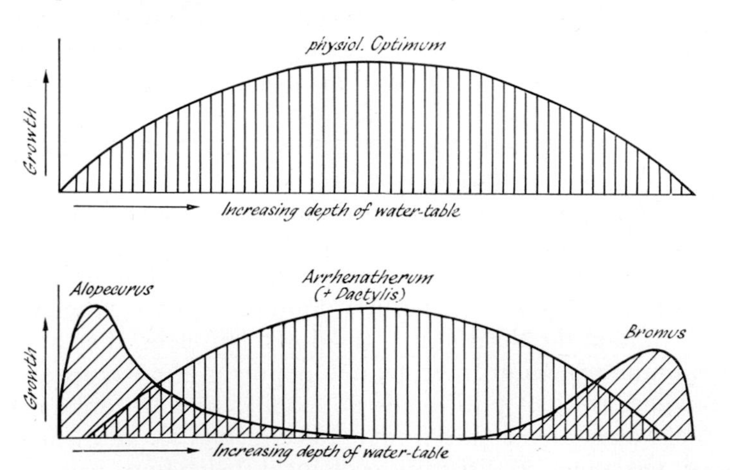

FIG. 4. Growth of meadow-grasses (ordinate) with increasing depth of water-table (towards right; abscissa). The upper sketch-graph applies more or less to all species investigated in pure seed-rows (optimum at intermediate depths of water-table); the lower applies to the situation in mixed seed-rows (*Alopecurus pratensis*—moist adapted, *Bromus erectus*—dry adapted).

competition with the grasses and amongst themselves, weeds became established according to their characteristic distribution in nature; some only in the moist part of the gradient, others in the centre and yet others in the dry part (Lieth, 1958). A few are indifferent with regard to the gradient of the water-table.

The pH value also has an indirect effect on the plants. Thus it is not surprising that a species competing under natural conditions may show a correlation with a very narrow pH range, whereas it loses this trend when competition is removed (Ellenberg, 1950).

We have to distinguish, therefore, between a physiological optimum, when there is no competition, and an ecological optimum, when the plant is growing in competition with others. This ecological optimum corresponds with the habitat conditions in which the species is most abundant and most vigorous in nature. However, since the ecological optimum is not only a function of the habitat, but, also, to an even greater extent, a function of other competing species, it can vary considerably with different competitors. For example, *Hypochaeris radicata* is found only in certain pasture communities in central Europe. When introduced to New Zealand, it apparently found no serious competitor, because there it grows in nearly all plant communities up to the timber line in the mountains.

I intend to refer to all inhibiting influences, which are the result of competition, as 'competition pressure'. This allows one to explain the relationship of the ecological optimum to the physiological optimum under different competition pressures by the following scheme (Fig. 5).

The range of growth under different environmental conditions is fairly broad for a species cultivated alone; that is, when not exposed to competition pressure. The species develops best in its physiological optimum (Fig. 5, *A*). However, its range of growth will be much more limited as soon as it occurs in mixtures with other species. Its ecological optimum, i.e. its area of dominance, will be pushed to one or the other side depending on the associated species and their particular competition pressure (Fig. 5, *B* and *C*). A species may even show two ecological optima (Fig. 5, *D* and *E*). Only if a species is very aggressive compared with its associated plants will it grow best in the habitat most favourable to its development. In this case the ecological and physiological optima are identical, but its range of growth will be somewhat reduced at its limits (Fig. 5, *F*).

An example may show how much we have to adjust our views in the light of the effects of competition. *Quercus palustris* is a stand-forming tree along rivers, around wet depressions and in swamps in Missouri (U.S.A.). *Quercus rubra*, on the other hand, predominates on the drier habitats in higher positions. Therefore, Sullivan and Levitt (1959) thought that *Q. rubra* must be more drought resistant and able to withstand a greater lack of water.

In order to test this, they compared the physiological responses of seedlings of both species. One group was grown at first under normal conditions, the other group was pre-conditioned by lack of water. Surprisingly, both species responded equally with respect to their drought resistance. The authors mentioned, as the only difference, that *Q. palustris* grew faster with large amounts of water than did *Q. rubra*. I believe that this difference is of great significance and provides a possible explanation for the distribution of both species. *Q. palustris* is more aggressive in moist habitats and pushes *Q. rubra* to the drier habitats, on which the latter is superior. It is not the drought resistance

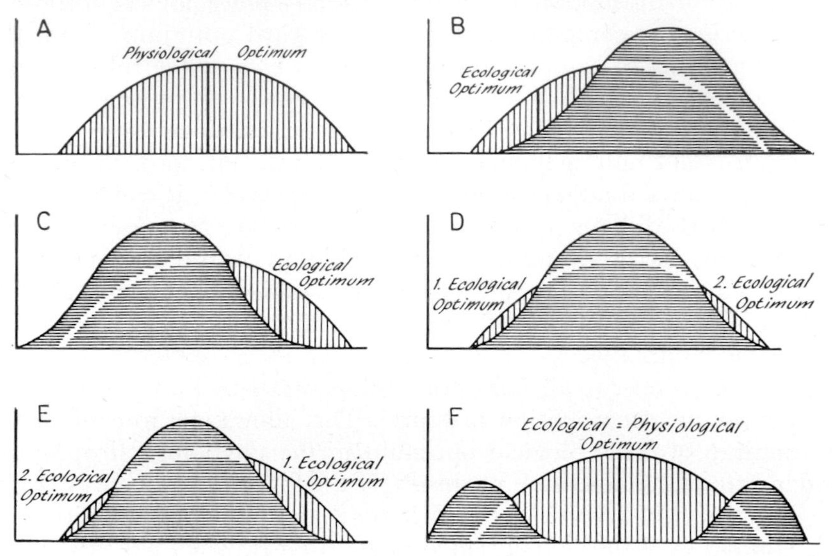

FIG. 5. Growth curves (vertical stippling) of a plant species without competition (*A*) and under competition pressure, which is indicated by dense horizontal stippling (*B-F*). After H. Walter: *Standortslehre* (i.e. Physiological ecology), 2nd ed. (Verlag Eugen Ulmer, Stuttgart).

which is decisive but the greater production of dry matter under the respective habitat conditions.

The aggressiveness of a species or its ability to compete is a very complicated phenomenon. It is determined by the entirety of all the morphological and physiological properties of a species, such as its rate of germination and growth, the kind of shoot and root system, its developmental rhythm and longevity, its reproductive capacity and kind of reproduction, the number of seeds and their mode of dissemination as well as the requirements for such factors as temperature, water, light and nutrients at the site (cf. Ellenberg, 1956, p. 120). In the following sections, I attempt to give a partial analysis of the phenomenon.

A decisive role is played by the production of dry matter, particularly amongst plants of the same life form. Therefore, it is essential to consider dry-matter production in some detail. I shall start with the dry-matter production of the individual plant and then consider that of entire communities.

4. The production of dry matter and the leaf-area index

Photosynthesis is the process by which the plant produces organic matter, that is the material of which it is comprised. Conversely, organic substances are used up during respiration.

Thus, physiological considerations often lead to the assumption that the dry-matter production of a plant increases with increasing CO_2 absorption per unit leaf area. However, this assumption is correct only for a given period. It is incorrect if one considers dry-matter production of a plant over the whole growing season; that is, the total amount of dry matter produced. This depends on the utilisation of the photosynthetic products ('Assimilathaushalt') by the plants (Walter, 1962, pp. 170ff; 1960, pp. 399ff). The larger the proportion of photosynthetic products used in the formation of new foliage, and the smaller the proportion utilised for non-green parts which only respire, the greater is the amount of dry matter accumulated in the end. Other differences show up if one deals with a single plant or a plant population (Takeda, 1961).

For example, the gain of dry matter by a single cereal plant is initially small immediately after germination as expected from the small leaf-area. The dry-matter increment increases with increasing leaf-area and reaches a maximum shortly before flowering, before the withering of the lower leaves and before the reproductive parts have respired a large proportion of the organic matter. If the plant receives applications of nitrogen (N), the production of dry matter, i.e. the yield, is much increased. Takeda has shown that under these conditions the increase in yield is largely due to the enlargement of the leaf-area. An 8-fold increase in yield corresponds to a 5-fold increase in leaf-area, while the rate of photosynthesis increases 1·5-fold. The N-content of the leaves increases 6-fold with a corresponding increase in respiration.

In an isolated plant all leaves receive sufficient light. Initially, its CO_2 absorption increases rapidly then less rapidly with increasing light intensity and reaches a maximum at about 50% full sunlight. Different relations obtain in a pure stand. The leaf-area expands as soon as the stand grows up and the lower leaves become shaded. Therefore, when the upper leaves attain their maximum photosynthetic capacity at 50% full sunlight, the lower leaves will not have reached this capacity because of the effects of shading. This is why the dry-matter production of a plant in a stand increases in a linear relationship with light intensity up to maximum daylight intensity. However, dry-matter production by a plant in a stand also depends, primarily, on the size of the leaf-area but the relationships are more complicated. The size of the leaf-area of a stand is expressed by the leaf-area index (LAI), which is the relationship of the total leaf-area to the unit area of land occupied by the stand. As long as this index is below 2·0, overshadowing of leaves is insignificant. However, light intensity near the ground decreases rapidly if the value increases above 2·0. The lower leaves will then have to carry out photosynthesis under increasingly unfavourable conditions. At the same time, respiration increases proportionately with leaf mass. It follows that the leaf-area index must attain a certain optimum at which dry-matter production is at maximum. This optimum increases with increasing daylight intensity. Takeda gives the following LAI optima for rice plants during the branching stage (radiation intensity in cal/dm^2 per day):

Radiation intensity	100	200	300	400-500
Leaf-area index	3·2	5·8	7·8	*c.* 10

If the LAI is 5·5 at 100 cal/dm^2/day, respiration losses are greater than the production of dry matter. This causes withering of the lower leaves. The LAI is also dependent upon the position of the leaves. Vertical leaves allow more light to penetrate between them than horizontal leaves. Therefore, one should consider only the vertical projection of the leaves on a horizontal area. However, this is not always easy to do. One can observe on many plants, for example broad-leaved trees, that the upper leaves, which receive an excessive amount of light, assume nearly upright positions, while the lower and poorer lighted ones spread horizontally.

The change of the LAI in a stand of *Trifolium repens* has been investigated by Brougham (1962) in New Zealand. The LAI increased from 0·7 in the winter months to 3·25 in the summer and was correlated with changes in the seasonal amount of incident light. An equilibrium was maintained by newly formed and dying leaves between the amount of light and the LAI, as a consequence of which only a little light penetrated to the soil surface in all seasons. The stems of this clover species creep along the soil surface and each leaf develops initially in low light intensity. However, in summer the newer leaves are pushed above the older ones by elongation of the petioles so that they are exposed to full daylight after unfolding. The LAI depends on the number of green leaves and their lamina size.

In contrast, the young leaves of herbaceous plants with upright stems always develop in full daylight, thus, in the summer they are exposed to much greater light intensities than in the spring. Therefore, they assume an increasingly xeromorphic appearance with the progress of the season, as was shown by Yapp (1912) for *Filipendula ulmaria*. The leaves receive decreasing amounts of light with advancing age, since they become shaded by the younger, higher placed ones.

In arid areas the leaf-area is regulated in relation to the water supply and often much reduced during the drought period (cf. Chap. XIV, 6).

On trees with dense crowns one can observe the formation of two kinds of morphologically and physiologically different leaves: sun-leaves and shade-leaves. Similar conditions were observed in herbaceous plants by Saeki (1959). Sun-leaves can utilise higher light intensities better, while shade-leaves work more efficiently under lower light intensities because of their low respiration rates. Trees without shade-leaves have open crowns as, for example, birch. In this connection, we must remember that the average daylight intensity varies through the growing season in higher latitudes. Light intensity increases in the northern hemisphere until the end of June, from then on it decreases again towards autumn. This implies that a LAI which is optimal in

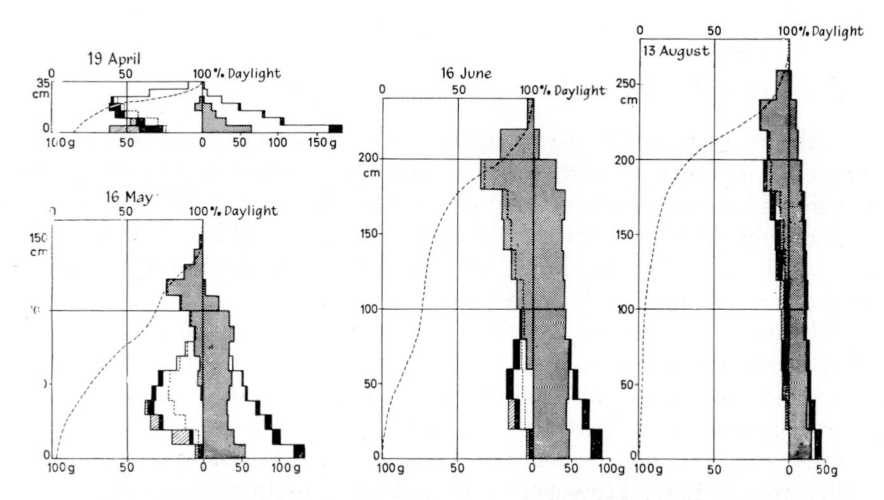

FIG. 6. Seasonal development of the *Phragmites* (shaded area)—*Sanguisorba* (blank area) community. Associated species black; withering leaves hatched lines. Fresh weight of the leaves (left) and of the non-green shoot parts (right) separated in 20 cm high strata. Dotted curve, light intensity as % of daylight (after Monsi and Saeki).

summer will be supra-optimal in autumn. This is reflected in a progressive withering of the lower leaves.

The conditions become even more complex in many-layered stands composed of several species. Here we find heliophytes and shade-plants in all gradations. The latter are pre-conditioned to lower light intensities. The smaller their proportion of non-green parts the lesser are their light requirements (e.g. ferns and mosses).

The LAI of such communities is made up of leaf-areas of ecologically very different species. Therefore, if possible, it should be estimated separately for each layer.

This was done by Monsi and his colleagues (Monsi and Saeki, 1953). As an example, we will use their observation of a *Phragmites-Sanguisorba tenuifolia* community on April 19, May 16, June 16 and August 13 (Fig. 6). The fresh weights of the CO_2-absorbing and non-absorbing parts are shown on the right and left sides, respectively, of the 100% light ordinate. The results are shown for successive 20 cm high strata above the soil. The leaf-area can be calculated from the fresh weights. One gram of fresh weight (68% water content) corresponds with 37-44 cm^2 in *Phragmites* and 64-80 cm^2 in *Sanguisorba*. The latter is a late-summer, not a spring plant; its leaves remain green into the autumn and long, white flower spikes are produced under favourable light conditions. But, in the *Phragmites* stand, *Sanguisorba* can only develop during the spring because it is not shaded. By the summer, however, the light conditions are so unfavourable that it cannot develop further and disappears before it has had a chance to flower (Fig. 6).

5. Growth capacity and competitive ability

The ability of a species to compete varies with the environment. Its competitive ability relative to other species is superior in some environments and inferior in others. Therefore, it would be desirable to evaluate competition quantitatively in ecological studies. This is extremely difficult because of the large number of morphologic-physiological attributes of plants involved in competition.

However, a necessary attribute for the ability to compete is the amount of dry-matter production. The larger the dry-matter production of a plant, the better and taller it will develop and, theoretically, it should produce more fruits and larger and more numerous seeds. Thus, greater production of dry matter is usually correlated with increased ability to compete. However, it is not the absolute quantity of dry matter produced that is decisive but the quantity relative to that produced by competitors.

If we designate the relation of dry-matter production (M) per individual of species A (M_A) to that of species B (M_B) as the growth capacity (G) of species A. Then, $G_{A/B} = M_A : M_B$ or $G_{B/A} = M_B : M_A$. It has been shown that growth capacity gives a fairly good indication of competitive ability among species of equivalent life form in certain environmental conditions (Bornkamm, 1961, 1963). If one compares, for example, the competitive ability between annuals such as *Sinapis alba*, *Avena sativa*, *Vicia sativa* and *Triticum aestivum* in different combinations with equal numbers of individuals (always a total of 100 plants/m^2) one observes that *Sinapis* grows fastest thereby dominating the others, next follows *Avena*, then *Vicia* and *Triticum*. The difference between the latter two is uncertain. If one knows the growth capacity of *Triticum* relative to *Sinapis*, and of *Triticum* relative to *Vicia*, one can calculate the growth capacity of *Vicia* relative to *Sinapis* for: $G_{T/S} = G_{T/V} \times G_{V/S}$ or $G_{V/S} = G_{T/S} \times G_{V/T}$.

Individual plants of *Sinapis* develop most poorly in a pure stand, i.e. when the other 99 plants on the 1 m^2 plot are of the same species.

But its development improves in mixtures with weak competitors. The greater the proportion of weak competitors among the other 99 plants, the better the development of the *Sinapis* individuals. By contrast, the weakest competitor, *Triticum*, develops its largest individuals in pure stands. The greater the number of strong competitors, the smaller the *Triticum* individuals. Therefore, intraspecific competition is less favourable than interspecific competition only when the species has a high competitive ability, otherwise the reverse is true.

However, growth capacity can be used as measure of competition only when the competing plants have a similar, upright growth habit so that light does not play a significant role in competition. If one plant species grows erect rapidly and the other spreads more or less along the ground in the shade of the former, the taller growing one always has the stronger competitive ability.

Iwaki (1959) carried out competition experiments with *Fagopyrum esculentum* and *Phaseolus viridissimus*. The production of dry matter was nearly equal in pure stands. The proportion of leaf material in the total dry weight was larger for *Phaseolus* (50-60%) than for *Fagopyrum* (20-60%). Thus the conditions were more favourable for *Phaseolus*. However, *Fagopyrum* has about twice the apical growth rate so that its leaves reach a more favourable position enabling it to photosynthesise more rapidly. The *Phaseolus* leaves, therefore, only obtain 16-35% of full daylight in mixed stands and their rate of photosynthesis declines rapidly. *Phaseolus* develops much better if sown 3-13 days ahead of *Fagopyrum*. In spite of this, *Fagopyrum* is not much affected since its stems rapidly outgrow the beans, thereby always receiving enough light.

It becomes obvious, therefore, that height growth, which is a species-characteristic feature that is only indirectly dependent on dry-matter production, is of very great significance in determining the competitive ability of a species. Height growth is decisive also in intraspecific competition, for example in a dense stand of sunflowers (*Helianthus annuus*). Seedlings from larger seeds grow a little faster in height (Kuroiwa, 1960). They receive more light and produce a correspondingly greater amount of dry matter so that their advantage over the smaller plants is increased even more. Since photosynthesis of the suppressed plants is reduced under the less favourable light conditions but not their respiration, they respire a greater proportion of their photosynthetic products. While net production is about 48% of the gross production among the fast growing plants, it declines to about 12% among the suppressed ones. The plants die once the net production reaches zero % under still less favourable conditions.

The competitive ability of different species always depends considerably on environmental factors. Shoots compete primarily for light, while roots compete primarily for water and nutrients. However, a clear separation of the relative influence of these factors is usually not possible, since the formation of the root system depends on production of dry matter by the leaves and, conversely, the photosynthetic rate depends on a sufficient supply of water and nutrients.

Bornkamm (1961) investigated the effect of light with regard to competition among *Sinapis alba*, *Agrostemma githago*, *Bromus secalinus* and *Anagallis arvensis* var. *azurea*. Light intensity (as % full daylight) amounted to: 100%, 60%, 30% and 15%. Each Mitscherlich-container was planted with 24 plants in pure stands, and with combinations of 2×12, 3×8 and 4×6, respectively. The dependency on light followed a logarithmic pattern if we consider total production (all combinations together). The largest proportion (80%) was produced by *Sinapis*, the smallest (1-2%) by *Anagallis*. A decrease in light intensity from 100 to 60% in pure stands had only a little effect in reducing the production of dry matter, while a further decrease to 15% of daylight reduced dry-matter production to $\frac{1}{4}$-$\frac{1}{3}$. The reduction was approximately one half in *Bromus* which, therefore, had the least light requirement. The situation in mixed stands was much more complicated. The least aggressive plant, *Anagallis*, benefited from reduced light that weakened its competitors. *Anagallis* produced approximately the same amount of dry matter in every 2-species combination, regardless of differences in light intensity. However, in 3-species combinations at 60% and 30% light it produced four times the amount of dry matter compared with that at 100% light, and in 4-species combinations even seven times the amount. Production in *Bromus* increased about twice in 4-species combinations at 30% daylight. This shows that the aggressiveness of *Anagallis* increased with decreasing light intensity when compared with its competitors. For example, its growth rate relative to *Sinapis* increased five times at 15% daylight, relative to *Agrostemma* it increased eleven times and relative to *Bromus* (at 30% daylight) three to four times, as compared with the growth rate at full daylight. Similar relations applied to *Bromus* in comparison with *Sinapis* and *Agrostemma*, while the competitive ability of the latter two species remained unaffected. However, one should not conclude that these relations are caused directly by light. The retarded development of the competitors at lower light intensities was not only derived from their reduced shoot systems but, in even greater proportion, from their reduced root systems, because the shoot/root ratio is known to increase with decreasing light intensity. Therefore, root competition especially is diminished at low light intensity, which gives a less aggressive competitor a better chance.

An investigation in the boreal spruce forest (Piceetum myrtillosum) by Karpow (1961, 1962a, b, c) illustrates these relations under natural conditions. The light conditions inside an old spruce forest (80% crown density) are very unfavourable. The undergrowth is composed predominantly of *Vaccinium myrtillus* and mosses. *Oxalis acetosella* grows with very low vigour. Light is usually considered to be the limiting factor for the development of the undergrowth. However, it was demonstrated that competition from spruce roots is much more important. When the roots were cut vertically to a depth of 50 cm around the outline of a one-metre square, the plants of the herb layer developed much better in spite of unchanged light conditions. *Oxalis*, in particular, covered the area rapidly suppressing the mosses. After a few years even *Rubus idaeus* appeared, which likewise spread rapidly. Investigation of photosynthesis in *Oxalis* showed that the rate of CO_2 absorption/leaf area was the same on the quadrats without root competition and on the control quadrats. In spite of this, the dry-matter production of the individual *Oxalis* plants increased tenfold. They became vigorous and increased their leaf-area (Fig. 7). The same was true for *Majanthemum bifolium* and *Trientalis europaea*, although here it was not quite so pronounced. This apparent discrepancy can be explained as follows. Both

FIG. 7. Development of *Oxalis acetosella* in a spruce forest without root competition (left) and with root competition (right). Half natural size (Karpow).

the roots of spruce and those of the herbs are restricted to the upper soil horizons; 90% of the root mass occurs in the upper 20 cm of the soil profile. Therefore, the roots of the herbs grow in the dense network of spruce roots. Soil water is not limiting in the moist, boreal climate; ground water is near the surface. The soil moisture contents were not significantly different in the experimental and control plots and were always much higher than permanent wilting point. The limiting factor was the amount of available nitrogen (N). Plants on the control plots exhibited clear symptoms of N-deficiency; a yellow-green foliage colour. In contrast, the leaves of the plants on the experimental plots were deep green. In addition, leaf analyses for *Oxalis* showed a N-content of only 2·06% on the control plots, but 3·62% on the experimental plots without root competition. The acute N-deficiency, from which the herbs usually suffer in these forests, causes the development of only a small leaf area relative to the non-photosynthesising organs.

Therefore, net CO_2 absorption is only sufficient to compensate for the respiration losses of the whole plant, there is no significant net production and thus no significant growth. However, elimination of root competition considerably improves the availability of N for the herbs, and they respond by increasing their leaf-area. This in turn leads to increased dry-matter production, which causes a further increase in leaf-area. That nitrogen really was the limiting factor in these cases was shown by the fact that where not only root competition was eliminated but also the whole herb and moss layers, the experimental plots eventually became invaded by nitrophilous species and those characteristic of improved mull-humus soils. Similar results were obtained through transplant experiments with more exacting species such as *Aegopodium podagraria*, *Ajuga reptans*, etc. However, one has also to remember that the cut tree-roots in the experimental plots will die and decompose in time so having a fertilising effect, particularly with regard to nitrogen. These experiments have shown that the poorly developed herb layer in boreal spruce forests is not primarily conditioned by low light intensity but by low N-supply resulting from competition with tree roots. The mosses are not affected by this. They receive their nutrients through drip water from trees and through needle litter. Thus, they are able to grow vigorously and cover the forest soil with their cushions.

Tree-seedling growth is also inhibited by root competition from the older trees. It was observed that spruce seedlings only grew very slowly in birch forests in spite of their favourable light conditions. They grew much better where root competition was eliminated in experimental plots by cutting through the birch tree roots. In the following experiment, Karpow has shown that mineral deficiency was the factor responsible. Control and experimental plots were uniformly fertilised with ^{32}P. The needles of spruce seedlings were found to be radioactive within the next few days. Radioactivity was 5 to 6 times greater in the experimental plots than in the control plots. No inhibiting substances or root excretions from the trees were found that could have had any influence in the controls.

It can be shown that soil water rather than mineral nutrients becomes the factor limiting understorey development through root competition, when investigations are done at the drought-limit of the boreal forest zone.

6. Life forms and competitive ability

So far we have discussed competition only with regard to dry-matter production of individuals during their period of development. However, the fact that different species usually maintain their position in a habitat for several generations is of particular significance for interspecific competition. Thus, successful reproduction of these species in a given habitat is a necessary prerequisite. From this it is evident that the generation time plays a role in competition. Short-lived species, e.g. annuals, must regenerate repeatedly from seed each year. Long-lived species, however, have ample time to suppress their competitors before reproduction from seed becomes a necessity. The ability to compete increases, as a rule, with age of the individuals. Here, I intend to discuss briefly the most important life form types from this viewpoint.

Annuals (Fig. 8, I)

Species in this group always complete their life-cycles within a year, sometimes within a few weeks. The germinating seedling (G) develops the primary, CO_2-absorbing leaves and the respiring stems and roots. About half of the dry matter present in the seed is lost in this process. However, carbohydrates are already accumulating after a few days of photosynthesis. Net production (PN) is obtained by subtracting the losses due to respiration. The growth increase (Δ) of new organs results from this net production. The leaf-area expands and net production increases until maximum dry-weight (Σ) of the whole plant is reached at flowering stage. Then certain parts die off and some of their carbohydrates are transferred to the ripening, usually small, but numerous seeds (m). This completes the life-cycle of one generation. Only the seeds persist through the unfavourable season. The seeds of the following generation (m′G) may germinate, but a certain percentage may die prematurely so that fewer (m′) plants may become established. Thus, in the following year, the number of plants may be $(m')^2$, etc. Annuals can, therefore, rapidly increase in density if they have no competitors. This may lead to strong intraspecific competition in a limited area. The ratio between CO_2 absorption and loss may decrease to a point where net production is insufficient for flowering and seed formation. Then the entire stand may disappear. This has been observed, for example, in the north-western states of North America, where *Bromus tectorum* forms temporary pure stands on overgrazed areas.

Biennials (Fig. 8, II)

Their development at the outset is similar to that of annuals, except that the biennials do not form a stem in the first year. Instead they develop a rosette which increases considerably in size throughout the year and they then over-winter in this form. The accumulated food reserves (F) are stored in the roots and the amount of food material is much greater than that stored in seeds. Thus, the plant has a better start in the second year which is further favoured by the rosette leaves, if they are not killed back by frost. The end result is a large-sized flowering plant with a capacity for forming abundant seed. Therefore, biennial plants have a much greater competitive ability than annuals in the second year, in spite of their somewhat slower increase in numbers.

Perennials (Fig. 8, III)

Their development through the first few years is similar to that of the biennials in the first year, except that there is usually a dying back of all above-ground organs, and food reserves (R) are deposited in special storage organs (rhizomes, tubers, bulbs). These are the only organs to survive the unfavourable season; the root system also remains partially intact. The food reserves increase from year to year. Therefore, new growth becomes increasingly more favourable and the plants increase in size from year to year until they produce flowers and seeds after a few years. However, simultaneously, they continue to increase their reserve food storage. Thus, the plants regenerate again in the following year together with the new individuals that originate from seed. Usually, there is also an increase in population size from vegetative reproduction, for example, development of several new bulbs which each support a new propagule in the spring. The longevity of perennial species results, in time, in their gaining a growing space of their own through competition with annuals and biennials. Seed production in perennials serves primarily for distribution over larger areas.

Perennial, evergreen plants (Fig. 8, IV)

They occur in zones without unfavourable seasons, for example, in the moist tropics. They do not need to develop storage organs. The entire plant remains intact from year to year and can continue to grow without serious interruption. It begins to flower and seed repeatedly after attaining sufficient vigour. Only certain monocarpic species die after one flowering period (e.g. *Agave* spp.). Their evergreen basal shoots may grow to several metres in height.

Woody, deciduous plants (Fig. 8, V)

They develop like perennial plants in the first year (scheme III). However, an important difference is that they do not shed the entire above-ground plant portion at the end of the growing season but only the leaves, while the woody stems and branches are retained together with the buds. At the same time, the stem is used as a storage organ. Formation of the stem utilises a large proportion of the net production in terms of dry matter. Therefore, not much is left over for enlarging the leaf-area. The small leaf-area in woody plants results in a low annual production of dry-matter during the first few years, for example in beech only 1·5 g. However, in time, woody plants gain an advantage over herbaceous plants due to the fact that the woody stem is perennial and increases in height from year to year rather than having to regenerate each year from ground level. The height attained by woody plants differs considerably. Chamaephytes and dwarf shrubs are usually shorter than 50 cm. Therefore, they occur in cold climates, where they receive a protective snow cover in the winter, or in arid areas, where the land surface is so poorly vegetated that larger height growth has no advantage in terms of light use. Nanophanerophytes or shrubs attain heights of about 4 m, whereas trees may grow up to 100 m high in extreme cases (*Eucalyptus regnans*, *Sequoia sempervirens*), thereby overtopping all other life forms.

The critical development period for trees is the first few years during which growth is slow, while they are subjected to competition with all other plant forms. Therefore, large seeds (*Castanea*, *Quercus*, *Fagus*) are favourable,[1] because they provide for more rapid, initial seedling growth and thus better reception of light. However, the ability of young woody plants to compete with herbaceous plants increases from year to year as they continue to gain height until they can unfold their leaves to full daylight, thereby producing an increasing shade effect on their former competitors. The woody plants gain absolute dominance when they form a closed crown canopy. Once the optimal leaf-area index is attained, there is little change in subsequent years. Consequently, there should be an increasingly unfavourable ratio between CO_2-absorbing and respiring parts resulting in a decreasing net production of dry matter. However, this is balanced by a continuous reduction of respiring plant parts. This results in part from the shedding of lower branches and from death of the inner woody portion. Non-respiring, dead, mature wood is formed in trunk and roots with increasing diameter growth, which does not interfere with the production balance. However, a complete equilibrium cannot become established, because the amount of living sapwood increases continuously with height and diameter growth. Therefore, net production decreases more and more until it becomes insufficient to maintain an optimal leaf-area. The crown then becomes more open in aging trees, some branches die off and finally the whole stem.

This point is attained sooner in light-demanding trees with relatively small foliage masses than in shade-tolerant trees. However, it is dependent also on the photosynthetic capacity and respiration intensity of the particular tree species.

Woody, evergreen plants (Fig. 8, VI)

There is no essential difference between these and the previous group. The leaves may last several years and leaf-shedding does not occur simultaneously.

[1] The size of one-year-old oak seedlings is proportional to the size of acorns (cf. Jarvis, 1963).

Evergreen foliage offers great advantages to the tree if temperature and water relations are favourable for tree growth throughout the year so that there is no interruption of photosynthesis. If evergreen plants are forced by drought, or cold, to undergo a dormant period each year, they have to develop leaves with a xeromorphic structure, which may not be superior in terms of all the year round productivity than thinner, larger, deciduous leaves. Both types occur in nature side by side in the same climate. We find deciduous *Acacia* species in Africa, while *Acacia* species with xeromorphic, evergreen phyllodes are found in Australia under the same conditions. The Arctic timberline is formed by deciduous birches, evergreen spruces and deciduous larches. The Alpine timberline is formed by a deciduous larch (*Larix decidua*) and by the evergreen *Pinus cembra*, and by two shrubs, the deciduous *Alnus viridis* and the evergreen *Pinus montana*.

The competitive ability of the different life forms under universally favourable conditions can be observed easily in a fallow field. Initially annuals dominate, they are soon suppressed by biennials. These are followed by perennial, herbaceous species, which become admixed with rapidly growing, woody tree and shrub species. After slow, but persistent gains, trees become absolutely dominant, so only shade-producing woody species dominate in the end. Therefore, the competitively-weak annuals (Therophytes) only persist as zonal vegetation in areas where climatic conditions are so unfavourable that other life forms cannot grow at all; for example, in the most extreme deserts with only episodical rains. They are often very numerous, forming 'vegetation buffer zones' in areas where each year a long period of drought restricts long-lived species to a great extent yet where enough moisture from rainy periods allows the development of a denser plant cover, i.e. in semi-deserts. There is no climate which especially favours biennial species. They maintain themselves in waste places, where the plant cover is not destroyed annually by man. Herbaceous, perennial species can predominate in semi-arid areas where trees cannot develop because of temporary lack of water as, for example, in the prairies and steppes.

Trees gain absolute dominance in climates that permit tree growth. Therefore, the larger portion of the earth's surface would be occupied by either deciduous or evergreen forests, if man had not cleared extensive areas for agricultural use or other purposes. Other life forms have a chance to regain secondary dominance on these areas (e.g. in meadows and croplands). They play a minor role in forest areas, where they only dominate restricted habitats that are unfavourable for tree growth (rocks, moors, etc.).

7. Energy cycles and production of dry matter in ecosystems

Natural vegetation is in equilibrium with its environment and together they form a complicated ecosystem. Within an ecosystem we can distinguish three major components:

1. The physical and chemical factors of the atmosphere and the soil.
2. The producers, i.e. the autotrophic organisms or green plants, which transform inorganic compounds into organic compounds and trap part of the radiant energy of the sun in the form of chemical energy.
3. The consumers, i.e. the heterotrophic organisms, which include the animals and micro-organisms. They return the decomposed organic compounds to the soil after having used them for their purposes. This brings about a circulation of matter.

It is the essential task of ecology to clarify the circulation of energy and matter in the different ecosystems of the world (Duvigneaud *et al.*, 1962; Ovington, 1962). However, this aim is still rather remote as we only know, in most cases, certain parts of the cycle. As an example, I shall refer to a study by Duvigneaud, who established some important data, on a hectare basis, for a west European mixed hardwood forest ecosystem. The annual energy that is radiated from the sun on one hectare of forest amounts to 9×10^9 kcal. The greatest part of this energy is changed into heat and re-radiated into the atmosphere. About $\frac{1}{3}$ is used for transpiration; that is, for transforming water from a liquid to a vapour phase. Only 1% of the radiated energy is used by the green leaves for photosynthesis, which is an extremely small proportion. Nevertheless, this amount is the basis for all life on earth.

About half of the organic matter formed from CO_2-absorption is lost immediately through the respiration of the plant. The remaining amount, i.e. the net production, in the form of dry matter, is about 12 ton/ha. Of this, about 4 ton/ha are formed in the foliage of the trees, about 5 ton/ha are laid down as wood and about 2 ton/ha are accumulated as root mass. The remaining 1 ton/ha is formed from the understorey.

The leaf surface is about 4·5 ha/ha. Therefore, the leaf-area index is 4·5. All leaves and herbaceous parts of the understorey die each year and are deposited as litter on the soil surface. The litter becomes decomposed and humified within a few years by micro-organisms.

Under natural conditions, of a constant reserve as wood per ha and a constant soil humus content, the annual amount produced must equal the amount recycled per year. In our example, this would therefore be 12 ton/ha. The greater part of this is decomposed directly through micro-organisms (fungi, bacteria) aided by the soil fauna. Only a very small proportion is first consumed by browsing animals, some of whom are eaten in turn by predators. But, in the end, even the dead bodies of these animals are returned to the soil, where they are decomposed.

The following values per hectare of total matter in a 120-year-old forest give an idea of how small the role of higher animals is in the circulation of matter in our forests:

Above ground:			
Producers		Consumers	
leaves	4 tons/ha	birds	1·3 kg/ha
branches	30 tons/ha	large mammals	2·2 kg/ha
trunk-wood	240 tons/ha	small mammals	5·0 kg/ha
understorey	1 ton/ha	predators	undeterminable
Total	275 tons/ha	Total *c.*	8·5 kg/ha
		plus invertebrates:	amount unknown
In the soil:			
root mass	several tons/ha	soil fauna*	*c.*1 ton/ha
humus	1-70 tons/ha	soil flora†	0·3 ton/ha

* Of these *c.* 0·2 ton are earthworms.

† Comprised of numerous bacteria [95 × 10^6/g soil], actinomycetes [36 × 10^6/g soil], and fungi [1 × 10^6/g soil].

These figures are only crude approximations. However, much would be accomplished if similar approximate calculations could be obtained for the zonal vegetation of different climatic zones. But this aim is still quite remote.

In my discussion of the different vegetation zones I will refer to such values wherever they have been established. However, in general, we have to confine ourselves to relationships between the composition of the plant cover and the environment, emphasising wherever possible any causal relationships.

8. The elucidation and significance of the ecological water relationships of plants[1]

Of the different environmental factors, temperature and water relations are those especially significant for vegetation. It is, nevertheless, important to consider the fundamental and different nature of the actions of both these factors on the plant. So far as temperature is concerned a plant is almost as exposed as if it were an unprotected dead thing.

The plant has no special temperature, and cannot protect itself against an increase or a decrease in temperature. Its temperature therefore simply changes with that of its surroundings. With increased radiation, the temperature of the leaves is raised above air temperature, but with more intensive transpiration as a result of the great heat consumption from the conversion of liquid water to vapour (at 20°C, 582 cal/g of water) their temperature decreases to below air temperature; these changes are clearly physically controlled. Measurement of the external

[1] See the recent publications by Walter and Stadelman (1968) and Walter and Kreeb (1970).

temperature generally enables one to estimate the temperature of the plant, especially that of the protoplasm in which all vital life processes occur.

The situation with respect to the water factor is quite different. In their water relationships, the higher plants—by contrast with lower plants—are more or less independent of environmental conditions, so that measurement of the humidity of the environment gives no information about the conditions under which the living processes occur in the plant.

In order to understand this, it is necessary to go into the physiological hypotheses of how the plants adapted to terrestrial life (Walter, 1967; cf. footnote p. 34). Life originated primarily in water and today it is wholly dependent on the presence of water. The living substance, the protoplasm, can only show the phenomena of life when it is hydrated. If it should dry out, it does not necessarily die and protoplasm can change to a dormant state in which no material changes are detectable.

Adaptation to a terrestrial life has only been successful with that branch of the plant kingdom which developed in fresh water.[1] It began with unicellular algae, such as *Chlamydomonas* and *Chlorella* and lead via green algae (Chlorophyta) and Characeae to those Cormophyta now living out of water, i.e. the mosses (Bryophyta), the ferns (Pteridophyta) and the seed plants (Gymnospermae and Angiospermae). Of course, in addition, a variety of lower green algae live out of water (Protococcales, *Trentepohlia*, etc., Fig. 9).

But one cannot strictly define these as land plants because they are only in an active living state when moistened with water or surrounded by an almost saturated atmosphere; otherwise they dry out like any other gel-like substance. This ability to withstand desiccation is a property common to the protoplasm of all plants. One must also consider, for example, unicells which are carried through the air; the spores of mosses and ferns and embryos in the seeds of higher plants. The question is, how do these cells without vacuoles survive, when their protoplasm is surely subjected to desiccation? Under these circumstances, the submicroscopic structure of the protoplasm seems to remain unchanged, so that, with re-exposure to water, the normal processes of life are renewed. Such organisms, which can exist without water but whose degree of hydration depends on the changing humidity of the environment, can be defined as 'wetness-followers' or *poikilohydric types*. The majority of them are associated with relatively damp habitats, but they can also be found in deserts.

[1] Many claim that the oceans were originally fresh-water basins, because marine salts are only the non-sedimentary part of the products of weathering and of volcanic gases which have reached the sea. NaCl, the chlorine of which probably came from volcanic gases, could easily have reached sea water in the course of time. This would mean that present-day marine algae must have come from fresh-water algae. It is only if this were so, that one can explain how these two elements in sea water, sodium and chlorine, in general, play no part in the life of non-halophytic plants.

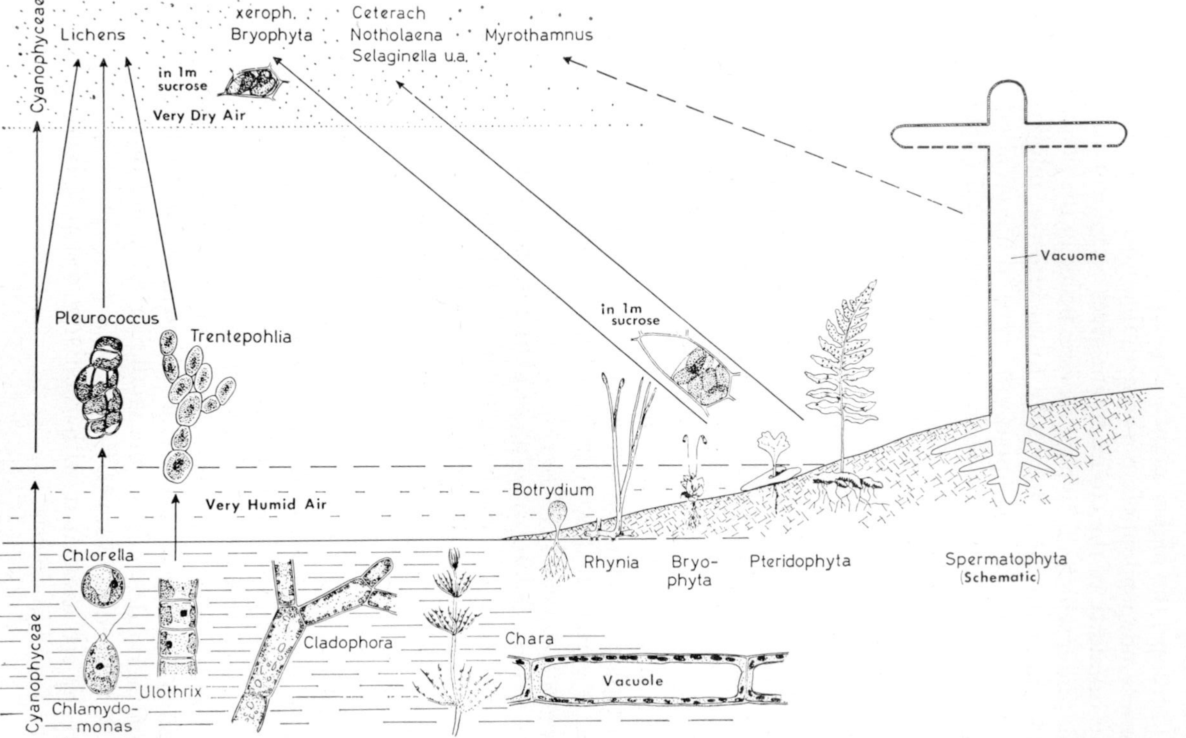

FIG. 9. Diagrammatic representation of the adaptation of plants for terrestrial life: Left: Lower algae with nonvacuolated cells evolved to the poikilohydric aerial algae. Middle: Vacuoles developed in the Chlorophyceae and Characeae, which live in water. Right: Homoiohydric cormophytes evolved. Bryophytes still rely upon high humidity of the air and the species living in arid areas revert to poikilohydric organisms. (Note also the difference in relative vacuolar volume: the large cells of the Bryophytes grown in humid habitats have large vacuoles and distinctly plasmolyse in 1 M sucrose solution; cells of Bryophytes grown in dry habitats are small and plasmolyse only slightly at this concentration because of the high volume of protoplasm and high cell sap concentration in the only very small vacuoles.

Poikilohydric forms are also known from the ferns, but none is found in the Gymnosperms. Poikilohydric plants are rare exceptions in Angiosperms (*Myrothamnus*).

The biomass of these poikilohydric types is not large, however, so that they play no important role on the earth; their cells, even when turgid, are small. Because of the restricted amount of light passing through their surface area, their production of organic material can only be very small, so much so, that they spend the greater part of the year in a dormant state.

However, in water, filamentous algae possessing relatively large, vacuolated cells which were low in protoplasm, comparable with types like *Spirogyra*, *Cladophora* or *Vaucheria* evolved at an early date (bottom of Fig. 9). *This vacuolation gave them the opportunity to develop a large surface area exposed to light, even though they only had a small protoplasmic mass*, and thus enabled organic material to be produced more easily. (This reduced the need for nitrogen as nutrient which is often a limiting factor in autotrophic plants.)

Of course, the capacity to dry out is lost with vacuolation. Peripheral protoplasm is subjected to severe mechanical strain, due to the drying out, between the reduced vacuole and the firm outer membrane so that its delicate structure is damaged and the cells die.

Vacuolation brings about a series of important changes. The protoplasm of living vacuolated algae is not only bounded by an external watery medium but also by an *internal watery medium—the cell sap*. Thus the degree of hydration of the cytoplasm is determined only by the internal medium.

This vacuolation has led, at the same time, to the conquest of the land by plants, in a different way from that of the poikilohydric types. It has proved to be a much more successful method.

I shall now make a big jump directly to the ultimate members of this line of development, to the particularly interesting *flowering plants*, which can be discussed in terms of their three basic organs—root, stem and leaves (right of Fig. 9).

The flowering plants are multicellular organisms of which one part, the root system, is embedded in a water-holding medium—the soil. By contrast, the shoot system stands in apparently dry air. Any drying out of the shoot system is prevented by a water-impermeable cuticle. If plants could develop a layer impermeable to H_2O but permeable to CO_2, then this would be the most *perfect protection* conceivable against water loss. A cuticle, however, prevents not only aqueous, but also gaseous exchange. Thus it was necessary to develop a 'pore-opening apparatus'—or stomata in the CO_2-assimilating organs of land plants.[1] The internal parts of the plants are filled for the most part by the internal medium to form a 'vacuome'—by which is meant the sum total of all the vacuoles.

[1] Petals which do not assimilate CO_2 are almost devoid of stomata. Their transpiration rate is very weak (Hellferich, R., 1942). The O_2 necessary for respiration can diffuse through the cuticle as a result of the high partial pressure of the atmosphere. In heterotrophic flowering plants the stomata are also reduced.

This inner water medium, the cell sap of the vacuome, first gave plants the chance to maintain the hydration of the protoplasm and also to support the growing organs in a dry atmosphere external to the watery medium. We can compare the cell sap of the vacuome to the body fluids of animals, although cell sap is divided amongst a vast number of isolated cells, in part highly specialised cells, which have a direct influence on the composition of the sap.

The average osmotic concentration of expressed cell sap can be estimated after the death of the cells of the whole organ.

The physiological basis of the plant's water balance, the complicated stomatal mechanism and all the other adaptations for water uptake, water conduction and water loss, can only be understood if one appreciates, in the first place, that they all serve to facilitate CO_2 uptake through the leaves for photosynthesis, together with maintaining plant water balance at the lowest possible concentration of the cell sap concomitant with the highest degree of hydration and activity of the protoplasm under varying extreme conditions. For *homoiohydric types* or 'self water-regulating' plants are, in some degree, 'water plants' but with an inner watery medium whose concentration is very different from that of the surrounding environmental conditions of the air.

To assess the water balance of homoiohydric plants it is more important to understand the state of the inner aqueous medium of protoplasm than to characterise the plant's environment. With regard to moisture, the cells are to some extent controlled environment chambers for protoplasm. When using a controlled environment chamber for research, only the humidity inside it is of interest, not that outside. Similarly, the osmotic cell sap concentration of various ecotypes from different climatic zones is the key to determining how much protoplasm is affected by environmental conditions (see osmotic spectra, Walter, 1960, pp. 238-258).

On this basis, I determined the osmotic value of the cell sap in every climatic zone, in order to see under which conditions of moisture (hydrature) the protoplasm was active and whether the dryness of the atmosphere outside the plants was affecting it or not. To demonstrate this—that the hydrature and therefore the degree of hydration of the protoplasm is dependent only on the osmotic concentration of the cell sap—I carried out the following simple thermodynamic model experiments (Fig. 10).

1. I put a gel-like material in a 'Pfeffer cell' (a clay cylinder with a semipermeable membrane) and placed the whole apparatus in water. The material showed maximal swelling when equilibrium was achieved.

2. I exerted a pressure of 14 atmospheres on the material. Some of the water was forced out and it continued to come out until the pressure inside was equal to the external pressure, i.e. until it had reached 14 atmospheres.

3. I replaced part of the material internally by a solution whose

osmotic potential was 14 atmospheres; it was separated from the material by a semi-permeable membrane. The equilibrium remained unaltered. This model represents a water-saturated cell with a suction potential (or DPD) of exactly 0 and a hydrostatic (turgor) pressure of 14 atmospheres—which is exactly equal to the osmotic potential of the solution. The material represents the peripheral protoplasm of a

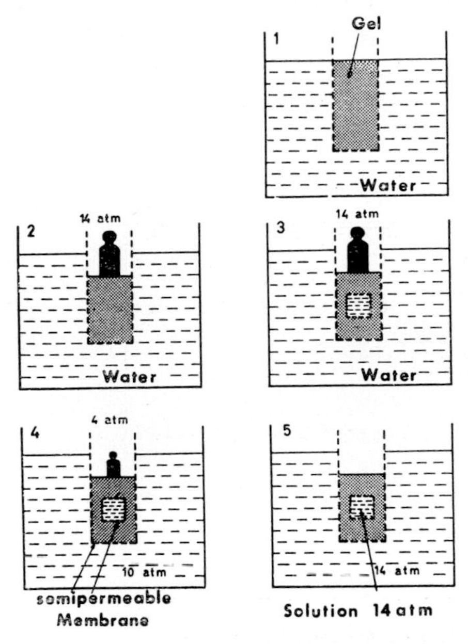

FIG. 10. A model demonstrating that protoplasm of a cell does not reach maximum imbibition in a water saturated cell. For details see text.

1. A gel or gel-like material is placed into a 'Pfeffer cell' (a clay cylinder with a semipermeable membrane) and immersed in water. After it has been maximally hydrated, the material has reached equilibrium.

2. Next, the gel is subjected to a pressure of 14 atm. Some of the water is forced out. The gel dehydrates until the pressure of the colloidal swelling and the external pressure are at equilibrium (at 14 atm).

3. Part of the gel is replaced with a solution of potential osmotic pressure (Π) equal to 14 atm, and is separated from the gel by a differentially permeable membrane. The equilibrium remains unchanged. This model corresponds to a water-saturated cell with a diffusion pressure deficit (DPD) of zero and an external pressure (here wall pressure) of 14 atm equal to the potential osmotic pressure (Π) of the inner solution (i.e. cell sap). The gel represents the protoplasmic lining of the cell and does not reach maximum hydration, although the absolute vapour pressure of the cell sap (as a consequence of the hydrostatic pressure upon the cell) is equal to the water pressure outside.

4. The degree of imbibition remains unchanged if the external pressure is lowered to 4 atm while at the same time the outer diffusion pressure deficit (DPD) is raised to 10 atm.

5. The degree of imbibition remains unchanged if the external pressure is removed and an outside solution, equal in concentration to that of the inner solution, is substituted for the latter. In all these cases the hydration of the gel (protoplasm) remains unchanged as long as the concentration of the inner solution (cell sap) remains the same. Maximum imbibition (as in 1) is not reached.

cell but it is not swollen to its maximum although the absolute vapour pressure of the cell sap (the inner solution) as a consequence of the hydrostatic pressure upon the cell is equal to the water vapour pressure outside.

4. The swollen state was not altered when I lowered the hydrostatic pressure and increased the external suction pressure correspondingly.

5. The swollen state remained unchanged if the hydrostatic pressure

was removed and an external solution, equal in concentration to the inner solution, substituted. In all these cases, the water content of the material capable of swelling (protoplasm) was unchanged, so long as the concentration of the inner solution (the cell sap) remained unchanged. Maximum imbibition (Experiment 1) was not reached.

Direct proof can be given that the protoplasm of a water-saturated cell is not in its maximally swollen state. For this purpose, the thin cell walls of fungal hyphae can be digested with enzymes from digestive glands of snails (Strunck, 1965). The naked protoplasm so formed retracts itself into a sphere which, when placed in pure water, swells and bursts. These spherical protoplasts can only remain alive in a 12·5%

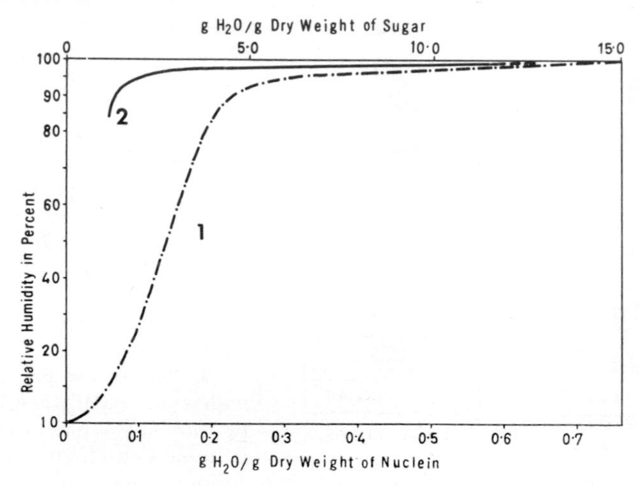

FIG. 11. Relation between colloidal swelling of Nuclein (1) and relative vapour pressure (after Katz, 1918). The relative water pressure of different concentrations of sucrose solutions (2) is included for comparison. For both curves the water content is calculated as grams of water per gram of dry substances.

sucrose solution, with an osmotic pressure of 10 atmospheres (corresponding to case 5 in the model experiments); they form vacuoles and those parts which contain a nucleus can grow a new hypha.

We come now to the question of how to estimate the degree of hydration of the protoplasm which is so important to all the vital processes of life. It is often supposed that water is so strongly bound by colloids that its hydration only varies a little with fluctuations in humidity. This view is not correct.

The curve showing the hydration of all dead materials capable of swelling plotted against relative humidity is in the form of an S (Fig. 11). If we proceed from a maximum swelling of 100%, we can see that with a decline in humidity, the greater part of the water is easily given up by the swollen material. When the humidity is 96%, the solution has an osmotic value of 55 atmospheres (or a depression of the freezing point of $-4 \cdot 5°C$) and the curve shows a bend, i.e. the remaining water seemed quite firmly held; but one should not refer to this as 'bound water'.[1] In cells lacking vacuoles, the volume of protoplasm can be measured, e.g. with the carpospores of *Lemanea*, it can be shown that the swelling curve of living protoplasm has the same shape as that of a mixture of equal parts of nuclein, casein and gelatine (Fig. 12) (Walter, 1923). The physiological range of protoplasmic swelling in homoio-

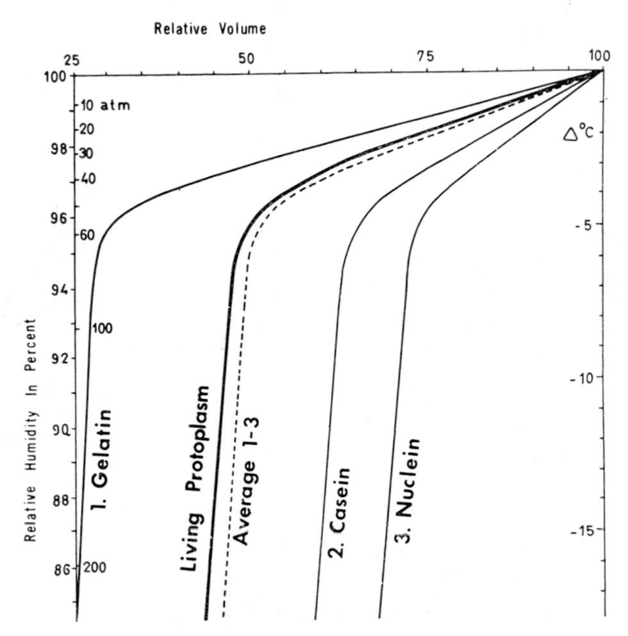

FIG. 12. Relation of relative vapour pressure (or potential osmotic pressure or freezing point depression) and hydration of protoplasm to various non-living colloidal materials. Ordinate: left scale, relative humidity in per cent and potential osmotic pressure in atm (at 20°C); right scale, freezing point depression. Abscissa: relative volume of the imbibant.

hydric plants lies between a relative humidity of 99·7 and 96%; the equivalent osmotic concentration being between 40 and 55 atmospheres. Higher osmotic concentrations are seldom found in plants other than halophytes and lie mostly in the sub-lethal range.

With vacuolated cells, it is not possible to estimate the volume of the protoplasm. We can, therefore, make no direct assessment of the variations in the degree of hydration of their protoplasm yet, although indirect, quite definite connections between these and the 'water activity' can be expressed by the use of the relative humidity (Fig. 12). In the range from 100%-96%, the slope is approximately linear.

[1] The same applies also to a sucrose solution (cf. Fig. 11 Curve 2).

Moreover, there is an equation connecting relative humidity and the osmotic concentration (= potential osmotic pressure Π^*) of the solution

$$\Pi^*_{20} = \frac{-1000\ RT}{Vw} \ln \frac{p}{p_0} = -3065 \log \frac{p}{p_0}$$

in which p/p_0 is the relative vapour pressure in the thermodynamic sense, i.e. measured under identical external conditions such as pressure, temperature, etc. The osmotic concentration of the cell sap, which determines the degree of hydration of the protoplasm, gives at the same time an idea of the hydration measured by the relative humidity. The minus sign means that an inverse relationship obtains, i.e. when the osmotic value increases, then the relative humidity drops and vice versa. I proposed the expression 'Hydrature' (= *Hydratur*) for hydration of the protoplasm as measured by the relative humidity. The ending 'ature' ('atur' in German) implies an activity state (compare, e.g. temperature in relation to the thermal state). This quantity is without dimensions and independent of external conditions; it is only dependent on the osmotic concentration of a solution, or on the degree of hydration of a material capable of swelling. Hydrature represents the thermodynamic magnitude a = the relative activity of the H_2O (in %). I return to consider this problem more exactly in Chap. VII, 5.

If we wish to understand the water conditions of a plant in nature we have to know how far the plant is able to keep the hydration of the protoplasm as high as possible. The osmotic concentration of the cell sap gives us the information. The higher this value, the lower the hydrature of the protoplasm and the less favourable the water relations. Many plants take this into account and give up CO_2 assimilation, close their stomata in order to maintain a lower osmotic concentration and hence a higher hydrature of their protoplasm. Others, on the other hand, keep their stomata open in order to facilitate CO_2 uptake for photosynthesis.

This antagonism between the maintenance of a water balance and the possibility of an active gas exchange to promote photosynthesis, forces plants to compromise and has led to complicated adaptations in connection with the water economy of plants.

Water absorption, the conduction of water and transpiration must be balanced so that the concentration of the watery medium which borders on the protoplasm, i.e. the cell sap, can remain low in order to ensure highly hydrated protoplasm and so guarantee its activity. Transpiration must be regarded in the first instance as an inevitable evil to the plant which it cannot avoid because of the gas exchange of the leaves. In order to transport mineral salts from the roots to the leaves it is not necessary to transpire so much water; even a weak guttation would be sufficient. It follows, therefore, that high production of dry matter is always associated with a great water loss. Plants which grow rapidly, transpire greatly at the same time. However, these considerations only apply to plants which do not grow in salty soils. If the soil contains soluble salts, e.g. NaCl, plants take them up until a particular concentration is reached in the cell sap of the leaves. Plants whose protoplasm is not resistant die. In the salt-resistant types which are called *halophytes*, the NaCl exerts a specific swelling effect on the protoplasm, so that the raising of the osmotic concentration does not decrease the hydration of the protoplasm. Halophytes grow well even with very high osmotic values of 50 atmospheres or more. Indeed, a certain salt content benefits the halophytes. In this case, the value of the osmotic concentration is unimportant but the salt content should be estimated (see Chap. VII, 6).

If we turn back to the more primitive Cormophytes, the mosses, it can be seen that they are much less protected against drying out; they grow best of all, therefore, where the atmospheric humidity is permanently high. So, if they appear in dry places, they must revert secondarily to a poikilohydric way of life. These xerophytic mosses are characterised by small almost non-vacuolated cells in which the protoplasm, as in the lowest algae, is capable of drying out (Fig. 9).

The ferns are much better adapted to life on land but in these also the water conducting system is less efficient, therefore, they favour wet places but, nevertheless, have relatively high cell sap concentrations (p. 121). Because of this, land plants were almost exclusively confined in the Permo-Carboniferous period—the age of Pteridophytes—to swampy areas, while the solid dry land was almost free of vegetation (Erhart, 1962). The ferns which today are found in arid areas (cf. pp. 314, 398), just like some mosses, have similarly reverted secondarily to the poikilohydric way of life. In these plants, cell reduction also occurred, and the vacuoles were reduced or filled with cell sap containing phloroglucin-tannins etc., which solidify when dry. Mechanical damage of the protoplasm by drying-out of such plants is also prevented by easy deformability of the cell-wall resulting in curling of leaves. There is no poikilohydric type in the gymnosperms, but since their conducting system is only composed of tracheids, they must be able to maintain the hydrature of their protoplasm in order to limit water losses greatly. They are characterised by a relatively xeromorphic habit.

Among the angiosperms, the poikilohydric types are rare exceptions. The most important example of this type—*Myrothamnus flabellifolia* (Rosales)—comes from South Africa (see Chap. IX, 3c).

Another type must also be added which is encountered in the arid regions of the temperate zone, the *Stenohydric xerophytes* which close their stomata immediately with the onset of drought, bringing photosynthesis to a full stop. With long continuing drought these plants eventually become starved but their leaves do not dry up, they turn yellow and the osmotic concentration, far from rising, on the contrary often even sinks. Many monocotyledonous geophytes belong to this group and also herbaceous *Euphorbia* spp. etc. A very typical representative of this group is the Sonoran-desert species, *Fouquieria splendens* (Chap. VIII, 3e).

The different ecological types of drought-resisting plants will be fully considered in Chap. VII, 5.

9. Zonal climatic types

Climate and vegetation show a very close relationship. In a way it is possible to regard vegetation as an indicator of climate and to utilise plant distribution for the delimitation of climatic zones. Therefore, we must consider the more important criteria used in classifying climates without attempting to give a complete outline of climatology (Köppen, 1931; Alissow, 1954; Alissow *et al.*, 1956).

Climate encompasses all weather factors in respect of their behaviour pattern over a given period of time (e.g. day, year, etc.). The ecologist is particularly interested in those parameters of climate that have a pronounced effect on plant response. These are temperature and water conditions.

In order to evaluate properly the effect of the climatic water factor on the plant it is necessary to consider potential evaporation together with precipitation.

With regard to the temperature factor we can recognise the following principles:

1. The annual mean temperature decreases in a constant manner from the equator to the poles, because of the diminishing amount of total solar radiation received per unit area correlated with the decreasing incident angle of the sun's rays.

2. By contrast, the annual temperature range between seasons increases from the equator to the poles according to the annual variations in radiation.

In calculating mean temperatures separately for corresponding north and south latitudes one notices that the annual mean temperatures decrease more rapidly in the southern hemisphere although the seasonal differences increase less rapidly. This can be attributed to the smaller land mass in the southern hemisphere and the buffering influence of the water mass; that is, to its more oceanic climate.

3. A comparison of temperature relations on land and over water shows a smaller decrease of annual mean temperatures polewards over water, because of its movement and its greater heat absorbing capacity. For the same reason, temperature differences between seasons are less pronounced over water and more pronounced on land. On land they increase with increasing distance from the coast. The climate becomes more continental, and the date of change-over times between seasons occurs somewhat earlier.

4. A thick layer of ice on water cools off in winter like a land surface, whereby the cooling effect of the ice is increased through a snow

cover, because of the strong re-radiation. In addition, an increase above the freezing point is retarded by ice and snow.

5. The temperature distribution is influenced by the direction of air and sea currents: they transport cold air and water masses from the polar regions and warm air and water masses from the equator. Cold sea currents move along the subtropical west coasts of the continents, such as the California, Peru, Canaries, and Benguela currents; fog belts form above these. Warm currents, such as the Kuroshio, Agulhas, Brazil and Gulf streams, are found on the east coasts of continents. Here, thunderstorms and torrential rains are common. Where warm and cold streams meet, e.g. near the Grand Banks of Newfoundland and the sea of Okhotsk, there is continual fog formation. Air currents from the sea alter the temperatures on the land to those over the water. Therefore, a less extreme climate is produced by predominantly west winds on the western parts of the continents in the temperate zones.

6. Local deviations in temperature relations, up or down, can be caused by the checking action on winds of mountainous areas.

7. Average temperatures decrease and temperature extremes increase with altitude in mountains. However, valley and peak positions may show deviations (periodic temperature inversions).

The water relations on the earth's surface are much more complicated. The distribution of precipitation is related to the planetary winds which originate from a particular circulation of the atmosphere. These winds can deviate from their course through the influence of barometric lows or highs over the mainland. They are then called monsoons.

One cannot understand the distribution of precipitation without a knowledge of the circulation of air masses. The seasons in the tropical and subtropical zones are conditioned less by the annual march of temperature than by periods of rain and drought.

Let us consider first the air circulation in spring or autumn (Fig. 13). At these times the sun is at its zenith-position at the equator. The moist air is heated considerably in this zone, it rises adiabatically causing low air pressure at the earth surface. The rising air cools off and its vapour condenses. This condensation slows down the cooling process and maintains a higher temperature in the air masses than is present in the surrounding air. The rise to great heights continues rapidly, the condensed water being precipitated as zenith-rain. At greater heights the air masses themselves flow off to the north and south and descend to the earth's surface at approximately the 30th parallels. In descending, the air warms up considerably, so becoming very dry. Consequently, there is no precipitation in this zone and the air pressure at the earth's surface is always high which results in calm air.

From these high pressure zones the air moves back to the low pressure zone at the equator. Because of the rotation of the earth, the winds are deflected to the right in the northern hemisphere and to the left in the southern hemisphere, resulting in the north-east and south-east trade winds. The doldrums are along the equator itself, where there is usually no wind. West winds prevail in higher latitudes (40-60°), moving with considerable regularity around the earth, particularly on the southern hemisphere, where they are known as the 'roaring forties'. An easterly direction prevails once again in the polar areas. Here, at the border between the polar front and the warmer air below is the area where the large, cyclonic storms that originate in the northern hemisphere mainly around Iceland and the Aleutians continuously develop.

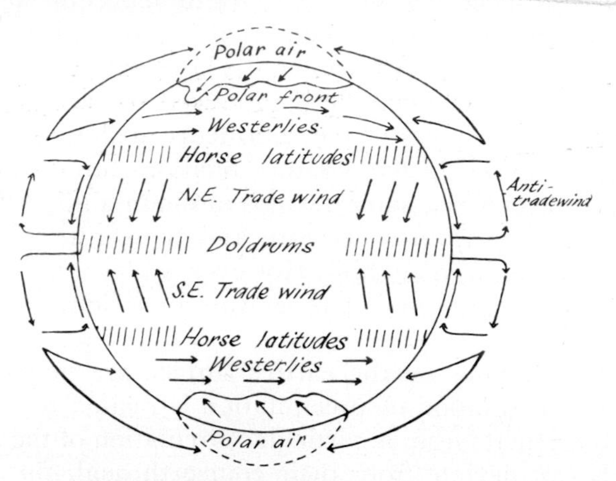

FIG. 13. Air currents at the equinoxes (after Gebauer). From Walter (1960).

These are the low pressure areas, the cyclones, which move from west to east forcing themselves more or less deeply southwards where they gradually disintegrate.[1] The cyclones are accompanied by rain because of the contact of cold and warm fronts of the air masses. These contacts are unrelated to the seasons and depend only on the movement of cyclones. In contrast to the cyclonic rains, zenith-rains occur only during the warm season, when the sun's position is at the zenith. Since the position of the sun changes throughout the year, there is, in addition, an associated change in the atmospheric circulation system (Fig. 14). In the northern summer, the sun's position at the solstice is over the Tropic of Cancer and, in the southern summer, over the Tropic of Capricorn. The two air mass systems shift in a corresponding manner from the equator northwards in the northern summer and to the south in the southern summer. Also the tropical doldrums, which separate the two systems of air masses, change their position about the equator with the movements of the air mass systems. However, the doldrums oscillate

[1] In the southern hemisphere, e.g., south of Australia, migrating anticyclones usually occur with areas of low pressure troughs in between.

mainly between 0° and 10° N because of the stronger air circulation in the southern hemisphere. The paths of the cyclones, with their frontal movements of warm air followed by cold air, undergo a similar deviation. In winter they extend farther south to nearly 40° in the northern hemisphere, while in summer their routes are farther to the north.

To understand the water balance, we have to know the potential evaporation, which depends to a great extent on temperature. If we consider temperature conditions and the amount of precipitation simultaneously, we obtain climatic zones that are of significance to vegetation. On this basis we can recognise the following major climatic zones:

I. Equatorial zone; about 10° N and S, with two rainy seasons (correlated with the two zenith-positions of the sun), which are usually interrupted only for a short period. Annual rainfall and humidity are very high and the monthly mean temperatures vary only a little.

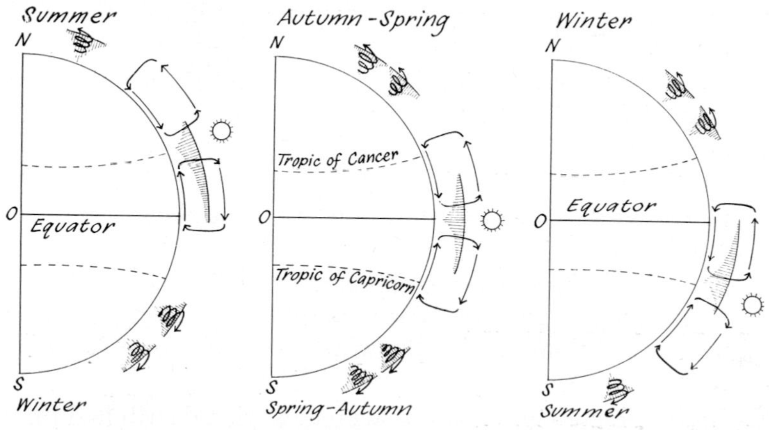

FIG. 14. Schematic representation of the oscillation of the precipitation zones in relation to the position of the sun. Amount of rain is indicated by length of parallel lines. After Walter (1960).

II. Tropical zone; north and south of the former, to about the 30° latitude. It has a moist summer-rain-season (coinciding or somewhat lagging behind the zenith-position of the sun) and a drought-season during the cool season. Mean annual rainfall decreases with distance from the equator; the cool season is, as a rule, still frost-free and the humidity of the air is low.

III. Subtropical arid zone; about the horse latitudes. Precipitation and air-humidity are very low; the yearly and daily temperature variations are large, because of the dryness of the air and the strong re-radiation. Night frosts occur.

IV. Transition zone with winter rains which, in the summer, comes within the range of the high air pressure area with its descending, dry air masses and, in the winter, is within the range of cyclones. Frosts are more common, but a pronounced cold season is absent. The summers are hot. This is the climate characteristic of the Mediterranean area, of California, south and south-west Australia, the Cape of Good Hope and the south-west of South America.

V-VIII. Temperate zone with cyclonic rains at all seasons. Annual precipitation decreases with increasing distance from the sea, so that a characteristic summer maximum becomes more and more evident. At the same time, the temperature range increases; the summers become increasingly hotter in the continental zones and the winters, particularly, become increasingly colder.

The temperate zone has a range of different climates in relation to vegetation types. Therefore, it is subdivided as follows:

V. Warm-temperate climate; in which all seasons are more or less moist with a moderate, not very pronounced cold period in the winter.

VI. Typical temperate climate; with cold but short winters, as for example, in Central Europe; or with nearly frost-free winters but very cool summers instead (extreme oceanic climate).

VII. Arid, temperate climate; with a continental character, strong temperature contrasts and low precipitation and low air-humidity.

VIII. Boreal or cold-temperate climate; which is characteristic of an extensive zone in North America and north Eurasia. The summers are cool and moist, the winters are cold for more than half the year.

IX. Arctic climate-zone; with low precipitation distributed throughout the year, mostly falling in the form of snow. The summer is characterised by uninterrupted daylight but it is short and cool, and the air-humidity is usually high. This zone grades into the polar region, where plant growth is absent. This last is true for the entire Antarctic.

This zonation is complicated by the monsoon winds, particularly on the south and east side of Asia, the largest continent. The south-west and south-east monsoons dominate throughout the summer months in the area between 10 and 23° north, from East Africa to the Philippines, where they are sucked in by the low pressure area over the Asiatic continent and further strengthened by the high pressure that prevails at this time (southern winter) over South Africa and Australia. The north-east trade winds are completely suppressed. They blow only in the

winter when the pressure distribution over the continents is reversed and, at this time, they extend south of the equator and are also called monsoons. Such monsoon winds, less pronounced than over the Indian ocean, however, occur also in north Australia, on the west coast of Africa (Liberia, Sierra Leone) and in the Gulf of Guinea (Cameroons) as well as on the east coast of North America. Along the west coast of Europe they are masked by cyclonic winds.

Wherever winds coming from the sea come up against mountains, there are intensive orographic rains on the windward side, while the lee-side is relatively dry. If the winds move constantly in one direction, as for example the trade winds, the climatic differences between windward and leeward side are very pronounced.

Rain that falls on land is partly evaporated and later re-precipitated as rain; all precipitated water by no means originates from the sea, therefore. This is the case only in the winter, when evaporation from the land in the temperate zone is slow because of low temperatures. At this time, precipitation is largely limited to areas neighbouring large bodies of water (maritime areas).

Water that does not evaporate from the land is returned to the sea via the rivers. This amount is estimated for the entire earth as 37,000 km^3 per year. About twice this amount evaporates from the land and about eight times from the sea. Three times this amount is precipitated on the land and seven times this amount on the sea. These are the approximate features of water circulation over the earth's surface.

The distribution of precipitation, shown in Fig. 15, results from the interaction of the factors mentioned and more local influences. In particular, mountains which intercept a vapour-laden wind track and force it to rise, cause strong precipitation. This applies to the trade winds on the coast of Guyana and Brazil (from Cabo de São Roque to Cabo Frio), the east coast of Madagascar, the east coast of Australia and south-east New Guinea. Similar conditions apply to the south-west monsoon during the warm season with regard to Liberia, Sierra Leone, Malabar, Arakan, Tenasserim; to the south monsoon with regard to the south-east coasts of China, Japan, and Kamchatka, to the north-west monsoon during the colder season in respect of north-west Japan; to the north-east monsoon for the east coasts of Taiwan, the Philippines and Indochina; to the north-west monsoon respecting the eastern part of the Malaysian Archipelago and to the north monsoon with regard to the north-east coast of New Guinea. This phenomenon is seen also on all mountainous oceanic islands within the range of the trade winds or the monsoons. Even large, gently rising plains have an inhibiting effect on fast-moving sea breezes, forcing their air masses to ascend to greater heights. This explains the high rainfall on the plains of the Amazon and on the lower Brahmaputra and its mountainous background-areas. In general, mountains are characterised by higher moisture than are their surrounding lowlands, so the mountains are often hidden in clouds.

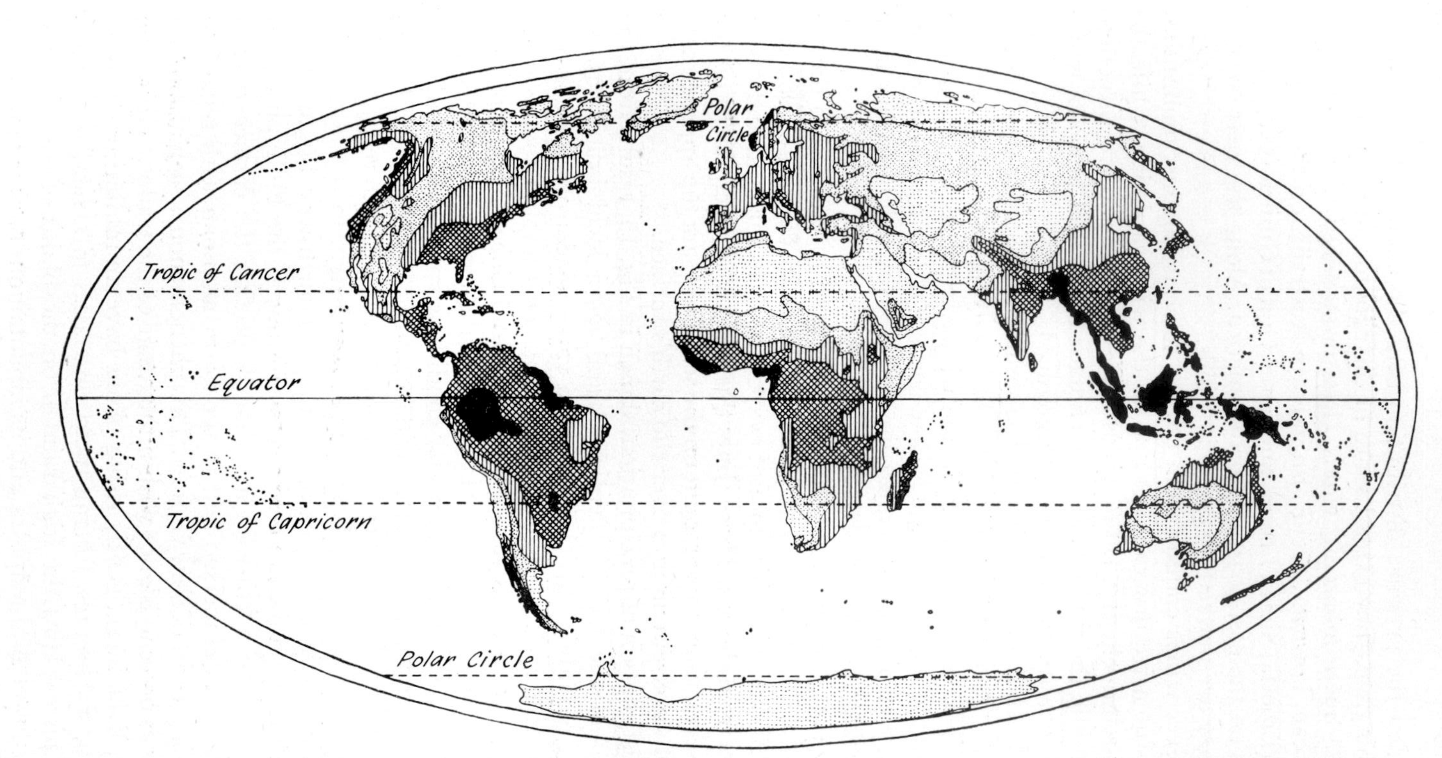

FIG. 15. Precipitation map: widely spaced dotted, less than 250 mm; densely dotted, 250-500 mm; vertical lines, 500-1000 mm; cross-hatched, 1000-2000 mm; black, more than 2000 mm. Walter (1960).

A decrease in precipitation becomes noticeable only on the higher mountain tops that reach above the lower wind tracks. Only the tallest mountains, often snow-covered, reach above the general cloud cover (Peak of Tenerife, Cameroons Mountain, Kilimanjaro, etc.), and they afford a beautiful view from a distance, particularly before sundown.

Of course, other climatic parameters enter into the evaluation of total climate although they are not of such immediate ecological significance and usually have only an indirect effect by influencing the temperature and water relations. The major climatic zones can be further subdivided, but the definition of more narrowly-defined climatic types is rather difficult. Accurate maps of various climatic parameters are available for entire continents or large geographic regions; for example, for mean annual temperatures, January and July temperatures, or for temperature ranges, for the duration of drought periods, the distribution of precipitation at different seasons, for indices of drought or evapotranspiration, etc. However, the specific climatic type of any particular place or area can be interpreted only with great difficulty from detailed comparisons of such maps.

Thus, any description of the total climate of particular areas which could go a step further than the classification of broad climatic zones would meet a long-felt need. Such an evaluation should also incorporate the local characteristics due to the distribution of continents and oceans and the influences of air movements.

The most detailed evaluation of climatic regions by means of temperature, precipitation and its seasonal distribution is still the classification of Köppen (1931). It is used for most geographical investigations. More recently, Thornthwaite (1948) has presented a classification of climates that is frequently cited. In this book I have used a different approach by not defining the climatic type by means of single climatic parameters, such as temperature, precipitation, evapotranspiration, length of drought period, etc., but by characterising it by means of appropriate climatic diagrams. These will be discussed next.

10. The representation of climatic types by climatic diagrams

'Weather' is understood to be the result of interaction of several meteorological factors at a given moment. 'Climate' represents the normal course of weather through a year. Therefore, two integrations must be done to obtain climate. However, no meteorological textbook explains how these are done! Even the most accurately tabulated climatic data provide no answer to this problem. Data-filled tables are difficult to comprehend and not very useful for rapid comparisons. Climate can be assessed adequately only if the yearly march of weather can be easily observed. For example, the duration and intensity of moist and dry, cold and warm seasons must be readily apparent. It was, therefore, necessary to develop a new method for ecological purposes. This method makes use of a diagrammatic representation of climate, and permits recognition at first sight of important similarities and differences between climates of different places.

The diagrams incorporate ecologically significant climatic parameters and clearly portray the seasonal variations in climate. When plotting mean monthly temperatures and precipitation graphically, a scale of 10°C = 20 mm precipitation is used, following an earlier suggestion of H. Gaussen (Bagnouls and Gaussen, 1953; Gaussen, 1955). It has been found empirically that a drought period occurs for plants when the precipitation (N) curve cuts the temperature (T) curve on this scale (T:N = 1:2). A comparison of the temperature curve with the precipitation curve is scientifically sound, because the former can be used as a measure for the yearly variation in evaporation, as mentioned earlier. The temperature curve indicates the debit side of the water relations, while the precipitation curve shows the credit side. Both together convey some idea of the water balance or the water conditions. Temperature-precipitation relationships are employed in most climatic formulae for characterising water balance relations, because direct measurements of evaporation are only available from very few stations. The limiting values used to separate arid from humid conditions differ depending on different assumptions. However, they vary about the value suggested by Gaussen; that is, 1:2. It may be worth mentioning here that there is never an absolutely abrupt change-over between arid and humid seasons, so any separation value is to some degree arbitrary. However, it is essential to portray the actual relations in a realistic way.

Not all climatic parameters significant to plant life are shown in representing the annual march of temperature and precipitation and a drought period. Equally important is the duration and intensity of the cold period, parameters not satisfactorily portrayed by monthly means. Frosts can occur in months with mean temperatures above freezing. Such early or late frosts and exceptional frosts as well are particularly important as selective influences on vegetation. Therefore, they must be shown also in a climatic diagram. In turn, the climatic diagram should not be overcrowded with information, particularly if it is to be used for comparisons on a world-wide scale. The method of representation shown in Fig. 16 has found general applicability (Walter and Lieth, 1967).

In investigating the drought period it was found that Gaussen's scale (T:N = 1:2) is realistic for all those areas where precipitation shows an abrupt decline for some period of the year. However, it is not quite satisfactory for areas with a more uniform precipitation curve. For example, the climatic diagram of Odessa (Fig. 16) hardly shows a drought period. Yet the summer in this steppe area is always a dry season for plant growth, even if it is not quite so extreme as in Anatolia or in the Mediterranean region. In such cases we have to distinguish a less extreme drought (= dry) period, which is obtained by emphasising evaporation through choosing a scale of T:N = 1:3. According to

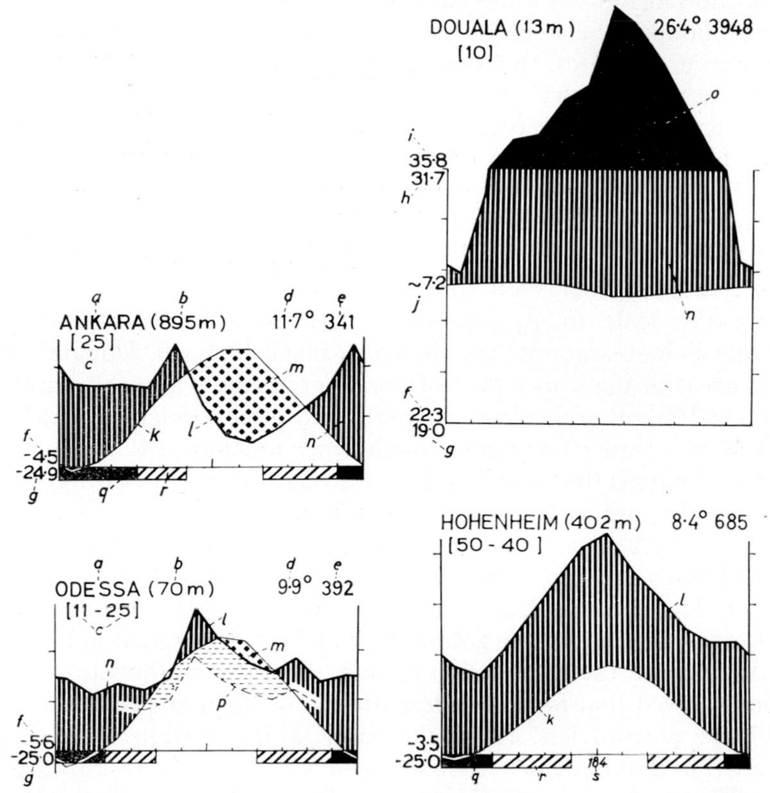

FIG. 16. Typical examples of climatic-diagrams with explanations. The symbols and figures on the diagrams have the following meanings (temperature in °C): *a*, station; *b*, height above sea level; *c*, number of observation years (occasionally first figure for temperature, second for precipitation); *d*, mean annual temperature; *e*, mean annual precipitation (mm); *f*, mean daily minimum temperature of the coldest month; *g*, absolute minimum temperature (the lowest ever recorded); *h*, mean daily maximum temperature of the warmest month; *i*, absolute maximum temperature (the highest ever recorded); *j*, mean daily temperature range; *k*, curve showing monthly mean temperatures (one scale interval = 10°C); *l*, curve showing mean monthly precipitation (scale 10°C = 20 mm); *m*, drought periods (dotted); *n*, humid period (vertical hatching); *o*, mean monthly precipitation exceeding 100 mm (scale reduced to 1/10), black area; *p*, precipitation curve lowered to 10°C = 30 mm, horizontally hatched area above, dry period; *q*, months with mean daily minimum less than 0°C (black); *r*, months with absolute minimum less that 0°C (diagonally hatched); *s*, average duration of mean daily temperatures above 0°C (i.e. frost-free period) in days (figure in italics).

These data are not shown for all stations. The appropriate areas on the diagram are left blank in such cases (cf. for example, Figs. 19 to 48).

Selyaninov (1937) this relation characterises the climatic conditions of the steppe region excellently. For this purpose I have retained the normal curve (at 10° = 20 mm) and plotted an additional precipitation curve (at 10° = 30 mm) on the diagram. The new precipitation curve (dotted line) is somewhat lower on the graph than the normal curve. The space between the new precipitation curve and the temperature curve indicates the duration (horizontally) and the intensity (vertically) of the 'dry' period, when it is not occupied by a drought period as well. It can be seen that a drought period occurs only in spring and late summer in the steppe area of Odessa. At both times it is only slight but the dry period lasts throughout the summer.

Monthly precipitation may amount to several 100 mm in the humid tropics. The diagrams would become cumbersome if the same scale were retained under such conditions. It was, therefore, expedient to reduce the scale to $\frac{1}{10}$, whenever the monthly rainfall exceeded 100 mm. These amounts are shown in black (Fig. 16, Douala). Thus, black areas in the upper part of the diagram show monthly rainfall in excess of 100 mm indicating a superhumid period. Such high rainfall is of little importance for plant growth, since much of it is lost as run-off.

Humid areas that show neither a drought nor a dry period in their climatic diagram and in which the temperature curve remains below the upper (1:2) as well as the lower (1:3) precipitation curve (Fig. 16, Hohenheim), include very different climatic types. For example, the forest climates of the hard- and soft-wood zone, the wooded tundra, the tundra proper and the greater part of the polar regions are included.

A check on the application of such diagrams to the Euro-Siberian region showed that further differentiation of the precipitation relations would be meaningless. Here, the temperature relations are of significance in determining whether hard- or soft-wood forests can grow, or where forest gives way to tundra. This is so despite the fact that these areas are humid throughout the year. Among the various temperature parameters, neither the annual mean temperature nor the average or absolute minima are of any use in delimiting vegetation types. Much better correlations were obtained with length of the growing season.

However, growing season is an obscure concept which needs to be defined more precisely. It may be considered to be the frost-free period, or the part of the year with daily means above 5°C, above 10°C, or above 15°C. The best results were obtained by defining the growing season as that part of the year with daily means above 10°C. I computed this period simply from the temperature curve of the monthly means since more exact information was not usually available. Temperature periods below −10°C are of importance only for the distribution of subarctic larch-pine forests, because they occur in the very cold continental areas with extremely severe winters. In these cases I have indicated in addition the length of the year with temperatures above −10°C. Examples of such diagrams are shown in Figs. 40-43.

One should avoid rigidity in interpreting climatic diagrams. Their particular advantage lies in permitting a total evaluation to be made of the climate. It is the interaction of the different climatic parameters that matters in ecological considerations. One should not emphasise one feature alone, for example, length of the drought period.

The climatic diagrams are based on average values appertaining to a specific number of years. The number of observation-years is written beneath the name of the station. Therefore, we obtain a climatic type from which single years may deviate considerably. Normal years are rare and almost every year shows some abnormalities that may deviate in any direction. The different seasons are either colder or hotter, or too moist or too dry compared with the norm. But the actual conditions, particularly if they are extreme, not the average, are of importance to the plant cover.

In spite of this, I use climatic diagrams to characterise climatic types in relation to vegetation, for I consider the following aspects to be important:

1. The natural plant cover, which is in a dynamic equilibrium with its environment, is the product of many hundreds of years. Its composition oscillates during short time-periods about an average equilibrium position, which has to adjust itself to average climatic conditions.

2. Within a climatic region there is some correlation between long-term averages and extreme values. For example, if low temperature extremes are common, the average minima will be lower as well. If extreme drought years occur, the mean annual precipitation will also be lower.

Table 1 may serve as an example; it shows the average daily minima of the coldest month (Mi_d) and the absolute minima (Mi_a) in °C for several stations in east Europe. With few exceptions, there is a close parallel within the different vegetation zones.

TABLE 1

	Mi_d	Mi_a		Mi_d	Mi_a
Forest zone:			*Steppe zone:*		
Kiev	−8·8	−30·0	Nikolajew	−6·8	−28·7
Tula	−14·3	−38·4	Melitopol	−7·5	−30·0
Forest-steppe zone:			Margaritowka	−9·3	−33·0
Uman	−9·5	−33·0	Nikolajewskoje	−17·1	−41·0
Charkow	−12·2	−36·9	Tschkalow	−19·1	−41·7
Woronesh	−14·1	−36·5	*Semi-desert:*		
Tambow	−15·3	−38·7	Berdjansk	−6·3	−26·5
Pensa	−16·2	−41·4	Astrachan	−10·6	−29·8
Uljanowsk	−17·2	−39·6	Wolgograd	−13·2	−34·6
			Kamyschin	−15·1	−37·2

It is also known that the relationship between the amount of rain in the wettest year to that in the driest year increases with decreasing annual precipitation, which implies an increase in severity of drought years. This ratio (max:min) of annual precipitation was given by Köppen for 30-50 year observations over the entire world (Table 2).

TABLE 2

Average amount of rain (mm)	200–400	400–600	600–800	1000–1300	>1600
Max:min ratio	4·4	3·3	2·9	2·4	2·1

We may conclude that the extreme values are reflected to some extent in the average values and that it is, therefore, quite sound to use average values in characterising climates for ecological purposes. However, a wrong impression can be conveyed in a few cases as will be shown later (Chap. XII).

These considerations apply only to natural vegetation. Agricultural crops and forest regeneration depend to a much greater degree on the weather conditions of individual years since these determine to a large extent the yield, or amount of damage in perennial stands.

It is not worth while to propagate a forest tree species if it is destroyed every 50 years by extreme weather conditions or if it is severely damaged every 10 years. Moreover, the farmer and fruit-grower can only tolerate a limited amount of crop failure. He should, therefore, be able to obtain information on the frequency of extreme years. Long-term averages are of little use to him and information on the interaction of significant weather parameters during a number of individual years is more helpful to him. Such information is provided by a sequence of graphs each showing information for one year only, a so-called climatogram. These permit a much closer insight into the climatic relations of a certain place than the climatic diagrams.

Climatograms are prepared by plotting the monthly means of temperature and precipitation separately for all observation years and supplementing these with certain additional information. Otherwise the graphs are prepared in the same way as the climatic diagrams. The climatogram can also be described as a 'climatic diagram-strip'.

The climatograms for Hohenheim near Stuttgart (Fig. 17) and Ankara (Fig. 18) are shown as examples. Hohenheim is in the central European climatic zone. The climatogram refers to the years 1907-56. Ankara is in the dry steppe of central Anatolia and its climatogram refers to the years 1936-53.

A climatogram enables extremely dry and extremely cold years to be recognised immediately. One can also draw conclusions regarding the suitability of certain crop plants.

Climatograms cannot be presented for a larger number of stations, because of the space problem. They are required only for special studies. In such cases one should acquire the data and prepare the climatogram for the area with which one is concerned.

If the geographical distribution of climate over a given area is required, it is best to construct a climatic diagram-map or a 'klimakartogram' in the terminology of C. Troll (1955). This is done by superimposing small climatic diagrams over the respective stations on a geographical map. The places with similar climates are at once clearly shown and corresponding geographical boundaries are easily drawn. With the exception of mountainous areas, one notices that geographical changes in climate are always continuous and that climatic zones are always connected by broad transition zones. This cannot be observed on maps which only show the borders of climatic zones. Such zones are always strongly marked, while a climatic diagram-map shows the places where the measurements were recorded. Such measurements are reproduced in unmodified form. Thus, it is obvious that my method has a number of advantages, and that it is extremely simple. More than 8000 climatic diagrams for stations in the world are presented in a climatic diagram-world atlas (Walter and Lieth, 1967).

The question arises, of course, whether the climatic diagrams or climatograms described here are applicable to all climatic regions of the world. It was found that climatic indices or climatic formulae are quite valuable for certain areas, while they fall short in climatically quite different regions.

The situation is quite different with regard to the climatic diagrams. No assumptions are involved in their representation. The meteorological values are shown graphically in the same form as they were measured at the stations. Choice of scale is a question of expediency. I shall show by some examples from all the continents that the scales used for representing temperature and precipitation correspond to reality in all climatic zones.

The mean climatic zones I-IX were discussed on pp. 47-48. I shall add a Xth climatic type. This refers to all mountainous climates, so that X (I) represents a mountain climate in the equatorial zone (e.g. the Andes around the equator), X (IV) one in the winter-rain zone (e.g. the Lebanon Mountains) and X (VI) one in the typical temperate zone (e.g. the Alps).

Figs. 19-48 show three rather different diagrams for each climate type, in part extreme examples from different continents.

11. Soil zones of the world

Comparisons between climatic and vegetation types have shown that the general outline of both distribution patterns is similar but not

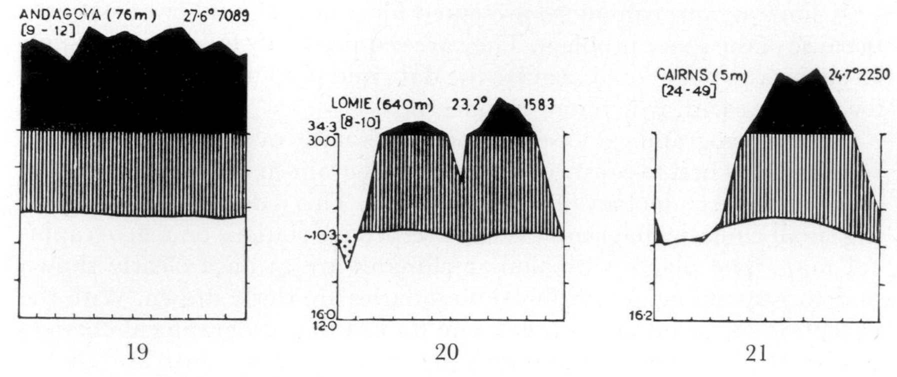

FIGS. 19-21. Climatic type I. Humid-equatorial climate in Colombia, Cameroons and Australia (transitional to II).

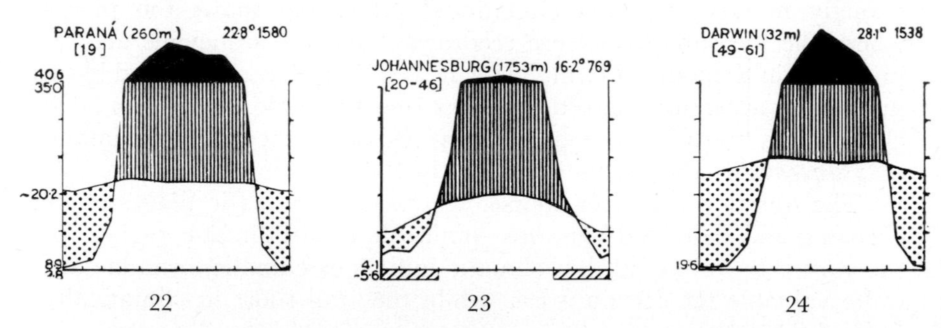

FIGS. 22-24. Climatic type II. Tropical climate with summer-rains in Brazil, South Africa (with frosts) and Australia.

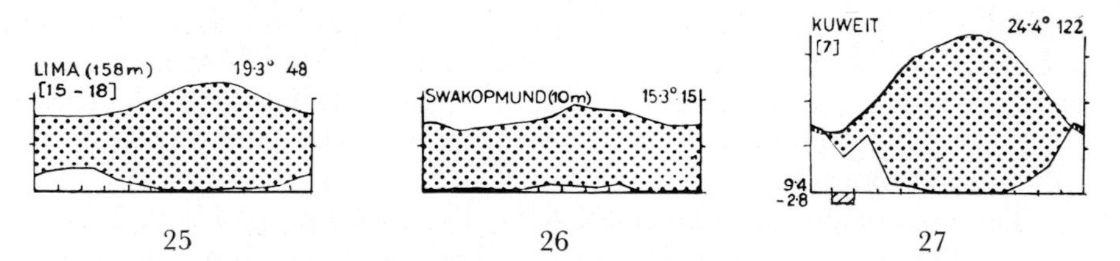

FIGS. 25-27. Climatic type III. Arid-subtropical desert climate in Peru, S.W. Africa and Arabia.

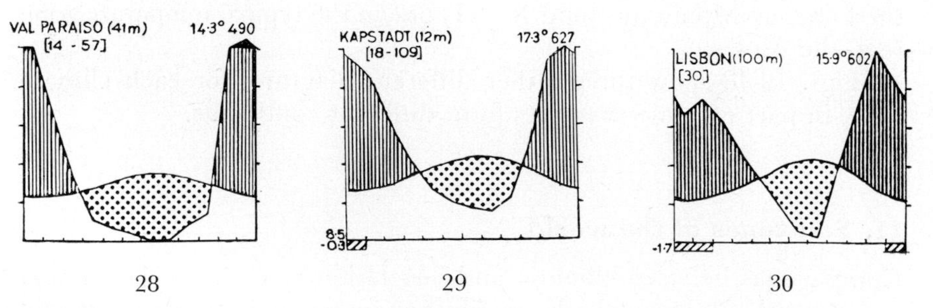

FIGS. 28-30. Climatic type IV. Etesian climate of E. Mediterranean type in Chile, South Africa and Portugal.

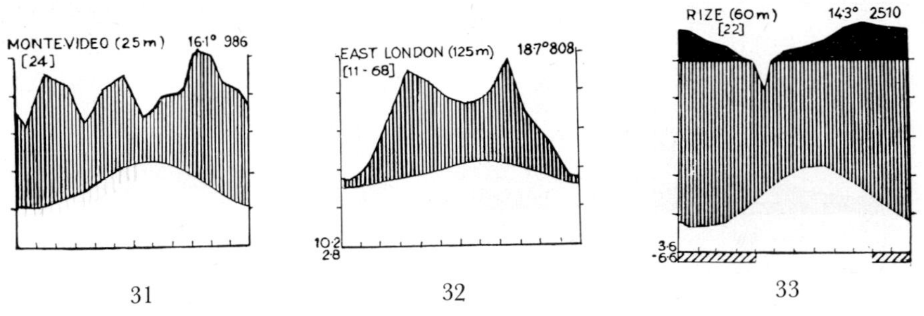

FIGS. 31-33. Climatic type V. Warm-temperate, humid climate in Uruguay, South Africa and North Anatolia.

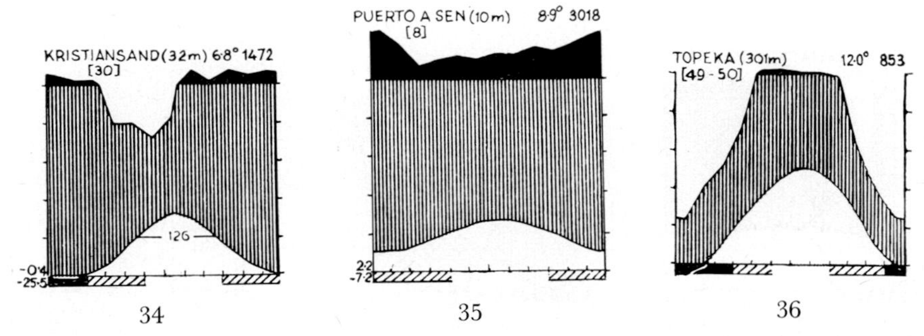

FIGS. 34-36. Climatic type VI. Temperate climate in Norway, Chile (very moist, mild winters, but cool summers) and U.S.A. (cold winters, but hot summers).

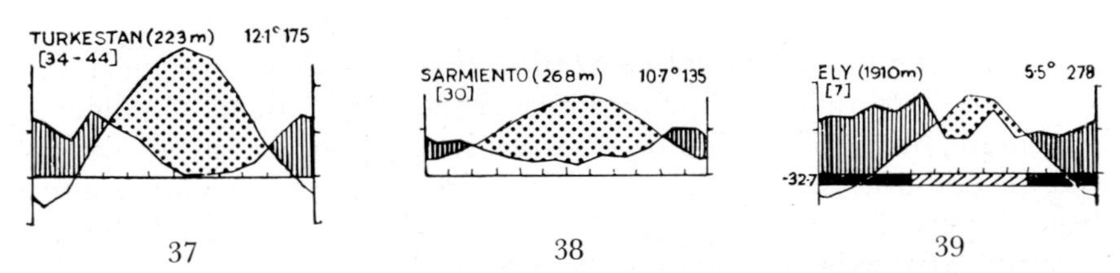

FIGS. 37-39. Climatic type VII. Arid-temperate climate in Central Asia (extreme continental), Argentina (less extreme) and U.S.A.

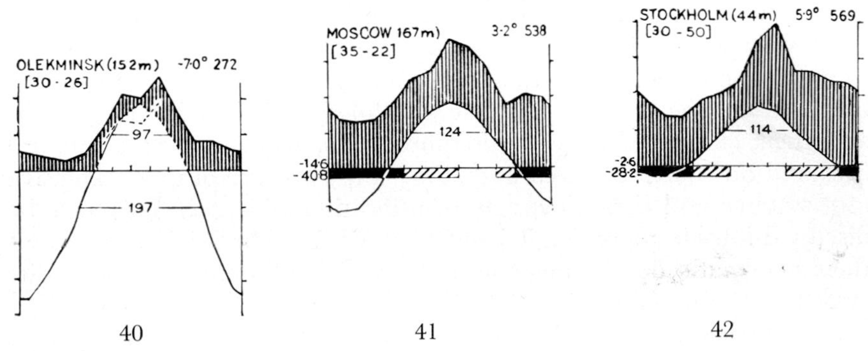

FIGS. 40-42. Climatic type VIII. Boreal-cold climate in Siberia (extreme continental), Central Russia and Sweden (less extreme).

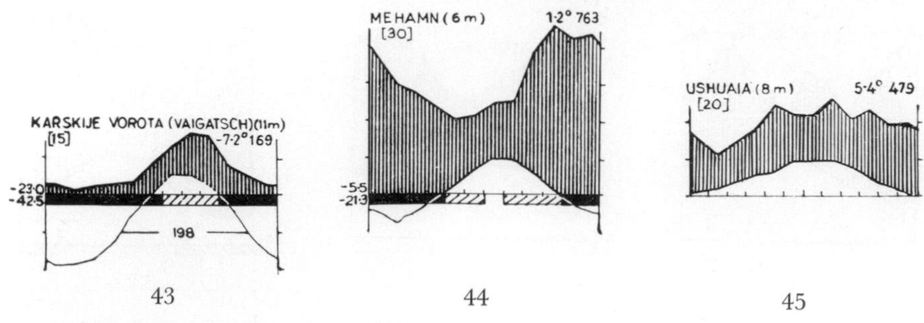

FIGS. 43-45. Climatic type IX. Arctic climate in North Russia (continental), Norway (moist) and Argentina (maritime).

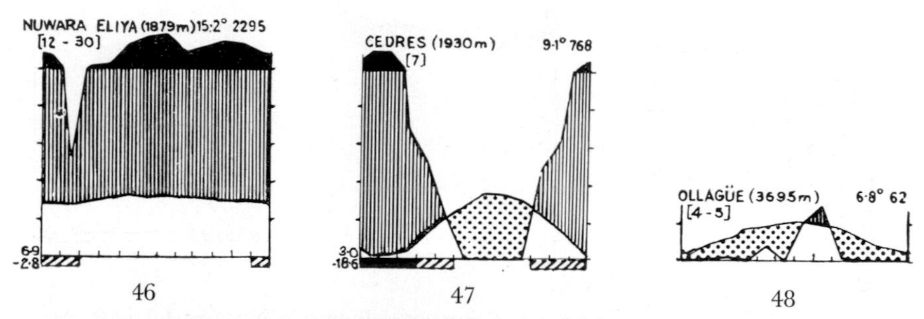

FIGS. 46-48. Climatic type X. Mountain stations in Ceylon (zone I), Lebanon (zone IV) and Chile (zone III).

identical. The boundaries of vegetation types are more irregular. The reason for this is that the distribution of vegetation is not only influenced by climate, but also by soil. For example, it was found in eastern Europe that the climatic diagrams in the steppe zone indicated, in addition to a drought period, a long dry period in summer; in the forest-steppe zone they indicated a dry period only and the forest zone lacked a dry period altogether. The borders of these climatic zones run from west-south-west to east-north-east (Walter, 1957). This is not the case for the vegetation types, which show large extensions in other directions. Forest extends southward on sandy soils and steppe vegetation penetrates into the forest zone on loess soils in the form of tongues and islands. Grasses are favoured on fine textured soils, because of their intensive root systems and their characteristic water requirements, while woody plants are favoured on coarse textured and stony soils, because of their extensive root systems and their slow rate of utilisation of water. This will be discussed later in more detail (see Chap. VI, 2). The origin of soils and their pedogenic development as influenced by climate has been discussed by Walter (1960, pp. 435-75; cf. Kubiena, 1948; Ganssen, 1957). Therefore, here the discussion will be restricted to a brief outline of the distribution of zonal soils in the world.

There are close, reciprocal relationships between climate, soil and vegetation, which may be schematised as follows:

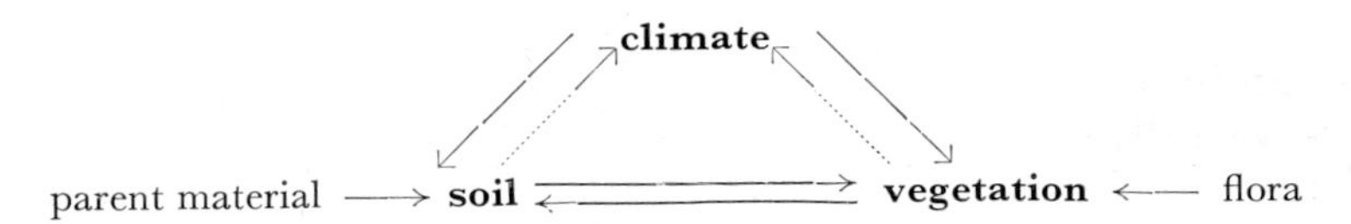

Climate influences not only vegetation but also soil. Temperature and water relations are the most important climatic parameters in soil formation. Conversely, vegetation and soil may influence climate but only to a minor degree and only with respect to the air layers near the soil surface. The interaction between soil and vegetation is so close that they form a distinct unit, the components of which can hardly be treated separately. Soil is part of the environment of a plant community, just as is the climate. Together with its environment, the plant community forms a unit—the ecosystem. However, both soil and vegetation have features of their own, which can be understood only from their historical relationships. The vegetation is derived from the flora of a particular region, while the origin of the soil goes back to the parent material. The properties of the parent rock predominate in young soils; and study of this material is an objective of geology. With increased soil formation, however, the properties of the soil become influenced more markedly by climate and vegetation.

When these relationships are recognised, one understands why there is such a close parallel between climatic zones, soil zones and vegetation zones. This is so in spite of the fact that the influence of parent materials on the nature of soil and, therefore, on vegetation also is never completely eliminated. The soils of the humid and arid climates are distinct. Water movement is mainly downward in soil profiles of humid climates, so characteristic leaching becomes apparent in the upper soil horizons. Such leaching is not shown in soils of arid regions. On the contrary, some enrichment by soluble substances may result from weathering. These substances may be deposited in particular horizons, or even on the soil surface, because of the strong evaporation, which causes a marked upward movement of the soil water. On the other hand, the vegetation is always much lusher in humid than in arid areas. Therefore, more organic matter is deposited in soils of humid areas and humus enrichment is much more pronounced, particularly when the temperature is lower as well.

Temperature differences are associated with differences in weathering and with subsequent changes in weathering products. They also influence the development and fate of humic materials. Thus, soil formation has its distinct characteristics in the tropics, the temperate and arctic zones. Table 3 shows the distribution of the most important global soil types as related to climate, summarised by Ganssen. The soils will be discussed in more detail with each vegetation zone in the following chapters.

12. Principal features of a three-dimensional classification of the earth's vegetation

The different vegetation zones coincide with the broad climatic and soil zones that extend from the North Pole to the equator and thence to the South Pole.

The unequal distribution of land and sea in the northern and southern hemispheres results in an asymmetrical distribution of climatic zones on the two sides of the equator. The southern hemisphere is not a mirror-image of the northern. The climate of the southern hemisphere is decidedly more oceanic because of the larger water surface. The broad desert zone and the temperate zone are absent. For example, the palm zone reaches as far as South Island of New Zealand and the forests on this island are still mostly of subtropical character, while only a little farther south the Auckland islands are beyond the limits of the polar timber line. The narrow, subantarctic zone shows many differences from the cold-temperate zone of the northern hemisphere and also from the subarctic zone. On the other hand, the polar zone of the Antarctic is much more pronounced than that of the Arctic. However, this difference is unimportant for our discussion as there is no plant cover in these areas.

The vegetation zones in the two hemispheres coincide in their asymmetry with the distribution of climates. One cannot simply transfer conditions in the northern to those in the southern hemisphere. This is important to emphasise. Since vegetation research had its origin in the northern hemisphere, it has led to unintentional extrapolation of conditions there to those in the southern hemisphere. This has often resulted in wrong theories.

The same is true also for altitudinal vegetation belts in mountainous areas. The view that changes in vegetation with altitude, from sea level upwards, coincide with small-scale repetitions of vegetation zones from south to north, is true only for the northern temperate zone, and even here with some reservations. The mountainous vegetation in the rest of the world shows much more complicated relationships.

The climate of alpine areas cannot be considered to be identical with that of the Arctic, even if this correlation is restricted to the alpine zone of central Europe. The only common feature is the short growing season. The Arctic climate is considerably more uniform during the growing season. The days are longer and midnight sun prevails for a short period. Consequently, the daily temperature ranges are small; that is, the days are not very warm and the nights not very cool. By contrast, days are much shorter in the alpine zone, daily radiation is far more intensive and nocturnal re-radiation is much stronger as well. This results in considerably greater daily temperature ranges than occur in the Arctic, whereas monthly mean temperatures show little differences. The generally level terrain of the tundra causes widespread bog formation in the Arctic in spite of lower precipitation there than occurs in the alpine zone, where bogs are restricted to more local depressions.

The climate of tropical mountains cannot be compared with that of the Arctic or that of the central European mountains (Troll, 1955). The tropics are characterised by very uniform temperature throughout the year. The difference between monthly means of the warmest and coldest month amounts to a few degrees only. Therefore, growing seasons are not related to temperature, a condition that applies not only to the lower altitudes in the tropics but also to the highest tropical mountains. A change is brought about only by a lower position of the temperature curve. Daily temperature ranges decrease with increasing altitude up to the cloud belt, after which they increase rapidly. Temperatures below freezing point occur daily at about 4000 m elevation, although the mean annual temperature may still be above 5°C. The night frosts hardly penetrate into the soil and the intermittent snowfalls melt during the day. Snow remains as a permanent cover only in certain places at about 5000 m height, where the air temperature rarely or never rises above freezing point.

The climate of the Arctic with pronounced annual temperature variations and small daily temperature ranges contrasts considerably with the daily climatic variations in the tropical mountains with their characteristically small, annual variations in daily mean temperatures. The mountain climates of middle latitudes show intermediate characteristics (cf. Walter, 1960, p. 65).

We may conclude from this that a discussion of the earth's vegetation should take into consideration not only the lateral, or horizontal, distribution of vegetation from north to south and from east to west, but also the vertical distribution. This implies a three-dimensional viewpoint, which was especially emphasised by C. Troll (1948, 1959), who illustrated this for the tropics in an excellent discussion.

The horizontal distributions of climate and vegetation zones in an idealised continent are shown in Fig. 49. The vertical distribution, i.e. the third dimension, is shown in a profile diagram extending from the North to the South Pole (Fig. 50).

The vegetation profile portrays clearly the asymmetrical distribution of vegetation on the two sides of the equator. It can be seen that in the northern hemisphere the montane-subalpine and alpine vegetation extends to higher altitudes towards the south to be replaced quite suddenly by different vegetation. The climatic conditions in tropical mountains show much closer similarities to those of lower altitudes in the southern hemisphere. This is reflected by closer similarities of the vegetation.

The extreme oceanic, cool climate of the subantarctic islands is

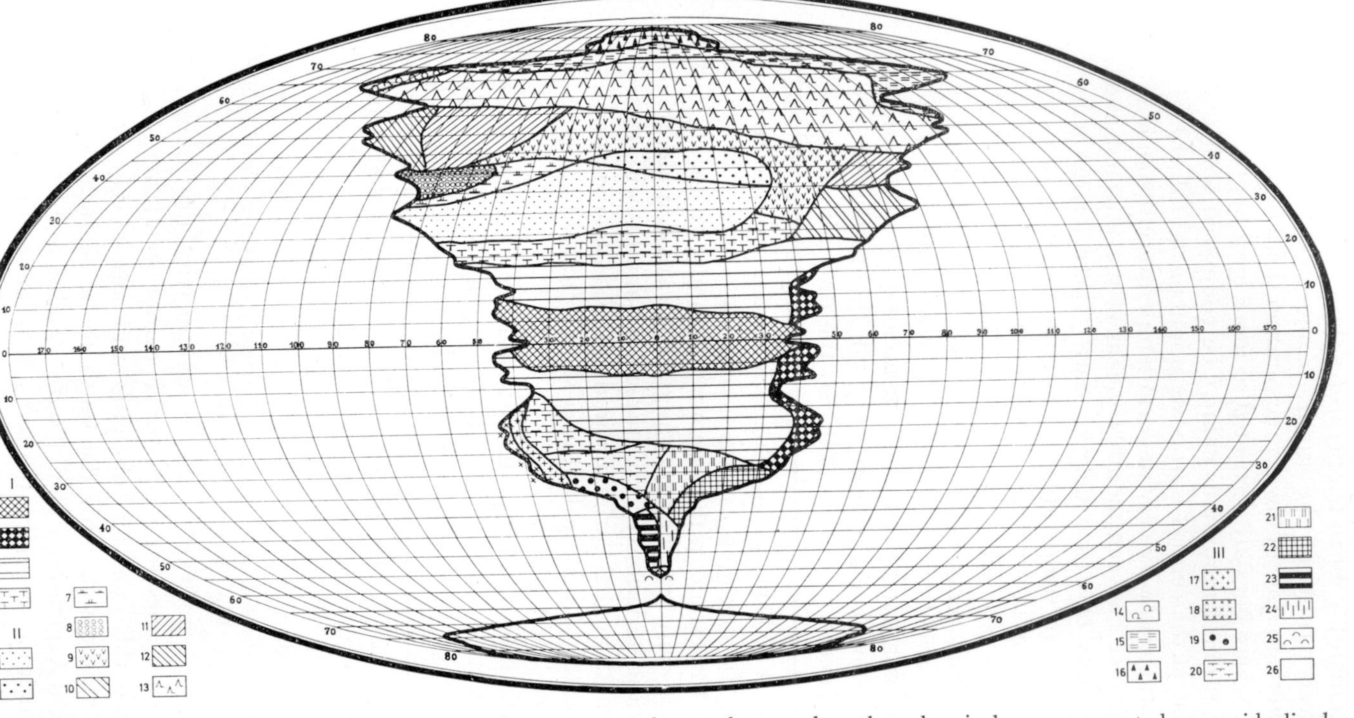

FIG. 49. The asymmetrical relation of vegetation zones on the northern and southern hemisphere, represented on an idealised continent (after C. Troll).

I. Tropical zones: 1. Equatorial rain forest; 2. Tropical rain forests with orographic rains from trade winds; 3. Tropical savannahs (moist savannahs) and raingreen tropical deciduous forests; 4. Tropical thorn-steppes and thorn-forests.

II. Non-tropical zones of the northern hemisphere: 5. Hot deserts; 6. Cold inland deserts; 7. Subtropical winter-green steppes; 8. Hot summer, rainy winter zones; 9. Cold winter grass-steppes; 10. Hot summer, moist monsoon and laurel forests; 11. Summer-green forests; 12. Oceanic summer-green deciduous and laurel forests; 13. Boreal conifer forests; 14. Boreal birch forests; 15. Subarctic tundras; 16. Arctic deserts.

III. Non-tropical zones of the southern hemisphere: 17. Coastal deserts; 18. Deserts with garua; 19. Winter-rain zones; 20. Subtropical thorn-steppes (Karroo, Monte); 21. Subtropical grassland; 22. Subtropical rain forests. 23. Cool-temperate rain

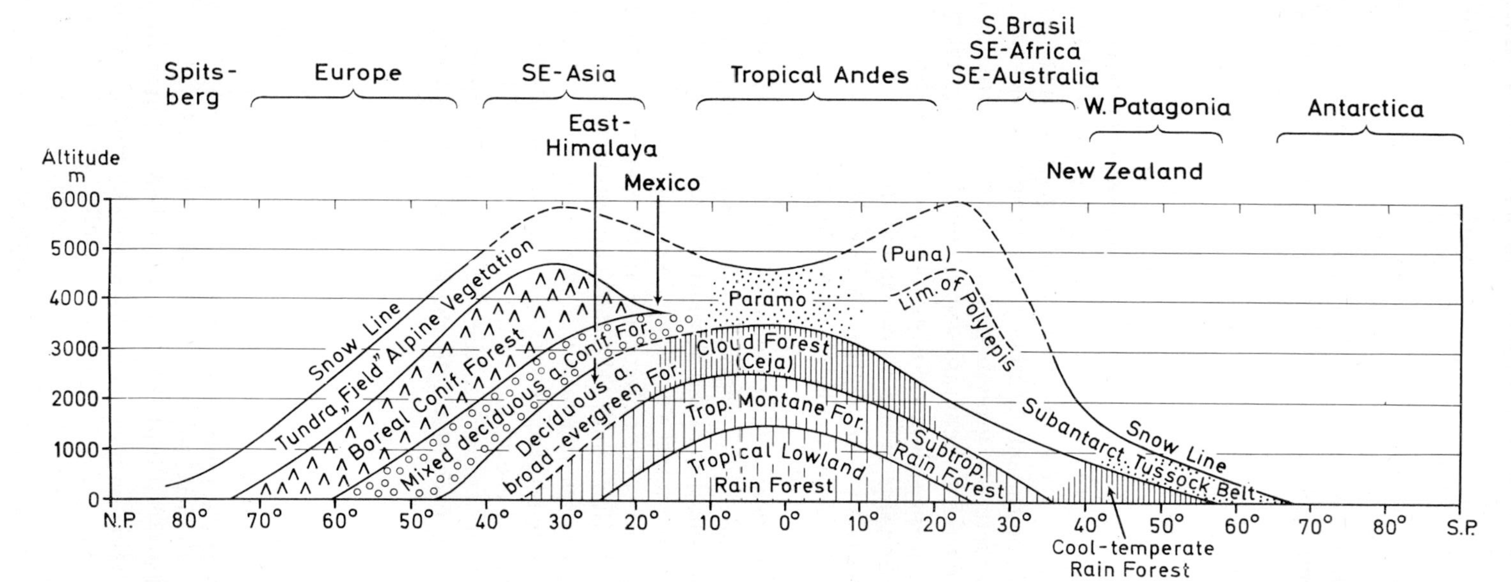

Fig. 50. Schematic vegetation profile of the world from the Arctic to the Antarctic, showing asymmetrical altitudinal belts in the humid regions (after C. Troll).

characterised, just as the climate of tropical mountains, by small annual ranges in temperature (Kerguelen Islands, 6·5°C; West Patagonia, 4·5°C; Macquarie Islands, 3·5°C). However, the daily temperature variations are also very small. The largest range of hourly mean temperatures throughout a year is only 4·9°C. If the temperature is near freezing, the number of days with alternate frost and thawing will be very large and there will be no penetration of frost into the soil. Many close floristic similarities can be observed between the subarctic islands and the tropical mountains of the Andes. This is so because there is a direct connection, whereas any resemblance is much less obvious with the African tropical mountains. Such similarities are shown by the abundance of cushion plants, which occur on boggy habitats. 64·1% of all cushion plants grow in the South American Andes and in the subantarctic; 50·5% of these in South America, including Tierra del Fuego and 13·6% on the subantarctic islands and New Zealand (Rauh, 1939). Other characteristic growth forms in these areas are the 'tussock-grasses', which are tall bunchgrasses with coarse blades that belong to different genera, or the strange-looking plants with tufted-leaved-stems, such as the *Espeletia* spp. of the South American Paramos, or the tree-like *Senecio* species of the alpine areas in East Africa, the huge lobelias, certain species of lupine or the Kerguelan cabbage, *Pringlea antiscorbutica* (Figs. 105, 106, 107, 118).

A similar difference in physiognomy and floristics to that existing between the tropical alpine vegetation and the northern hemisphere vegetation is also found between the latter and the tropical subalpine and tropical montane forest vegetation. Vegetation analogous to these tropical mountainous types only occurs in the southern hemisphere at lower elevations farther south. We can cite similarities between such characteristic genera as *Podocarpus*, *Araucaria*, and *Nothofagus*, or the different tree ferns, to representatives of the family Proteaceae, or to the genus *Gunnera*, which are found in the cloud forests of Columbia and Costa Rica. Among epiphytes, the delicate Hymenophyllaceae are restricted, with few exceptions, to the humid forests of the southern hemisphere.

It is remarkable that all these forests are evergreen, even those of higher latitudes. Only a few species of *Nothofagus* are deciduous. The eco-physiology of evergreen trees with laurel-like foliage which form forests down to the southern tip of Africa have not yet been investigated in any detail. In the tropical mountains of the southern hemisphere these evergreen forests extend as far as the alpine timberline in spite of their pronounced subtropical characteristics.

Troll's scheme of altitudinal vegetation belts, shown in Fig. 50, refers only to humid climates. This was especially emphasised by Troll. Stocker (1963) has pointed out that Troll's profile diagram cuts through moderately humid areas in the northern hemisphere, such as are found from north Sweden to the Alps, whereas it cuts through extremely humid areas in the southern hemisphere. Stocker believes that the asymmetrical relationships of vegetation belts in the northern and southern hemisphere would disappear by drawing the northern hemisphere profile section through extremely humid climates, for example, from Iceland through Ireland. However, one has to bear in mind that the vegetation of the extremely humid part of western Europe is strongly degraded and that it differs considerably from the vegetation in the extremely humid climates of North America along the Pacific seaboard.

The importance of such schemes should not be over-estimated, because they always incorporate a certain arbitrary element; that is, they vary with regard to the examples chosen. The different floristic relations of the northern and southern hemisphere complicate comparisons of parallel vegetation types.

The question of the uniqueness of the mountain vegetation was much discussed at the 1954 International Botanical Congress in Paris. If they are notably unique it would suggest comparison only in relation to the particular vegetation zone and would rule out any logical paralleling of mountain vegetation with vegetation zones of higher latitudes, e.g. looking at the alpine spruce belt in relation to the boreal vegetation zone, etc. This question cannot be answered in any general way.

The scheme in Fig. 50 appears to substantiate the second point of view. However, this is true only if there is a pronounced increase in precipitation with decreasing temperature in the mountains, as is the case in humid areas. The situation is quite different in arid areas. Let us consider for example, the Mediterranean zone. Wherever there is a pronounced humid, cloud belt throughout the summer in the mountainous part of this zone we find, above the Mediterranean vegetation, a belt of typically central European vegetation, e.g. a beech forest belt in the Apennine mountains. On the other hand, wherever air humidity is low enough to cause summer-drought, even in montane-subalpine altitudes as in the Sierra Nevada of southern Spain, we find vegetation belts that have no counterpart at lower elevations to the north. This applies, for example, to the subalpine thorn-cushion vegetation belt.

A *Cedrus* and *Juniperus* forest belt occurs in the Taurus and Atlas mountains, and there is no analogous vegetation zone farther north. The increase of humidity with altitude is usually so small in desert mountains that altitudinal vegetation belts are only poorly discernible and are related merely to temperature.

The amount of precipitation decreases rapidly near the peaks of the highest mountains which extend beyond the general cloud layer, as will be shown later. This produces a desert-like alpine climate without seasons in the equatorial zone. Such conditions are nowhere repeated at lower elevations (cf. Chap. IV, 4-5).

Thus, the question discussed at the Congress cannot be answered without reservations. Altitudinal vegetation belts may resemble latitudinal vegetation zones in certain situations, while there is no resemblance in others. Therefore, I shall discuss the various altitudinal belts together with the individual vegetation zones, pointing out their uniqueness or any analogies wherever applicable.

References

AHLGREN, H. L., and AAMODT, O. S. 1939. Harmful root interactions as a possible explanation for effects noted between various species of grasses and legumes. *J. Amer. Soc. Agron.*, **31**, 982-5.

ALISSOW, B. P. 1954. *Die Klimate der Erde.* Berlin.

ALISSOW, B. P., DROSDOW, O. A., and RUBINSTEIN, E. S. 1956. *Lehrbuch der Klimatologie.* Berlin.

BAGNOULS, F., and GAUSSEN, H. 1953. Saison sêche et Indice Xérothermique. *Docum. Cartes Prod. Végét. Série: Général*, Tome III, vol. 1, 1-48.

BÖRNER, H. 1960. Excretion of organic compounds from higher plants and its role in the soil-sickness problem. *Bot. Rev.*, **26**, 393-424.

BORNKAMM, R. 1961. Zur quantitativen Bestimmung von Konkurrenzkraft und Wettbewerbsspannung. *Ber. Dtsch. Bot. Ges.*, **74**, 75-83.

BORNKAMM, R. 1963. Erscheinungen der Konkurrenz zwischen höheren Pflanzen und ihre begriffliche Fassung. *Ber. Geobot. Inst. Rübel*, **34**, 83-107.

BOYSEN-JENSEN, P. 1949. Causal plant ecology. *Danske Vidensk. Selskab., Biol. Medd.*, **21**, No. 3.

BROUGHAM, R. W. 1962. The leaf growth of *Trifolium repens* as influenced by seasonal changes in light environment. *J. Ecol.*, **50**, 449-59.

BRAUN-BLANQUET, J. 1951. *Pflanzensoziologie.* Second ed. Vienna.

CAIN, S. A., and DE OLIVIERA CASTRO, G. M. 1959. *Manual of vegetation analysis.* New York.

CANNON, J. R., CORBETT, N. H., HAYDOCK, K. P., TRACKEY, J. G., and WEBSTER, L. J. 1962. An investigation of the effect of dehydroangustione present in the leaf litter of *Backhousia angustifolia* on the germination of *Araucaria cunninghamii*—an experimental approach to a problem in rainforest ecology. *Austr. J. Bot.*, **10**, 119-28.

CHAPMAN, V. J. 1959. Studies in salt marsh ecology, IX. Changes in salt marsh vegetation at Scolt Head Island. *J. Ecol.*, **47**, 619-39.

CLEMENTS, F. E., WEAVER, J. E., and HANSON, H. C. 1929. *Plant competition.* Carnegie Inst. Washington Publ. No. 298.

DONALD, C. M. 1946. Competition between pasture species, with reference to the hypothesis of harmful root-interactions, I. *Comm. Sci. Ind. Res., Austr.*, **19**, 32-37.

DUVIGNEAUD, P., *et al.* 1962. *L'écosystème, l'écologie, science moderne de synthèse*, Vol. 5, Brussels.

ELLENBERG, H. 1950. Kausale Pflanzensoziologie auf physiologischer Grundlage. *Ber. Dtsch. Bot. Ges.*, **63**, 24-31.

ELLENBERG, H. 1954. Über einige Fortschritte der kausalen Vegetationskunde. *Vegetatio*, **5-6**, 199-211.

ELLENBERG, H. 1956, 1963. *Einführung in die Phytologie*. Vol. IV, *Grundlagen der Vegetationsgliederung*. Part 1, *Aufgaben und Methoden der Vegetationskunde*. Part 2, *Die Vegetation Mitteleuropas mit den Alpen*. Stuttgart.

ERHART, H. 1962. Témoins pédogénétiques de l'époque Permo-Carbonifère. *C.r. Séanc. Soc. Biologéogr.*, **335-37**, 21-35.

GANSSEN, E. 1957. *Bodengeographie*. Stuttgart.

GAUSSEN, H. 1955. Expression des milieux par des formules écologiques; leur représentation cartographique. *Coll. Int. Cent. Nat. Rech. Sci.*, **59**, 257-69.

GOODALL, D. W. 1961. Objective methods for the classification of vegetation. *Austr. J. Bot.*, **9**, 162-69.

GOODALL, D. W. 1963. The continuum and individualistic association. *Vegetatio*, **11**, 297-316.

GRÜMMER, G. 1955. *Die gegenseitige Beeinflussung höherer Pflanzen. Allelopathie*, Jena.

HELFFERICH, R. 1942. Über die Transpiration von Blüten. *Planta*, **32**, 493-516.

IWAKI, H. 1959. Ecological studies on interspecific competition in plant community. I. *Jap. J. Bot.*, **17**, 120-38.

JARVIS, P. G. 1963. The effects of acorn size and provenance on the growth of seedlings of sessile oak. *Q. Jl For.*, **57**, 11-19.

KARPOW, W. G. 1961. On the influence of tree root competition on the photosynthetic activity of the herb layer in spruce forest. *Dokl. Akad. Nauk SSSR*, **140**, 1205-08 (in Russian).

KARPOW, W. G. 1962a. Some phytocoenological questions about spruce forests from an experimental viewpoint. *Akad. d. Wiss. Mitt. Labor. Forstwiss. Mosc.*, **6**, 35-61 (in Russian).

KARPOW, W. G. 1962b. Experiments in the utilisation of ^{32}P for investigating root competition of trees and undergrowth in the southern Taiga forests. *Dokl. Akad. Nauk SSSR*, **146**, 717-20 (in Russian).

KARPOW, W. G. 1962c. Some experimental results on the composition and structure of the lower strata of the *Vaccinium*-rich spruce forests. *Die Probleme der Botanik*, **6**, 258-76 (in Russian).

KLAPP, R. 1955. *Experimentelle Soziologie der höheren Pflanzen*. Stuttgart.

KOLB, F. 1961. *Experimentelle Untersuchungen zur gegenseitigen Beeinflussung von Kulturpflanzen und Unkräutern*. Diss. Hohenheim.

KÖPPEN, W. 1931. *Grundriss der Klimakunde*. Berlin and Leipzig.

KREEB, K. 1967. Thermodynamische Betrachtungen zum Wasserhaushalt der Pflanze. *Zeit. Pflanzenphysiologie*, **56**, 186-202.

KUBIENA, W. 1948. *Entwicklungslehre des Boden*. Vienna.

KUROIWA, S. 1960. Intraspecific competition in artificial sunflower community. *Bot. Mag. (Tokyo)*, **73**, 300-09.

LIETH, H. 1958. Konkurrenz und Zuwanderung von Wiesenpflanzen. *Zschr. Acker-u. Pflanzenbau*, **106**, 205-23.

MARTIN, P. 1957. Die Abgabe von organischen Verbindungen, insbesondere von Scopoletin aus den Keimwurzeln des Hafers. *Z. Bot.*, **45**, 475-506.

MONSI, M., and SAEKI, T. 1953. Über den Lichtfaktor in den Pflanzengesellschaften und seine Bedeutung für die Stoffproduktion. *Jap. J. Bot.*, **14**, 22-52.

ORLENKO, E. G. 1955. Interrelationships of oak under conditions of dense stocking. *Dokl. Akad. Nauk SSSR*, **102**, 841-44 (in Russian).

Ovington, J. D. 1962. Quantitative ecology and the woodland ecosystem concept. *Adv. Ecol. Res.*, **1**, 103-92.

Poore, M. E. D. 1962. The method of successive approximation in descriptive ecology. *Adv. Ecol. Res.*, **1**, 35-68.

Rademacher, B. 1959. Gegenseitige Beinflussung höherer Pflanzen. In *Handbuch der Pflanzenphysiologe* (ed. Ruhland, W.), **11**, 655-706.

Rauh, W. 1939. Über polsterförmigen Wuchs. *Nova Acta Leopoldina*, **7**, 268-508.

Saeki, T. 1959. Variation of the photosynthetic activity with ageing of leaves and total photosynthesis in a plant community. *Bot. Mag.* (*Tokyo*), **72**, 404-08.

Schmid, E. 1944. Kausale Vegetationsforschung. *Ber. geobot. ForschInst. Rübel*, **1943**.

Schmithüsen, J. 1959. *Allgemeine Vegetationsgeographie*. Berlin.

Selyaninov, G. T. 1937. *Agro-climatic reference book* (in Russian). Leningrad and Moscow.

Stocker, O. 1963. Das dreidimensionale Schema der Vegetationsverteilung auf der Erde. *Ber. Dtsch. Bot. Ges.*, **76**, 168-78.

Strunck, O. H. 1965. Über die Entstehung und Reversion enzymatisch erzeugter Protoplasten. *Biol. Rundschau*, **3**, 242-44.

Sulliven, C. Y., and Levitt, J. 1959. Drought tolerance and avoidance in two species of oak. *Plant Physiol.*, **12**, 299-305.

Takeda, T. 1961. Studies on the photosynthesis and production of dry matter in the community of rice plants. *Jap. J. Bot.* **17**, 403-37.

Thornthwaite, C. W. 1948. An approach towards a rational classification of climate. *Geograph. Rev.*, **38**, 55-94.

Troll, C. 1948. Der asymmetrische Aufbau der Vegetationszonen und Vegetationsstufen auf der Nord- und Südhalbkugel. *Ber. Geobot. Forsch-Inst. Rübel*, **1947**, 46-83.

Troll, C. 1955. Der jahreszeitliche Ablauf des Naturgeschehens in den verschiedenen Klimagürteln der Erde. *Stud. Gen.*, **8**, 113-33.

Troll, C. 1959. Die tropischen Gebirge. *Bonner Geogr. Abh.*, **25**.

Walter, H. 1923. Plasma- und Membranquellung bei Plasmolyse. *Jb. Wiss. Bot.*, **62**, 175-322.

Walter, H. 1954a. Klimax und zonale Vegetation. Angewandte Pflanzensoziologie. *Festschr. Aichinger*, **1**, 144-50.

Walter, H. 1957. Die Klimadiagramme der Waldsteppen und Steppengebiete in Osteuropa. *Stuttgart. Geogr. Stud.*, **69**, 253-63.

Walter, H. 1960, 1962, 1954b. *Einführung in die Phytologie*. Vol. I, *Grundlagen des Pflanzenlebens*. 4th ed. Vol. III, *Grundlagen der Pflanzenverbreitung*. Part 1, *Standortslehre*, 2nd ed. Part 2, *Arealkunde* (2nd ed. 1970). Stuttgart.

Walter, H. 1967. Die physiologischen Voraussetzungen für den Übergang der autotrophen Pflanzen vom Leben im Wasser zum Landleben. *Zeit. Pflanzenphysiologie*, **56**, 170-85.

Walter, H., and Kreeb, K. 1970. Die Hydration und Hydratur des Protoplasmas. *Protoplasmatologia*, **2**, C/6, Vienna.

Walter, H., and Lieth, H. 1967. *Klimadiagramm-Weltatlas*. Gustav Fischer Verlag, Jena.

Walter, H., and Stadelmann, E. 1968. The physiological prerequisites for the transition of autotrophic plants from water to terrestrial life. *Bioscience*, **18**, (7), 604-701.

WALTER, H., and WALTER, E. 1953. Einige allgemeine Ergebnisse unserer Forschungsreise nach Südwestafrika 1952/53: Das Gesetz der relativen Standortskonstanz; das Wesen der Pflanzengemeinschaften. *Ber. Dtsch. Bot. Ges.*, **66**, 228-36.

WEAVER, J. E., and CLEMENTS, F. E. 1938. *Plant ecology*. 2nd ed. New York.

WHITTAKER, R. H. 1962. Classification of natural communities. *Biol. Bot.*, **28**, 1-239.

YAPP, D. 1912. *Spiraea ulmaria* L. and its bearing on the problem of xeromorphy in marsh plants. *Ann. Bot.*, **26**, 815-70.

II. *The Continuously Wet Tropical Rain Forest*

1. General features

BIOLOGICAL sciences had their beginnings in the northern temperate zone. Thus, we are used to considering the vegetation relationships in this area as normal, while those of other climate zones are regarded as deviations from this norm. However, the vegetation of the temperate zone is really a very specialised type, resulting from its adaptation to a climate with strongly pronounced annual seasons, characterised by a relatively warm summer and a relatively cold winter.

Flowering plants originated most probably under tropical conditions. A tropic-subtropical forest still covered much of the northern hemisphere during the Tertiary. This vegetation type represents the purest form of development of the plant world under conditions most favourable to plant growth. All other plant communities are more or less depauperate: they are composed of much smaller numbers of species. Moreover, these species have become specialised in different ways so that they are capable of surviving either unfavourable water or unfavourable temperature conditions. Therefore, it is advantageous, in an outline of the vegetation of the earth, to begin with the tropical rain forest, and to discuss other vegetation types which are adapted to more extreme conditions later. The special characteristics of the temperate vegetation are better understood when treated in this more logical order.

The tropical rain forest is restricted mainly to the equatorial climatic zone, which extends from about 10° N to 10° S. This zone is characterised, in particular, by extremely small annual temperature ranges. Annual precipitation is very high and may reach several metres. Rain occurs during the entire year and, in the most typical cases, it shows two maxima that coincide with the two zeniths. However, the humid tropical climate does not form a continuous zone around the equator. The eastern part of equatorial Africa, influenced by monsoon winds, is characterised by relatively severe drought at certain periods of the year. The same is true for the eastern part of Ceylon, the southern tip of India and also for certain areas of north-western South America (Fig. 51). On the other hand, the continuously humid climate extends over the eastern part of Central America to the Greater and Lesser Antilles and along the west coast of India, over Burma and China up to the Tropic of Cancer and even somewhat beyond this line. The same applies also to the mountain ranges of eastern South America and east Australia. It is, therefore, understandable that the distribution of rainfall does not always correspond to the equatorial climatic type in this

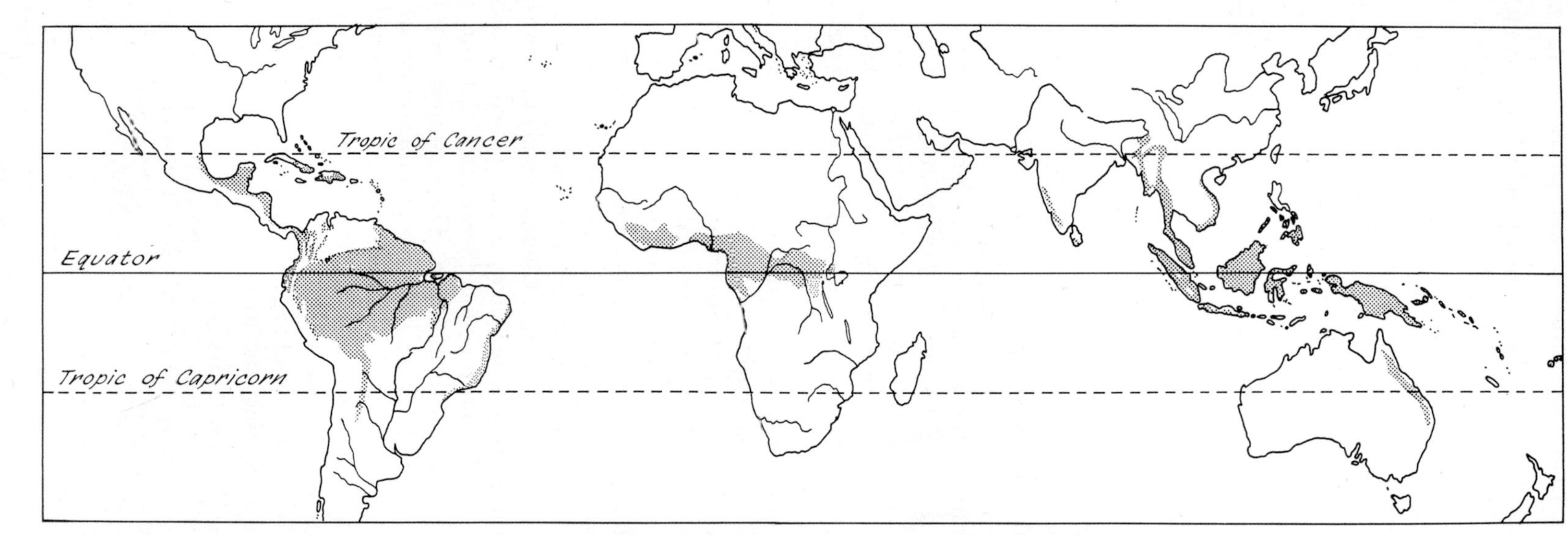

FIG. 51. Distribution of the tropical rain forest (dotted area). From Richards, 1947.

area, and that we often find a long rainy period interrupted by a short drought period.[1] The distribution of tropical rain forests is correlated with these climatic conditions (Fig. 51). The largest, continuous forest area occurs in America in the Amazon basin on the eastern slope of the Andes, the second largest occurs in the Indo-Malaysian area and a smaller one in West Africa around the Gulf of Guinea, extending into the Congo basin. Outside these three large areas, tropical rain forest is only found scattered along the high rainfall slopes of mountain ranges lying within the Tropics of Cancer and Capricorn (eastern Brazil, north-east Australia, India, etc.). Tropical rain forest is also found in smaller, mountainous areas, for example, in East Africa on the slopes of the large volcanoes (Kilimanjaro, Meru Mt., etc.) and in the Usambara and Uluguru mountains. At higher altitudes, we find montane forests that differ considerably from the typical rain forests of lower elevations. It should be emphasised at this point that the tropical rain forest not only shows floristic differences in different areas but also many structural peculiarities which do not readily permit over-all generalisations.

The uniformity of the climate in the equatorial zone is in part illusory, resulting from the use of monthly means. The tropics are characterised more by daily than yearly variations in climate (see Chap. II, 2). Variations in temperature and humidity during a day are much greater than the variations of monthly means throughout a year. This is even more pronounced at higher altitudes.

Everyone who has stayed in the humid tropics has experienced the magic mood of the cool early morning hours when, shortly before sunrise, life reawakes with numerous bird calls; when the first rays of sunlight appear and the dew-drops on the water-impregnated leaves glisten like numerous pearls. At that time the air is fresh and cool. But, soon after, when the sun has risen to a higher position, the air loses its freshness, one looks for shade and is more or less forced to use some head protection. The birds too become silent. Around noon, the atmosphere becomes uncomfortable and muggy. Clouds quickly increase in size and density and a thunderstorm develops, accompanied by driving rain, which only lasts for a short period. Soon after, the sun reappears. Everything is drenched with moisture. The freshness only lasts for a short period. The humidity again becomes unbearable, as in a hot-house. Relief comes only with the evening, when one can regain strength after sundown. The mild tropical nights are a wonderful experience. A vast number of bizarre forms of insects gather in the beam of a lamp, which in turn attract the insatiable geckos. However, simultaneously the mosquitoes appear, forcing one to look for shelter in a stuffy room

[1] A drought period of one month has hardly any effect on the vegetation as shown by my investigations (UNESCO, *Humid Tropics Res. Progr.*, **106** B). The tropical rain forest in India requires an annual precipitation of at least 2250 mm in areas with two months drought period and 3500 mm in areas with four months drought period (Fig. 123). At the same time the drought period must coincide with the cool season of the year.

beneath a mosquito net where it is difficult to find sleep. The same daily sequence is repeated virtually unchanged throughout the whole year. This makes a prolonged stay in the humid tropics nearly unbearable for people from the temperate zone.

It is also one reason for our limited knowledge about the tropical rain forest. We have many good travelogues but hardly any detailed long-term measurements. These require suitable stations and a prolonged stay in the tropics. Such a station is the Botanic Garden of Bogor on Java (Indonesia), which has a well equipped laboratory. Treub and von Faber have worked there for a number of years and the German Tropical Fellowship provided an opportunity, in former years, for many German botanists to spend a year at this station. However, the station is situated in the heart of a cultivated landscape and the distance to the nature reserve at Tjibodas, located at a higher altitude, is rather far. Therefore, only a small number of ecological investigations have been done in the area so far. The Biological Experimental Station at Amani in the Usambara mountains (East Africa), which has now been discontinued, was situated in a very favourable location. The station was in the midst of an extensive virgin forest at 900 m elevation (Moreau, 1935a, b). I had an opportunity to carry out ecological work at this station for a period of six months in 1934. Otherwise, work at Amani was mostly of an applied nature, as is the case also at other experiment stations in the tropics.

The 'Institut d'Enseignement et de Recherches Tropicales' in Adiopodoumé near Abidjan on the Ivory Coast (West Africa) has existed for a number of years. The station was intended for operation at an international level. However, its further development must be awaited. The same applies to the experimental stations in the Congo.

2. The climate of the region of the humid tropics

The climatic diagrams of stations in the humid tropics provide us with an idea of the average climatic conditions. Here, I will present a few examples (Figs. 52-56, see also Figs. 19-21). The annual rainfall may attain very high values.

Some 12,500 mm of rain falls on the peak of the Hawaiian Island of Kauai and 10,500 mm at the south-west foot of the Cameroons mountain; 11,630 mm is reported for Cherrapundja in the Chasia mountains (Assam), which are somewhat cooler. However, drought periods may occur even at locations showing such high rainfalls. In Cherrapundja, December and January are particularly dry with only 10, and 12 mm of rainfall, respectively.

The climatic diagrams are insufficient to show the ecologically significant parameters of the tropical climate. There is a strongly fluctuating daily cycle, in which the extremes are repeated every 24 hours (note the temperature extremes shown for Douala on Fig. 16). I use as an example the conditions at Bogor (Indonesia), which were discussed by Stocker (1935a).

The monthly means of temperature vary only by 1° (February 24·3°C, October 25·3°C). Annual precipitation amounts to 4370 mm. January is the wettest month with a rainfall of 430 mm and August the driest with 230 mm. This apparent uniformity in climate, however, is

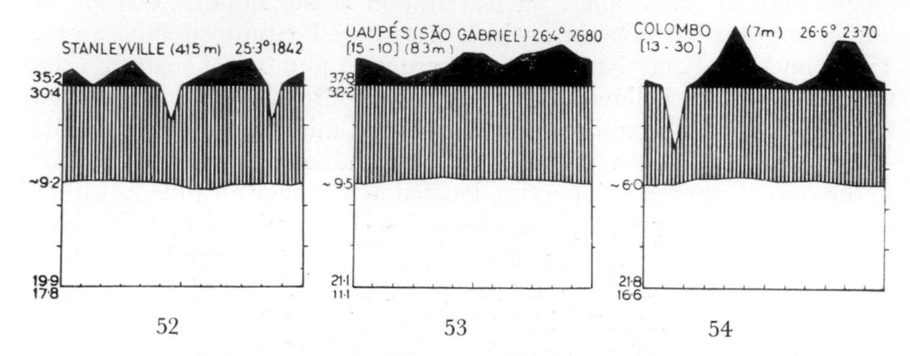

Figs. 52-54. Climatic diagrams of stations in the tropical rain forest: Congo, South America and Ceylon (from *Klimadiagramm-Weltatlas*).

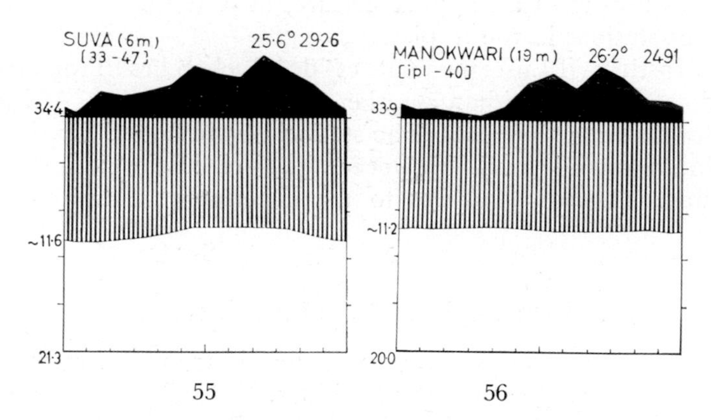

Figs. 55-56. Climatic diagrams of stations on New Guinea and the Fiji Islands.

due only to recording the means. Daily variations of these climatic parameters are much greater.

The temperature can vary on sunny days in November between 23·4°C (at 0600) and 32·4°C (at 1400), i.e. by 9°C. Variations of 6-7°C within 24 hours even occur during the rainy season but they may be reduced to 2°C on cloudy days. The daily variations in temperature cause fluctuations in relative humidity between 100% and 40% (to 25%) (Fig. 57). The relative humidity remains above 90% only on very rainy days (Fig. 58). The saturation deficit can be very great at such

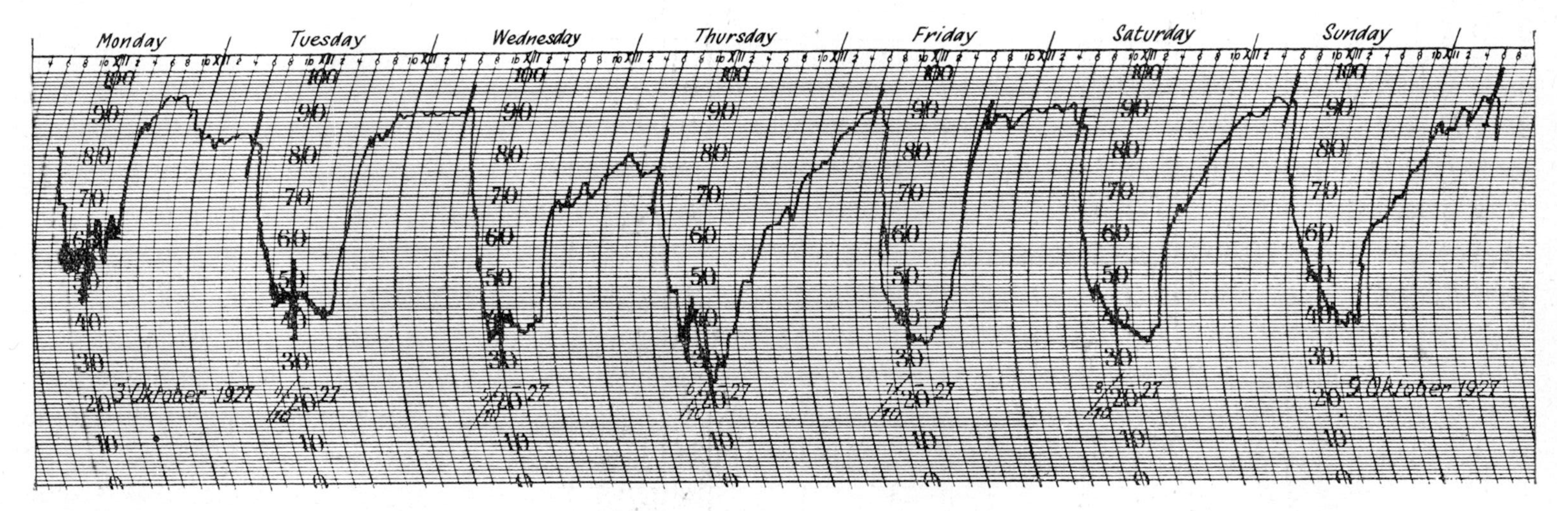

Fig. 57. Daily march of relative humidity during the dry season in Bogor from October 3 to 9 (after Schimper-Faber 1935).

high temperatures; 50% relative humidity at 32·3°C is equal to a saturation deficit of 18·3 mm. Such a value is characteristic for steppe areas. However, the atmospheric dryness only lasts for 4-6 hours in the humid tropics, coinciding with an air temperature increase between 0800 and 1400. The air is always vapour-saturated at nights resulting in regular dew formation. In contrast, the dryness can last for weeks in arid areas where the relative humidity is so low, even at nights, that there is no dew formation.

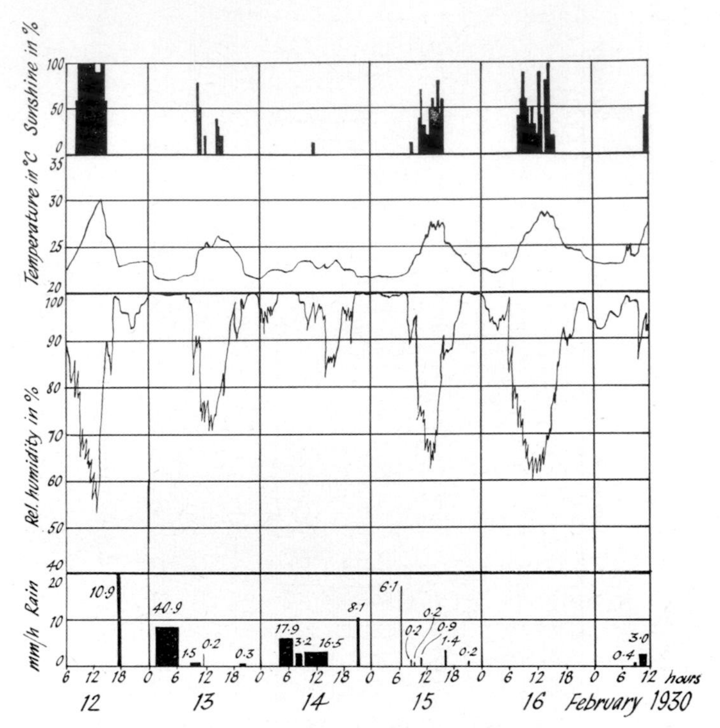

FIG. 58. Daily march of climate at Bogor (Indonesia) during the rainy season (compare the sunny Feb. 12 with the rainy Feb. 14). Figures on the rain-histogram show absolute rainfall in mm (after Stocker, 1949).

The real magnitude of daily climatic variations in the tropics becomes apparent from the fact that the monthly means of daily temperature ranges at Bogor lie between 10·2°C and 6·9°C, corresponding to those at Vienna, with 10·2°C-4·7°C and far exceeding those at Hamburg. Bernard (1945) reported an average daily saturation deficit of 13 mm for the virgin forest in the Congo basin. He emphasised in this connection that this value is greater than that recorded for stations in northern France. A variation of monthly temperature means of only 1-2°C is contrasted with average daily temperature variations of 10-12°C (absolute maximum 36°C, absolute minimum 18°C).

Relative humidity increases with altitude in the mountains corresponding with cloudiness, while temperature variations decrease. However, the daily temperature fluctuations become more pronounced at still higher altitudes corresponding with decreasing mean annual temperatures (compare Fig. 54 with Fig. 46).

The differences between the daily maximum and minimum become greater. The decreasing minimum is particularly important. Forest composition changes only slowly with increasing altitude. Certain species are limited to the lowest vegetation belt, others occur only at higher elevations. Distinct altitudinal vegetation types are not apparent. Lianas decrease in abundance with increasing altitude and, among the epiphytes, orchids become less abundant. Instead, certain groups of ferns increase in abundance, e.g. Hymenophyllaceae. Boreal taxa, such as *Quercus* and *Pinus* are found only in the northern hemisphere.

Precipitation decreases rapidly above the cloud zone which is characterised by almost continuous fog formation. The forests become more open and exhibit a less humid character. It is often customary, therefore, to divide the tropical rain forest into three altitudinal levels: lower slope, montane, and subalpine.

A sharp vegetation boundary is shown only where the daily minimum drops below 0°C. Since there are no pronounced annual seasons, the temperature drop below zero occurs nearly every day; that is, the number of days with temperatures above and below freezing increases rapidly at a particular altitude. The evergreen tropical rain forest ends abruptly, thereafter it is replaced by alpine vegetation. This change occurs at elevations much above the normal cloud zone. Therefore, the tropical alpine belt is characterised not only by the large number of days with alternate freezing and thawing but also by relatively low precipitation. The relative humidity varies greatly because of the large daily fluctuations in temperature. For example, Schimper-Faber (1935) measured relative humidities ranging from 8% to 100% at the top of Pangarango on Java on September 18.

In contrast to the common belief, radiation intensity is less in the humid tropics than in several other climatic zones. This even applies to the temperate zone when radiation is measured on an area perpendicular to incoming radiation.

The high relative humidity in the tropics, which is here correlated with a high absolute vapour pressure, screens out much of the radiation before it reaches the earth's surface, in spite of the more vertical position of the sun (Walter, 1960, p. 31). Clear indications of this are the constantly damp atmosphere and the light, grey-blue colour of the sky.

Instrumentation is very necessary to give an objective evaluation of habitat relations.

There are reasons other than radiation intensity for the uncomfortable feeling created by radiation in the tropics. The absolute vapour pressure regulates, to a large extent, the evaporation of water from the human body (which has its characteristic temperature), because it determines the vapour-pressure gradient between the body surface and the surrounding atmosphere. Therefore, cool air always feels dry, since it contains only a little water even when saturated. However, the vapour pressure gradient at the body surface is very low when the air temperature approaches the body temperature while at the same time it has a high relative humidity. Only a little water evaporates in spite of the body being covered with sweat and there is no cooling effect from evaporation. The danger of overheating and, therefore, of sun stroke, becomes very great on further heating from radiation. Wind provides the only cooling effect. Therefore, tropical sun can be endured relatively well without head protection in areas with sea breezes, whereas the sun becomes uncomfortable in areas protected from wind. Experience in tropical mountains has shown that the belief regarding the extreme intensity of tropical radiation is only imagined. Radiation intensity increased constantly during the ascent of Kilimanjaro above the tropical mountain forest. Yet conditions were quite comfortable above 4000 m elevation as a result of decreased air temperature.

One should not project one's own feeling into judgements of plant environments. The humid tropics are very far from being as moist a plant environment as is usually assumed. Plants have no temperature of their own and their foliage can show excessive temperatures where exposed to direct radiation. It is, therefore, not surprising that sun-leaves possess a rather tough structure often even giving the impression of xeromorphy.

The day length is important for plant development which, in the tropics, hardly changes through the year. The longest day is only 13·5 hours and the shortest 10·5, even at the Tropics of Cancer and Capricorn. Since the sun sets at a perpendicular to the horizon in the tropics, dusk is shorter than in higher latitudes, where the sun cuts the horizon at an angle. However, most descriptions exaggerate the brevity of the dusk. According to my determinations of dusk lengths in the tropics, it takes 30 minutes from the disappearance of the sun beyond the horizon to the moment when reading in the open is just still possible; in southern Europe at 41° N it took 40 minutes on April 18 and 48 minutes on April 22, north of Le Havre at the Channel. The dusk period lasts a considerably longer time only farther north. Long-day plants cannot flower in the tropics. For example, lettuce can be grown throughout the year without danger of bolting.

The morning is usually sunnier in the tropics than the afternoon because of the frequent afternoon rains. Fig. 59 may serve as an example. Evaporation, which depends largely on radiation, shows a daily rhythm in the tropics similar to that in the temperate zone on a humid summer day; that is, it shows a pronounced daily fluctuation (Fig. 60).

An important factor in many areas of tropical rain forest is the cyclone activity (hurricanes). Forest is much affected along their paths; many large branches break off, trees become defoliated, smaller stems of the intermediate tree layers are subjected to wind breakage but larger

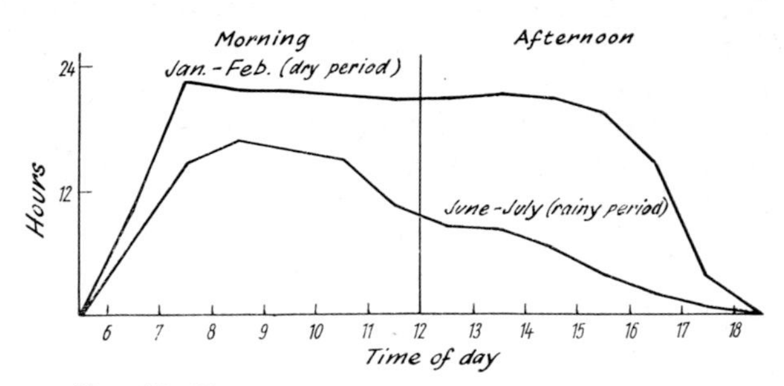

FIG. 59. Frequency (hours) of sunshine at different hours of the day in San Jose (Costa Rica) 9°56′, 1135 m above sea level (data from Nann-Süring).

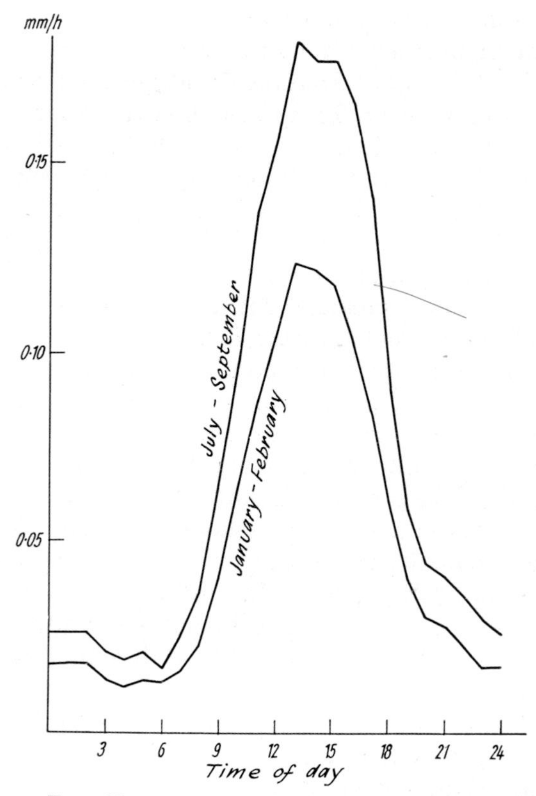

FIG. 60. Daily march of evaporation in Djakarta after data from Braak.

trees are rarely uprooted. The damaged and disturbed areas are soon re-vegetated, particularly with lianas. However, the dead wood so accumulated increases the fire hazard. If plantations are in a neighbourhood in which burning is practised, as, for example, in sugar plantations, fire can easily get loose into the forest. This results in further floristic changes. In east Australia, *Rubus*-lianas and *Flagellaria* invade by occupying the place of the fire-sensitive rattan-palm (*Calamus*). Also, *Acacia* species are known to follow fire in north Queensland (Webb, 1958).

The rain forest cannot attain its final equilibrium in areas with very frequent cyclone activity. Such forests show a very heterogeneous structure.

3. Microclimate in the tropical rain forest

The macroclimatic features mentioned have a direct effect only on plants fully exposed to the sun. This only includes plants cultivated in open fields and the rain forest trees of the upper crown canopy. Quite a different climate exists in the forest stand itself. It is much more uniform compared with the regional macroclimate (Fig. 61).

Of particular significance is the radiation-screening effect inside the forest. One obvious difference from the deciduous forests of the temperate zone is that it remains uniform throughout the year because of the absence of seasonal defoliation.

Typically, a relatively large amount of light penetrates into the forest interior, since the tropical rain forest has usually an interrupted crown canopy. However, much light is intercepted by the abundant lianas, epiphytes and trees of the lower layers. This effect is especially intensified by the almost closed shrub and herb layers, which often attain heights of several metres. Man is forced to move beneath these layers. The impression from this position is that of very low light intensity. Very little light penetrates to the forest floor. Therefore, a ground vegetation layer is usually absent.

About 0·1% of daylight reaches the forest floor in the East African rain forest at Amani in the East Usambara mountains (5° S, 900 m above sea level), where the annual rainfall amounts to 2000 mm. I observed flowering plants usually at only 1% daylight (on overcast days). It is very difficult to determine light intensity on sunny days because of the presence of sun flecks that constantly change their position, causing sudden increases in light intensity wherever they strike a leaf surface.

Orth (1939), working in the central African virgin forest at the Virunga volcanoes, came to a similar result, namely, that the minimum light intensity for shade plants was determined as 0·5%, the same as found in European hard-wood forests. The light intensity near the forest floor in forests of south Nigeria is likewise between 0·5-1% (Evans, 1939, 1956). The light flecks may reach 5 times this intensity. However, they affect only a very small area. Similar relations were found near Santarem, Brazil (Ashton, 1958).

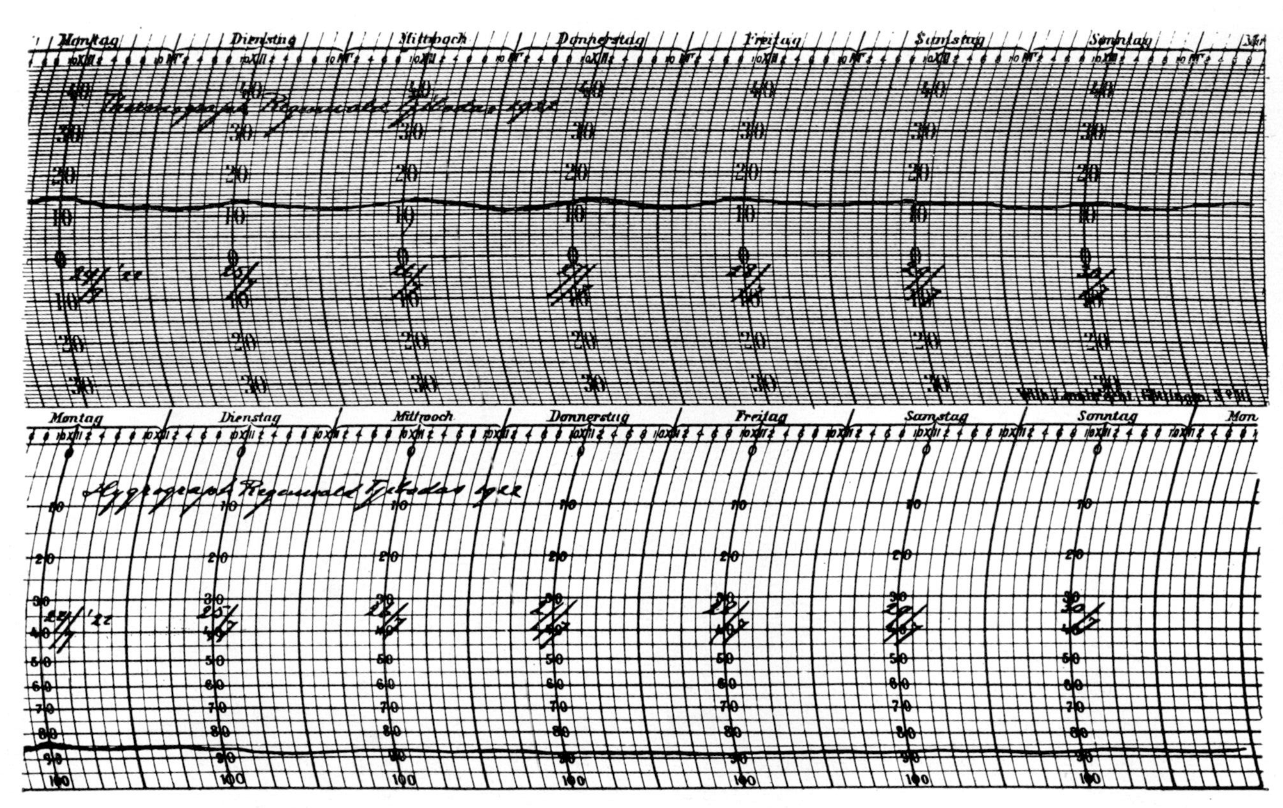

Fig. 61. Daily march of temperature (above) and relative humidity (below) in the rain forests of Tjibodas recorded at about 1 m height in a closed forest during the dry season (from Schimper-Faber, 1935).

More light penetrates to the forest floor in the mountain forests at about 2400 m on Kilimanjaro. They have only a thin herb layer, while the forest floor is covered with a bright green carpet of *Selaginella*.

However, these findings may not be of universal application. The light relations in the Dipterocarp forests of the heavy rainfall areas on north Sumatra (4500-6000 mm rain) are much less favourable according to travel descriptions by Bünning. These tall tree species form a nearly closed crown canopy, which effectively obstructs penetration of light into the forest interior, resulting in the prevalence of extremely dim light conditions throughout the day. Measurements near the forest floor gave a mere 0·1% daylight. Only weakly developed moss protonema are found in such places. *Selaginella* and liverworts were found at about 0·2%; Hymenophyllaceae, other ferns and clubmosses (*Lycopodium*) can survive with 0·25-0·5%. The mosses usually grow in little raised places, on dead stems and buttressed roots and not at the soil surface. Other plants found growing at such low light intensities include species of *Impatiens*, Begoniaceae, Rubiaceae, Commelinaceae and Zingiberaceae.

Conditions are less extreme in the rain forest at Tjibodas in west Java (1500 m above sea level). Schimper-Faber (1935, p. 344) reports on this forest: 'Investigations of light distribution in the tropical rain forest have shown that the variations in light intensity are very large and that nearly uniform light conditions prevail only in the herb layer of completely closed forests. The large variations in light intensity are caused mainly by the interrupted canopy of the upper tree layers and by the leathery, shiny surface of the foliage. The contrasts between light and dark are greatest in the lowest part of the forest and decrease with height above the ground. In this respect, light shows an opposite relationship to relative humidity and air temperature.'

A constantly vapour-saturated atmosphere occurs near the soil in the densest forests of north Sumatra. There are no temperature variations in this air layer. In the herb layer of the rain forest at Tjibodas, Schimper-Faber observed an average minimal daily variation of 1·4°C during the most moist month and an average maximal variation of 3·2°C during the driest month (Fig. 61). At 18 m height the corresponding values are 7·2°C and 12·4°C.

Other examples are given by Richards (1952). They refer to average daily temperature variations at different heights in a virgin forest (Table 4).

TABLE 4

	During the dry season (after Allee)			In the dipterocarp forest—Philippines (after Brown)	
Height in m	0·2	17	26	Undergrowth	18
Daily variation	1·1°–2·9°	6·8°–7·3°	8·6°–11·8°	5·7°	6·0°

Fig. 62 gives information about the march of relative humidity at different heights above the soil. It can be seen that the relative humidity frequently drops to 60-40% at 18 m height, even during the rainy season. The conditions are probably still more extreme in the crown canopy, where most of the sun's radiation is absorbed and converted into heat. The climatic factors are probably much less uniform there than in the instrument shelters used by meteorologists. According to Evans (1939), the highest saturation deficit values in Nigeria did not exceed 10 mm Hg. The variations in absolute humidity were at the most 2-4 mm.

Table 5 determined by Brown at different altitudinal levels on Mt. Maquiling (Philippines), refers to temperature extremes and relative humidity in the crown canopy of tall trees (from Richards, 1952).

TABLE 5

Height above sea level	Vegetation	Mean max.	Mean min.	Relative humidity Av. (%)	Daily min. (%)
80 m	second-growth forest	38·2°	24·0°	82·2	68·6
300 m	dipterocarp rain forest	32·4°	20·0°	—	—
450 m	dipterocarp rain forest	31·3°	19·6°	87·9	75·5
740 m	submontane rain forest	30·4°	18·1°	—	—
1050 m	montane, moss-rich rain forest	28·1°	17·3°	usually 100·0	91·0

Schimper-Faber concluded (p. 346): 'The climate in the tropical rain forest is relatively uniform in closed stands only near the ground; at higher levels the rain forest is characterised by considerable contrasts in intensity of the different climatic elements.'

My Piche evaporimeter readings (green paper, diameter 3 cm) in Amani (East Africa) often showed no measurable water loss for several days near the ground and at 1 m height. During a dry period the evaporimeters lost 1 ml in 24 hours on 13 out of 70 days, when the paper was placed on the prothallia of *Marattia*. The highest daily value was 1·7 ml, the highest hourly values near 0·1 ml. Double the amount evaporated at higher levels above the soil and in more exposed places.

In contrast to this, an evaporimeter placed in the open at 70 cm above a lawn frequently lost 6 to nearly 10 ml in 24 hours, whereas no evaporation occurred at nights. The maximum hourly value was 1·3 ml. Such values compare closely with those obtained in central Europe on hot, sunny summer days.

The high humidity in the virgin forests, which persists even after a number of days without rain is, according to my observations, to be attributed to the large quantities of dew that condense on the crown canopy. Dew drips from leaf to leaf thereby providing constant moisture

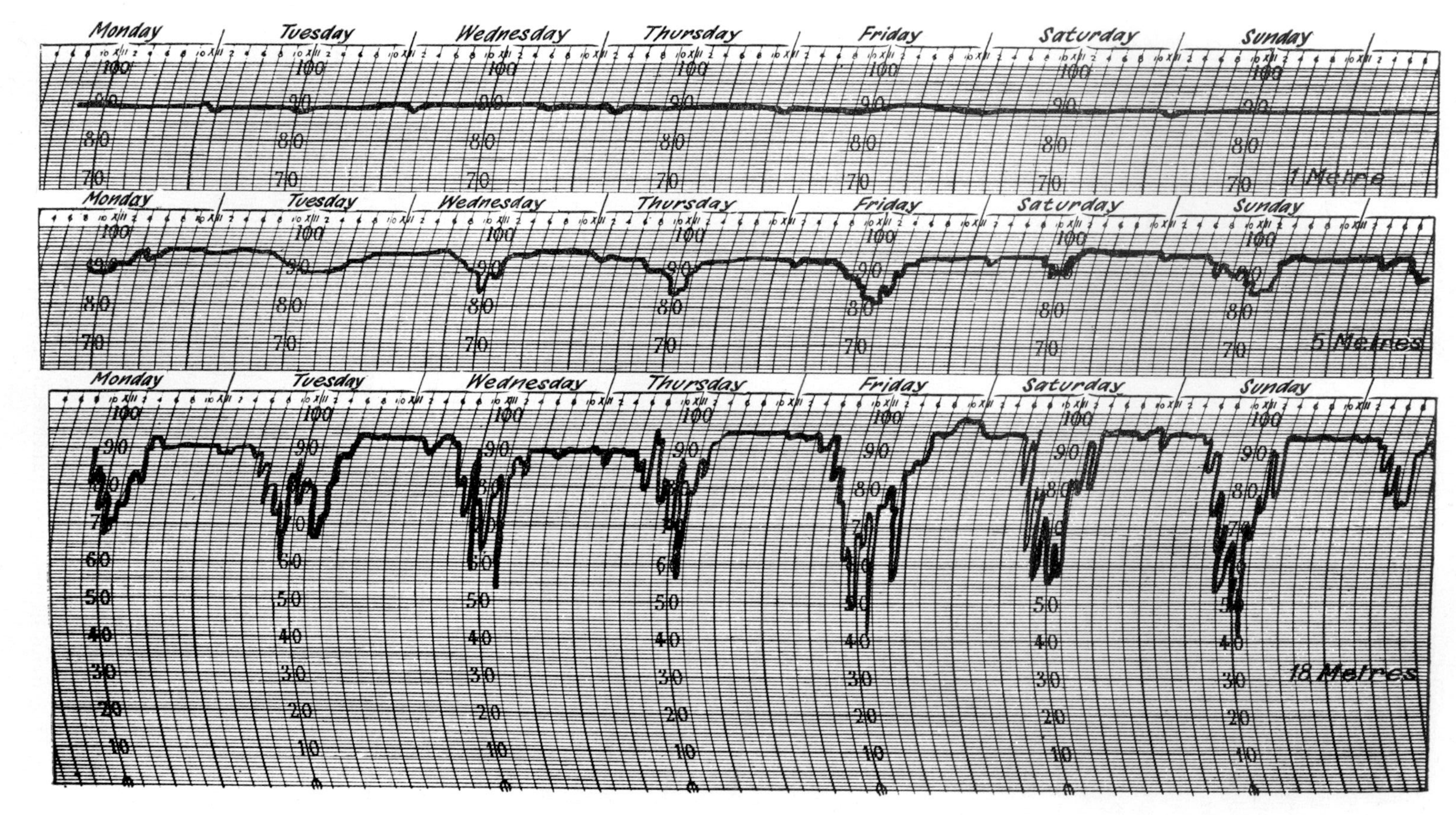

FIG. 62. Relative humidity at 1 m, 5 m and 18 m height above soil in the rain forest at Tjibodas during the rainy season (after Schimper-Faber, 1935).

to the lower vegetation layers. Dew occurs every night in the humid tropics.

The amount of dew was measured (with Leick's dew plate) in a forest-opening at Amani during 40 nights without rain. The maximum amount recorded was 0·26 mm. The amount was over 0·15 mm 14 times, 12 times between 0·15 and 0·1 mm and only during 3 nights was it hardly measurable.

A dew precipitation of 0·1 mm corresponds to 100 ml per m^2. This quantity can quite possibly drip from the leaves, which usually show no residual wetness. The amount of dew condensing on the crown canopy is probably much greater.

Little is known about the CO_2 concentration in the tropical rain forest. Soil respiration must be relatively intense because of the high rate of dry-matter production and the rapid decomposition of all dead organic matter. Soil respiration under natural conditions is known to be a function of the net production (dry weight) of the vegetation (Walter and Haber, 1957). Therefore, one might expect the CO_2 content of the air layer next to the soil surface to be higher than in the general atmosphere. However, the very high values (up to 10 times normal) given by MacLean (1919) were not supported by Evans (1939) in Nigeria. The highest values, occurring during morning hours (up to 1000) did not exceed twice the normal atmospheric content. Also Stocker's (1935b) values from Tjibodas deviated only little from the normal. Lemée (1956b), working in cacao plantations (Ivory Coast), found 0·7 mg CO_2 per litre of atmosphere in the morning and 0·4 mg in the afternoon. The CO_2 concentration is probably rapidly normalised with increasing height above ground, because of the increasing air turbulence.

According to Allee (1926), working in the rain forest of Panama, the wind speed at 25 m height is about 10 times that near the soil and it is, of course, much greater still at the crown canopy. The giant trees are exposed to the full force of the tropical storms. Therefore, uprooting is common.

The sheltering effect of the tropical forest ceases once larger openings are created. Today, large areas are cleared of forest, in part by natives and in part by planters. Thus, it is also important to understand the environmental conditions in cultivated areas. Kirkpatrick (unpublished) deserves credit for a detailed study of a coffee plantation near Nairobi, East Africa (1° S, 1722 m above sea level).

The region is relatively dry (1062 mm) for the tropics but dew formation occurs every night.

The coffee trees, planted in rows, were 20 years old and were maintained at a height of 2 m. They were covered with dense foliage. A comparison was made with climatic values from a meteorological station located 100 m away from the plantation. Within the plantation, temperature was read between the rows at 1·3 m height. The daily temperature variations recorded at the station were 5-8°C on cloudy days and 12-16°C on clear days. In the plantation, the variations were 10-25% greater on cloudy days and 25-50% greater on clear days (maximum variation 22·4°C).

Temperatures in the plantation were 3-4° higher around noon (max. 5°C) on clear days and 2-3° lower at nights (max. 4-5°C). The difference was up to 6·2° at noon on days with clouds that did not cover the sun. The cloudier the day, the smaller was the difference, but it did not usually decrease below 1°C. Lower temperatures were recorded only when the leaves were wetted by a shower at noon. No difference was shown during complete cloud-cover at nights.

Maximum temperatures were from 0·3-5·9°C higher and minimum temperatures were from 0·0-4·2°C lower. Wind had little influence on these relations. The lowest minimum temperatures were not measured at the soil, but at the upper surface of the crowns. The soil surface became more heated between the rows than in open areas. The difference was 5-8°C at 5 cm depth. The soil was cooler beneath the plants (due to a shorter duration of insolation).

Leaf temperature is of great importance. On clear days it was 10-15°C above air temperature. Maximum leaf temperature was 46°C. The slightest air movement lowered the temperature by several degrees. Occasionally a passing cloud lowered the temperature as much as 10°C. But if the sun was covered for more than an hour, the leaf temperature was still 3-5°C above air temperature. It will be seen that coffee has only a small rate of transpiration.

Absolute humidity was lowest at sunrise. It increased rapidly till 0800. After that it decreased slightly, reaching a second minimum at 1630 in the plantation (turbulence of the air). Relative humidity showed a decrease correlated with increasing temperature up to the daily maximum. In the bushes, relative humidity was only 3-4% higher than between the rows.

Wind was noticeable in the plantation only when exceeding 3 m/sec outside the plantation.

Evaporation between the rows, with wide spacing (3 × 3 m), was 15-18% higher than in the meteorological shelter; in the bushes it was 30% lower. In case of narrow spacing (2·4 × 2·4 m) evaporation was 15-20% higher between the rows and from 30-50% lower in the bushes.

Light intensity at the soil surface between the rows was 49% of full daylight under *Grevillea* on sunny days and 15-20% on cloudy days. The corresponding values under *Albizzia* were 50% and 26-33%, respectively. The mass of foliage of the coffee plants was more dense under shade trees and the leaves were larger, thicker and darker. The mean temperature was about 1° higher than in the open. It was lower only at noon and the maximum was attained at a later time.

If evaporation from an unshaded plantation is considered to be 100%, evaporation of a shaded plantation was:

On cloudy days	On sunny days	On very dry days
60–65%	75%	90%

Shelter-belt hedges against wind raised the temperature during the day and lowered it during the night. Relative humidity was lower during the day. The positioning of the plantation in a valley increased the extremes. Temperature was higher on an east slope in the morning, while in the afternoon it was higher on a west slope, in accordance with the path of the sun in the tropics. North and south slopes are less extreme near the equator. Dew remained for 2·5 hours after sunrise on west slopes and for only 1-1·5 hours on east slopes.

4. Soil relations in the tropical rain forest[1]

The soils of the tropical rain forest, in contrast to those of the temperate zone, are very old. Only volcanic soils may be of more recent origin. The Pleistocene period had no influence in the equatorial zone. Several pluvial and drier periods may have occurred in Africa but there was no complete destruction of the soils. The origin of the soils probably goes back to the Tertiary in most cases.

Therefore, depth of weathering is extremely deep. It can be deeper than 20 m in certain places. Under forest, there is no soil erosion even on rather steep slopes. Thus, the weathering products remain *in situ*.

High temperature, moisture and CO_2 production favour a far-reaching destruction of silicate rocks and leaching of bases and silicic acid. The remaining substances are mainly Al_2O_3 and Fe_2O_3. This process is called laterisation. There is no humus enrichment in the rain forest areas because of the rapid decomposition of organic matter. However, creek water is often a light brown colour, indicating the presence of some humus colloids. It is otherwise very poor in minerals. Its conductivity corresponds to that of distilled water.

As a typical example, we will discuss the tropical virgin forest soil of Amani (East Africa), which has been investigated most closely by G. Milne. His data, given to me personally at Amani, are cited here.

Very deep, reddish brown lateritic loam soils overlying gneiss, granulite and pegmatite are found even on steep slopes. The sesquioxide ratio $[SiO_2:(Al_2O_3+Fe_2O_3)]$ amounts to 0·68-0·75. An overlying humus layer is entirely absent. A litter layer composed of dead leaves and branches is present, but it only needs to be pushed aside with the foot in order to see the red colour of the mineral soil. Termites participate in the destruction of dead wood, although they do not erect mounds in the forest and thus are not easily noticed. Earthworms are present, although not very abundant. Most decomposition is probably through fungi. The soil profile shows no visible stratification into horizons. Lateritic concretions or compacted horizons are absent. The soils are quite permeable and thus strongly leached. Soil reaction is acid

[1] See Vageler (1938), Robinson (1939), Ganssen (1957).

(pH = 5·3-4·6). Litter with woody remains is very acid (pH = 4·05-3·55). The pH values vary little down the profile to a depth of 2 m (here, for example, pH = 4·95-5·00). There is no humus-enriched horizon in spite of the fact that the organic substances may attain 2·5-4% (wet combustion). Organic colloids are present in dissolved form but may only cause a somewhat duller colouration of the upper few centimetres of the soil. The composition of the organic colloids is unknown. They exhibit a rather constant C/N ratio, as shown in Table 6.[1]

TABLE 6

Carbon and nitrogen content (%) of soils at Amani (East Usambara Mts)

Depth (cm)	C	N	C/N
0–8	4·55	0·364	12·5
10–26	2·85	0·215	13·4
31–47	1·55	0·117	13·2
52–63	1·07	0·089	12·0
70–86	1·09	0·084	13·0
94–104	1·08	0·074	14·2

Table 7 gives information on the mechanical composition of the soils.

TABLE 7

Kwamkoro near Amani. Virgin forest soil of an old coffee plantation. Altitude 1000 m; rainfall 2000–2200 mm, slope 1:3. Soil texture gravelly sandy loam throughout the profile. Lime entirely absent (figures in %).

Depth (cm)	Colour	Gravel larger than 2 mm	Coarse sand larger than 0·2 mm	Fine sand larger than 0·02 mm	Silt larger than 0·002 mm	Clay less than 0·002 mm	Field capacity	Water content of air-dry soil
0–13	dull light brown	14·1	24·4	20·5	11·15	18·0	32·4	3·0
15–31	light brown	31·5	22·6	19·8	11·1	42·7	33·3	3·2
31–47	orange brown	21·9	20·3	18·8	11·6	45·3	32·7	3·8
78–94	reddish brown	22·2	24·2	24·7	13·0	35·1	27·0	3·45
125–157	orange brown	16·3	28·9	24·7	14·5	30·1	25·7	3·05
157–188	purple brown	23·4	32·7	22·7	15·9	29·7	27·7	2·6
220–250	dull heliotrope	22·0	32·5	22·8	18·0	25·2	30·5	2·4

[1] Similar values were found also for non-podzolised soils in the Amazon region by Klinge (1962a).

Weathered gneiss is abundant below 2·5 m and a few rock fragments occur above.

The water content of two soil samples that were extracted directly in the forest, from 40 cm, was 26·8-27·2%.

The nutrient content of the soils is usually extremely low: $CaCO_3$ and N, in ammonium form, are entirely absent. N exists only in an organic form (see Table 6). Nitrate nitrogen can be found only in traces. Equally low is the phosphorus content. Soluble phosphorus in diluted H_2SO_4 (pH = 3) amounted to:

at the surface	18 ppm
farther down	11–12 ppm
at the bedrock	10 ppm

In contrast, recent volcanic soils in the Kilimanjaro region are 5 to 25 times richer in phosphorus.

This shows that tropical, virgin-forest soils lack the most important nutrients essential for plant growth. However, they carry the most lush of all rain forests. In the temperate zone, we commonly observe degradation to dwarf-shrub heath vegetation in similarly extreme humid climates. Progressive rock weathering in the humid tropics cannot provide nutrients for the plants since the unweathered rock material lies beyond the reach of tree roots. The upper layers are relatively the richest in nutrients. The direction of water movement is always downward because of the high rainfall. No data are available on rooting depths of trees and associated plants. Uprooted trees only show laterally extended root systems. Bünning estimated maximal rooting depth to be 75 cm. Ellenberg (1959) gave an estimate of only 10-20 cm for the Amazon lowland forest in Peru, which appears unbelievably shallow. Structural support for the stems is obtained through the buttressed roots.

The seemingly contradictory relationship of low soil nutrient content and extreme vigour of the vegetation is however quite explicable (Walter, 1936a, b). A natural virgin forest, in which there is no exploitation by man, is a self-sustained, closed system, including the soil. Such a system is in a dynamic equilibrium. The nutrients are absorbed by the plants from the soil and are returned again to the soil through decay of organic matter. There is a continuous cycle of matter. The more mineralisation of dead plant materials in the soil is favoured by climatic factors the more rapid is the availability of nutrients from dead organic matter for reabsorption by living plants. These nutrients are immediately absorbed by the roots. Therefore, the amount of nutrients in the soil at any one time can be practically zero. The entire nutrient capital necessary for the continued growth of this lush type of vegetation is tied up in the living matter itself and in the small amount of undecomposed, dead, plant fragments. Decomposition is much slower in a cooler climate. This results in humus accumulation which increases with decreasing temperature. The nutrient capital is then found partly in the living plant material and partly in the soil humus. The proportion of nutrients tied up in living plants and the plant mass itself decreases for equal rainfall with decreasing temperature. Thus, we find only sparse vegetation associated with deep organic soils in moist arctic and alpine areas.

The soil relations in the tropics are understandable only if one considers the constant cycle of matter. There are no nutrient losses under natural conditions. The roots act as filters; they absorb the substances resulting from the mineralisation of organic matter and allow only humus colloids to pass by. This is shown by water issuing from springs; in Amani spring water was used in place of distilled water, its conductivity was equally low or even lower. If there were to be a slow but continuous leaching of soil nutrients, virgin forest would degrade gradually, for addition of nutrients to the system can occur only in respect of nitrogen (N-fixation through electric discharge during thunderstorms or through micro-organisms). Yet the tropical virgin forests have grown unchanged for many thousands of years, and only recently have large areas of it been destroyed by man.

This destruction is probably permanent, because logging and burning of slash causes a sudden mineralisation of the entire nutrient capital. Since such cleared and burned areas are denuded of plants that could absorb the suddenly-released nutrient capital, it is leached out instead by rain. Only a small remaining proportion is held by cultivated plants and weeds. Therefore, cleared forest areas are infertile in the tropics and their peoples are forced to practise shifting agriculture. They leave their fields in fallow after a few years to clear new areas from virgin forest. A second-growth forest can become established on the fallow fields. But it will hardly attain the original vigour of the virgin forest as a result of the reduced nutrient capital. After repeated use, the cleared areas are often invaded by bracken (*Pteridium aquilinum*) or, after repeated burning, by alang-alang grass (*Imperata*).

These observations, which apply to African virgin forest, are fully supported by observations of Sioli (1958) who worked for many years in the Amazon area. Plantations on cleared virgin forest soil have to be abandoned after three years, because of exhaustion of the soil-nutrients. A second-growth forest (Capueira) follows. This can be cleared again after 10 years for re-use of the land for another 1-2 years of cropping. 'The original, beautiful, tall forest does not regenerate for decades and instead of a prosperous cultivated landscape we observe on the Amazonian terra firma the continuous spread of poorly vegetated, scrubby Capueiras, resulting from human practices of land clearing and agricultural use.' In contrast to this, Ellenberg believes—following some observations in the Amazon area of Peru—that the second-growth forest can re-establish the original soil fertility. He assumes that it is largely a question of nitrogen supply and that nitrogen can be rapidly re-accumulated, particularly when the second-growth forest is rich in legumes. More detailed investigations of this question would be desirable. So far they are not available.

An important question relates to the origin of the nutrient capital that is tied up in the plant mass of the virgin forest. For example, where does the essential phosphorus come from? The answer to this fundamental question is probably that the nutrient supply had accumulated gradually during the period when weathering had not yet penetrated too deeply and when roots were still in contact with the undecomposed rock. This period must have lain in the more distant past in most tropical areas.

However, not all soils of the tropical rain forest zone are lateritic red-loams. They are formed particularly on silicate rocks and limestone. Because of the high rainfall, lime is leached out and only the clay-like impurities remain. Thus, the remaining material rapidly assumes an acid reaction. Only areas with outcropping limestone have a selective action on the vegetation. Such substrates are very permeable to water thereby supporting vegetation characteristic of the semi-evergreen forest of drier tropical climates.[1]

Red-loams cannot result from weathering of sandstone, or from very acid volcanic rock rich in silicic acid, or from alluvial sands, since such parent materials do not contain any clay-forming minerals. Such soils are characterised by bleached sand horizons and can be described as podzols (see Walter, 1960, p. 454). Richards points out that podzols are found not only in boreal climates but that they are also widely distributed in the humid tropics. The following profile is from Borneo with 26°C mean annual temperature and 3000 mm precipitation (no month with less than 100 mm rain). The vegetation on this soil is a poor forest with heath shrubs and differs strongly from the typical tropical rain forest. This poor vegetation type appears to form a climax community on this soil. The parent rock was probably already so poor in nutrients that the vegetation could not accumulate a larger nutrient capital during weathering. Information on soil structure is given in Table 8.

Similarly, another profile from east Borneo showed a substantial, bleached sand horizon (pH = 6·1) underlying a 20 cm thick raw humus layer (pH = 2·8). Beneath the bleached horizon occurred a yellow-brown illuviated (B) horizon with a pH value of 5·4. In both cases the soils were formed on alluvial sands but the same profiles develop in soils from sandstone. Such podzols are also found in Thailand and on the Malay peninsula. Bleached sandy soils, covered with savannah, *Humiria*-scrub and *Eperua*-forest, are also known from Guyana. These are sands that originated from weathered granite and were deposited in a shallow basin during the Tertiary. Podzol soils with bleached sand

[1] Lötschert, W., 1959 has described these relations for San Salvador. See also his work on Cuba (cf. Chap. III, 4).

are also widely distributed in the Amazon basin (Sioli and Klinge, 1961; Klinge, 1965). These are covered with Campina forests, which attain heights of 25 m and are composed predominantly of tree-legumes (*Eperua*, etc.) with little undergrowth (Takeuchi, 1962). The black water of the Rio Negro gets its colour from the high content of humus colloids that are leached from the raw humus. Similar soils are mentioned from Mafia Island (East Africa). They are covered with a heath-type vegetation with *Philippia* (Ericaceae).

TABLE 8

Profile of a tropical lowland podzol (after Hardon, cited from Richards, 1952). The gravel (>2 mm) and silt (6·005-0·05 mm) content was omitted.

Horizon	Depth (cm)	Description	Organic Matter (%)	% Sand (2–0·05 mm)	% Clay (0·005–0·0005 mm	Ratio $\frac{SiO_2}{Al_2O_3 + Fe_2O_3}$ of Clay	pH	CaO (%)
A_0	0–10	Semi-decomposed, black organic substances	much	—	—	—	2·7	—
A_1	10–25	Loose, grey-black humus-enriched quartz sand	—	95·6	1·6	7·17	3·9	0·022
A_2	25–40	Loose, grey-white quartz sand	—	94·0	0·6	8·64	6·1	0·032
B_1	40–70	Hardpan, dark brown, very compacted quartz sand	5·2	86·9	7·2	4·02	3·9	0·029
B_2	70–100	Loose, light brown quartz sand	—	92·1	4·6	0·31	4·6	0·035

Not only podzols are found in the tropics but also hydromorphic soils. They are conditioned by high ground water, as in the temperate zone. These are either humus-poor, mineral soils or peat soils. The former originate under eutrophic conditions. They consist of grey or yellowish clays and are similar to gley soils. The overlying humus layer is very thin and the soils are poorly aerated. The peat soils are less widely distributed. Ombrogenous peats are covered with tropical bog forests, while topogenous bogs are usually covered with grasses. *Sphagnum* occurs at higher altitudes but it does not play an important role in the tropics. Peat depths of 7 m have been measured. A dark-coloured water seeps out from ombrogenous bogs having a pH of 3·0.

These bogs are oligotrophic. The ash content of the peat is very low. They correspond to the raised bogs of western Europe and are probably conditioned not only through high rainfall, but also through nutrient-poor rock substrates. The topogenous bogs are mostly eutrophic, as in the fens of Europe.

The low nutrient content of the tropical soils does not apply to areas with recent volcanic rocks. Here, the supply of nutrients may be sustained through weathering after forest clearing. Such regions are, therefore, the most important settlement areas in the tropics, often occupied by a dense agricultural population.

Cultivation of the nutrient-poor tropical soils can cause severe soil erosion so that unweathered parent rock may reappear at the surface. The whole process of nutrient accumulation can start over again provided there is no further interference by man. However, this must involve a very slow process even under the rapid weathering conditions of the tropics. Re-establishment of the primary virgin forest in this manner is conceivable.

The differentiation of soils in the tropics is reflected in the composition of the vegetation. The typical rain forest is usually correlated with red-brown lateritic loams, while plant communities of different composition are found on the other soils. Thus, the relationships between soil and vegetation are very close in the humid tropics. However, they are usually detected only upon more exact evaluation of the floristic composition.

Probably the most important edaphic factors in relation to vegetation are water permeability, soil aeration and soil depth. In contrast, soil acidity and the content of calcium (which is practically always absent) appear to be of lesser importance. It is still doubtful whether there are truly basiphilous species in the tropics. The Dipterocarps are considered acidophilous, although this view is not universally accepted. The Dipterocarp forests in Burma and Thailand are supposed to occur on limestone, and one species, *Hopea ferrea*, is even said to show a preference for lime. However, all these questions, for example, the role of aluminium, are insufficiently investigated as yet and it is of little value to discuss in detail certain unproven hypotheses or assumed possibilities.

In summary, we may emphasise that the lushness and vigour of the tropical rain forest is no indication of great soil fertility, as was assumed by the first planters. The living plant mass is rather a function of the humidity of the climate. This refers not only to the tropics but to all areas of natural vegetation that are not completely modified by man. Since, however, the lushness of the vegetation on deeply weathered, mature soils is inversely related to the nutrient capital in the soil, we find that cultivated soils degrade most rapidly in humid areas, where the second-growth vegetation is characteristically depauperate.

5. Structure of the tropical rain forest

This section gives a general description of the tropical rain forest. There are as many, if not more, different types of rain forests as there are types of temperate forests. However, the tropical rain forest has not yet been studied sufficiently for adequate classification. Therefore, the description presented here will not necessarily fit all cases. Richards (1952) gave an excellent description of a typical tropical rain forest.

Tree heights in the typical rain forest usually do not exceed 50-55 m, although certain trees may become 60 m tall. As a rule, the number of tree species in a forest stand is large. Commonly one may find 40 species per hectare, occasionally more than 100. However, the tree layer can be very uniform. Usually the shrub and herb layers, including lianas and epiphytes, are rich in species.

In spite of this wealth of species, the physiognomy of the rain forests in different parts of the world is very similar. The trees are straight and slender and begin to branch only just below the apex. Consequently, the crowns are small. Commonly, buttress-like roots are developed at the base of the trunk. The bark is thin and smooth; the leaves are usually large, leathery, dark-green and entire. (They resemble, in general, *Prunus laurocerasus* rather than *Laurus nobilis*.) The leguminous trees with their feathery leaves are the only exceptions, but even their individual leaflets are usually large. The green colour is typically dull and flowers are not easily noticed. Three tree strata are often distinguishable, but in other types only two, or stratification may be absent. Bizarre tree forms such as palms, *Pandanus* or Dracaeneae do not occur in typical tropical rain forest; they are more common on waterlogged low-lying areas. The undergrowth in the interior of the rain forest is not as impenetrable as has often been described, but the slippery, loamy soil makes walking difficult. However, borders of the forest stand present an impenetrable green wall.

A striking characteristic of the mountain forest is the abundance of epiphytes and lianas which hang down from the crests of the trees in ropes often thicker than a man's arm. The greater the humidity the more numerous they are, but they prefer more open woodland.

In spite of many different plant forms, the tropical rain forest has a rather monotonous appearance, which is enhanced by the absence of seasonal change. The high humidity inside the forest gives a feeling of being imprisoned which is further emphasised by the closeness of the vegetation. The deep silence is interrupted only occasionally by the cries of a bird or monkey.

The vertical stratification of the tropical rain forest is best seen along edges of strip-cut areas.

Fig. 63 shows a profile of a virgin forest in Guyana. Only trees above 4·6 m height are drawn. They are arranged in three strata. The upper stratum is 35 m high and is composed of members of the families Lecythidaceae, Lauraceae and Araliaceae. Their crowns are umbrella-shaped. The intermediate stratum, which is about 20 m high, is composed predominantly of representatives of many different families. These two strata are not very dense and not always easy to distinguish. The largest trees with heights above 42 m occur as scattered individuals. The dense, closed, lower stratum is usually a little over 10 m high and consists of immature trees of the upper strata and tree species belonging to the families Anonaceae and Violaceae. Their crowns are elongated. All lower trees are connected by lianas, which only rarely extend into the tallest tree tops. Beneath the tree layer are shrubs, tall herbs and a ground layer consisting of seedlings, low herbs, ferns and *Selaginella*. There is no moss layer.

The second profile (Fig. 64) shows a Dipterocarp forest from Borneo. The family Dipterocarpaceae is represented by several species. Here also one can distinguish three tree strata at 35, 18 and 8 m height. The

FIG. 63. Profile through the rain forest at the Morabilli River, Guyana (from Davids and Richards) along a strip-cut area 41 m long and 7·6 m broad. All trees larger than 4·6 m are shown.

upper stratum of Dipterocarps does not form a closed canopy, but it is clearly separated from the second stratum. The two other strata are less easily recognised, and are composed of members of several different families. Below these is a shrub layer, 4 m tall, and a herb layer reaching 1-2 m in height. The herb layer is dominated by tree seedlings (184 per 10 m^2 as opposed to 135 herbs). The density of this layer varies considerably, but is nowhere very great. As in the former example there is no moss layer.

The third profile (Fig. 65) represents a rain forest in West Africa where rainfall is less uniformly distributed. The mean heights of the three tree strata are 42 m, 27 m, and 10 m or less, respectively. Only the lowest tree stratum is very dense and intertwined with lianas and for this reason the shrub and herb layers are weakly developed. It is not impossible that a few tall trees had been cut and removed from this forest.

FIG. 64. Profile through a mixed Dipterocarp forest at Mount Dulit, Borneo (from Richards). Only the trees above 7·6 m height were drawn on a transect 61 m long and 7·6 m broad.

FIG. 65. Profile sketch through the Shasha forest Reserve (Nigeria). The forest strip shown was 61 m long and 7·6 m broad. All trees above 4·6 m height are sketched (from Richards, 1952). *Ab*, *Pausinystalia* or *Corynanthe* sp.; *Ak*, Ako ombe (undetermined); *Eb*, *Casearia bridelioides*; *Ek*, *Lophira procera*; *Ep*, *Rinorea* sp. (cf. *dentata*); *Er*, *Picralima umbellata*; *Es*, *Diospyros confertiflora*; *Ip*, *Strombosia* sp.; *It*, *Strombosia pustulata*; *Od*, *Scottelia kamerunensis*; *Om*, *Rinorea* sp. (cf. *oblongifolia*); *Op*, *Xylopia quintasii*; *Os*, *Diospyros insculpta*; *Te*, *Casearia* sp.; *Y*, *Parinarium* sp. (cf. *excelsa*).

While most tropical rain forests have a large number of tree species, there are a few in which the tree layer is dominated by one species. An example of such a forest is shown in the following profile sketch (Fig. 66), which represents a *Mora excelsa* forest on Trinidad. Here the crown canopy is closed, just as it is in the central-European beech forest. The two lower strata are less well represented. The *Eperua* forests in Guyana and the *Dacryodes-Sloanea* forests on the Lesser Antilles are similar. Such types are also found in Africa and Asia. In these forests the shrub and herb layers may be entirely lacking.

A detailed description of a tropical rain forest, in the Amazon area 42 km north-north-east of Manaus is given by Takeuchi (1961). At higher elevations (60 m above sea level) the trees attain heights of more than 30 m. Mixtures of species of legumes with Sapotaceae and

FIG. 66. Profile through the evergreen *Mora excelsa* forest (trees marked with M) on Trinidad (from Beard) representing a forest strip 61 m long and 7·6 m broad.

Lecythidaceae occur at higher locations and, at lower ones, species of Moraceae. Three tree strata can be distinguished. However, in depressions (40 m above sea level) the crown canopy is often interrupted, no layering is discernible and the trees are usually shorter. The presence of densely stocked stemless palms is very striking in the lower stratum, while in the lowest parts there are palms with stems. Trees with buttress-like roots are common in all stands. Epiphytes (Araceae, Orchidaceae, Bromeliaceae, ferns) are more abundant in depressions.

These few examples give some idea of the kinds of structural variations that one may encounter in virgin forests. The tropical rain forest only appears homogeneous on superficial observation.

It has been suggested that virgin forests are a type of vegetation that has remained unchanged for thousands of years, but this is probably applicable only to broad geographic categories. On a smaller scale, one can notice signs of continuous change—for instance the replacement of older trees by younger ones. These replacements may also be associated with temporary floristic changes. No definite information is available on the age of trees that do not have annual rings, but on a basis of stem increment measurements, ages of 200 to 250 years have been estimated in some cases.

The collapse of trees past maturity results in openings. If these are relatively large they are invaded initially by species that are dominant in the secondary forest, e.g. by *Cecropia* in South America, by *Musanga* or *Schizolobium* in Africa and by *Macaranga* in Malaysia. Later they are replaced by regeneration of species of the upper tree stratum. These have differing light requirements when they are immature. Some species require much light and these germinate and grow only in the larger openings. Others may have been present already in the undergrowth thereby indicating their early development in deep shade.

Seed production and numbers of seedlings vary considerably both in different developmental stages and between species. Tree species of the upper stratum may be entirely absent or extremely rare in the understoreys. The same species is only found abundantly in the undergrowth of forests dominated by one tree species. Germination usually occurs immediately after seed fall, although seed dormancy has been noted. Only a few of the many seedlings survive the first critical years under the poor light conditions near the forest floor. Richards mentions that seedlings of some palms typical of swampy habitats are found throughout the entire forest, where they have been brought by birds. The appearance of seedlings, therefore, cannot always be accepted as evidence for the occurrence of a temporal succession. The growth rate during development depends entirely upon the local environment. In deep shade annual growth may be very slow. For example, mean annual diameter increments of only 0·7 mm were measured on young trees 5 cm in diameter in a Dipterocarp forest, while the larger trees showed diameter increments of nearly 1 cm. In addition to lack of light, root competition is probably another limiting factor. On the other hand, extremely fast growth has been observed in forest openings under favourable light conditions. For example, an annual diameter increment of 2·5 cm has been recorded for *Flindersia* growing in forest openings in north-east Queensland, while diameter increments were hardly measurable inside the forest.

The fact that saplings of the tree species of the upper strata are often lacking was also noticed by Aubréville (1938a), who studied the rain forests of the Ivory Coast. He concluded that the virgin forest has a mosaic structure. This implies that any tree species becomes replaced at one place by another tree species, and the first species will only recur at this site after a lapse of time and several more generations, i.e. the tree species are replacing each other on small areas in periodic cycles. A similar trend has been observed on very small areas among grassland communities. Here, also, the individual species interchange their growth-locations (Lieth, 1958).

As noted by Richards, this conclusion may be applicable in certain situations but should not be over-generalised. A pattern applicable to one type of tropical rain forest may not apply to another.

Old trees may either fall to the ground full length and thus create large openings, or they may be gradually hollowed out by termites or fungi and then collapse in stages leaving only a small gap. Termites are especially abundant in the virgin forest of the Congo. Fig. 67 shows the distribution of termite mounts (after Achtnich) on a cut-over area.

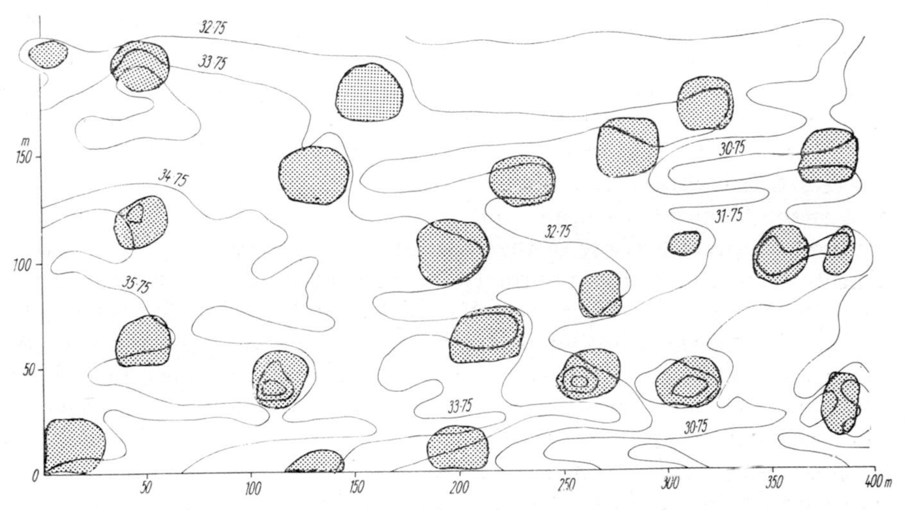

FIG. 67. Very numerous termite mounts (dark) on a clearing in the virgin forest of the Congo basin in Fangarubi (0°49′ N, 24°29′ E) after Achtnich (unpublished). Isohypses are 1 m apart.

Seventy per cent of all plant species in the tropical rain forest are phanerophytes (= trees and shrubs). They represent not only the most dominant life-form in terms of species number, but also in terms of numbers of individuals. However, other biological types are present which often form certain unions or synusiae which exhibit rather peculiar ecological adaptations.

The following life forms can be distinguished:

1. Trees and shrubs
2. Herbs
3. Lianas
4. Hemi-epiphytes
5. Epiphytes.

Heterotrophic life forms, such as saprophytes and parasites, only play a minor role in closed virgin forests, although there are very interesting species among them, e.g. *Rafflesia*; Loranthaceae are not uncommon.

6. Ecology of the trees in the tropical rain forest

The trees of the virgin forests commonly show certain peculiarities which are characteristic of many species, but not all. As mentioned before, their trunks are smooth and slim; the bark is usually thin and a striking light colour. Corky bark does not form because the trunks are hardly ever exposed to a dry atmosphere or to direct insolation. *Lophira procera*, a typical tree of the rain forest in Africa, only forms very thin bark, while *L. alata*, which grows in savannahs, forms very thick and deeply furrowed bark (Richards, 1952). These trees are often treated as one species. Species of thorns, which are so typical of arid areas, occur only very rarely in the rain forest. For example, prickly stems are found on *Cylicomorpha parviflora* in Africa.

Buttressed roots are a striking feature (Figs 68 and 69) of many tree species which attain great heights. These provide a firm support for the stem, a matter of some importance because of the shallow root system. Cambial activity of the roots at the stem base is quite normal during early development of the trees; later, however, it is restricted to the upper side only where wood is added and the buttress-roots are formed in this manner. They may reach 9 m up the stem and spread the same length radially, decreasing in height towards their outer edge.

Buttressed roots seem to be correlated with wet soils. It is possible that cambial activity on the lower side of the roots is inhibited, because of poor aeration. Buttressed roots are also developed in central Europe on tree species in moist habitats, e.g. in *Ulmus effusa* on alluvial bottomlands or in *Populus italica*.

Stilt-roots are less common in trees of the tropical rain forest. When they occur, they belong mostly to species of the lower tree layer.

The buds of the rain forest trees are usually only slightly protected, often merely by hairs, a sticky fluid or through specialised stipules or petioles. Wherever bud scales are formed, they are green and full of moisture.

The development of nodding leaf-shoots is remarkable (Fig. 70). Formation of supporting tissue often lags behind during the rapid elongation of young shoots, resulting in their nodding position during early development. Initially, they are frequently white or they may be coloured red by anthocyans. Later they turn green and become strengthened with supporting tissue. The development of young shoots is usually slower in less humid climates, where it is associated with the differentiation of tissues. Nodding shoots are poorly protected against desiccation so that they only occur in humid climates. Similarly, nodding shoots are only observed in the temperate zone in beech where the young shoots also have rather light green foliage.

In regard to leaf shape and size, the trees show, in most cases, rather large, undivided, entire leaves, often with a drip-tip. Such a drip-tip is found, for example, in 90% of the 41 species from 20 families of di-

Fig. 68. *Piptadenia africana* with plank buttresses. Entebbe, Uganda, Africa (Phot. E. Walter).

Fig. 69. Tropical rainforest tree with plank buttresses. Yingabara, North Queensland, Australia (Phot. E. Walter).

cotyledonous trees occurring in the Sinjara forest in Ceylon (Baker, 1938). The same situation applies to the tree species of Borneo and Nigeria. Trees with lobed leaves are the exceptions. Leaf-size usually decreases with the height of the tree species and, at the same time, the leaves become more leathery and the drip-tip shorter.

FIG. 70. *Brownea arhiza*, with nodding foliage. Entebbe, Uganda, Africa (Phot. E. Walter).

Differences in leaf size in the same tree species are very striking, either when the leaves of young individuals in the lower tree stratum are compared with those of taller individuals in the upper strata, or when comparing, on the same tree, the leaves on lower, shaded branches with those on the upper branches exposed to insolation.

In the rain forest of Amani (East Africa), I found, on the same tree, the relationships shown in Table 9.

Leaf lengths of 55-95 cm were measured on a fallen tree of *Polyscias polybotrya* belonging to the upper layer, while leaf lengths of a young tree in the shrub layer were 170 cm.

These differences in leaf size underline the microclimatic differences in the upper and lower tree (and shrub) layers.

TABLE 9

Length × width (cm)	Smallest leaf	Largest leaf	Ratio
Myrianthus arboreus	16 × 7	48 × 19	1:8
Anthocleista orientalis	22 × 10	162 × 38	1:28

The great humidity and constant adequate water supply of the rain forest trees are reflected also in favourable hydration conditions (compare Chap. I, 8); that is, the cell sap concentration is always low.

The following measurements in Table 10 were made by me in the rain forest of Amani (East Africa) from October to December.

[Note. It was not possible to make determinations on leaves from the crowns of the tallest trees. However, their osmotic values are probably not much higher than those of the sun-exposed leaves of the lower branches. A similar relationship applies to European trees (Walter, 1931)].

TABLE 10

Species of the upper tree stratum	Cell sap concentration (atm)
Allanblackia stuhlmannii with leathery, medium-sized leaves	
Sun-leaves of a large tree (lower branches)	16·4
Oldest leaves of another tree	16·5
Young, but already full-grown leaves of the same tree	14·7
Medium-sized leaves of a medium-sized tree	15·2
Shade-leaves of a young tree	13·3
Leaves shed from the same tree on the ground, still fresh	16·3
Piptadenia buchanani	
Leaves of a young tree, but at a rather sunny location	13·7
Fully grown leaves of a 3 m tall sapling	11·1
Young, very delicate leaves of the same sapling	8·8
Berlinia scheffleri with thin, leathery leaves, similar to beech	
Old leaves	9·7
Nodding shoots, leaves flaccid and light coloured	7·1
Parinarium holstii	
Leaves of a branch broken off through a storm	13·8
Albizzia fastigiata	
Fully grown, dark green leaves of a young tree	12·1
Young, nodding, light green leaflets	10·7
Paxiodendron usambarense	
Large leathery leaves, but not very thick	13·8

TABLE 10 (*ccntd.*)

Species of the upper tree stratum	Cell sap concentration (atm)
The following tree species either belong to the lower stratum or can be considered as colonising species during regeneration.	
Anthocleista orientalis (mid-rib was cut out)	
Mature leaves of a large tree	8·2
Young leaves of the same tree	8·1
Very large leaf of a young tree	7·7
Not quite fully developed leaves, at forest stand border	10·2
Macaranga usambariensis	
Old leaf without mid-rib	9·4
Mature leaves without mid-rib	9·0
Youngest leaves hardly 10 cm long	6·1
Polyscias polybotrya	
Mature leaves, with scales on the underside	9·5
Meristem and leaf primordia, densely covered with scales	7·2
Whitfieldia longifolia, shrub-like	
Leaves leathery, as in *Aucuba*	17.9
Cylicomorpha parviflora with prickly stem	
Leaves rather shaded	8·0

Table 11 gives some data for trees introduced to Amani.

TABLE 11

Tree species	Cell sap concentration (atm)
Ficus elastica	
Mature leaves	10·8
(For comparison a greenhouse-grown plant from Stuttgart):	9·8
Young, 2 mm thick tips of aerial roots	7·7
Ficus nitida (= *retusa*), very large tree	
Leaves of the lower branches, small and hard	14·3
Arctocarpus integrifolia	
Leaves of suckers	10·9
Cinnamomum ceylanicum	
Old leaves, leathery and strongly aromatic	9·7
Young leaves, entirely flaccid and brilliant red	6·6
Ilex paraguajensis	
Mature leaves, hard	13·5
Young leaves, soft, just emerged	11·2

Forty-five determinations were done on coffee trees. No important differences in water balance were shown between *Coffea arabica*, *C. liberica* and *C. robusta*.

The following values were found:

TABLE 12

Cell sap concentration (atm) of leaves of *Coffea* species

	Young	Optimum value	Highest value
Coffea arabica	10·4–12·0	14·4–15·5	18·9
Coffea liberica	9·7	15·8	19·5
Coffea robusta	8·8	13·5–16·6	20·6

The optimal values for shade-leaves were between 12·4-13·0 atm.

Schweitzer (1936) working in Java has shown that coffee trees flower only upon attaining a specific cell sap concentration. Normal flowering does not occur if the climate is very humid.

These facts were substantiated by Alvim (1960). While buds are formed regularly, they do not flush until a temporary decline occurs in the water balance. In a coffee plantation in Peru, a plot was watered weekly to eliminate water stress in the plants. None of the buds flushed and the plants did not flower. However, a second plot was watered when the soil dried almost to the wilting point. The buds flushed 10-11 days after re-watering and the plants were then in full bloom. The exact causal relations between decrease in water balance and flowering are not yet known. Since *Coffea* is a short-day plant, bud formation merely requires exposure to short days.

Schweitzer measured the highest osmotic pressure for different *Coffea* varieties during an abnormally dry period in 1935. The photosynthetic efficiency decreased with increasing osmotic pressure. But the leaves, which had suffered slight damage from drought, showed a positive photosynthetic balance 24 hours after a rain.

All osmotic values of tree species in the humid tropics are below those of woody plants in the temperate zone.[1] Their uniformity in water economy is also shown by the small range in suction tension. Such measurements had already been carried out by Blum (1933) in Java. The method applied by Lemée (1956a) to the shrub layer in the virgin forest of the Ivory Coast is probably more accurate. His values ranged between 3-8 atm. The suction tension increases as soon as plants become exposed to more intense radiation. It may exceed 20 atm on cut-over land. The values increase farther north in the savannah areas (10-25 atm). However, absolute values from suction tension measurements are not very reliable (Walter, 1963).

[1] The tough structure of the leaves is, in part, a function of the high SiO_2 content which was emphasised by Faber, referring to Stahl. This supposedly applies especially to the leaves of lianas with active internal water movement. Silicic acid is especially soluble in tropical soils and should accumulate in the leaf epidermis, if it is carried along in the transpiration stream.

The problem of transpiration in the tropics has been discussed since 1892, when Haberlandt presented the view—on the basis of investigations in Djakarta and Graz (Austria)—that the transpiration rates in the humid tropics are about 2-3 times below those in central Europe (Haberlandt, 1892; Giltay, 1897, 1898a, b, 1900; Holtermann, 1907). Stocker summarised the earlier literature on this subject in 1935. At the same time he presented results on the daily march of transpiration of three species obtained with the short-period, weighing method. These species were not, however, representative of the rain forest proper. He found relatively low values for transpiration and observed partial closing of stomata around noon but found no sign of a disturbed water balance.

I have made a large number of short-period transpiration experiments with large-leaved forms or with whole, cut-off twigs by using a sensitive indicator-dial precision balance (loading capacity 10 kg). When the fresh-weight transpiration values are related to a standard Piche evaporation of 0·48 ml/hr, it can be shown that the seven tree species investigated in Amani gave values between 2·6-22 mg/g/min (Table 13).[1]

Thus, transpiration intensity varies considerably between species. It shows approximately the same range of variation as plants from near Innsbruck which were investigated by Pisek and Cartellieri under identical evaporating conditions and unrestricted transpiration (see Walter, 1960, p. 310). The relatively low transpiration values and the decline in transpiration around noon become understandable, if we consider the significance of the high temperatures of the leaves exposed to insolation in the tropics.

We have seen that leaf temperatures of 10-15°C above air temperature were measured on clear days in a coffee plantation. The danger to the plant of such high temperatures increases with increasing air temperature, since the vapour pressure at the leaf surface increases to a much greater degree.

Let us assume that a relative humidity of 50% occurs at an air temperature of 10°C in one case and at 30°C in another case. Assume further that, in both cases, the transpiring leaf surface acquires an excess temperature of 10°C as a result of strong radiation, i.e. the leaf surface temperatures are 20° and 40°C, respectively. Then, the vapour pressure gradients in Table 14 would occur in relation to a vapour pressure of 100% at the leaf surface.

Thus, evaporation would be three times greater in the second case for the same excess-temperature of the leaves and the same relative humidity. Such high water losses from the leaves are prevented through thick cuticles and closing of stomata. These leaf properties effectively decrease the rate of transpiration and the vapour pressure at the leaf

[1] Stocker's values would amount to about 5-9 mg/g/min at the same evaporation rate.

TABLE 13

Transpiration of tree species (in mg/g/min) belonging to the lower and upper tree strata. The water content of the transpiring leaves, expressed in per cent of fresh weight, is shown in brackets:

Species	mg/g/min
Macaranga usambarica (58·6%)	22·0
Albizzia fastigiata (62·1%)	14·5
Piptadenia buchanani (45·5%)	9·0
Anthocleista orientalis (82·1%)	8·0
Allanblackia stuhlmannii (71·3%)	3·2
Tabernaemontana holstii (70·1%)	3·0
Myrianthus arboreus (62·7%)	2·6

Transpiration of *Coffea* species and of the tea shrub from whole branches and individual leaves (in mg/g/min):[1]

Species	mg/g/min
Coffea robusta (54–66%), branches, av. from 5 tests	3·3 (3·0–4·0)
Coffea robusta, individual leaf	7·3
Coffea liberica (52–69%) branches	3·0
Coffea arabica, individual leaves, av. from 10 tests	7·6 (5·5–12·8)
Tea shrub (*Thea sinensis*), branches, average	1·8 (0·9–2·6)
Tea shrub (*Thea sinensis*) (67–72%), individual leaves, average	1·5 (0·7–2·7)

Transpiration of other planted trees (in mg/g/min):

Species	mg/g/min
Breadfruit tree (*Artocarpus integrifolia*) (74%)	5·2
Orange (*Citrus auranticum*) (59%)	4·1
Bambusa vulgaris (49%)	5·8
Acacia saligna (phyllodes) (31%)	5·3
Eucalyptus citriodora (42–48%)	3·4
Casuarina equisetifolia (54%)	1·8

[1] Transpiration of individual leaves was determined with the Bunge travel-balance. Individual leaves of the tea shrub gave the same results as whole branches on calculation per g fresh-weight of the leaves; however, in the coffee species individual leaf values were twice as high.

TABLE 14

Vapour pressure of the air at 10°C (50% of 9·2 mm)	4·6 mm
Vapour pressure at the transpiring leaf surface (t= 20°C)	17·5 mm
Difference	12·9 mm
Vapour pressure of the air at 30°C (50% of 31·8 mm)	15·9 mm
Vapour pressure at the transpiring leaf surface (t= 40°C)	55·3 mm
Difference	39·4 mm

surface, which in turn, however, increases the leaf temperature because of the absence of the cooling effect of evaporation. A high noon-tide water loss would be catastrophic for trees with crowns exposed to the sun. Therefore, we can say that the trees of the tropical rain forest are only exposed temporarily to drying conditions for a few hours. These, however, are just as extreme as those in arid regions.

The common, xeromorphic leaf structure of rain forest trees becomes understandable in relation to these conditions. This leaf structure always allows the trees to maintain a uniform water balance with highly hydrated protoplasm (low cell sap concentration) in spite of the temporarily high saturation deficit at the leaf surface.

One should not be deceived by the annual rainfall curve based on monthly means derived from long-period observations. Dry periods occur in some years in all equatorial areas. However, a temporary water deficiency can occur quite frequently in rain forest stands, because of the very large transpiring leaf-area, which promotes considerable loss of water and because of the relatively shallow root system for which only a limited amount of storage water is available in the soil. The stenohydrous species of the tropical rain forest can endure such conditions only if they are enabled to reduce water loss temporarily to a minimum.

The same conclusion was reached by Coutinho (1962) on the basis of his detailed investigations at the Alto da Serra Station between Santos and São Paulo on the east coast of Brazil. There, tropical rain forest occurs at an annual rainfall of 3600 mm. However, very little rain may fall in June and July and, in some years, there are even dry periods of 60 days. After closure of the stomata, cuticular transpiration can be reduced to 0·3% of free evaporation because of the xeromorphic structure of the leaf epidermis. The transpiration values are low under normal conditions, when they do not exceed 6 $mg/dm^2/min$, and the water balance is very constant.

Thus, it is not surprising that irrigation is practised in tree plantations even in the most humid parts of the tropics to prevent damage from short dry periods. This is necessary because the most important cultivated tropical trees, the rubber tree (*Hevea brasiliensis*), the cacao tree (*Theobroma cacao*) and coffee (*Coffea* spp.), are stenohydrous species with low cell sap concentration and low transpiration rates.

Coffea closes its stomata as soon as the leaves are exposed to the sun. Therefore, coffee plantations have to be raised in partial shade, provided by nurse trees. CO_2 absorption on cloudy days is about twice as high as on sunny days (Nutman, 1937). It reaches a maximum of 4·5 mg $CO_2/dm^2/hr$ and shows an average of 1·9 mg $CO_2/dm^2/hr$. The same applies to cacao trees (Alvim, 1959). Without shade, the temperature of the leaves increases in the sun from 28°C to 48°C in two minutes. Transpiration increases by 3-5 times this amount and the stomata close within 6 minutes at a leaf-water deficit of only 4·5%. This also impedes photosynthesis. Cacao requires favourable soil moisture, not high air humidity. Even a decrease in soil water to $\frac{2}{3}$ field capacity lowers transpiration and at the same time strongly reduces photosynthesis. Cacao trees can be cultivated with irrigation even in the Peruvian desert where they attain very good yields since they do not suffer from fungal diseases. A tree uses about 40 l water per day, 14,000 l per year and the entire plantation an amount of water equal to a rainfall of almost 900 mm. The tree also behaves as a shade plant with respect to photosynthesis (Lemée, 1956b). The young, delicate leaves are very sensitive to dry air. This shows that the tough structure of the mature active leaves is a necessity in the humid tropics for those plants that are exposed to the sun.

There are only a few CO_2 assimilation values available from other plants. These investigations were done more than 30 years ago (Giltay, 1898b; McLean, 1920; Dastur, 1935; v. Guttenberg, 1931; Stocker, 1935b; and in greenhouses: Hartenberg, 1937; Sander-Viebahn, 1962) and the methods used at that time were not as accurate as now. Moreover, the investigations were done with detached leaves and in artificial light. So far, of typical rain forest trees, only *Stelechocarpus burahol* (Anonaceae) has been investigated (Stocker). This tree has large, long, metallic leaves with drip-tips. All other species investigated are either trees of dry tropical areas, herbs or cultivated species, respectively (see Table 15).

Values obtained in the tropics are neither higher than those from other climatic zones nor are there other peculiar features in the response of tropical plants. Respiration is not much different from that in the temperate zone in spite of the much higher temperatures. A peculiarity of many tropical trees is the vertical position of the highly light-reflecting leaves. Through this feature and because of the vertical position of the sun, the broad side of the leaf receives light of decreased intensity, which is advantageous, as pointed out earlier. According to Stocker, maximum CO_2 assimilation in *Stelechocarpus* is attained at relatively low light intensities. In any case, all rain forest species must be able to get along with low light intensities, at least when young.

However, the total leaf-area per hectare rather than the intensity of CO_2 absorption per unit leaf-area is responsible for the dry-matter production of the tropical virgin forest. The total leaf-area is probably not much larger than that of the hardwood forests of the temperate zone but the leaf-area of the tropical forest is photosynthetically active through the whole year without interruption. In spite of this, a yearly net production of 100 ton/ha for optimal conditions, as estimated by Vageler, is much exaggerated. Although exact investigations are not yet available, recent estimates are much more conservative (Weck, 1963). The annual timber yields of tropical forest plantations amount to about 13 ton/ha. Thus, they are about 2·5 times as high as in central Europe (cf. p. 33). The total primary dry matter production should be about 30-35 ton/ha. This value also probably applies to the natural tropical rain forest. However, yield increments would be much lower in natural stands than in plantations and would probably correspond to those of central European hardwood forests.

Almost the entire organic matter produced is, in due course, decomposed in the soil because there is no humus accumulation. Thus, measurements of soil respiration could give some indication of the actual rate of dry-matter production. However, such investigations are not yet available.[1] One could easily apply the simple absorption method (Walter, 1952; Haber, 1958).

A peculiarity of flower and fruit formation of trees of the tropical rain forest is cauliflory, i.e. the formation of flowers and fruits on leafless, woody stems. Latent, axillary buds only burst through the bark after

TABLE 15

CO_2-assimilation: maximal values in mg $CO_2/dm^2/hr$

Species	Position with regard to light	Value	Author
Stelechocarpus burahol	sun-leaves	6·3	Stocker
Stelechocarpus burahol	shade-leaves	2·6	Stocker
Cassia fistula	sun-leaves	11·0	Stocker
Cassia fistula	shade-leaves	8·6	Stocker
Calophyllum inophyllum	sun-leaves	7·3	Stocker
Abutilon darwinii	sun-leaves	7·5	Dastur
Sparmannia africana	sun-leaves	7·5	Dastur
Cocos nucifera	sun-leaves	0·9	McLean
Musa textilis	sun-leaves	1·0	McLean
Sugar cane	sun-leaves	5·0	McLean
Cassia timorensis	sun-leaves	19·2	Giltay
Cedrela serrulata	sun-leaves	9·0	Giltay
Acalypha tricolor	sun-leaves	7·0	Giltay
Helianthus annuus	sun-leaves	8·0	Giltay
Nicotiana rustica	sun-leaves	9·0	Giltay

many years, thereafter the flowers unfold. This phenomenon is known in the temperate zone only from the sub-mediterranean species *Cercis siliquastrum* (the Judas-tree), and the mediterranean *Ceratonia siliqua* (the Carob-tree). The phenomenon is widespread in the tropical rain forest. The best known example is the cacao tree, which belongs to the lowest tree layer of the American rain forest. In the entire tropics there are perhaps about 1000 species with cauliflory.

Mildbraed (1922) mentions 278 such species for Africa alone. He distinguished several types of cauliflory on the basis of flower position: (i) on leafless twigs, (ii) only on the main stem, (iii) at the stem base, (iv) on long sprouts with scale leaves, which originate from the stem or

[1] Some results of investigations in Venezuela have since been published by E. Medina (1968).

from the bases of branches, and (v) on long, whip-like twigs, which lie on the ground.

Several hypotheses have been advanced concerning the origin of cauliflory but they are very vague. For example, Klebs (1911) assumed a concentrated accumulation of photosynthates in the stem to be responsible. It is interesting that cauliflory occurs almost exclusively among tree species of the lower strata, which are frequently chiropterophilous; i.e. they are pollinated by bats (Porsch, 1941; van der Pijl, 1957a; Vogel, 1958). Chiropterochory is also very common among these species; i.e. the fruits are eaten by flying foxes, which distribute the seeds with their droppings. The strange appearance (for Europeans) of many tropical fruits, their unattractive colours and the frequently bad odour and

FIG. 71. *Spathodea campanulata.* Entebbe, Uganda (Phot. E. Walter).

taste—to which one has to become accustomed at first—and their peculiar position on the tree is understandable as an adaptation to flying foxes that only hunt for fruits at night (van der Pijl, 1957b). This mode of distribution is unknown in Europe. The fruits often contain large seeds which supply sufficient food material for the young seedling. This is a necessary adaptation to the low light intensities near the forest floor and enables the seedling to carry on until photosynthesis takes over after sufficient initial height has been attained.

Another peculiarity of some tropical trees are 'water-calyces'; that is, the phenomenon of a flower bud being maintained within a closed calyx filled with water excreted by hydathodes, e.g. the case of the beautiful *Spathodea campanulata*, amongst others (Figs. 71-72). Von Faber refers to the relevant literature. Nothing can be said about the ecological significance of this phenomenon. Pascher (1960) has shown that such 'water-calyces' are also present in some non-tropical Solanaceae.

The inner floral parts in Bromeliaceae frequently also develop in water which accumulates after rains in the funnel-shaped leaf-arrangement.

Finally, we must consider the phenomenon of periodicity in the tropical rain forest.[1] Seasonal periodicity is absent because of the year-round, uniform climate. The forest is continuously clothed with green foliage. A few trees flower at any time of the year, their flowers are often beautiful but they are not conspicuous amongst such a mass of green foliage. More often the light- or brilliantly red-coloured young foliage is conspicuous (Fig. 70). Such colour indicates a change in foliage. The temporary leafless trees are less conspicuous.

It is said of the New Hebrides that the annual seasons vary so little that the natives do not count their age in years. Instead they plant their cultures at regular intervals: yams, when *Erythrina indica* flowers, and batatas, when *Alphitonia zizyphoides* is in fruit.

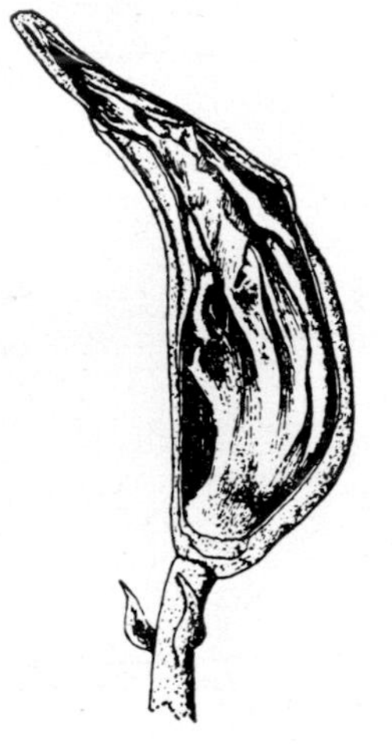

FIG. 72. Opened water calyx of *Spathodea campanulata* (after von Faber).

This shows that individual tree species have a marked periodicity. However, their development phases do not occur simultaneously and even those of individuals of the same species are not always synchronised. There are trees that shed all their leaves and remain leafless for some time. Others shed the older leaves only when the young ones unfold, or even somewhat later. No trees are known that grow continuously. This may just be the case in the juvenile stage. There are certain species that renew their leaves on different branches at different times. This is characteristic, for example, of the mango tree (*Mangifera indica*): while the old foliage appears dark green, some individual twigs with light coloured, young foliage protrude from the crown. Commonly, trees of the same species can be seen side by side with and without foliage.

Some species renew their foliage at regular intervals with a periodicity of 12 months, others of 3-4 months (Holttum, 1953). However, there are periodicities also of 9 months and even of 32 months.

[1] The rather numerous references are cited by Schimper-Faber (1935), pp. 401-2.

It appears probable that species with such regular periodicities are very sensitive to annual variations in climatic factors, such as small variations in daylength—which could perhaps have an influence on flower formation—or to small changes in temperature or rainfall.

The periodicity in all such cases is autonomous; i.e. it can adapt itself to the periodicity of the environment. It is, therefore, understandable that the same species responds differently in different tropical climates with a more-or-less pronounced periodicity. For example, *Tectona grandis* (teak-wood or djatti-tree) never defoliates completely when transplanted to west Java, while it sheds its entire foliage during the dry period in the periodically dry east Java. In the tropical part of east Australia, which has a cooler annual season, all mango-trees burst into foliage at the same time. The ease with which trees adapt themselves to environmental periodicity varies with species and even with individuals of a species. Bünning pointed out, therefore, that periodicity in tropical plants is often related to different biotypes.

There is probably no tropical area without deciduous trees, i.e. trees that are without foliage at some time or other. However, they are inconspicuous in the humid tropics and only become predominant in tropical climates with pronounced dry seasons.

Most interesting are acclimatisation experiments with deciduous trees from the temperate zone. Fruit trees, *Quercus pedunculata* and *Liriodendron tulipifera*, have been planted in Ceylon at an elevation of 1800 m, and the botanic garden at Tjibodas on Java, situated at 1500 m, contains deciduous trees from Europe, North America and temperate Asia. A beech tree (*Fagus sylvatica*) was planted on the top of the Pangerango volcano (3020 m). Regular observations of this tree over 4 years have shown a very irregularly phased growth pattern. Von Faber reported, 'the twigs show very different periodicities. During the dry period there are quiescent as well as shoot-forming branches. Even during the very dry year of 1914, there were twigs that shed their leaves and others that sprouted. There is no sign as yet of unlimited bud-growth' (Schimper-Faber, p. 387).

A similar reaction was also shown by other woody plants brought to Tjibodas. Initially a prolonged reaction effect, related to the annual seasons of the country of origin, persisted. Leaf emergence and shedding only became independent of the annual seasons gradually, until finally each individual branch system showed its own pattern. According to Schimper, who visited the planted trees in December and January, the individual branches of *Liriodendron tulipifera* and *Quercus pedunculata* indicated winter, spring and summer simultaneously. Responses typical of all four annual seasons were displayed on the same individuals by apple and pear trees that showed little success in the tropical climate. *Amygdalus communis* was in full bloom. Similar observations were made in Ceylon.

In this connection we must consider the fact that days are short in the tropics, yet long-day plants can develop flowers at higher altitudes, where they are subjected to lower temperatures. Thus, *Pyrethrum* plantations in Kenya are situated at 2000-3000 m elevation where the temperature drops below 15-16°C. *Primula veris* develops well at 2400 m in Indo-Malaya where it flowers and fruits abundantly; it grows only vegetatively up to 1400 m (Van Steenis, 1962).

Peach trees were grown from European seed on Reunion Island. Initially, they shed their leaves every year but, in time, the leafless periods became shorter and shorter until finally some foliage was retained at all times of the year.

These observations support a theory of an endogenous rhythm,

FIG. 73. Individual twigs of *Coffea robusta* with flowers and mature fruits in Amani, East Africa (Phot. E. Walter).

which adapts itself to climatic periodicity only when the latter is pronounced. Under these circumstances, the endogenous rhythm appears to be induced by the environmental rhythm.

I have already mentioned that periodicity is also noticeable with regard to flowering. However, the flowering period can last for an extremely long time so that the number of flowering individuals exceeds that of the non-flowering ones at any time of the year. Flower emergence starts at different times so that trees in full bloom occur side by side with those having ripe fruits, or flowers and even mature fruits can occur on the same branch (Fig. 73). However, at certain times mass-flowering can occur as well. In cases of short flowering periods, there is commonly the striking phenomenon of a sudden flowering of all members of a species which begins on the same day and extends over several square kilometres. A well known example is the epiphytic orchid, *Dendrobium crumenatum* on Java.

This orchid, which flowers for only one or a few days, has developed buds that undergo a dormant period. An environmental stimulus is required for them to burst, probably a sudden cooling after a thunderstorm. According to Holttum the blossom of *Fagraea fragrans* always appears in Singapore 4 months after the rainy season if it has been followed by a sudden dry spell.

Other plants flower at distinct, equally spaced, intervals. Two species of the genus *Hopea* and four *Shorea* species flower every 6 years. *Strobilanthes cernuus* is supposed to flower every 7 years but this is not always the case. Several *Bambusa* species and some palms, such as *Corypha umbraculifera*, flower after a certain number of years and then simultaneously over a certain large area. This time-interval amounts to 13 years in the *Bambusa* species of south Brazil, and to 30-32 years in *Bambusa arundinacea* on the west coast of India. The Bambusae die after flowering.

I had an opportunity of witnessing this phenomenon in north Australia, south of Darwin. The rainy season did not occur in 1951-52. Following this, the *Bambusa* stands flowered in the cooler dry period of 1952 along the rivers and streams of the area. According to W. Arndt, they all died in 1952-53. However, the seeds germinated, and I was able to observe the young stands in 1958. The dry period was not effective in the lower part of the river valleys which are always supplied with sufficient water. Here, the *Bambusa* stands neither flowered, nor died. This indicates that flowering was triggered off by unfavourable water relations. Such periodicity would be understandable if the environmental factors were correlated with a solar rhythm. The long intervals between flowering periods might be connected with the necessity to store greater food reserves for flower formation. Von Faber points out in this connection that leaves are often shed from flowering branches. The flowering twigs later become purely vegetative in some deciduous trees. 'Some trees are evergreen during their young stages of development, before flowering, while in their later stages they become deciduous, shedding their leaves regularly before the flowering period. This behaviour is shown, for example, in *Schizolobium excelsum*, at least on Java' (Schimper-Faber, p. 398).

7. Other ecological types of plant of the rain forest

(*a*) *Shrub layer*

The tree layer is the absolutely dominant layer of vegetation in the tropical rain forest. But there are also other associated plants dependent on the tree layer. Apart from trees in an advanced stage of regeneration, actual shrubs play only a minor role in the shrub layer. Instead, tall herbaceous plants are represented much more abundantly in this layer.

These species and, of course, the smaller herbs as well receive only as much light as can penetrate through the tree canopy. Thus, they are closely dependent on the trees. To what extent root competition plays a role in the soil is uncertain. Competition for water can hardly be considered important in the water-saturated atmosphere near the forest floor; competition for nutrients is a possibility. Coster has thrown some light on this question with his root studies in the tropics (see Richards, 1952). He observed a lack of undergrowth under certain species of trees with relatively high light intensity at the soil surface, although undergrowth occurred in deeper shade under other tree species. He related this to competition between roots for oxygen in the moist tropical soil. More probable explanations appear to be either an excess CO_2 concentration in the soil atmosphere or root competition for the minimal amount of nitrate or ammonium. Adequate investigations of this question are not yet available from the humid rain forest.

(*b*) *Herb layer*

The number of herbaceous species in the rain forest is smaller than that of the woody species. This may be related to the decreased light near the soil, which acts very selectively. A peculiarity that contrasts with the temperate zone is that the parts of herbs above the ground remain alive for an indefinite period because of the absence of any extreme season. Thus, the herbaceous species can attain considerable heights, up to 6 m. They often reach into the shrub layer and occasionally even into the lower tree stratum, e.g. bananas and several large Scitamineae. The herbs are often equipped with underground perennating organs, such as rhizomes or tubers. But these serve less as storage organs for reserve food than as means for vegetative reproduction. Flower formation in herbaceous plants is correlated even less with annual seasons than it is in woody plants. However, there is often a developmental periodicity, e.g. in *Amorphophallus*, in which the flowers appear before the leaves. The herbaceous plants include many ferns and *Selaginella* species.

The leaves show much diversity in shape and size and the frequency of coloured leaves is rather peculiar. Red and white leaf-areas occur as well as leaves with metallic lustre. Velvety leaves are also found but, usually, the leaves are delicate and thin, although they may also be somewhat succulent, as for example in most of the begonias.

Adaptation to low light intensity appears to be their most conspicuous ecological feature. There are species that occur exclusively in deep shade and others which grow under more favourable light conditions, where they develop better. In deep shade, reproduction is mostly vegetative.

Shade plants transpire only very slightly, as was demonstrated by von Faber (1915) with potted plants. They usually guttate strongly at night so that the water lost in liquid form is often greater than that in vapour form. I observed uninterrupted streaming of water droplets from the leaf tips of *Colocasia* starting immediately after sundown in Amani. Shade plants often close their stomata immediately on being illuminated by the sun. They may wilt, even if the exposure only lasts for a short time. This is true, for example, for *Impatiens* species.

Growth measurements have shown the great sensitivity of herb species to a small decrease in water balance. Plant growth in the forest slows down immediately at the onset of a slight decrease in relative humidity during the day. Thus, growth is more intense at nights than during the day (Smith, 1906).

Stocker investigated transpiration by a few shade plant species using the short-period weighing method. It proved to be very low in accordance with the low evaporation. In addition, I also measured, at Amani, transpiration of large-leaved herbaceous species such as banana, *Colocasia*, and *Ravenala*. My data are given in Tables 16-21.

The figures in brackets in Tables 16-18 refer to the water content, as a percentage fresh weight of the transpiring organs. The transpiration values in mg/g/min are converted to correspond to a Piche evaporation of 0·48 ml/hr (provided there is a proportionality between evaporation and transpiration).

TABLE 16

Transpiration of obligatory shade plants at Amani (East Africa)

	mg/g/min
Impatiens vallerina (96%)	1·8
Costus subbiflorus (85%)	1·2
Pleomele (*Dracaena*) *papahu* (68%)	0·2
Begonia zimmermannii (92%)	0·023
(This species transpired only 7 mg/g in 24 hours in its natural habitat.)	

TABLE 17

Transpiration was higher for facultative shade plants, viz.

	mg/g/min
Banana (*Musa paradisica*) (83·3%), sunny day	3·3–3·4
Banana (*Musa paradisica*) (83·8%), dull day	8·0
Traveller's tree (*Ravenala madagascariensis*) old leaf	3·7
Traveller's tree (*Ravenala madagascariensis*) young leaf	1·0
Arrowroot (*Maranta arundinacea*) (74·5%) sunny	4·3
Arrowroot (*Maranta arundinacea*) (74·5%) shady	9·0
Taro (*Colocasia antiquorum*) (81·4%)	12·0
Cassava (*Manihot utilissima*)	7·4
Sugar cane (*Saccharum officinarum*((71%)	2·8
Papaya (*Carica papaya*), leaves slightly wilted	2·8
Cardamon (*Elettaria cardamomum*) (73·5%)	1·6

TABLE 18

Some measurements carried out with ferns.

	mg/g/min
Cyathea usambariensis (47·5%)	
Overcast day, bright locality	7·6
Overcast day, shady	16·0
Sunny day, bright locality	5·4
Sunny day, shady	15·0
Marattia fraxinea (77%), wilts fast and dries up	4·8
Pteridium aquilinum (70%)	3·3
Asplenium nidus (82%), epiphyte, remains fresh for days	0·32

In this connection it is worth emphasising that transpiration in shade—converted to equal evaporation—was 2-3 times that in the sun in the case of ferns, the banana and the arrowroot. Thus, the stomata of these species probably close partially in the sun. In the case of arrowroot, the leaves became somewhat curled-up in the sun.

A few measurements with other species are shown for comparison in Table 19.

TABLE 19

	mg/g/min
Phoenix humilis (55%)	3·4
Carludovica palmata (66%)	4·2
Cycas circinalis (71·8%)	5·1
Encephalartos hildenbrandtii (54·3%)	3·3
Monstera deliciosa (79·3%)	3·7
Pineapple (*Ananas sativa*) (85%)	0·035
Agave sisalana et al. as well as *Fourcroya gigantea* } exposure time up to 4·5 hours	0·004–0·025

All these data refer only to single measurements. They show, however, that the values are very low in general; that is, in respect both of non-xeromorphic and of xeromorphic species (*Colocasia* appears to be an exception). Succulent species lose extremely small amounts of water.

The internal water balance must be very stable at such low transpiration values and in such humid habitat conditions. In fact, the cell sap concentrations of herbaceous species are even lower than those of trees.

All values which I obtained at Amani were below 10 atm, commonly even below 5 atm (Table 20). An exception was *Brillantaisia madagascariensis* with 15·6 atm (shoot-tips of trailing adventitious shoots) and *Pleomele* (*Dracaena*) *papahu* with 10·5 atm (at a rather exposed locality in a forest opening).

TABLE 20

Cell sap concentrations of species of the herb stratum

	atm
Cylicomorpha parviflora, with milky juice	8·0
Piper umbellatus, large leaves, shady	8·9
Piper capensis, very delicate leaves	6·6
Clinogyne ugandensis	8·5
Calathea zebrina, shady	8·2
Calathea zebrina, sunny	8·9
Aframomum or *Kaempferia*	7·5
Lobelia kumuriae	7·3
Costus subbiflorus	5·7
Fleurya lanceolata (Urticaceae)	8·0
Fleurya lanceolata, trailing shoot at ground	6·0
Coix lacrimans, exposed, swampy place	8·6
Synadenium (*Euphorbia*) *carinatum*	5·6
Notenia amaniensis (Crassulaceaen-like composite)	5·6
Impatiens parviflora, fresh	4·3–4·7
Impatiens parviflora, wilted plants	5·4
Streptocarpus ovata	3·6
Streptocarpus ovata, swollen base of stem	4·8
Lissochilus (soil-orchid), sunny position among grasses	4·4

TABLE 21

Sample values were obtained also for ferns

	atm
Cyathea usambariensis (tree fern), sunny	10·5
Cyathea usambariensis (tree fern), shady	7·5
[The same, but young leaves (still rolled-in)	4·6–5·4]
Marrattia fraxinea, centre leaflets with synangia	10·7
Microlepis speluncae, at a creek	9·7
Pteris biaurita	8·6
Lonchitis hirsuta	9·0–11·9
Asplenium cicutarium, leaves with sporangia	8·9–9·1
Asplenium unitum	7·7
Adiantum capillus-veneris	11·3
[The same in Arizona (spring in desert)	13·9]
Pellaea viridis	13·3
Pellaea viridis, young	10·5
Pellaea viridis, yellowing and of dry appearance	11·0–11·2
Selaginella sp.	10·6
Pteridium aquilinum, sunny	16·2

This shows that the ferns tend to have higher values than the phanerogamic herbs, in spite of their rather delicate leaves (e.g. *Adiantum*).

No experimental data are available on the photosynthetic activity of the herbaceous ground vegetation in the tropical rain forest. Some dry-matter production is indicated by the fact that the herbs grow at low light intensities. In relation to their fresh weight, this may not amount to very much, since the organs consist largely of water (frequently more than 80%). Their low respiration rates facilitate some net production. This applies to *Aspidistra*, which is, therefore, able to grow in the darkest corner of a room. Also Usambara-violet (*Saintpaulia usambariensis*)—which I saw growing in deep shade in East Africa—can get along with very little light as an indoor-plant.

Size and number of stomata are often used as a quantitative ecologic-morphological measure. I present, therefore, a few such data although the measurements were done on greenhouse plants in Stuttgart. The number of stomata are usually less than 100/mm² in shade plants. However, there are, similarly, not very many in tropical heliophytes (see Walter, 1960, pp. 322 ff).

TABLE 22

Number of stomata per mm² on the under-side of leaves (number of stomata on the upper-side of the leaves are shown as denominators; length of stomata in μ shown in brackets).

Tradescantia zebrina	15–20	
Begonia hybrida	24	
Aspidistra elatior	28	
Cycas circinalis	36	(66)
Ananas sativa	42	(27)
Asplenium nidus	45	(42)
Monstera deliciosa	46	
Encephalartos lehmannii	52	(79)
Colocasia odorata	52/6	(36)
Colocasia cochleata	91/6	(35)
Costus afer	95	
Elettaria cardamomum	100/4	(27)
Angiopteris thysmanniana	111	(45)
Ficus elastica	100–130	(18)
Coffea arabica	148	(23)
Phoenix canariensis	89/91	(22)
Musa sapientium	194	(30)
Saccharum officinarum	133/60	(40)
Bambusa arundinacea	233	(20)
Alsophila australis	266	(33)
Citrus aurantium	293	(22)
Carica papaya	370	(20)
Acacia melanoxylon—		
phyllodes	130/125	(25)
feathery leaflets	238/174	(21)
Acacia latifolia	295/268	(36)
Manihot utilissima	400	(23)

(*c*) *Lianas*

In contrast to herbs, which are only indirectly dependent on the trees, the lianas and the two following groups, the hemi-epiphytes and epiphytes, are directly dependent on the trees in a mechanical way for they need them for support.

Lianas germinate on the soil but their stems grow very rapidly in length without forming sufficient supporting tissue and they attach themselves to other plants in order to gain height above ground. This is accomplished in very different ways:

1. *Spreading climbers* are usually shrubs with spreading branches, which grow for support into the branch-system of other plants. A tendency to slide downwards may be avoided by the development of spines, thorns or hooks. Climbing palms, such as *Calamus*, are particularly characteristic in this respect. The thin but tough and spiny stems of this palm are known as 'rattans'. Their leaf-rachis is elongated into a flexible flagellum bearing hook-shaped spines. The young shoot attaches itself by this flagellum to a supporting tree and then climbs up to its crown. When growth ceases at a later stage, leafless, older shoot-like parts detach themselves and lie on the ground as sling-like structures. Part of a stem measured by Treub was 240 m long (see Faber). The rattan palms are particularly common in south-east Asia and also in north-east Australia. Prickly *Rubus* lianas, the diameter of an arm, are commonly found in New Zealand.

2. *Root-climbers* form thin roots, with which they attach themselves to the bark of a tree trunk; or the roots may be thicker and longer, encircling their support as ring-like structures, as in the case of *Vanilla*. Many *Araceae* belong to this group (Fig. 74) and also *Freycinetia*, which often covers the trees up to their tops.

3. *Winding plants* have fast-growing apical tips with very long internodes where the leaves at first remain underdeveloped, as in an etiolated plant. The stem winds round a vertical support (geotropism playing some role in this process) to which it becomes tightly attached. The stem may then become woody and increase in diameter. The numbers of this type of liana are very great in the tropics. They correspond to the Chinese Kidney bean (*Wistaria sinensis*) and *Aristolochia* species cultivated in Central Europe.

4. *Tendril plants* are equipped with special attachment-organs that are sensitive to touch and which may also become woody (Fig. 75).

Lianas with woody stems are very characteristic of the tropics and there are a very large number of species. They range all over the trunk from the soil to the crowns of the trees often giving the impression of intertwined ropes. They also dangle occasionally with their growing points hanging downwards from the crowns, some continue along the ground, from whence they can ascend again by becoming attached to another tree. In this way they often attain lengths of 70 m or more.

According to Schenck—who studied lianas in considerable detail (1892, 1893)—more than 90% of all liana species are restricted in their distribution to the tropics. About 8% of the flora of the West Indies are woody lianas.

FIG. 74. Epiphytes and climbers in the more open canopied rain forest of St. Lucia. Left, *Philodendron*; right, *Carludovicia*. (Reprinted by Courtesy of J. S. Beard and the Clarendon Press, Oxford.)

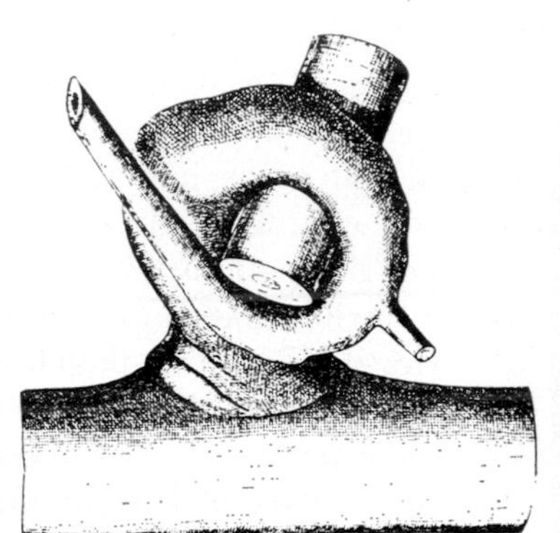

FIG. 75. Old, strongly thickened twined branch of *Dalbergia variabilis*, $\frac{2}{3}$ natural size (after H. Schenck).

The lianas can reach the crowns that are fully exposed to light without much investment in supporting tissue. However, they have to get along with low light intensities near the ground during their juvenile stage. It is, therefore, understandable that they are particularly frequent where the forest is more open. They form impenetrable thickets along the margins of rivers and at forest borders and they can become pests in plantations. When they have reached a position in the crowns of the trees where they are fully exposed to light they develop a crown of their own that may become so vigorous as to kill the supporting tree. In open areas, lianas remain short and their structure resembles that of shrubs. In the forest, however, the main axis is initially unbranched. After becoming woody, the stem retains its flexibility through anomalous secondary thickening. This does not result in compact wood but in segments separated by parenchymatous tissue (Figs. 76 and 77). The water supply of the rapidly expanding crown takes place through the long thin stem. This is possible only through a

FIG. 76. Cross-section through the stem of the liana *Anisosperma passiflora*, 3·2 times enlarged (after H. Schenck).

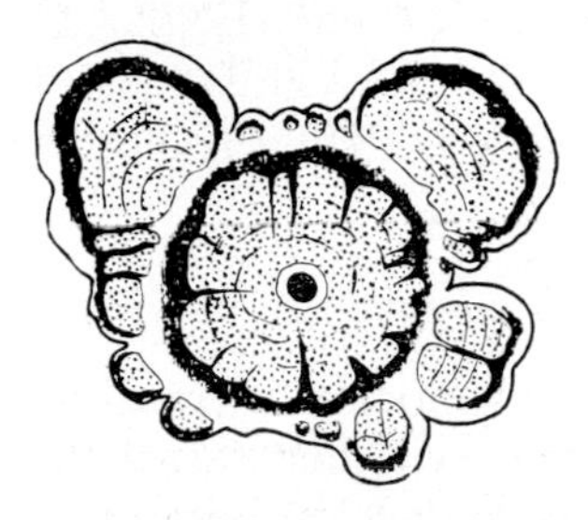

FIG. 77. Cross-section through the stem of the liana *Dalechampia ficifolia*, natural size (after H. Schenck).

reduction to a minimum of the resistance of the vessels to water movement. These vessels are very large in diameter and cross-walls are absent. When a liana stem is cut off at the base at one swoop, the vessels empty and, in this way, it is easily possible to obtain a glass of sterile drinking water from certain species. I determined the concentration of such water obtained from an unknown species at Amani as 0·3 atm; for *Ampelocissus volkensii* I obtained 0·7 atm.[1]

The water columns in such vessels are maintained by cohesion and can easily be torn apart by strong suction tensions resulting in an interruption of the transpiration stream. This may explain the distribution of woody lianas, which occur almost exclusively in the humid tropics, where strong suction tensions are practically never developed in the leaves. The transpiration rates of three species investigated by Coutinho (1962) were less than those of trees. There are often very pronounced differences between shade- and sun-leaves and between primary and subsequent leaves, respectively, and the leaves are morphologically and anatomically very variable.

[1] According to Gessner the conductivity of water in the vessels of lianas is about 20-100 times higher than that of soil water (Gessner, 1965).

The cell sap concentrations that I determined at Amani are similar to those of trees (Table 23).

TABLE 23

	atm
Smilax kraussiana, tough leaves with spines	12·9
Culcasia sp. (Araceae)	12·2–12·3
Cucurbitaceae	10·0
Adenia schweinfurthii, delicate shade-leaves	7·7
Monstera deliciosa, planted	6·7–7·5
Philodendron sp., planted	6·3–6·7
Ampelocissus volkensii, young leaves	5·8

The lianas play an important role in the structure of the crown canopies of the middle and lower tree strata in the tropical rain forest. They contribute greatly to the low light intensity near the forest floor. In forest plantations they are regarded as weeds and are removed by cutting through their stems. Wherever liana stems are suspended directly from the tree canopy, it indicates that the tree which formerly supported them and up which they have climbed has died and rotted away.

(*d*) *Hemi-epiphytes*

These represent a transitional category between lianas and epiphytes. They may either start their life cycle like the former or the latter. Many Araceae develop initially as normal lianas but, later, the lower parts of the stem may die off so that they lose direct connection with the soil and become epiphytes. (See Table 19 for the transpiration rate of *Monstera*.)

An opposite type of development has been observed in some epiphytes, which germinate on the trees but then develop aerial roots that become so long as to reach the soil. Upon contact, their supply of water and nutrients takes place in the normal way from the soil.

This group includes certain Araceae, the American *Coussapoa fagifolia* and the Malayan *Pyrus granulosa*, which occurs sometimes as an hemi-epiphyte and sometimes as an ordinary tree.

Finally, the strangest hemi-epiphytes are the stranglers. The textbooks usually mention only the strangling figs (*Ficus* sp.). But in New Zealand examples occur from several families, e.g. *Metrosideros robusta* (Myrtaceae), which nearly always germinates on the conifer *Dacrydium* only to strangle it later. Others are *Griselinia littoralis*, *G. lucida* (Cornaceae), *Nothopanax simplex*, *N. colensoi* (Araliaceae), etc., the latter two genera can develop equally well into normal trees. Sometimes one strangler is even found growing on another. In South America *Clusia* spp. are stranglers.

The stranglers, e.g. the figs, germinate as epiphytes on forked branches. Here they develop a small shoot system and aerial roots, which in part align themselves along the stem where they grow downward and in part dangle freely in the air. The roots that grow down along the stem form anastomosing branches around the supporting

FIG. 78. *Pseudospondia microcarpa* with strangler fig, Uganda (Phot. E. Walter).

stem before they reach the soil. After this, the plant gains in vigour, the crown becomes fully developed and the root system thickens around the stem of the host-tree. The latter cannot increase in diameter further and eventually dies. However, in the meantime, the root reticulum has changed into a stem so that the strangler becomes a normal tree (Fig. 78). These *Ficus* trees are often amongst the largest trees of the virgin forest, and, when confronted with them, it is hard to imagine that they germinated as epiphytes high above in the tree canopy on a forked branch. *Ficus retusa* and *F. bengalensis* also belong to this group; their aerial roots, that have grown into branch-supporting stems, can convey the impression of a stand of trees. The largest known individual covers an area of over 2 ha at a canopy height of only 26 m; the crown circumference amounts to 530 m, the diameter to 170 m.

The cell sap concentrations of *Ficus* species at Amani were around 10 atm (Table 24).

TABLE 24

	atm
Ficus ulugurensis, as epiphyte, but already rooting in the soil	9·6
Ficus subcalcarata, only one root had reached the soil	10·0
Ficus rhynchospora, normal tree	10·6
Ficus rhynchospora, sun-leaves	11·8

(*e*) *Epiphytes*

Among the epiphytes I shall first discuss those that belong to the higher plants, namely, the ferns and flowering plants. They do not root in the soil and represent perhaps ecologically the most interesting group of tropical rain forest plants. Their distribution is almost exclusively restricted to this vegetation type.[1]

Epiphytes use other plants only as substrates. They germinate high above on the branches of the trees and thereby receive the advantage of greater light intensity from the first. This advantage, however, is offset by the difficulty of obtaining an adequate supply of water and nutrients. Only in the tropical rain forest is it posssible for them to maintain the water balance necessary for their life-processes and, even here, this is accomplished only by special adaptations of the shoot and root system.

The epiphytic forms have evolved in part from species that grow on the soil, particularly from those with thickish leaves (Begoniaceae, Piperaceae, Orchidaceae, etc.) and low transpiration rates, and in part from groups that are endemic to arid areas (epiphytic cacti, for example, *Rhipsalis* spp. and probably also Bromeliaceae). Rock-plants can also become epiphytes. Soil is absent in both these situations and, therefore, the danger of desiccation is common to both substrates. On the other hand, epiphytes are often found on rocks, when trees are absent.

Epiphytes are good indicators of certain microclimatic relations, particularly high air humidity. They hardly have any influence on the substrate-plant. It has not yet been established with certainty whether there are any specific relationships between epiphytes and their substrate-plants. Many epiphytes are supposedly found only on living trees, often restricted to particular tree species. The properties of the bark may play a role in this connection. The bark is easily removed from

[1] Bünning (1956) has described a number of the strangest forms and also other ecological types in a very informative manner.

dead branches. On the other hand, abundant epiphytes mean a considerable load on the substrate tree (Fig. 79). It is possible also that fungal infections are established more easily through the presence of epiphytes that maintain the bark in a continuously moist condition. Pockets of organic matter can accumulate on larger branches and

FIG. 79. Tropical rainforest with epiphytes near Amani, E. Usambara (Phot. E. Walter).

branch-forks, which may serve as soil substrate for the epiphytes. Klinge (1962b, 1963) has estimated the total quantity of epiphyte-humus—which shows acid reactions—to consist of several tons per hectare. The organic content of humus-soil pockets is high but they also contain mineral particles from dust deposition (in El Salvador from volcanic dust and ash spray). Snail shell fragments also occur. Nitrogen is supplied either by rain, which in the tropics appears to have a higher bound-nitrogen content, or by drip-water in the form of ammonia. Drip-water contains three times the amount of NH_3 found in rain water. In this manner, a new, continuously moist biotope is created in the tropics that offers suitable environments even for tree frogs.

The number of ants living above the soil surface in the tropical rain forest is very large, particularly in the detritus-accumulations around the epiphytes and even in the inner chambers of the swollen stems or tubers of the *Myrmecodia* and *Hydnophytum* species or in the urn-shaped leaves of the Dischidiaceae. Certain kinds of ants even build real nests on the trees, in which the seeds of epiphytes germinate. The ant nests then appear to form a unit with the epiphytes and their root systems. Involved in such relationships are several Bromeliaceae, Gesneriaceae, Araceae, Solanaceae, *Ficus myrmecophila*, *Peperomia*, *Phyllocactus*, etc. Ule (1905), referring to such epiphytes in the Amazon forest, speaks of the 'flower gardens' of the ants that are suspended like flower pots from the tree branches. The accumulated water in the funnels of the Bromeliaceae are breeding grounds for mosquito larvae and several water insects. A similar habitat is formed by the leaf axils, in which there is always some stagnant water.

Outside the tropics, only lower plants are found as epiphytes. The only exception among the ferns is *Polypodium vulgare* in the more humid European deciduous forests. Otherwise, there are only mosses and particularly lichens and certain algae. These are all poikilohydrous species that are able to endure complete desiccation. In the tropics, however, there are great numbers of epiphytic, homoiohydrous flowering plants.

The Orchidaceae, which in terms of species numbers represent the richest family, include far more epiphytic than soil-dwelling species. The Bromeliaceae include virtually only epiphytes or rock-inhabiting plants.[1]

The high light requirements of most epiphytes and the extremely poor light relations near the forest floor in the rain forest make their distribution on the trees understandable. On the other hand, epiphytes cannot compete with the fast-growing soil-inhabiting plants in forest openings.

Among the epiphytes one can distinguish two groups that are not always well differentiated:

1. *Shade-epiphytes*, which occur in the lower parts of the trees where they usually occupy the larger branch-forks. They are mostly ferns.

2. *Sun-epiphytes*, which occur in the upper parts of the crowns, where they are found almost to the tips of the branches. The thinner the branches, the smaller are the epiphytes. Among these smaller ones are commonly orchids, which encircle the branches with their aerial roots. In Amani I found a minute orchid that had a height of scarcely a single centimetre including its inflorescence. However, there are also very large epiphytic orchid species.

[1] The family is restricted to America. Only one soil-inhabiting species, *Pitcairnia feliciana*, Harms et Mildbr., which was formerly included in the Liliaceae, occurs in West Africa (Hutchinson, 1959).

Water relations are often probably more decisive in the separation of these two groups than are the light relations. The higher up the location of the epiphytes, the farther away they are from the ever-humid atmosphere of the forest interior and the more extreme become the dry intervals between showers that they have to endure. Epiphytic orchids are found even in relatively dry savannah-forests.

Many different adaptations to less favourable water conditions have been developed. Particularly important is the accumulation of soil at the site of growth. This soil is formed from dead plant materials, therefore, it consists entirely of humus. The large bird's nest fern, *Asplenium nidus-avis*, collects the detritus between its leaves; the roots hold it together. The slowly decaying veins of the dead leaves have a similar effect. *Platycerium*, *Drynaria*, etc., develop large 'nest-leaves', which adhere closely to the substrate. They live only for a short period but hold the humus in place for a long time after their death. Humus

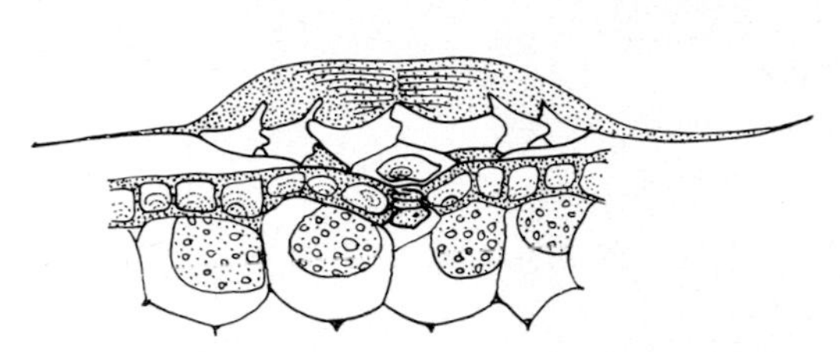
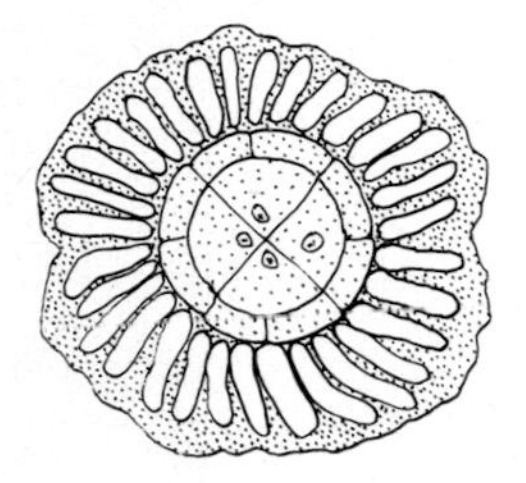

FIG. 80. Suction-scales of Bromeliaceae (after Schimper) viewed in cross-section (*Tillandsia*) and from above (*Vriesea*). Enlarged 340 and 375 times respectively.

formation is favoured by ants. They probably supply the epiphytes with nitrogen and other nutrients. It has not yet been established whether any nitrogen fixing bacteria are involved. The humus has an acid reaction (Boyer, 1964).

Spanner (1939) has shown that *Myrmecodia* utilises nitrogen better in the form of ammonia than as nitrate.

Epiphytes that do not collect humus in their rooting area must have adaptations that permit rapid absorption or storage of water from rain or dew, whereby water loss through transpiration may be kept to a minimum.

The best known examples in this respect are the aerial roots with velamina of orchids and the 'suction-scales' on the leaves of Bromeliads, which collect the water in funnels (Fig. 80). The funnels are formed through the densely interlocking rosette-leaves (Fig. 81).[1] The significance of the 'urn- and shell-leaves' of *Dischidia*—into which even the

[1] More precise eco-physiological experiments have been carried out on two Tillandsias of the drier tropical region. They take up droplets only of liquid water very quickly, especially the capillaries of the scale-hairs; later it is absorbed by the living tissues. Their osmotic pressures are about 6-8 atmospheres, much like that of succulents, and the intensity of transpiration is also very low so that they can endure a considerable water deficit (cf. Biebl, 1964).

roots may grow—has not yet been established, but epiphytic *Nepenthes* species can utilise the water that accumulates in their pitcher-like leaves.

The water absorbed is often stored in relatively large quantities, e.g. in the tubers and succulent shoots of many orchids, similarly in *Rhipsalis*,

FIG. 81. Epiphytic *Vriesea* (Bromeliaceae) in the primeval forest in the Cerra da Mar, near Paranapiacaba, Brazil (Phot. E. Walter).

etc. The leaves of Bromeliaceae are also mostly succulent. In ferns, the thick rhizomes often contain considerable amounts of water.

The water relations of shade- and sun-epiphytes are further analysed through the following four examples investigated in Amani.

1. *Shade-epiphyte—Asplenium nidus-avis* (*see also Table* 17)

The 'nest' had an upper diameter of 2·5 m and consisted of 24 mature leaves (largest leaf 104 cm) and 3 immature leaves. The following weights (g) were obtained during the dry period:

leaves	958
rhizomes with roots and humus	5099
total weight	6057

The fern lost 1287 g water in 20 days on the veranda of my cabin located right at the margin of the virgin forest (protected from rain). It suffered somewhat, but did not die. The plants are never exposed to such prolonged dry periods in the virgin forest, where the atmosphere is always much moister.

The water contents of humus soil samples of two other bird's nest ferns were as follows (in per cent dry weight):

Date	I	II
23. XI. 1934 (dry period)	152	96
7. XII. 1934 (rainy period)	220	317

Thus, the ferns do not seem to suffer from any lack of water since there is always sufficient stored in the humus soil.

2. *Sun-epiphyte—Drynaria laurentii*

This fern is found on the upper branches; its thick rhizome grows in a spiral around the branches, where it also forms, temporarily, living 'nest-leaves'; it has in addition long-petioled, pinnate leaves for photosynthetic activity.

The distribution of water in a plant with a 50 cm long rhizome was as follows:

TABLE 25

	fresh weight (g)	water (g)	% fresh weight
photosynthetic leaves	24·0	13·0	54
nest-leaves	75·5	28·5	38
rhizome	616·5	422·0	68
humus and roots	222·0	150·0	70
Total weight	938·0	613·5	—

Here, the reserve amount of available water in the humus is small. The largest proportion of water is stored in the rhizomes. The reserves are even smaller in other ferns, such as *Asplenium theciforme* and *Polypodium lanceolotum*. Their leaves were rolled up during the dry period. They represent poikilohydrous forms, an adaptation even further developed in the Hymenophyllaceae. These ferns, which have leaves that consist of only one layer of cells, behave almost like mosses. They are particularly characteristic of the montane tropical forests. Their physiology was investigated by Gessner (1940, 1956) and Härtel (1940) in a greenhouse at Munich. However, Coutinho (1962) found in the rain forest of Brazil that *Hymenophyllum polyanthos* is also able to withstand water deficits of 94%.

3. *Sun-epiphytes—orchids without humus accumulation*

Two plants of *Bolbophyllum* sp. with tubers gave the following distribution of fresh weights (as per cent).

TABLE 26

	Leaves	Tubers	Rhizomes	Roots
I	46	40	8·0	6·0
II	60	30	5·5	4·5

The water content of the leaves amounted to 92% of the fresh weight, that of the tubers to 93%. Thus, both organs serve for water storage. Upon wetting, the aerial roots absorbed 60% of their own weight in water very rapidly.

In a minute species with five leaves and 10 small tubers, the weights were: 1·1 g for the plant, 0·16 g for the leaves, 0·87 g for the tubers, and 0·07 g for the rootlets. In *Angraecum*, the roots amount to only 5% of the fresh weight, while the shoot, consisting of succulent leaves, contains 92% water.

Other orchids have a strongly developed root system that amounts to 70% of the total weight, including air-containing velamina. Such root systems absorb 30-170% of their own weight in 15 minutes upon wetting.

In epiphytic orchids, transpiration takes place from all organs, including the roots. Therefore, transpiration experiments have to be done with the whole plant after they have been carefully removed from their substrates. Such a test was done with three plants that were hung up in a rain-protected location on my cabin-veranda for 9 days.

The water loss is shown in Fig. 82.

In orchids without tubers, the curves have the shape of typical desiccation curves. The water loss after 9 days was 20-25% of the original weight. In spite of this, the plants were undamaged. The first damaging effects appeared only after 10-12 days. In their natural habitats, these orchids can hardly remain without absorbing water for so long a time, since they are wetted daily at least, through dew condensation.

In species with tubers, the water loss remained unchanged for the first 12 days and the daily loss amounted to 2·5%.

If only the aerial roots of the orchids are dipped into water for an hour, the plant can gain about 15% of its fresh weight, i.e. the water loss of several days is rapidly replaced.

In tropical virgin forest areas with a pronounced dry, rainless period of about a month, the orchids are found to shed their leaves, so that only the thick and tuber-like shoots remain. These show practically no water loss. New leaves emerge from the shoots after resumption of the rain.

4. *Epiphytic cacti*

Rhipsalis cassytha occurs in East Africa. A large individual was tested for its transpiration in a similar way to the orchids. It had lost 28% of its

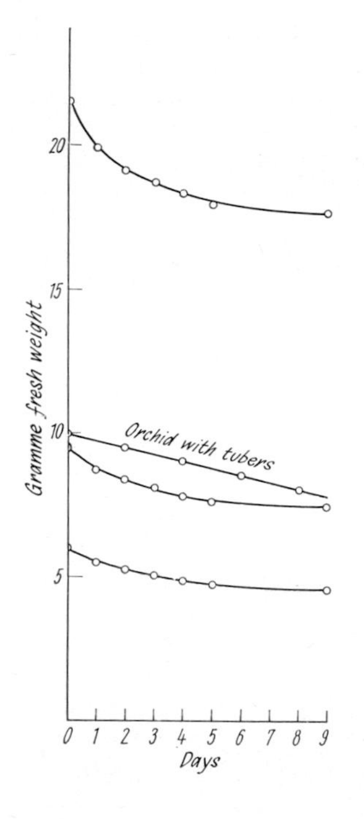

FIG. 82. Decrease in fresh weights of epiphytic orchids (whole plants with aerial roots) hung up in an open, but rain-protected location at Amani, East Africa.

Three orchid species had no tubers, but tubers were present on an individual of *Bolbophyllum*. Since the plants did not flower, their taxonomic identity could not be established exactly.

fresh weight after 28 days, when it showed visible damage. In contrast to the cacti of arid areas, *Rhipsalis* cannot endure a longer dry period.

Coutinho's short-period transpiration tests, carried out in the virgin forest of Brazil, resulted in maximal values of 1-2 mg/dm²/min for *Codonanthe gracilis* and *Hypocyrta radicans* (Gesneriaceae). Transpiration was hardly measurable in the orchid *Maxillaria picta* and the bromeliads *Vriesea altodaserrae* and *V. inflata* (see also *Ananas* (Table 19)). Recently Coutinho has obtained information concerning the 'De Saussure effect', i.e., the absorption of atmospheric CO_2 during the night and its assimilation in daylight (cf. p. 325), in a variety of epiphytes. In this phenomenon the acid content of the cell sap rises during the night and the pH falls. A positive result was found with the epiphytic Bromeliads, *Aechmea pectinata* and *Neoreglia concentrica* as well as the terrestrial

Bilbergia amoena, *Quesnelia testudo* and *Cannistrum cyathiforme*. The orchid *Epidendrum ellipticum* also showed the effect, whereas the epiphytic *Maxillaria picta* (Orchidaceae) and *Vriesia inflata* (Bromeliaceae) and all the Gesneriaceae showed the normal course of photosynthesis. The 'De Saussure effect' is only associated with those succulents which lack a sharp distinction between chlorenchyma and water storage tissues (Coutinho, 1964, 1965).

In conclusion it can be said that the behaviour of these epiphytic orchids and *Rhipsalis* is similar to that of non-extreme succulents, such as the European *Sedum*- and *Sempervivum* species from stone walls. The water balance is also reflected through the low cell sap concentration given in Table 27.

TABLE 27

		atm
orchids with tubers	leaves	3·1
	tubers	2·0
orchids (without tubers), with large aerial roots, but without thickened stems	I	4·3
	II	6·1
	III	6·6
Rhipsalis cassytha		5·2–5·6
Asclepiadaceae containing caoutchouc	leaves	10·8
	tubers	3·8

In ferns, the cell sap concentration was again higher than in flowering plants (Tables 28-30).

TABLE 28

	atm
Asplenium nidus, young leaves, rolled-in	6·1
Asplenium nidus, mature leaves	10·2–10·9
Asplenium nidus, senescent leaves	11·1–11·7

TABLE 29

Drynaria laurentii: plant I in drier location than plant II (values in atmospheres)

	Photosynthetic leaves	'Nest-leaves'	Rhizome
I	12·6	7·0	5·7
II	10·2	5·7	4·8

TABLE 30

Ferns that endure desiccation; values of fresh leaves (atmospheres)

Vittaria lineata (or *isoëtifolia*), sunny	15·7
Vittaria lineata (or *isoëtifolia*), shady	10·6
Polypodium phymatodes, sunny	15·0
Polypodium phymatodes, shady	11·6
Polypodium phymatodes, thin rhizome	10·4
Oleandra africana	9·4
Asplenium sinuatum	9·0
Asplenium macrophyllum	11·7–12·7
Asplenium protensum, leaves wilted	11·5
Asplenium theciferum, leaves rolled-in	33·3
Lycopodium cernuum, leafy shoot	11·6
Lycopodium cernuum, main stem	8·4

Went (1940) has studied the epiphyte communities of Tjibodas on Java. He found a close dependence of epiphytes on light relations and humus accumulation.

The bark plays a role also, so that Went considered not only the physical but also the chemical properties of bark, including those of the down-running water. For example, in *Castanopsis* the bark-tannins could be important. According to Richards, *Utricularia schimperi* in Guyana occurs only in the thick cushions of liverworts which are located on the stems where the water runs down after rain.

The seeds of the epiphytes have to get on to the trees. Therefore, they are either as fine as dust (Orchidaceae) or they have other adaptations to efficient wind dispersal (Asclepiadaceae, Gesneriaceae, etc.). Distribution by birds plays a role also (Cactaceae, Bromeliaceae); they may even transfer vegetative parts of *Tillandsia usneoides* for nest building. Seed distribution by ants could likewise be of importance. No epiphytes have evolved from families with heavy seeds.

Ferns have no difficulties in reaching epiphytic habitats, since they are distributed by spores. The prothalli develop on the trees.

As well as the higher-plant epiphytes, there are also lower-plant epiphytes in the tropics, e.g. numerous mosses and algae. They usually live in association with the former. Mosses are often found enclosing tree branches entirely. They may even hang for several metres from the branches. Lichens are less common, because of their slow growth. They are found only on exposed sites together with drought-resistant mosses or near the alpine timber line, where they are quite abundant.

A special group is formed by the epiphyllous plants. These grow on the surface of very old leaves. Nevertheless, even these older leaves are still so short-lived that such epiphyllous growth is found only in extreme habitats, e.g. along forest creeks, where lower plants grow relatively rapidly. Moreover, light relations and the wettability of the leaves are important. Lichens and algae, among them *Trentepholia* and

Phycopeltis are the pioneer colonists. Foliose liverworts are later invaders.

Amongst algae and lichens are also those that penetrate into the leaf tissue or just beneath the cuticle (Fitting, 1910). Recent evidence has shown that there are also nitrogen fixing bacteria (*Azotobacter*, *Beijerinckia*) involved in addition to actinomycetes, yeasts and fungi. The former are of particular importance for this biotope (Ruinen, 1961).

Renner (1933) has studied the physiology of the water budget of epiphyllous algae, mosses and lichens. A strange, regressive development is shown in an epiphyllous moss, *Ephemeropsis tjibodensis*. The protonema develops vigorously, but the moss plant containing the sex organs is so small that the capsule appears to sit directly on the protonema. This is, therefore, an analogous case to *Taeniophyllum* among the higher epiphytes. Only the flattened, green and photosynthesising roots are developed in this orchid and the flower-bearing shoot with a few scale-like leaves develops directly from these roots.

(*f*) *Saprophytes and parasites*

Heterotrophic higher plants are not very conspicuous in the tropical rain forest. Larger forms are rare, most of them are very small. Amongst monocotyledons are saprophytes belonging to the families Burmanniaceae, Liliaceae, Orchidaceae and Triuridaceae; among the dicotyledons are members of the Gentianaceae and Polygalaceae. Of course, they are only found in places where dead organic matter has accumulated, i.e. on dead tree stumps, in local depressions filled with dead foliage or between buttressed roots. They commonly occur in the deepest shade where the soil is always moist. They are seldom found, but where they occur they usually form small colonies.

Surprisingly, also, phanerogamic parasites are not frequent in the tropical rain forest, especially the root-parasites, which are all holoparasites. They belong to two families only, the Balanophoraceae and the Rafflesiaceae, with the genera *Mycetanthe* and *Rafflesia* that are confined to the roots of Vitaceae in Indonesia. The vegetative body of *Rafflesia* is apparently reduced to cellular-threads that penetrate the wood of the host plant. A larger tissue accumulation occurs only before flowering. This tissue is formed at the base of the stem or on the surface roots of the host, where it gives rise to a huge flower. In *Rafflesia arnoldi* (in Sumatra) the flower is bright red to red-brown, it assumes a diameter of 1 m and releases a carrion-like odour. *Cuscuta* spp. which parasitise trees in the tropics of the West Indies, where they creep along the crowns, as well as *Cassytha*, which belongs to the Lauraceae, are not endemic members of the rain forest. Much more common are the epiphytic hemi-parasites, to which the Loranthaceae belong; that is, the mistletoes. They can form huge balls in the tree crowns, where they often become conspicuous because of the brilliant colours of their flowers.

Very strange are the 'hyper-parasites', which have been observed in the tropics. For example, one *Loranthus* species may parasitise another one (Koernicke, 1910). There are even hyper-parasites of the second degree. In such cases one parasite is attached to another, which in turn is attached to a third parasite that was the first to become attached to a host plant.

8 Conclusion

Finally, when considering once more the different ecological types with their differing morphological adaptations, it should be emphasised that the greatest variety of plant forms is found in the tropical rain forest. Selection of forms under such favourable growing conditions has not been so severe as in other climatic zones. The humid tropics have room for many types, for extreme hygromorphic forms as well as for those that appear more xeromorphic or succulent. Thus, the tropical rain forest can be viewed as a great reservoir which has provided material for natural selection of various sorts to develop forms and ecological types with adaptations to extreme conditions: to climates with pronounced periodicity, with cold or drought periods.

It will probably never be determined with any certainty whether plant forms which migrated as a result of such adaptations to other climates eventually returned to the rain forest (e.g. Bromeliaceae), where they conquered extreme sites as epiphytes on trees, or whether they evolved originally in the rain forest itself.

A more detailed description of the vegetation of individual areas of tropical rain forests showing great floristic differences will be given in appropriate vegetation monographs (cf. p. xv).

References

ALLEE, W. C. 1926. Measurement of environmental factors in the tropical rain-forest of Panama. *Ecology*, **7**, 273-302.

ALVIM, P. DE T. 1959. Water requirements of Cacao. *FAO Tech. Cacao Meeting. Accra (Ghana)*.

ALVIM, P. DE T. 1960. Moisture stress as a requirement for flowering of coffee. *Science*, **132**, 354.

ASHTON, P. S. 1958. Light intensity measurements in rain forest near Santarem, Brazil. *J. Ecol.*, **46**, 65-70.

AUBRÉVILLE, A. 1938a. La forêt coloniale. *Ann. Acad. Sci. Colon. Paris*, **9**, 1-245.

AUBRÉVILLE, A. 1938b. La forêt equatoriale et les formations forrestières tropicales africaines. *Scientia (Como)*, **63**, 157.

BAKER, J. R. 1938. Rain-forest in Ceylon. *Kew Bull.*, 9-16.

BERNARD, E. 1945. *Le climat écologique de la Cuvette Centrale Congolaise*. Brussels. (Referred to in *Ecology*, **30**, 265, 1949).

BIEBL, R. 1964. Zum Wasserhaushalt von *Tillandsia recurvata* L. und *T. usneoides* auf Puerto Rico. *Protoplasma*, **58**, 345-68.

BLUM, G. 1933. Osmotische Untersuchungen auf Java. *Ber. Schweiz. Bot. Ges.*, **42**, 550-680.

BOYER, Y. 1964. Contribution a l'étude de l'écophysiologie de deux fougères epiphytes: *Platycerium stemaria* et *P. angolense*. *Ann. Sci. Nat., Paris, Ser.*, 12, **5**, 87-228.

BÜNNING, E. 1947. *In den Wäldern Nordsumatras*. Bonn.

BÜNNING, E. 1956. Der tropische Regenwald. *Verständliche Wissenschaft*, **56.**

COUTINHO, L. M. 1962. Contribuicão ao conhecimento da ecologia da mata pluvial tropical. *Fac. Fil., Univ. de São Paulo Bolet*, **257**.

COUTINHO, L. M. 1964. Untersuchungen über die Lage des Lichtkompensationspunkts einiger Pflanzen zu verschiedenen Tageszeiten mit besonderer Berücksichtigung des 'de-Saussure-Effektes' bei Sukkulenten. *Beitr. zur Phytologie* (Walter-Festschr.), 101-08.

COUTINHO, L. M. 1965. Algunas informacoes sobre a capacidade ritmica diaria da fixacão e acumulacão de CO_2 no escuro em epifitas e erbacaes terrestre da mata pluvial. *Bolet. Fac. Fil.Cience. Letras. Univ. São Paulo Botan.*, **294**, 1-21.

DASTUR, R. H. 1925. The relation between water content and photosynthesis. *Ann. Bot.*, **39**, 769-86.

ELLENBERG, H. 1959. Typen tropischer Urwälder in Peru. *Schweiz. Zschr. Forstwesen.*, **3**, 169-87.

EVANS, G. C. 1939. Ecological studies in the rain forest of southern Nigeria. *J. Ecol.*, **27**, 436-82.

EVANS, G. C. 1956. An area survey method of investigating the distribution of light intensity in woodlands, with particular reference to sun flecks. *J. Ecol.*, **44**, 391-428.

VON FABER, F. C. 1915. Physiologische Fragmente aus einem tropischen Urwald. *Jb. Wiss. Bot.*, **56**, 197-220.

FITTING, H. 1910. Über die Beziehungen zwischen den epiphyllen Flechten und den von ihnen bewohnten Blättern. *Ann. Jard. bot. Buitenz. Suppl.*, **3**, 505-18.

GANSSEN, R. 1957. *Bodengeographie*. Stuttgart.

GESSNER, F. 1940. Die Assimilation der Hymenophyllaceen. *Protoplasma*, **34**, 102-16.

GESSNER, F. 1956. Wasserhaushalt der Epiphyten und Lianen. In *Encyclopedia of Plant Physiology* (ed. W. Ruhland). Vol. 3, pp. 915-50. Springer-Verlag, Berlin.

GESSNER, F. 1965. Untersuchungen über den Gefässaft tropischer Lianen. *Planta*, **64**, 186-90.

GILTAY, E. 1897, 1898a, 1900. Vergleichende Studien über die Stärke der Transpiration in den Tropen. *Jb. wiss. Bot.*, **30**, 615-44; **32**, 447-502; **34**, 405-24.

GILTAY, E. 1898b. Über die vegetabilische Stoffbildung in den Tropen und in Mitteleuropa. *Ann. Jard. bot. Buitenzorg.* **15**, 43-72.

VON GUTTENBERG, H. 1931. Beiträge zur Kenntnis der Laubblattassimilation in den Tropen. *Ann. Jard. Buitenzorg*, **41**, 105-84.

HABER, W. 1958. Ökologische Untersuchungen der Bodenatmung. *Flora*, **146**, 106-57.

HABERLANDT, T. 1892. Anatomisch-physiologische Untersuchungen über das tropische Laubblatt. *Sitz. ber. Akad. Wien*, **101**, 785-816.

HÄRTEL, V. 1940. Physiologische Studien an Hymenophyllaceen II. *Protoplasma*, **34**, 489-514.

HARTENBERG, W. 1937. Der Wasser- und Kohlensäure haushalt tropischer Regenwaldpflanzen in sommerlicher Gewächshauskultur. *Jb. wiss. Bot.*, **85**, 641-97.

HOLTERMANN, K. 1907. *Über den Einfluss des Klimas auf den Bau des Pflanzengewebes. Anat.-physiol. Untersuchungen in den Tropen.* Leipzig.

HOLTTUM, R. E. 1953. Evolutionary trends in an equatorial climate. *Symp. Soc. Expl. Biol.*, **7**, 159-173.

HUTCHINSON, J. 1959. *The families of flowering plants.* Vol. 2.

KIRKPATRICK, T. W. *Studies on the ecology of a coffee plantation in East Africa.* Unpublished.

KLEBS, G. 1911. Über die Rhythmik in der Entwicklung der Pflanzen. *Sitz. ber. Heidelb. Akad. Wiss., math-nat. Kl.*, **23**.

KLINGE, H. 1962a. Beiträge zur Kenntnis der tropischen Böden. *Z. Pfl. Ernähr., Düng. Bod.*, **97**, 106-18.

KLINGE, H. 1962b, 1963. Über Epiphytenhumus aus El Salvador (Zentralamerika), I and II. *Pedobiologia*, **2**, 1-8; **2**, 102-7.

KLINGE, H. 1965. Podzol soils in the Amazon Basin. *J. Soil Sci.*, **16**, 95-103.

KOERNICKE, M. 1910. Biologische Studien an Loranthaceen. *Ann. Jard. bot. Buitenz. Suppl.*, **3**, 65-97.

LEMÉE, G. 1956a. La tension de suction foliaire, critère éco-physiologique, des conditions hydriques dans la strate arbustive des groupement végétaux en Côte d'Ivoire. *Naturalia Monspeliensis, Ser. Bot.*, **8**, 125-40.

LEMÉE, G. 1956b. Recherches éco-physiologique sur la cacaoyer. *Rev. gén. Bot.*, **63**, 41-94.

LIETH, H. 1958. Konkurrenz und Zuwanderung von Wiesenpflanzen. Zschr. *Acker.-u. Pflanzenbau*, **106**, 205-23.

LÖTSCHERT, W. 1959. Vegetation und Standortsklima in El Salvador. *Botan. Studien*, **10**.

MCLEAN, F. T. 1920. Field studies of the carbon dioxide absorption of coconut leaves. *Ann. Bot.*, **34**, 367-90.

MACLEAN, R. C. 1919. Studies in the ecology of tropical rain forest. *J. Ecol.*, **7**, 15-54; 121-72.

MEDINA, E. 1968. Bodenatmung und Streuproduktion verschiedener tropischer Pflanzengemeinschaften *Ber. Deutsch. Bot. Ges.*, **81**, 159-168.

MILDBRAED, J. 1922. *Wissenschaftliche Ergebnisse der II. deutschen Zentral-Afrika-Expedition. 1910 to 1911.* Leipzig.

MOREAU, R. E. 1935a. A synecological study of Usambara, Tanganyika Territory. *J. Ecol.*, **23**, 1-43.

MOREAU, R. E. 1935b. Some ecological data for closed evergreen forest in tropical Africa. *J. Linn. Soc.* (*Zool.*), **39**, 285-93.

NUTMAN, F. J. 1937. Studies in the physiology of *Coffea arabica* L. Photosynthesis of coffee leaves under natural conditions. *Ann. Bot. N.S.*, **1**, 353-67.

ORTH, R. 1939. Zur Kenntnis des Lichtklimas der Tropen und Subtropen sowie des tropischen Urwaldes. *Gerlands Beitr. Geophys.*, **55**, 52-102.

PASCHER, A. 1960. Über Wasserkelche von *Datura* und *Arisodus*, etc. *Flora*, **148**, 517-28.

VAN DER PIJL, L. 1957a. De bloem gezien tegen de achterground van de bloembiologie. *Vakblad voor Biologen*, **47**.

VAN DER PIJL, L. 1957b. The dispersal of plants by bats (Chiropterochory). *Acta Bot. Neerland.*, **6**, 291-315.

PORSCH, V. 1941. Ein neuer Typus Fledermausblume. *Biologia Generalis*, **15**, 283-94.

RENNER, V. 1933. Zur Kenntnis des Wasserhaushalts javanischer Kleinepiphyten. *Planta*, **18**, 215-87.

RICHARDS, P. W. 1952. *The tropical rain forest.* Cambridge.

ROBINSON, G. W. 1939. *Soils.* 3rd ed. London.

RUINEN, J. 1961. The phyllosphere I. An ecological neglected milieu. *Plant and Soil*, **15**, 81-109.

SANDER-VIEBAHN, G. 1962. Der winterliche Kohlensäurehaushalt tropischer Gewächshauspflanzen. *Beitr. z. Biol. d. Pfl.*, **37**, 13-53.

SCHENCK, H. 1892, 1893. Beiträge zur Biologie und Anatomie der Lianen. *Bot. Mitt. Tropen*, **4**; **5**.

SCHIMPER, A. F. W. and FABER, F. C. VON 1935. *Pflanzengeographie auf physiologischer Grundlage.* 3rd ed. Jena.

SCHWEITZER, J. 1936. *Handl. 7 Ned. Ind. Naturwat.*, **436**.

SIOLI, H. 1958. Die Fruchtbarkeit der Urwaldböden des brasilianischen Amazonasgebietes und ihre Bedeutung für die zukünftige Nutzung. *Staden-Jb.*, **5**, 23-36.

SIOLI, H., and KLINGE, H. 1961. Über Gewässer und Böden des brasilianischen Amazonasgebietes. *Die Erde*, **92**, 205-19.

SMITH, A. M. 1906. On the application of theory of limiting factors to measurements and observations of growth in Ceylon. *Ann. Roy. Bot. Gard. Peradeniya*, **3**.

SPANNER, L. 1939. Untersuchungen über den Wärme- und Wasserhaushalt von *Myrmecodia* und *Hydnophytum*. *Jb. Wiss. Bot.*, **88**, 243-83.

VAN STEENIS, C. G. G. J. 1962. Die Gebirgsflora der malesischen Tropen. *Endeavour*, **21**, 183-94.

STOCKER, O. 1935a. Ein Beitrag zur Transpirationsgrösse im javanischen Regenwald. *Jb. wiss. Bot.*, **81**, 464-96.

STOCKER, O. 1935b. Assimilation und Atmung Westjavanischer Tropenbäume. *Planta*, **24**, 402-45.

STOCKER, O. 1949. *Grundlagen einer naturgemässen Gewächshauskultur.* Verlag Eugen Ulmer, Stuttgart.

TAKEUCHI, M. 1961. The structure of the Amazonian vegetation II. Tropical rain forest. *J. Fac. Sci. Univ. Tokyo*, **8**, 1-26.

TAKEUCHI, M. 1962. The structure of the Amazonian vegetation IV. *J. Fac. Sci. Univ. Tokyo*, **8**, 279-88.

ULE, E. 1905. Blumengärten der Ameisen am Amazonenstrome. *Veg. Bilder*, **3**, 1-6.

VAGELER, P. 1938. *Grundriss der tropischen und subtropischen Bodenkunde.* 2nd ed. Berlin.

VOGEL, S. 1958. Fledermausblumen in Südamerika. *Österr. Bot. Zschr.*, **104**, 491-530.

WALTER, H. 1931. *Die Hydratur der Pflanze.* Jena.

WALTER, H. 1936a. Nährstoffgehalt des Bodens und natürliche Waldbestände. *Silva*, **24**, 201-05, 209-13.

WALTER, H. 1936b. Zur Frage nach dem Endzustand der Entwicklung von Waldgesellschaften. *Naturforsch.*, **13**, 151-55.

WALTER, H. 1952. Eine einfache Methode zur ökologischen Erfassung des CO_2-Faktors am Standort. *Ber. Dtsch. Bot. Ges.*, **65**, 175-82.

WALTER, H. 1960. *Einführung in die Phytologie.* Vol. III, *Grundlagen der Pflanzenverbreitung.* Part 1, *Standortslehre.* 2nd ed. Stuttgart.

WALTER, H. 1963. Zur Klärung des spezifischen Wasserzustandes im Plasma und in der Zellwand bei der höheren Pflanze und seine Bestimmung. *Ber. Dtsch. Bot. Ges.*, **76**, 40-70.

WALTER, H., and HABER, W. 1957. Über die Intensität der Bodenatmung mit Bemerkungen zu den Lundegårdschen Werten. *Ber. Dtsch. Bot. Ges.*, **70**, 275-282.

WEBB, L. J. 1958. Cyclones as an ecological factor in tropical lowland rainforest (North Queensland). *Austr. J. Bot.*, **6**, 220-28.

WECK, J. 1963. Möglichkeiten und Grenzen der Zuwachssteigerung bewirtschafteter Forstflächen in verschiedenen Klimazonen. *Scientia* (*Como*), **57**, 1-4.

WENT, F. W. 1940. Soziologie der Epiphyten eines tropischen Urwaldes. *Ann. Jard. bot. Buitenz.*, **50**, 1-98.

III. *Other Vegetation Types of the Humid Tropics*

1. Swamp and aquatic vegetation

Because of the high rainfall in the equatorial zone and the relatively low evaporation, excess water collects in depressions, where it gives rise to swamps and lakes. From here, the water leads into streams that run into the ocean. Forest vegetation does not usually invade such habitats that are continuously covered with water.

Such swamp areas are very extensive in tropical Africa as can be noticed easily from a plane. Here I briefly mention the largest of them. I saw the huge area of the Lukanga swamps in Zambia, west of Broken Hill, on the flight from Livingstone to Nairobi. From Living-

Fig. 83. Papyrus-swamp (background) near Masaka, Uganda, Africa (Phot. E. Walter).

stone the flight continues across Lakes Kampolombo and Bangweulu, where the explorer of Africa and discoverer of the Victoria Falls, David Livingstone, died in 1873. Both lakes are surrounded by vast swamps, which can extend over approximately 150 km. In the northern part of this swamp a strange formation, similar to contour bogs, was noticeable but the orientation was perpendicular to the shores of the lake. The western part of Lake Bengweulu is almost separated from the rest of the lake by a sand-spit. Many huts were visible on this peninsula but otherwise the lake shore is virtually inaccessible as it is surrounded by floating grass mats and swamp. The swamp areas around Lake Mweru and to the west of it were not on the flight route. Swamp areas of greater extent were again observed north-west of Lake Rukwa and south of Lake Eyasi. Lakes Eyasi, Manyara and Natron as well as others of the East African rift valley occur in a relatively dry area with high evaporation. They have no river outlet and show accumulation of sodium carbonate (Na_2CO_3).

According to Wasawo (1964), the swamps in Uganda extend over an area of 12,800 km² (Fig. 83). This represents 6% of the total land surface. They are primarily *Cyperus papyrus* communities (pH 5·5-6·4), which allow little penetration of light to the ground so that there is no undergrowth. Between these and the *Phoenix reclinata* zone occur pure stands of *Miscanthidium*, in which one may find *Sphagnum* because of the strongly acid soil reaction (pH 4·6-5·5).

However, the largest swamp areas are produced by the Nile. The Victoria-Nile, which begins at Lake Victoria flows 80 km north of it through Lake Kioga. This lake is surrounded by swamps, which may extend across an area of 200 km. Farther to the north, the Nile runs through a small valley, where it plunges as the Murchison Falls into the Central African rift valley (Graben). It then runs through the north end of Lake Albert and from there it is named the Albert-Nile. Above Nimule, shortly before the Sudanese border, the Nile is obstructed by a mountain ridge and becomes embedded in extensive swamp-flats. From here it cuts through to a deeper bed, where it is known as Bar El Jebel or the White Nile and empties farther north into a huge lowland, about 400 m above sea level which it floods extensively. With its left tributary, the Bar el Ghazal, the Nile here forms probably the largest swamp area, the 'Sudd', which according to Mensching (1963) covers an area of 150,000 km². This swamp extends for more than 600 km from north to south and similarly from east to west. I had a chance to look over this area from the air during high flood level. Almost the entire area was covered with water, for the sun was evenly reflected from water standing in wide, green, often circular patches. These are mostly floating grass mats and floating islands, known under the name 'Sudd'.[1] They are formed from the grass *Vossia cuspidata*, which has floating shoots, from *Cyperus papyrus*, *Eichhornia* and *Pistia*. The swamp flats are interrupted by many open water surfaces, which are connected through a network of meandering streams. Only a small part of the total area was not flooded. Scattered huts and fields could be seen on the dry land, which was distributed as islands. These are described as a palm-savannah with *Hyphaene thebaica* and *Borassus aethiopica*, and as *Acacia-Balanites* savannah, respectively, particularly on clayey soil. The water level drops during the drought season and a large part of the flooded land surface reappears. This is a grassland covered with *Hyparrhenia rufa* or *Setaria incrassata*. On the moister parts *Echinochloa stagnina* and *E. pyramidalis* dominate together with the rush (*Phragmites communis*) and *Veteveria nigritana* (cf. the map in Harrison and Jackson, 1958).

[1] Migahid has made ecological studies in the Sudd-area, which include investigations of soils, water relations and the transpiration of *Cyperus papyrus*, *Vossia cuspidata* and *Pistia stratiotes* (for supplementary information see Migahid, 1948, 1952a, 1952b; Migahid *et al.*, 1955).

The Sudd-area evaporates so much water that half of the inflow is lost from the Nile farther down. If a project, already designed, which provides for channelling the water through the present swamp area directly into the White Nile at about the present confluence of the Sobat river were to be carried out, not only would the swamp area be drained, but the capacity of the White Nile would also be doubled. In addition, the swamp area pollutes the Nile farther down with the free-floating, South American introduction, *Eichhornia crassipes*, which not only hinders river boat traffic but also accumulates at the retaining dam above Khartoum in such enormous quantities that it is possible to walk over the plant mat.[1]

If one considers that large areas of swamp occur also in South Katanga and in the Congo Basin proper, one begins to appreciate the importance of the water-plant communities in relation to the total plant cover of tropical Africa.

Grassland with periodically moist habitats in depressions often changes rather gradually into swamp communities. However, a different flora occurs in those areas, in which the rainfall during the rainy period remains below the evaporation rate, particularly during the drought period. Many undrained areas have been developed as a consequence of tectonic movements which caused rift fractures. Here one finds an accumulation of bases that have originated through weathering. Since more recent marine sediments are absent and since also the sea coasts are rather distant, there is no chloride accumulation (see pp. 293, 304, 443). Instead, the accumulated base is sodium carbonate. Thus, most of the undrained lakes in the dry East African rift valley are sodium carbonate lakes, as mentioned before.

A zonation of sodium-tolerant species can be observed at the shore of the moderately sodium-enriched Lake Nakuru (Kenya): here one finds *Scirpus* sp. in the water, then *Juncellus* and, as the most important grass species, *Sporobolus spicatus* and a little higher, *Cynodon dactylon*. Outside the actual brackish zone grow *Cyperus rotundus* and *Pennisetum clandestinum*, the Kikuyu grass, with strange basal inflorescences and filaments 6 cm long. A swamp forest around the lake is formed by *Acacia xanthophloea* with its characteristically yellow trunks. Flamingoes, which occur in flocks of thousands, particularly prefer such sodium carbonate lakes.

One also finds transitory stages to communities on saline soil in other places, frequently in depressions in grassland. In addition to the plants above I should mention *Sporobolus robustus*, *Diplachne fusca*, etc.

[1] It is extremely difficult to keep this plant in check. In spite of spending 50 million Belgian francs for control in the Congo during 1956 and 1957, 150 tons/hour of *Eichhornia* float downwards on the Congo river near Leopoldville (after Lebrun, in Wild, 1961; Gay, 1960; Chadwick, 1960). Similarly dangerous, locally, is *Salvinia auriculata*, which is also introduced from South America. A 25 cm thick mat originating in the dammed Lake Kariba (Rhodesia) provided a substrate for the formation of floating grass carpets of *Vossia* and *Scirpus cubensis*.

Whenever the lowest parts of the depressions are completely dried up, they are barren and covered with a white crust. More exact analyses of these salt-soils are not available.

The aquatic-plant communities of the tropics are floristically different from those of the temperate zone, but the differences are not so great as between the terrestrial communities. Many cosmopolitan species occur also in the tropics (Ruttner, 1940). Stages of succession in aquatic habitats are here shown by the same vegetation zonations as are found in temperate regions: (*a*) submerged and free-floating species, (*b*) species with floating leaves, (*c*) swamp communities.

(*a*) This zone is characterised, among others, by *Hydrilla verticillata*, *Ceratophyllum demersum*, *Myriophyllum* spp., and *Potamogeton* spp. Of particular significance are the free-floating *Eichhornia* populations. Associated plants are *Lemna*, *Azolla*, *Salvinia* and in the New World tropics also *Jussiaea repens*.

(*b*) Among the species with floating leaves are several *Nymphaea* spp., also the Egyptian Lotus, *N. lotus* and the blue *N. coerulea*, both of which occur in Africa. Other plants in this group are *Limnanthemum* spp., *Trapa* spp., *Euryale*, and the famous *Victoria regia*[1] that is found in the Amazon area. The latter species has large floating leaves with upturned margins. These leaves have many small perforations (stomatodes), which serve as drainage holes for rain water. Such unwettability is generally characteristic of the leaves of tropical swamp plants. In *Nelumbium*, for example, run-off is facilitated through papillae on the lower side of the leaves (Gessner, 1955, 1959).

The Podostemaceae are a particularly strange tropical family that occur only in the foaming waters of waterfalls and rapidly flowing streams. The family belongs to the order Rosales, but has vegetative organs that are thalloid and crustose, resembling those of algae and mosses. Their ecology has only recently been investigated by Gessner and Hammer (1962), who studied this group at the waterfalls of the Caroni river, a major tributary of the Orinoco. There are four species in this area. The insect-pollinated flowers originate at the upper holdfast point, that is above the water. The vegetative organs float on the water. They have neither conducting elements nor protection against transpiration and desiccate quickly when only partly immersed in water. Their osmotic pressure during plasmolysis was 6·6 atm. They die in stagnant water and prefer tropical 'black-waters'; i.e. those that are low in electrolytes and are brown-coloured from humus soils containing little CO_2. Therefore, to investigate their photosynthesis, it is necessary to add CO_2 to the water. In their natural habitats, CO_2 is constantly absorbed from the air through the foam of the rapids. The Podostemaceae are obligatory heliophytes; only one species is found in shaded creeks.

[1] Now, more appropriately, *V. amazonica*.

The plants are so strongly fastened to rocks that they are never torn loose. They only live for short periods (hardly longer than two months), because of the violent fluctuations of the water level. New growth results rapidly from germinating seeds. These are very small (only 0·4 mm long) and they are carried to current-free interfaces where they adhere to the rocks. The outer layer of the seed coat swells up in less than a minute. The seed is glued to the rock by this gelatinous and pectin-like substance and cannot then be removed even by a strong water-jet from a hose.

(*c*) Representatives of swamp plants are: *Typha*, *Cladium*, *Phragmites*, *Nelumbium*, *Sagittaria*, *Pistia*, *Limnophyton* and many Cyperaceae amongst which *Cyperus papyrus* can attain several metres in height.

Floating islands, as already mentioned, are also very characteristic of tropical waters. They consist of grassy vegetation mats, which are primarily formed through plants rooting on the shores and which often extend their floating stems several metres along the open water surface. These include, for example, *Ipomoea reptans*, *Jussiaea repens*, *Polygonum barbatum* and also Gramineae and Cyperaceae. The floating vegetation mats that only receive rain water can show very acid reactions. They even form habitats for such plants as *Sphagnum* with *Rhynchospora*, *Eriocaulon*, *Xyris*, and for several insectivorous plants, such as *Utricularia* and *Nepenthes* spp. in south-east Asia.

Lind and Visser (1962) have described the zonation of the swamp vegetation as related to water level near Entebbe on Lake Victoria. *Papyrus* communities form extensive floating mats at the outer margin towards the lake. Tree-like monocotyledons join the association in the terrestrial zone. Among them are *Pandanus*, *Raphia* and *Calamus* spp. and other palms, for example, *Phoenix reclinata* in Africa. In most situations palms appear to grow on soils that are at least periodically flooded. They endure poor soil aeration better than most of the dicotyledonous trees and they have a greater competitive capacity on wet or periodically moist soils.[1] Therefore, they are commonly found along the margins of rivers. Here we find forest communities that correspond ecologically to the alluvial bottom-land forests of the temperate zone; that is, they are subject to periodic flooding, which occurs particularly in tropical areas whose rainfall is not distributed quite uniformly.

The distribution of vegetation across the lower Amazon basin is shown in Fig. 84. In this region one can distinguish three river types:

1. 'Agua branca' = white water. These are opaque rivers that contain many suspended particles which colour the water loam-yellow. They originate in the Andes.
2. 'Agua preta' = black water. These rivers contain clear water,

[1] The particularly favourable conditions for the cultivation of coconut palms at tropical sandy beaches is probably related also to the continuously seaward moving ground water that here comes close to the surface.

which, however, assumes a coffee-brown colour through enrichment with colloidal humic substances. They originate from extensive swamp-forest areas, e.g., the Rio Negro.

3. Clear-water rivers with yellow-green, transparent water, which originate in areas of lower relief, e.g. from the Matto Grosso area.[1]

The strongest sedimentation occurs from the white water rivers (Sioli, 1947). These rivers are bordered by alluvial flats extending from a few metres to 100 km across. When the river level rises these flats are flooded from about the end of the year on. The velocity of the flood water is thus reduced on the landward sides, resulting in sedimentation. The gallery forest along the margin of the river also acts as a sieve causing accelerated sedimentation, which results in the formation of

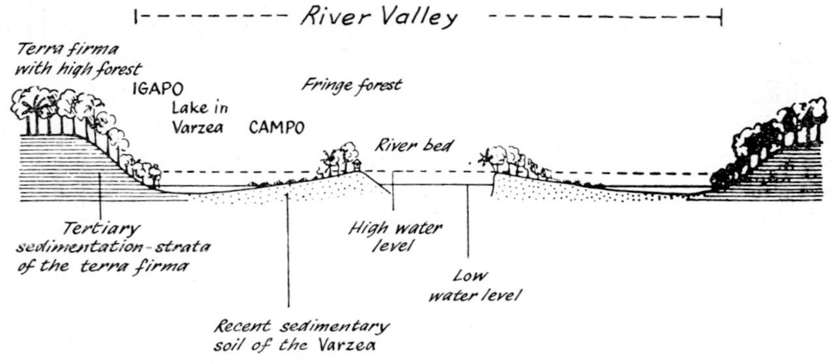

FIG. 84. Schematic profile across a valley section of the lower Amazon river (white-water river) (after Sioli).

elevated levées whose height gradually decreases towards the flood plain—the 'Varzea'. The height of the gallery forest decreases in the same way until it becomes replaced by grassland vegetation—the 'Campo' of the Varzea. The lowest parts are occupied by shallow lakes that are often very large. Beyond these lakes rises the 'Terra firme', which consists of Pliocene sedimentation from the Amazonian inland lake and which is covered with forest. The lowest regions at the edge of the Terra firme are occupied by a swamp forest, 'Igapo'. Here, the high water level may rise up to 15 m. The Igapo-forests are often under water for several months (Gessner, 1959). They are made up of but a few species. A dominant tree is *Eugenia inundata*, which sheds its leaves

[1] In the Matto Grosso just north of Paraguay in the heart of South America is the great Patanal, a very level plain extending over almost 100,000 km² and drained by many rivers. It has long been considered to be a swampy wilderness similar to that of the African Sudd (cf. p. 145). However, this is not the case. The region is only flooded extensively during the rainy season (December-March) when 80% of the annual rainfall of 1,200-1,400 mm falls. In the dry season very numerous 'ring lakes' remain only where the shore-like banks of the forest have hemmed them in. Otherwise it is a question of 'Floods on termite-savannah' (cf. p. 229) with grassy plains which serve for cattle grazing. Spontaneous salt lakes without outlets are lacking. Cf. Wilhelmy (1957), or his review (1958a), and also Wilhelmy (1958b).

below water. Other species are evergreen. The shrub-woodland attains a height of 5 m. Aquatic vegetation becomes established in the lower areas with *Polygonum hispidum*, *Jussiaea*, *Neptunia*, *Victoria regia* and the grasses *Paspalum repens* and *Oryza perennis*. A floating community is formed by *Eichhornia azurea*, *Phyllanthus fluitans*, *Ceratopteris pteridoides* and grasses, as well as *Pistia*, *Salvinia*, *Azolla*, *Lemna*, *Ricciocarpus*, etc. (Takeuchi, 1962). The clear-water rivers deposit their sediments, consisting largely of sand, in their upper reaches. The black-water rivers show little sedimentation.

Flood-water forests of unusual composition are also distributed widely in the Congo basin. However, not much is known about these.

The alluvial forests grade gradually into the mangrove forests near the river mouths where tidal influences become effective (cf. next Section).

In addition to such forests with widely fluctuating water levels, there are also true swamp forests in certain lowland areas, which are continually wet and so correspond to the swamp forests of the temperate zone. Peat formation occurs where acid reactions prevail. In such situations we may speak of peat forests (Anderson, 1961). They often resemble the rain forest but are usually much sparser in species numbers and not so luxuriously developed. They are also more open communities allowing invasion of swamp plants beneath the trees.

Bünning (1947) described such forests on the east coast of North Sumatra. The pH values of the soil are there around 3·0. The roots of the trees are crowded beneath the surface of the soil. Buttressed roots and stilt-roots are frequent; aerial roots and knee-like roots are also common. Species of bamboo are more frequent in such stands than in the rain forest. Peat depths may attain several metres (see also Straka, 1960). 'Switch-like' roots are mentioned as a special characteristic of these forests. They originate from the trunks of certain tree species several metres above the soil and grow out at right angles to the stem, often reaching a length of 0·5-1 m or more. Bünning assumed that these roots served for water absorption during periods when the surface of the peat soil was dry. However, these roots could only be wetted by rain, which would also moisten the surface soil, so that the shallow-growing surface roots would be able to absorb sufficient water from the soil. An absorption of water vapour from the air, or of dew in large quantities, appears rather improbable.

2. Mangroves

One of the botanically most remarkable vegetation types are the mangroves, which occur on the sea shores in salt water. They are forests whose crowns form canopies which alone extend above the water level at high tide, while their peculiar breathing roots become exposed during low tide. The slender tufts of coconut trees are seen only inland, next to the mangrove zone. These forests form a world of their own in the tidal zone. They show interesting peculiarities with regard to their plant forms and associated animal species.

Mangrove forests grow only on shores, where the vigour of the surf is broken as a result of outlying coral reefs or islands. They can exist only in habitats that are periodically flooded by sea water and they cannot endure frost. These conditions occur far beyond the equatorial zone. Therefore, mangroves extend beyond even the Tropics of Cancer and Capricorn but they are best developed in the humid tropics and the numbers of their species declines outside this area. The most northern sites are in the Bermudas (32° N), at the north tip of the Red Sea in the Gulf of Aqaba (30° N) and in South Japan (32° N). To the south, mangroves extend beyond Durban in South Africa (33° S), around Australia (38° S) and to the northern end of New Zealand. They occur even on Chatham Island, east of New Zealand (44° S).

Like all seashore communities, the mangroves have a very wide distribution and are not limited to the equatorial zone. Floristically, one can distinguish the species-rich mangroves of the Indian Ocean and Pacific from the species-poor mangroves of the Atlantic. The genera are mostly the same, but the species are different. Even among the eastern mangrove forests, the entire group of 20 species is found only on Java, New Guinca and the Philippines. In East Africa only 14 species occur, in North Australia likewise 14, in China and on Formosa only 4, and in New Zealand only one.

Tidal levels vary on different coast lines. But periodical, correlated changes occur even on the same coast in phase with the moon. One distinguishes between mean high and mean low water. The high tides increase and the low tides decrease at new and full moon. These are the spring tides. In between occur the neap tides with the least tidal variations during the first and last quarter of the moon. Finally, the tidal fluctuations are at maximum twice a year during the spring and autumnal equinox. Mangroves occur throughout the whole range up to the highest high-water mark although only a barren zone is often found at the low-water mark (Fig. 85).

Mangrove trees are obligatory halophytes. Some of them can grow also in fresh water, but they develop better in salt water. Like all obligatory halophytes, they store much sodium chloride in their cell sap (Fig. 92, p. 156) and their leaves are succulent.

Three environmental factors are important in determining the zonation of mangroves through influencing, in turn, the competitive ability of the individual species:

1. Frequency and duration of flooding with sea water.
2. Consistency of the soil: sandy or clayey mud-deposits.
3. The degree of admixture with fresh water at the river mouths, and the concentration of brackish water. The latter is not constant but

depends on the particular state of the tides, wind direction, the particular amount of fresh water and velocity of the rivers.

These factors determine a distinct zonation in the mangrove stands whereby each zone is represented by only one, or occasionally two or more species. This zonation is the result of competition among the mangrove species, which in turn is influenced by site factors. The larger the number of mangrove species, the more complicated are the competitive relationships. This likewise produces greater difficulties in recognising the zones and their causal relations. A change in the zonal sequence of

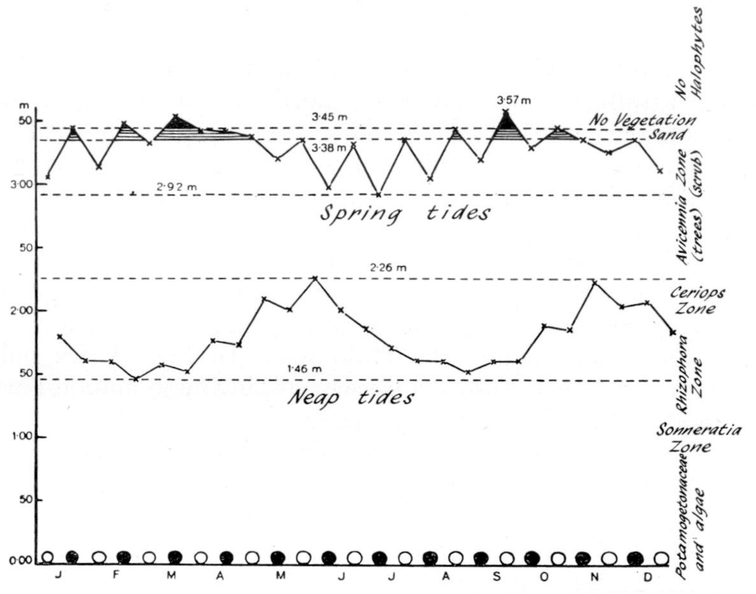

FIG. 85. Seasonal change of tidal levels at Tanga (East Africa) and distributions of mangroves (after Walter and Steiner). Black—zone reached only by equinoctial tides. Below—months and moon phases.

mangrove species can result from the absence of certain competitors. The relations of one area, therefore, cannot be extrapolated without qualification to another, floristically different area. A further factor is the rainfall, which influences the salt content, particularly that of the rarely flooded, higher regions in the mangrove terrain.

I will first explain, therefore, the ecological relations of an example that shows a relatively simple pattern and which I have investigated in some detail, the coastal mangroves of East Africa (Fig. 86) (Walter and Steiner, 1936).

In contrast to the mangroves of the river mouth, the coastal mangroves grow away from the river inlets in the shelter of coral reefs. Therefore, they are subjected only to sea water with an osmotic concentration of 24 atm. Their zonation is shown in Figs. 87-88. *Bruguiera*

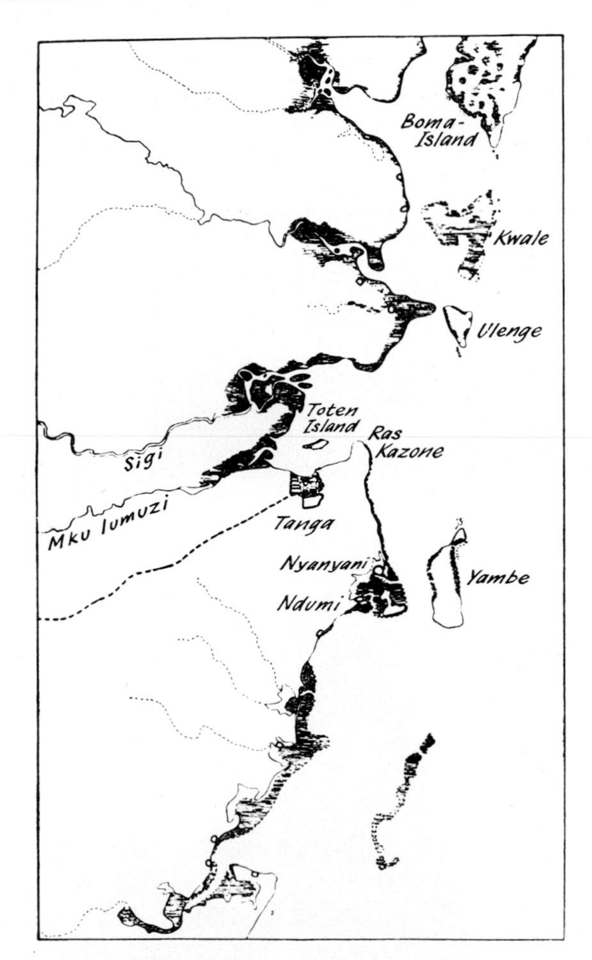

FIG. 86. Mangroves (hatched) on the East African coast near Tanga sheltered by outward-lying coral islands and reefs. River mouth mangroves, coastal mangroves and coral reef mangroves are distinguished (after Walter and Steiner).

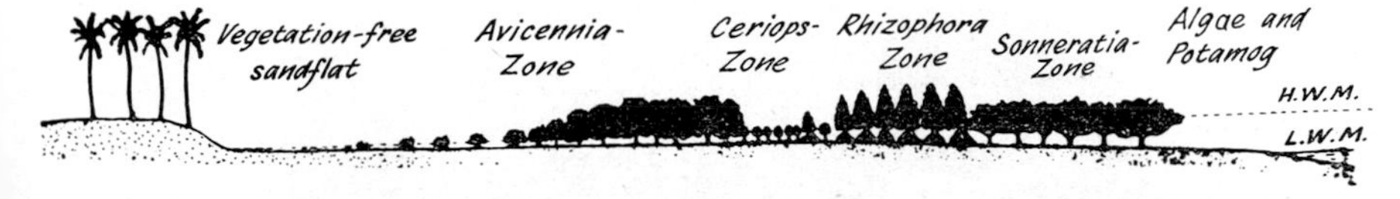

FIG. 87. Zonation of the East African coastal mangroves (after Walter and Steiner). H.W.M.—high water mark; L.W.M.—low water mark.

does not here form a zone of its own but is mixed with *Rhizophora* and *Ceriops*. *Bruguiera* only extends up into the brackish water along rivers where it comes together with the fern *Acrostichum aureum*.

FIG. 88. Outer *Sonneratia* zone at high tide on the Island of the Dead near Tanga (after Walter and Steiner). In the background the hospital of Tanga. (Phot. E. Walter)

FIG. 89. *Rhizophora mucronata* on barren rock on the coral island Ulenge near Tanga, East Africa. (Phot. E. Walter.)

Mangroves occur either isolated or in groups on coral reefs (Fig. 89). Here there is no competition and so no zonation.

The climate of Tanga (5° S) is relatively dry for tropical conditions (see Fig. 90). The two rainy periods are well pronounced. The monthly means of relative humidity at 1400 hr are between 68-76%. Evaporation (standard Piche) amounted to 9·2-12·0 ml/24 hours during the dry period, the hourly values at noon were above 1 ml, and the maximal value was 1·25 ml/hour. These values correspond to those in central Europe on hot summer days. Evaporation in the mangrove stand itself is only half this amount but, in spite of this, one must realise that some concentration of the soil water takes place at low tide. It has been shown, in fact, that the salt concentration in the soil increases inland, so that a value equivalent to nearly 40 atm is reached in the

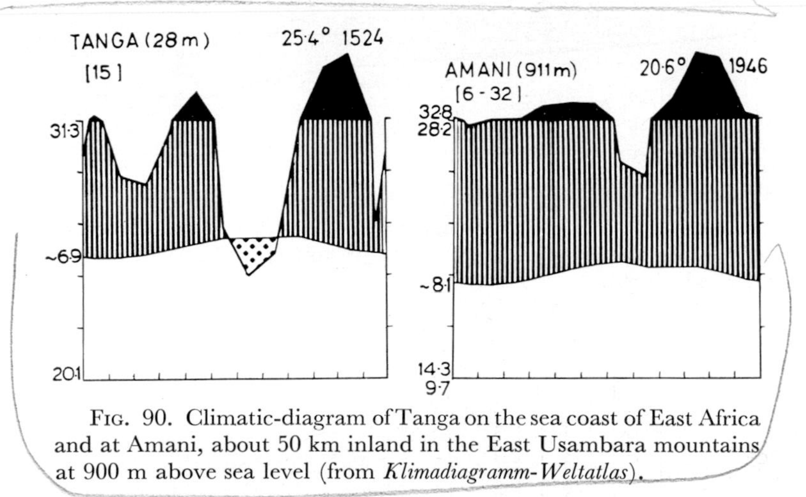

FIG. 90. Climatic-diagram of Tanga on the sea coast of East Africa and at Amani, about 50 km inland in the East Usambara mountains at 900 m above sea level (from *Klimadiagramm-Weltatlas*).

Avicennia zone. A corresponding increase in cell sap concentration is also shown by the leaves of the plants. The difference from the soil value amounts to 8-9 atm and only in the case of *Avicennia* was it about 20 atm (Fig. 91). In this connection, it was found that the chloride content of the cell sap was about the same as in the soil solution and that the difference was made up by non-chlorides (sugar, organic acids, etc.) (Fig. 92).

The difference of 8-9 atm corresponds approximately to the normal osmotic values of the leaves of the tropical rain forest trees (see Tables 10 and 11, pp. 105-106). This shows that the mangroves—disregarding the salt concentration of the sea water—suffer no disadvantage in regard to water absorption when compared with trees on the salt-less tropical soils. Indeed, *Avicennia* rather shows an advantage in this respect. It is also the species of mangrove which extends the farthest into non-tropical areas, up into the desert at the Gulf of Aqaba.

The water balance of the mangroves is well regulated. The leaves are markedly succulent and equipped with water-storage tissue (Fig.

93). Transpiration is very low (Walter and Steiner, 1936). With an evaporation rate of 0·53-0·62 ml/hr, transpiration amounted in the different species to 1·8-2·6 mg/g fresh weight/min or 6·1-8·8 mg/dm²/min. The amount was only about twice as high in *Avicennia* and *Loranthus* (higher in *Lumnitzera*). Thus, the mangroves transpire even less than

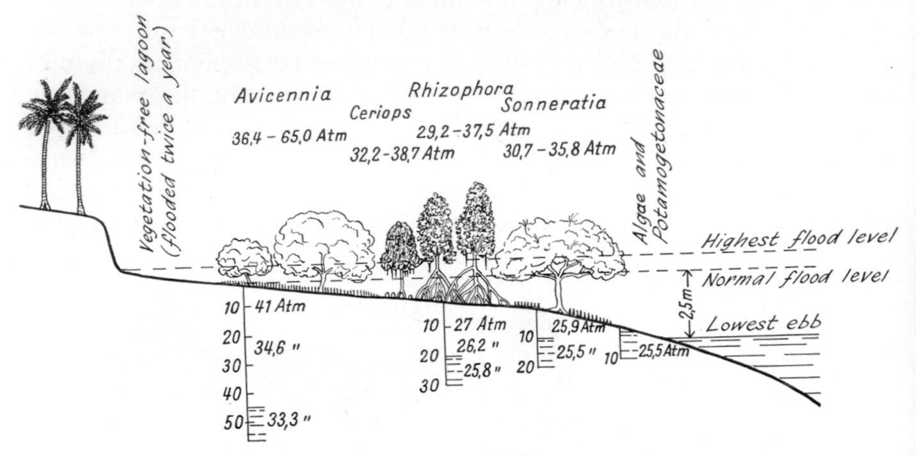

FIG. 91. Osmotic values of the mangrove species (lowest and highest value) and of the soil solution at different depths (in cm) (after H. Walter, *Standortslehre*, 2nd ed., 1960).

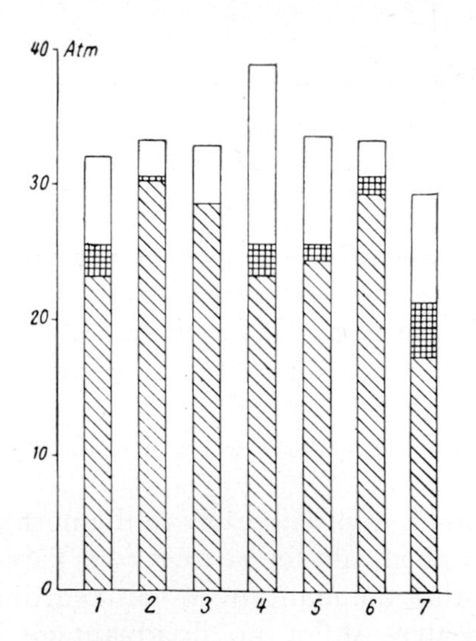

FIG. 92. Osmotic values (total length of histogram-blocks), chloride content (obliquely hatched) and total sugar content (cross-hatched) in East African mangroves: 1. *Sonneratia alba*; 2. *Rhizophora mucronata*; 3. *Ceriops candolleana*; 4. *Avicennia marina*; 5. *Bruguiera gymnorrhiza*; 6. *Lumnitzera racemosa*; 7. *Xylocarpus obovatus* (after H. Walter, *Standortslehre*, 2nd ed., 1960).

the trees of the tropical virgin forest. Transpiration on a fresh weight basis is below that of *Sedum maximum*; only that of *Avicennia* equals that of *Sedum*. However, in spite of this very low transpiration rate, salt accumulation would occur in mangrove leaves if the roots were to absorb sea water in unchanged form. The problem in mangroves is not the water—but the salt relations.

I differentiate between non-salt-excreting and salt-excreting mangroves. The latter belong to the genera *Avicennia*, *Aegiceras* and *Aegialitis*. The underside of the leaves of a salt-excreting *Avicennia* tree is densely

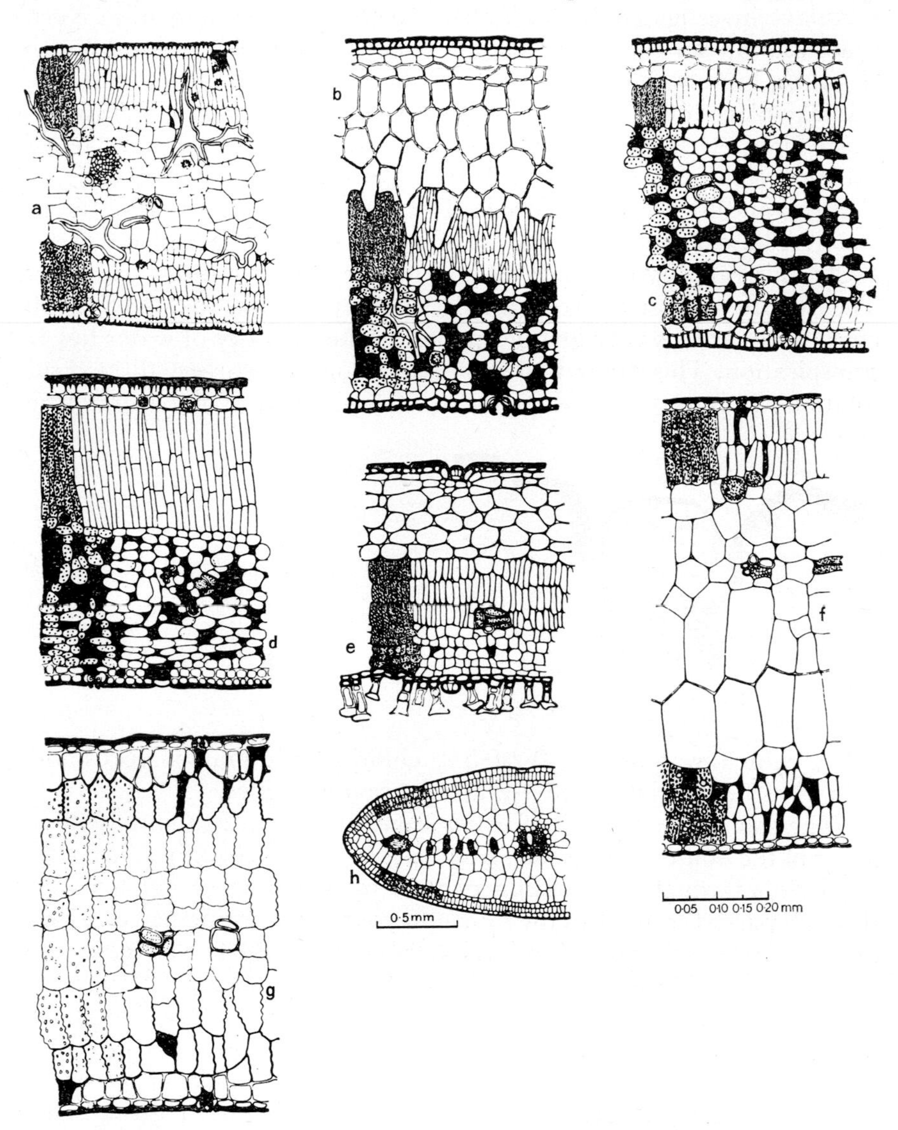

FIG. 93. Leaf cross-sections of different mangrove species (after Walter and Steiner): *a*, *Sonneratia alba*; *b*, *Rhizophore mucronata*; *c*, *Ceriops candolleana*; *d*, *Bruguiera gymnorrhiza*; *e*, *Avicennia marina*; *f*, *Lumnitzera racemosa*; *g*, *Loranthus* sp.; *h*, *Suaeda monoica*. All cross-sections (with exception of *h*) drawn at equal scale (shown on figure).

covered with sodium chloride crystals in the dry climate of East Africa (Fig. 94). At night these absorb water hygroscopically from the atmosphere, and so dissolve. But this has nothing to do with an atmospheric water absorption by the leaves. No sodium chloride crystals are formed in humid climates. Here, the salt is simply washed off by the rain.

Scholander (Scholander *et al.*, 1962, 1965, 1966) and his collaborators have investigated glandular salt excretion by mangroves in North Queensland. It shows a definite daily periodicity, especially in *Aegialitis*, but also in *Aegiceras*. It is highest at noon. Salt excretion does not take place if the leaves are shaded. One can observe the emission of single droplets from the glands under oil. The concentration of the excreted solution amounts to 1·8-4·9% in *Aegialitis*, to 0·9-2·9% in *Aegiceras* and to 4·1% in *Avicennia*. Thus, it is often twice that of the sea water. The excreted salts consist of up to 90% NaCl and up to 4% KCl. This ratio corresponds roughly to the composition of sea water. The quantity excreted during a day (9 hours) amounts to 0·2-3·5 mg/10 cm^2 leaf-area. It is very low in relation to the quantity of water lost by transpiration. This shows that the roots absorb only a very diluted salt solution. This was supported by an investigation of the water from

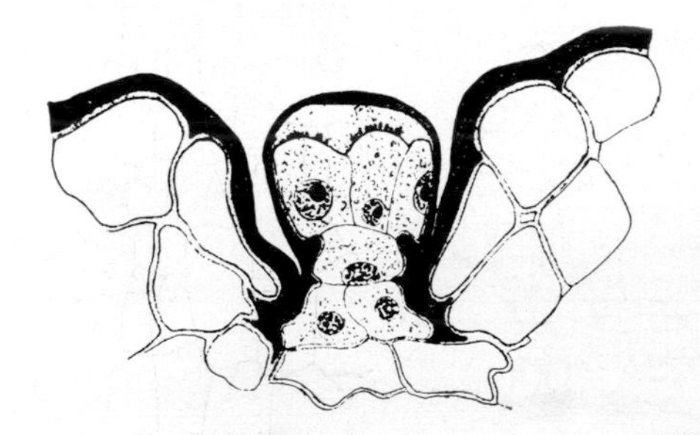

FIG. 94. Salt-gland of *Avicennia marina* (after von Faber).

xylem which was only a 0·2-0·5% chloride solution whose osmotic value was further increased 1-2 atm by non-chlorides.

The salt concentration of water from the xylem is about 10 times lower in the non-excreting mangrove species. They can only achieve salt regulation through elimination of salts by shedding of old leaves or through partial transfer of the salts into younger, growing leaves. It was found that the salt concentration increased relatively fast in relation to the low transpiration rate in young, developing leaves (Walter and Steiner, 1936). In a more recent paper Scholander (1968) has shown how mangroves desalinate sea water. The roots act as an ultrafilter permeable to water and nearly impermeable to the salts of sea water. Pure water is sucked into the vascular system by the high cohesion stress in the vessels due to a suction tension in the leaves equal to at least 24-26 atm.

The low concentration of the xylem sap makes it understandable how the seedlings of the viviparous mangrove species with a very low transpiration rate show low cell sap concentrations of 13-18 atm and low chloride contents (Walter and Steiner, 1936). Since there are no direct connections between the vessels of the mother plant and those of the viviparous seedlings, it appears probable that transfer of substances is mediated through certain glandular cells located in the cotyledons and surrounding tissue. This question was investigated by Kipp-Goller (1939/40; Biebl and Kinzl, 1965), and more recently in greater detail by Pannier (1962), who worked with seedlings of *Rhizophora mangle*. The cell sap concentration of the pericarp is 30·0 atm and that of the cotyledons 24·4 atm, but that of the tissue layers surrounding the cotyledons is only 16·4 atm; also, the chloride content is especially low. However, the sugar content is relatively high. In addition, these tissue layers and the cotyledons are also histochemically characterised by accumulating labelled phosphorus (^{32}P), an increased respiration, etc. All these

FIG. 95. *Bruguiera parviflora* with seedlings hanging from the branches (Phot. von Faber).

suggest a specific glandular function of importance for the nutrient supply of the viviparous seedling.

The large seedling of the non-viviparous *Avicennia* shows no variations in osmotic values throughout the seedling tissue. But it also contains practically no chlorides.

Soon after the seedlings have taken root in the soil, they show the same behaviour with regard to osmotic value and chloride content as the mother plant. Such seedlings are shown in Figs. 95-97. The respiration of mangrove seedlings was thoroughly investigated by Chapman (1962). It does not show any peculiarities.

A very conspicuous phenomenon of the coastal mangrove stands in East Africa is the vegetation-free sand flat that borders the mangrove forests on the landward side. This is related to the dry climate. The barren sand flat is flooded only twice a year for a few days during the equinoctial spring tides. Subsequently, the salt concentration can

Fig. 96. Detached branch of *Rhizophora mucronata* with viviparous seedling from a mangrove stand near Tanga (Phot. E. Walter).

Fig. 97. Seedlings of *Rhizophora conjugata* at low tide; in between pneumatophores of *Bruguiera gymnorrhiza* (Phot. von Faber). Mangrove stand at Segara Anakan (formerly 'Kindersee'), South Java.

increase considerably through evaporation.[1] During the remaining period, the salt is leached out of the soil by rain. It appears that neither halophytes nor glycophytes can grow under such conditions (Figs. 87 and 98).

In summary it can be stated that the regular zonation of the East African coastal mangrove forest is apparently to be attributed to the salt relations which, in turn, determine the competitive capacity of the individual species. Under different climatic conditions and different floristic compositions, different zonations would be expected and other factors may become decisive.

FIG. 98. Vegetation-free flat at the landward border of the mangrove forest (background), which is flooded only twice a year, Ndumi near Tanga (after Walter and Steiner).

For example, *Sonneratia* does not occur farther south than the coast of Mozambique. Here, *Avicennia* forms not only the landward zone but also the seaward zone on sandy soils. The southern limit for *Bruguiera*, *Rhizophora* and *Avicennia* is near 33° S.

Each mangrove zone also has its own fauna. However, Macnae and Kalk (Macnae, 1963; Macnae and Kalk, 1962) assume this association to be of an indirect nature. They believe the depth of the ground water table, duration of flooding and the salt content to be the factors controlling the distribution both of animals as well as plants.

Avicennia is the mangrove species that not only penetrates farthest into the arid areas but which is also the most cold resistant. *Avicennia marina* var. *resinifera* is the only mangrove species which occurs on the north island of New Zealand, near Auckland, where it is liable to be exposed to frosts. Its behaviour in this area has been investigated in

[1] On Jamaica, Harris (1917) determined the concentration of a water puddle on such an area as 69 atm.

more detail by Chapman and Ronaldson (1958). *Avicennia* cannot grow in localities where temperatures of −2°C recur every 5-10 years. It forms only fringe-forests on the outer margins of marshes with mud-soil and is replaced inland by salt marsh species, such as *Salicornia australis, Juncus maritimus, Samolus, Triglochin, Aster*, etc. *Avicennia*—which always prefers sandy soils—is apparently very sensitive to the poor aeration of mud-soils. It still achieves heights of 2-3 m along the channels in the mud flats, while its height decreases sharply to 30 cm towards the centre of mud flats. This decline in height is correlated with increasing water stagnation. It is interesting that the seedlings of *Avicennia*, which are completely flooded for a much longer period, can carry out anaerobic respiration.

Further consideration will now be given to other mangrove areas. Lötschert (1959) has presented a vegetation profile through the coastal zone at El Salvador, which includes mangroves (Fig. 99). A more exact

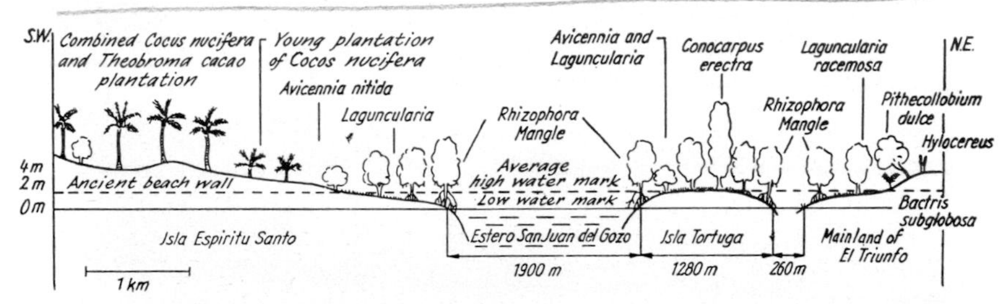

Fig. 99. Zonation of the mangrove forest in El Salvador (after Lötschert).

analysis of the salt relations was not carried out. Temperature and evaporation relations were similar to those in East Africa. Lötschert recorded 0·8 ml/hr evaporation by comparison with my maximal values of 0·9 and 1·25 ml/hr, which occurred only infrequently. A vegetation-free zone is absent on the landward mangrove border in El Salvador. The outer zone is formed of *Rhizophora mangle*. This is the case in all situations where *Sonneratia* is absent and frequently also in the case of river-mouth mangroves. According to Pannier (1959), *Rhizophora* develops best in the brackish water zone of river mouths in Venezuela near Caracas (in this connection see Fig. 100). In support of this observation, water cultures with *Rhizophora mangle* have shown their best development in 25% sea water. The dry weight produced at this concentration exceeded that produced in undiluted sea water by 5·6 times for the foliage and by 3·4 times in the case of roots. Unfortunately, the salt concentration of the cell sap was not determined. These findings do not agree with those of Stern and Voigt (1959) that germinating seeds of *Rhizophora mangle* grow best in 100% sea water.

The zonation of mangroves and the salt content of the soil profile was investigated in a river mouth area in Sierra Leone by Gledhill (1963).

Rhizophora mangle likewise grows at very low salt water contents (osmotic value 5·1 atm, that attributable to chlorides, 4·4 atm) near Duala in Cameroon. The cell sap concentration is correspondingly low (19·8-24·5 atm), but the chloride concentration in the cell sap (10·8-14·6 atm) is much higher than in the soil solution. Therefore, *Rhizophora* stores salt in the cell sap as is the case in all obligatory halophytes. Even where cultured in fresh water, I still found a chloride content of 41% and a chloride concentration responsible for 4·7 atm in leaves of *Rhizophora*.

FIG. 100. Mangrove (*Rhizophora mangle*) with stilt roots near Marina Bahia de Buche, Venezuela (Phot. E. Walter).

Osmotic values of mangroves were determined by Harris (1932) in Jamaica, by Sen-Gupta (1935) at Calcutta and by Bharucha and Navalkar (1942) at Bombay. They support my findings. A considerable admixture with fresh water occurs during the monsoon rains at Bombay, which results in a decrease of the cell sap concentration during this period.

Particularly complicated relations exist in south-east Asia. These were investigated by W. Troll and O. Dragendorff (1931), and described in more detail by Bünning (1947). Here also, *Sonneratia alba* forms the outer zone of mangrove forests but extends farther into the water. However, *Avicennia marina* appears to occur exclusively on sandy soils and can be traced far up-river, an observation that coincides with mine made in east Australia. *Rhizophora* is represented by several species and forms forests 20-30 m (up to 35-40 m) tall, which are deeply shaded, allowing only 1-5% of the light to penetrate to the ground. In addition, there are other species such as *Sonneratia acida*, which, according to Bünning, 'require a strongly acid soil water'. But such conclusions can be tested only by culture experiments, because the species may simply have a high competitive ability in this habitat, although its physiological optimum may be in a different pH range. As mentioned before, the zonation of mangroves is probably not related directly to the habitat factors but only indirectly via competition. A different ecological arrangement of mangrove zonation to that found in East Africa is not surprising, when one considers the different floristic and climatic relations of the optimal mangrove area in south-east Asia. By comparison with this area, the coastal mangroves in East Africa can be considered, so to speak, as a marginal case within the over-all mangrove area.

A peculiar feature of the mangroves is their breathing roots or pneumatophores. These had already been noted by Rumphius in 1660, but Goebel (1886) was the first to recognise them as breathing organs on the basis of their morphology. Proof of their function in respiration was not obtained by Westermaier (1900), or by W. Troll and O. Dragendorff (1931). On the basis of careful morphological investigations the latter emphasised, in the case of *Sonneratia*, that they were post-like organs for the formation of nutrient-absorbing roots in an increasingly dense medium, caused by sedimentation and up-growth of the soil. Only Scholander and his collaborators (1955) have clarified their function experimentally. The lenticels of these roots have such small openings that they are permeable to air but not to water. The oxygen in the intercellular spaces is used up when the breathing roots are completely covered by water. This results in a certain tension, because of the escape of CO_2 that readily dissolves in water. The pressure is balanced as a result of oxygen absorption from the air as soon as the roots emerge during low tide. The oxygen content in the breathing roots changes periodically, therefore, with the tides. It amounts to just under 20% at low tide and may decrease to less than 10% during high tide (Gessner, 1959, pp. 168-85).

Xylocarpus and *Bruguiera* have peculiar breathing roots. Similar structures exist in *Rhizophora*. The function of root respiration is accomplished here by a system of widely spreading stilt-roots. A bundle of sponge-like, air-containing roots originate at their lower ends in the mud. These are connected to intercellular spaces in the stilt-roots which connect with the outer atmosphere through lenticels.

The development of breathing roots on bare coral rock indicates that there is no direct relationship between their formation and soil sedimentation (Fig. 89, p. 154).

A more detailed analysis of zonation in mangroves is planned for the vegetation monographs of the different continents. Here I will only present, in conclusion, a description, from Schimper, of a river-mouth mangrove:

'The lagoon-like bay in South Java, known under the name of "Kindersee" (now Segara Anakan) is separated from the Indian Ocean to the south by the hilly island Nusa-Kambangan while the other sides are enclosed by the shores of the main island which are here very low-lying. Several rivers flow into this bay. Their outflow moves very slowly as a result of the low relief, and the tidal movements have an influence reaching far inland. The rivers split into many deltaic arms. This delta area is submerged during high tide and only emerges a little above sea level during low tide. A better substrate for mangroves can hardly exist and they are extremely well developed here.

'When travelling in a canoe along the shore of this bay or into one of the many river arms, one does not always observe the same picture. By contrast with the more exposed, steeper coastlines, where *Rhizophora mucronata* is the only species able to withstand the beating of the surf and where it is able to regenerate in relatively disturbed water, one finds on these low-lying, level shores, which are never reached by surfs, equally favourable growing conditions for several species where, through competition, one or the other occasionally becomes dominant. The shore is occupied alternately with a dense belt of *Rhizophora*, then with a clump of silver-grey, willow-like *Avicennia marina* var. *alba*, then with a stand dominated by the pale green of *Sonneratia acida*. At other places, the outlying vegetation is formed by a narrow hedge of *Nipa fruticans*. Here and there one observes the peculiar *Xylocarpus granatum*, whose head-sized brown-yellow fruits shimmer through their small crowns, or one sees a bush of *Aegiceras corniculatum* recognisable by its snow-white flowers and horn-like fruits. The two *Bruguiera* species (*B. gymnorrhiza* and *B. parviflora*) found in this part of the world, are distributed more sporadically at the margins of the mangrove forests, although they are more frequent in the forest interior. Here individuals of *B. gymnorrhiza* are conspicuous through their tall crowns and the large tubers, the size of children's heads, of *Hydnophytum montanum* that grow on their stems. The much smaller *B. parviflora* with its inconspicuous flowers is less easily spotted.

'During low tide one observes the tangle of *Rhizophora* stilt-roots or the stubble of pneumatophores of *Avicennia* and *Sonneratia* among which one can see swarms of fishes (including the strange *Periophthalmus* species) and many crabs. At other places one notices the sharply keeled roots of *Xylocarpus granatum* winding along the muddy surface in many directions.

'At high tide, the entire root system is invisible, even the lowest leaves of *Rhizophora* and *Sonneratia* remain below the water surface for a while. I was able to see from the canoe younger individuals of *Rhizophora mucronata* deep below the water surface.

'Epiphytes are not so frequent in mangrove stands. They are usually entirely absent on *Rhizophora mucronata* at the seaward forest border. Apparently, they cannot cope with the salty surface. They are more common along narrow channels that penetrate deeper into the interior of extensive mangrove stands, where the branches are not affected by salt spray. Here are, for example, Platyceriaceae, and at the "Kindersee" much *Hydnophytum montanum*, occasionally also Polypodiaceae and one *Loranthus* sp. In addition small lichens are always present but never any mosses. The latter are very salt sensitive' (Schimper-Faber, 1935, pp. 568-71).

3. Strand vegetation

Mangroves are almost entirely absent on all unsheltered sea coasts. This applies, for example, to the whole west coast of Africa with the exception of the large river inlets and the lagoons. Here, a more or less

FIG. 101. Flat Beach with *Ipomoea pes-caprae* on the southern coast of Java (Phot. von Faber). In the foreground many fruits and seeds of strand plants that have been washed ashore.

broad sandy beach—occasionally with dunes—is formed, because of the strong surf that prevails along the open coast line. It is vegetated by characteristic plants, which show an equally characteristic zonation that is found along all tropical beaches. As an example, I will discuss the conditions typical for the Indo-Pacific area.

The lower part of sandy beaches is usually unvegetated, because here the sand is shifted back and forth during storms. There follows immediately the first vegetation zone formed by an *Ipomoea pes-caprae*—*Canavalia* community (Fig. 101). This beautiful flowering creeper forms long horizontal runners along the surface. The same growth habit is exhibited by the legume *Canavalia*. These plants bind the sand. In addition, *Sesuvium portulacastrum* occurs, some grasses, such as *Sporobolus virginicus* and, in the eastern part, amongst others the strange grass *Spinifex*. This zone may still be affected by sea water. *Sesuvium* is a halophyte. *Ipomoea pes-caprae* supposedly has salt-excreting glands on its leaves according to Bünning but should be further investigated. *Ipomoea* usually grows on sand dunes beyond the salt-soil zone. However, the sand near the shore still contains some sodium chloride.[1]

The cell sap of leaves from plants growing on the sandy beach of the Island of the Dead near Tanga gave the following values:

Ipomoea pes-caprae, determined on November 9 :

osmotic value atm	NaCl component atm	chloride proportion %
12·7	7·6	60
10·7	2·5	23

Hibiscus tiliaceus (2 m tall shrub) growing a little higher on the beach, determined on November 25:

osmotic value atm	NaCl component atm	chloride proportion %
25·5	5·2	20

Sesuvium portulacastrum absorbs considerably more sodium chloride. It also grows somewhat lower on sandy beaches, where it is occasionally inundated by sea water. It is a typical halophyte, as seen from the values obtained on November 11 and 25 after a dry period:

Date	osmotic value atm	NaCl component atm	chloride proportion %
November 11	27·5	21·4	78
November 25	32·3	19·0	59

The values vary considerably with the salt content and dryness of the habitat as shown by some examples of cell sap concentrations of *Sesuvium* determined by Harris (1932):

Jamaica	sea coast	43·0 and 52·6 atm
	cacti- and *Acacia*-desert	19·5 and 49·9 atm
Florida (15 samples):	values between	11·6 and 34·2 atm
	chloride proportion in 2 samples	41% and 44%
Hawaii (4 samples):	16·5 (at a lake), 26·1, 33·5 and 40·6 atm (coral beach).	

Individual shrubs that contribute to dune formation become readily established on sandy tropical beaches. However, large dune areas are not known from the tropics, because of the common occurrence of *Barringtonia* communities. These are open forests, which prevent further sand movement. Other tree species of this zone are *Calophyllum inophyllum*, *Terminalia catappa*, *Hibiscus tiliaceus*, *Tournefortia argentea*, *Pandanus tectorius*, *Casuarina equisetifolia*. The latter can form large forests. The fruits of these species are equipped for floating, therefore, they are

[1] The values on this and the following page are cited from Walter and Steiner (1936), p. 171.

found washed ashore on all tropical sea coasts. The woody plants are often covered with the parasite *Cassytha*. *Ipomoea pes-caprae* also occurs on the Atlantic coast. In addition, there are other strand plants and in the next zone other tree species, which form forests up to 20 m high, or open stands. Very widely distributed along the coast of the West Indies is the shrub *Coccoloba uvifera* (Fig. 102) as well as *Hippomane mancinella*,

FIG. 102. *Coccoloba uvifera* (sea raisin) with fruits on the beach of the Caribbean Sea, Laguna de Tacarigua, Venezuela (Phot. E. Walter).

and *Chrysobalanus icaco* amongst others. Only a few species are pantropical, e.g. *Terminalia catappa* (Combretaceae) and *Thespesia populnea* (Malvaceae).

Most species in this zone are probably facultative halophytes. The salt concentration in the soil has already become very low here. The values in Table 31 are from Gooding (1947, after Richards, 1952).

The proportion of chlorides contributing to the osmotic value of the cell sap is still often considerable. Even in the coconut tree, this proportion is still 40% of the osmotic value of the leaf cell sap of 15 atm

(Walter and Steiner, 1936). However, coconut trees can also grow in salt-free habitats 50-100 km inland, but larger plantations are usually absent there. The yields are probably too low. I even noticed well fruiting coconut trees on an island on the Nile near Assuan, but, in this extreme desert climate, the trees need to be watered every week.

Behind the coastal zone there follow the forests characteristic for each climate respectively, if they have not been modified or entirely destroyed by man, as is usually the case near the sea.

However, the vegetation of tropical coasts can also be destroyed by natural means, due to periodical hurricanes or typhoons. In as much as the damage is restricted to wind breakage, regeneration follows rapidly. More lasting consequences result from the tidal waves that are commonly associated with these storms. These may not only destroy

TABLE 31

	salt concentration of the salt water (%)	av. water content in sandy soil (%)
Pioneer zone (50 m inshore)	5·3	0·26
Ipomoea pes-caprae zone (100 m inshore)	1·8	0·48
Coccoloba zone (150 m inshore)	0·4	1·65

the vegetation, but may simultaneously erode the soil (Sauer, 1962; Blumenstock *et al.*, 1961).

Very detailed microclimatological and ecological investigations have been started on the coastal sand-flats of Brazil, on the so-called 'Restingas' with their mosaic vegetation near Cabo Frio (east of Rio de Janeiro) (Dau, 1960; cf. Dansereau, 1947).

4. Vegetation of dry habitats

Dry habitats are rare in the humid tropics. Only areas of bare rock dry rapidly after each rain. However, it is surprising that even rocky slopes are nearly all covered with closed forests so that rock outcrops are hardly visible. In such habitats, species may be present that are not otherwise found in the forests. For example, the xeromorphic cycad *Encephalartos* occurs frequently on rocky outcrops in the virgin forest area of Usambara, East Africa.

The rock surfaces that are only wet during rains form a substrate for epiphytes similar to that on tree stems and branches. Many phanerogamic epiphytes, therefore, can grow directly on rock surfaces, if they receive sufficient light; especially those epiphytes that prefer acid substrates, e.g. *Nepenthes* and *Rhododendron* in Malaya. Such terrestrial attachments of epiphytes occurs in forest clearings and on mountain tops, where the light conditions are more favourable. On a rocky ridge of the highest range in the Binna-Burra reserve (Queensland) I found, for example, a whole colony of *Asplenium nidus-avis* growing on the ground.

Permeable limestone rocks are particularly dry, so that one can expect vegetation that deviates from that of the surroundings in such habitats. Lötschert (1958) was able to demonstrate this for the poorly vegetated, cone-shaped mountains (Mogotes) of west Cuba. But even these dry habitats are covered in the tropics with an impenetrable, primary vegetation. Their zonation is shown schematically in Fig. 103. The summit is occupied by that strange member of the Rutaceae,

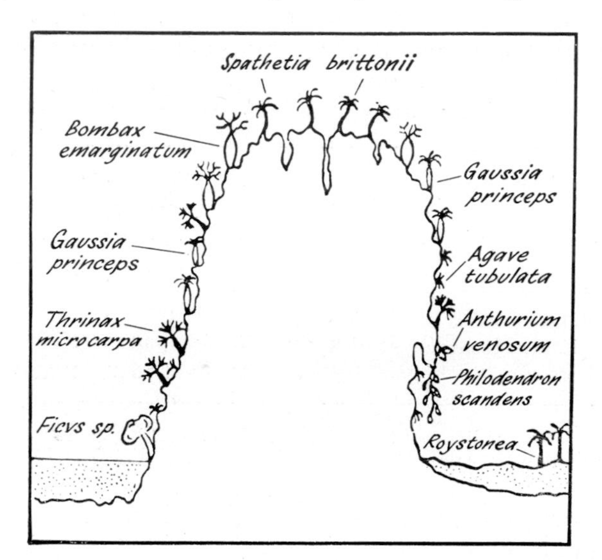

Fig. 103. Scheme of vegetation distribution on a cone-shaped mountain consisting of permeable limestone (Mogotes) in the Sierra de los Organos, West Cuba (after Lötschert).

Spathelia brittonii, which looks like a palm. At inaccessible places below the rounded tops *Bombax emarginatum* occurs, *Gaussia princeps* and *Thrinax microcarpa* grows on the flanks, and on the sheer cliffs *Agave tubulata*, *Anthurium venosum* and *Philodendron lacerum*. *Philodendron scandens* forms shoots 10 m long that hang over the stalactites. A cycad also occurs, the endemic *Microcycas calocoma*.

The dryness of these habitats is indicated by the development of bottle-shaped, water-storing stems (*Bombax*, and the endemic palm *Gaussia*), the presence of succulents (*Agave*, *Leptocereus*, *Selinicereus*, *Rhipsalis*) and the extensive root systems of most species. The soil is alkaline (pH above 7·7), and the weathering of the limestone is very pronounced.

Completely different from this vegetation is that on acid sandy soils (pH 5·0-5·5) in Cuba. It consists of oak and pine forests with many associated species of tropical origin. The dependence of vegetation upon geological substrates is rarely shown so clearly as in this area of west Cuba.

Successions, starting on bare rock, can be observed in the tropics on volcanic surfaces after a volcanic eruption. They occur relatively rapidly and can reach the forest stage within 100-150 years. Even the Island of Krakatoa, which was completely devoid of vegetation after the eruption in 1883, has become revegetated with forest in 50 years, in spite of its isolated location. This forest resembles a secondary forest.

5. The landscape of the cultivated tropics

The region of evergreen tropical rain forest has been disturbed by man over a wide area. Undisturbed forest is absent from the more densely populated areas, as for example at lower altitudes in Java. In other places it has been considerably modified. It still persists in less accessible areas only.

Richards, therefore, considered it necessary at the end of his book, *The Tropical Rain Forest*, to make a plea for an effort to establish virgin forest reserves in different areas of the humid tropics, particularly in the lowlands, following the example of the Albert National Park in the Congo. Only in this way would it be possible to maintain this virtually untapped resource of scientific knowledge, which is of such great importance for practical questions of management.

This requirement is of particular urgency in the humid tropics, where the same kind of rain forest does not regenerate after cutting, because of the extreme poorness of the soil. A secondary forest becomes established in its place. This forest can assume considerable lushness after the initial clearance. Frequently it is even more dense and more impenetrable than virgin forest and it is often confused with it.

Today, large areas of virgin forest have already become markedly modified and devalued through irrational timber utilisation, which has eliminated only the most valuable trunks. However, of even greater consequences are the effects of shifting agriculture. For use for cropping, the natives clear a forest area, when the larger trees, which are more difficult to cut, are often left. The felled timber is burned, the ash worked into the soil and then the area is seeded with corn, yam, cassava or mountain rice. The yields decline after several years, since the soils are very low in nutrients to begin with and fertilisers are not subsequently added. Then a new area is cleared for agriculture and the old one abandoned. The abandoned fields are usually re-afforested very rapidly particularly by pioneer species. If such an area is left to itself for a long time, a markedly disturbed forest community develops, which is not in equilibrium with its habitat factors. Here, lianas are particularly common, because of the more favourable light conditions.

The tree species of the secondary forest grow especially rapidly; they can reach 12 m in 3 years. This is related in part to the favourable light conditions and in part to their often very soft wood. However, they are usually only short lived, so that the secondary forest always remains less tall than the virgin forest. The actual virgin forest species occur only in later stages of succession. Their seedlings either cannot grow up without some protection against insolation, or they become suffocated by the rapidly growing species.

In many areas, the entire virgin forest land has already been under cultivation once. Here, secondary forest has to be cleared for new agricultural land. The denser the population, the faster is the turnover to re-use land and the more obvious becomes the degradation of the soil. Finally, only a scrub vegetation develops, in which bracken (*Pteridium*) or *Gleichenia* is often dominant. This shifting cultivation is a devastation of resources of the worst kind. It presents an ever increasing threat, because of the rapidly increasing population density in those areas and because suitable measures for a rational utilisation of the soil and protection of the forests require both a radical change in popular thinking and a long-term educational programme. Many areas in the tropics are already overpopulated relative to the current state of soil utilisation. But modern farm and plantation management has also greatly contributed to the reduction of the virgin forest areas. Mechanised equipment permits the clearing of larger areas and the destruction of the forest is complete (Hessmer, 1966).

A change-over from virgin forests to rationally managed forests has recently begun in most tropical countries (Teak forests, *Eucalyptus*, etc., e.g. Lamprecht, 1961, 1964a, b). A more common form of planting is that of tree plantations for rubber, palm oil and quinine; or the establishment of plantations for cacao, coffee, tea, sugar cane and bananas amongst others. The coastal forests are replaced almost everywhere by coconut groves that often cover considerable areas. The native settlements are mainly distributed within these groves.

Where the grass-covered fallow lands are burned repeatedly, open grassland areas may develop in which only fire-resistant trees are able to maintain themselves. In south-east Asia, these are, for example, *Tectona grandis*, *Pinus merkusii* or *Melaleuca leucadendron*.

In time, a cultivated landscape results in the tropics, which differs as much from the virgin landscape as is the case in central Europe. This landscape often impresses the foreigner as more primitive because it usually shows less regularity. The cultivated areas of the natives represent on one hand a mosaic of smaller forest compartments and groves of trees and, on the other hand, larger, often grass-covered, open areas; in addition, there are the irregularly distributed crop-patches and fields. The whole landscape has a savannah-like appearance; but one should not confuse this with natural savannah, for it really represents a cultivated landscape. It may be considered a secondary moist-savannah, in which species dominate that are foreign to the virgin forest. The tree strata are eliminated through human intervention. The herb layer of the original forest cannot utilise the high light intensity on the cut-over land, because they are adapted to low light intensities. Consequently, they grow only very slowly and cannot compete with species adapted to summer-rain climates. The latter species are adapted to utilising the short vegetative season and therefore show very rapid growth under favourable light conditions. If repeated burning is practised after cutting, grasses are especially favoured, particularly those grass types that branch only at their short-bases and in which the meristems are located at the soil surface so that they are not destroyed by fire (see p. 239). Among trees, those with thick bark are fire-resistant and also palms which, lacking cambium, are not easily destroyed by fire. Therefore, palms are much more frequent in such secondary communities. They are also preferred as ornamental trees on farms and plantations and are planted in town parks. In the absence of competition they grow well in many different habitats. This explains the popular opinion of tourists that palms are especially characteristic of the tropics, since tourists usually only see such cultivated landscapes.

Rapidly growing, herbaceous and shrubby weeds are spreading in the populated areas in a similar way to that in the temperate zone. The most widely spread species in south-east Asia—such as *Eupatorium pallescens*, *Ageratum conyzoides* and *Lantana camara*—have been introduced from tropical America. Also, on the Fiji Islands, there are to-day extensive areas covered with *Psidium guajava*, *Lantana*, and *Melastoma malabaricum*, amongst others, in place of the original forests.

Thus, the study of secondary forests and that of the different replacement communities in the humid tropics is an important task, since it is essential to stop the devastation of resources and the degradation of soils if the productive capacity of the tropical zone is to be preserved for the future. This is of the utmost importance to the entire human race, because the humid tropics, in particular, still represent an undisturbed reserve for nourishing the rapidly increasing world population. The highest yields per hectare of cultivated land could be attained in this ideal growth climate through rational management and appropriate use of fertilisers. However, land use and conservation should maintain a close relationship.

Up to the present, shifting cultivation has been the most widely spread form of land use in all forested and little-populated areas.[1] A stable cyclical equilibrium position can be developed, so long as the total area is rather large and the cultivated area involved at any given time rather small. Three factors are important determinants of the continual shifting of cultivated areas:

1. The degradation of the soil, 2. the abundance of weeds, 3. the occurrence of crop-pests (Popenoe, 1964; Phillips, 1964). Degradation of the soil is the main factor in areas with very acid, poor soils, e.g. in case of the white sandy soils in Guyana, but not in areas with young volcanic soils. Among weeds, the grasses play an important role. They

[1] Local terms for shifting cultivation are: taungya (Burma), chena (Ceylon), ladang (SE Asia), kaingin (Philippines), milpa (Central America), chitimene (Zambia).

become particularly prevalent in those forest areas that were more open to begin with. The effort of weeding an infertile soil is an inefficient procedure which bears no relation to the low crop-yields obtained. So, it is of greater advantage to clear and cultivate a new area than to weed the old area. The grass *Imperata cylindrica* is particularly troublesome in south-east Asia. In time, pests like birds, rodents, insects, etc., begin to congregate in the small patches of crops in the wilderness, where they destroy a large part of the crop. Therefore, once again it becomes expedient constantly to change and extend the cultivated area. Whenever shifting cultivation is discontinued because of increasing population density a changeover to modern agricultural practices, which includes soil cultivation against weeds, application of fertilisers and crop protection against pests, is a necessary requirement. At the same time, precautions against soil erosion have to be taken.

References

ANDERSON, J. A. R. 1961. *The ecology and forest types of peat swamp forests of Sarawak and Brunei in relation to their silviculture.* Ph.D. Thesis, University of Edinburgh.

BARUCHA, F. R., and NAVALKAR, B. S. 1942. Studies in the ecology of mangrove. *J. Univ. Bombay*, **9**, 93-100.

BIEBL, R., and KINZL, H. 1965. Blattbau und Salzhaushalt von *Laguncularia racemosa. Österr. Bot. Z.*, **112**, 56-93.

BLUMENSTOCK, D. I., FOSBERG, F. R., and JOHNSON, C. G. 1961. The resurvey of typhoon effects on Jaluit Atoll in the Marshall Islands. *Nature, Lond.*, **189**, 618-20.

BÜNNING, E. 1947. *In den Wälden Nordsumatras.* Bonn.

CHADWICK, M. J. 1960. Some observations of the ecology of *Eichhornia crassipes* Solms. *Univ. Khartoum 8th Ann., Rep. Hydrobiol. Res. Un.*, 23-28.

CHAPMAN, V. J. 1962. Respiration studies of mangrove seedlings, I and II. *Bull. Marine Sci. Gulf Carib.*, **12**, 137-67, 245-63.

CHAPMAN, V. J., and RONALDSON, J. W. 1958. The mangrove and salt marsh flats of the Auckland isthmus, New Zealand. *D.S.I.R. Bull.*, **125**.

DANSEREAU, P. 1947. Zonation et succession sur la restinga de Rio de Janeiro. *Rev. Canad. Biol.* (*Montreal*), **6**, 448-77.

DAU, L. 1960. Microclimas das Restingas do Sudeste do Brasil. *Arquiv. Mus. Nat.*, **50**, 185-236.

GAY, G. A. 1960. Ecological studies in *Eichhornia crassipes* Solms in the Sudan. *J. Ecol.*, **48**, 183-91.

GESSNER, F. 1955, 1959. *Hydrobotanik.* Vols 1 and 2. Berlin.

GESSNER, F., and HAMMER, L. 1962. Ökologisch-physiologische Untersuchungen an den Podostemonaceen des Caroni. *Int. Rev. ges. Hydrobiol.*, **47**, 497-541.

GLEDHILL, D. 1963. The ecology of the Aberdeen Creek Mangrove Swamp. *J. Ecol.*, **51**, 693-703.

GOODING, E. G. B. 1947. Observations on the sand dunes of Barbados, British West Indies. *J. Ecol.*, **34**, 111-125.

GOEBEL, K. 1886. Über die Luftwurzeln von *Sonneratia*. *Ber. Dtsch. Bot. Ges.*, **4**, 249-55.

HARRIS, J. A. and LAWRENCE, J. V. 1917. The osmotic concentration of the sap of the leaves of mangrove trees. *Biol. Bull. Mar. Boil. Lab.*, Woods Hole, **32**, 202-11.

HARRIS, J. A. 1932. *The physico-chemical properties of plant saps in relation to phytogeography*. Minneapolis.

HARRISON, M. N., and JACKSON, J. K. 1958. Ecological classification of the vegetation of the Sudan. *Forest. Bull., Khartoum*, **2**.

HESSMER, H. 1966. *Der Kombinierte land- und forstwirtschaftliche Anbau.* I. *Tropisches Afrika.* Stuttgart.

KIPP-GOLLER, A. 1939/40. Über Bau und Entwicklung der viviparen Mangrovenkeimlinge. *Z. Bot.*, **35**, 1-40.

LAMPRECHT, H. 1961. Tropenwälder und tropische Waldwirtschaft. *Beih. Z. Schweiz. Forstver.* (*Zurich*), **32**, 1-110.

LAMPRECHT, H. 1964a. Europäischer Waldbau und Waldbau in den Tropen. *Der Forst- und Holzwirt*, **19**, 433-6.

LAMPRECHT, H. 1964b. Über Waldbau in tropischen Entwicklungsländern. *Schweiz Z. Forstwesen.*, **4**, 211-27.

LIND, E. M., and VISSER, S. A. 1962. A study of a swamp at the north end of Lake Victoria. *J. Ecol.*, **50**, 599-613.

LÖTSCHERT, W. 1958. Die Übereinstimmung von geologischer Unterlage und Vegetation in der Sierra de los Organos (Westcuba). *Ber. Dtsch. Bot. Ges.*, **71**, 55-70.

LÖTSCHERT, W. 1959. Vegetation und Standortsklima in El Salvador. *Botan. Studien.*, **10**.

MACNAE, W. 1963. Mangrove swamps in South Africa. *J. Ecol.*, **51**, 1-25.

MACNAE, W., and KALK, M. 1962. The ecology of the mangrove swamps at Ithaca Island, Moçambique. *J. Ecol.*, **50**, 19-34.

MENSCHING, H. 1963. Nordafrika. In *Die Grosse Illustrierte Länderkunde.* Vol. 2.

MIGAHID, A. M. 1948. *Report on a botanical excursion to the Sudd Region*, I and II. Cairo Univ. Press.

MIGAHID, A. M. 1952a. *Velocity of water current and its relation to swamp vegetation in the Sudd Region of the Upper Nile.* Cairo Univ. Press.

MIGAHID, A. M. 1952b. *Further observations on the flow and loss of water in Sudd swamps.* Cairo Univ. Press.

MIGAHID, A. M., ABDEL RAHMAN, A. A., and EL SHAFFEI ALI, M. 1955. An ecological study of a swampy island in the Upper Nile. *Arabic Congress*, pp. 677-718.

PANNIER, G. 1959. El efecto de destintas concentraciones salinas sobre el disarrollo de *Rhizophora mangle* L. *Acta Cient. Venezolana*, **10**, 68-78.

PANNIER, F. 1962. Estudio fisiologico sobre la viviparia de *Rhizophora mangle* L. *Acta Cient. Venezolana*, **6**(13), 184-97.

PHILLIPS, J. 1964. Shifting cultivation. *I.U.C.N. publ., N.S.*, **4**, 210-20.

POPENOE, H. 1964. The pre-industrial cultivator in the tropics. *I.U.C.N. publ., N.S.*, **4**, 66-73.

RICHARDS, P. W. 1952. *The tropical rain forest.* Cambridge.

Ruttner, F. 1940. Die Wasservegetation einiger Seen Niederländisch-Indiens. *Veg. Bilder.*, **25**, Plates 37-42.

Sauer, J. D. 1962. Effect of recent tropical cyclones on the coastal vegetation of Mauritius. *J. Ecol.*, **50**, 275-90.

Schimper, A. F. W. and Faber, F. C. von 1935. *Pflanzengeographie auf physiologischer Grundlage.* 3rd ed. Jena.

Scholander, P. F., van Dam, L., and Scholander, S. I. 1955. Gas exchange in the roots of mangroves. *Amer. J. Bot.*, **42**, 92-98.

Scholander, P. F., Hammel, H. T., and Garey, W. 1962. The salt balance in mangroves. *Plant Physiol.*, **37**, 722-29.

Scholander, P. F. *et al.* 1965. Sap pressure in vascular plants. *Science*, **148**, 339-46.

Scholander, P. F. *et al.* 1966. Sap concentrations in halophytes and some other plants. *Plant Physiol.*, **41**, 529-32.

Scholander, P. F. 1968. How mangroves desalinate seawater. *Physiol. Plant.*, **21**, 251-261.

Sen-Gupta, J. 1935. Die osmotischen Werte bei einigen Pflanzen in Bengal (Indien). *Ber. Dtsch. Bot. Ges.*, **53**, 783-95.

Sioli, H. 1947. Sedimentation im Amazonasgebiet. *Geolog. Rundschau*, **45**, 608-35.

Stern, W. L., and Voigt, G. K. 1959. Effect of salt concentration on growth of red mangroves in cultures. *Bot. Gaz.*, **121**, 36-39.

Straka, H. 1960. Über Moore und Torf auf Madagaskar und den Maskarenen. *Erdkunde*, **14**, 81-98.

Takeuchi, M. 1962. The structure of the Amazonian vegetation, VI. *J. Fac. Sci. Univ. Tokyo, Sec. III*, **8**, 297-304.

Troll, W., and Dragendorff, V. 1931. Über die Luftwurzeln von *Sonneratia* und ihre biologische Bedeutung. *Planta*, **13**, 311-413.

Walter, H., and Steiner, M. 1936. Die Ökologie der Ostafrikanischen Mangroven. *Z. Bot.*, **30**, 65-193.

Wasawo, D. P. S. 1964. Some problems of Uganda swamps. *I.U.C.N. Publ., N.S.*, **4**, 196-204.

Westmaier, M. 1900. *Zur Kenntnis der Pneumatophoren.* Freiburg-Schweiz.

Wild, H. 1961. Harmful aquatic plants in Afrika. *Kirkia*, **2**, 1-66.

Wilhelmy, H. 1957. Das Grosse Pantanal in Mato Grosso. *Verh. dt. Geogrtags*, **31**, 45-71.

Wilhelmy, H. 1958a. *Dtsch. Geogr. Tag Würzburg*, **18**, 555-59.

Wilhelmy, H. 1958b. Umlaufseen und Dammuferseen tropischer Tieflandflüsse. *Ann. Geomorph.*, **2**, 27-54.

IV. *Cooler Rain Forests of Higher Altitudes in Tropical Mountains*

1. General features of the altitudinal vegetation belts in the tropics

In mountainous areas the tropical evergreen rain forest develops most richly on the lowest slopes. Here the climate is uniformly warm and, at the same time, continually humid because of cloud immobility, yet the soil is well drained. On the level terrain of the lowlands the relative humidity may decrease much more and between showers waterlogging can occur because of limited run-off during periods of high rainfall.

The forest on level terrain near mountains differs, therefore, from the lower forest on steep slopes by a notable reduction in lianas and epiphytes; it also differs because of the appearance of palms, which are well adapted to alternately moist and dry soils, and periodic flooding. With increasing soil moisture, *Pandanus* species become more frequent.

To-day, these lowland forests have almost entirely disappeared and the land has been taken into cultivation for sugar cane, rice or other crop plants. More precise ecological investigations, especially of the water relations and soil aeration during the rainy period, are not available.

The rainfall increases with increasing altitude beyond the submontane belt up to the cloud zone, while the mean annual temperature simultaneously decreases. The cloud zone is the wettest belt and also shows the least daily fluctuations in temperature. In all tropical mountain ranges we find this characteristic altitudinal belt of 'mist forest' (Nebelwald) (Lamprecht, 1954). These evergreen tropical montane forests are physiognomically not very different from the typical tropical rain forests, but they differ floristically, in part, by the composition of their tree species. The number of tree ferns, which at lower altitudes are restricted more or less to deeply cleft valleys, increases. The number of lianas and epiphytes is very great but among the latter ferns and lycopods predominate with increasing altitude. Hymenophyllaceae and mosses are especially well represented here and in more open places the soil is covered with *Selaginella*. Commonly, *Bambusa* species form an impenetrable undergrowth in the wettest altitudinal belt.

Rainfall decreases at still higher elevations above 2500-3000 m. The average daily temperature decreases further, the daily temperature fluctuations become more pronounced and the minima approach the freezing point gradually. The forest now undergoes a pronounced change.

The foliage of the evergreen trees becomes more xeromorphic; orchids disappear from the epiphytes and the tree crowns become covered with bearded lichens instead. This is the subalpine tropical rain forest belt. A little higher up, the temperatures fall below freezing point. Since it is still warm during the day at this altitude and since the climatic conditions remain uniform throughout the year, the number of days with alternating frost increase suddenly (at about 3500-4000 m). Temperatures of 1-4°C are, after all, deadly for tropical forest plants (Biebl, 1964b).

This altitude is well beyond the tree-line and we are now in the alpine belt. It has not been determined with certainty whether the tree-line is controlled by the occurrence of alternating frost days in tropical mountains, or whether the rainfall, which rapidly decreases here, is also responsible. The fact that, in tropical mountains, forest reaches higher up in corries than on ridges seems to indicate that moisture also plays a role; otherwise, one would expect that drainage of cold air in the corries would depress the tree-limit, as is generally found in the Alps. The soil types also change with increasing elevation. The lateritic loams of the lower belt, without a humus horizon, first change in the submontane belt into reddish and then into yellow loams with pronounced mull horizons, as the clay content decreases. The soil is already somewhat podzolic in the montane belt and higher up real podzols with raw humus horizons develop. The bleached and leached $\mathbf{A}_2$ horizon is usually very deep and the **B** horizon is but little developed. In the upper montane, perhumid cloud zone, the podzols are replaced by peaty gley-soils. The peat can attain a thickness of 25 cm and more on level sites. Its reaction is very acid ($pH < 3$).

Raw humus and peat formation are also common in the drier, but cooler subalpine belt. Such formations extend far up into the alpine belt. Stony rankers occur only at altitudes where the plant cover is reduced to isolated cushions. Here, solifluction is noticeable. Soil type formation differs of course with the parent material and slope.

Quite a different change occurs in the tropical evergreen rain forest at low altitudes with increasing distance from the equator. It is related to the change of climate either from the constantly humid, equatorial zone, or a zone with two short, dry periods, respectively, to the tropical summer-rain zone. Here, a period of drought becomes increasingly pronounced at latitudes of more than 10°. One observes that the evergreen tree species of the upper stratum, which are most strongly exposed to drought, become replaced by tree species which shed their leaves during the dry period. However, evergreen species still dominate the lower tree stratum. Accordingly, these are called tropical semi-evergreen forests. With increasing duration of the period of drought the forests become composed entirely of species which remain barren during the dry period. They resemble in appearance the European deciduous forests in winter. These are called tropical rain-green (or deciduous) forests or dry forests. These types will be discussed in more detail in Chapter V.

Here I will discuss only the rain forests which occur at higher latitudes in climates that remain humid, but become cooler. Such conditions are found, for example, on the east coast of Australia and South America where a certain parallelism is shown to the altitudinal vegetational types of the equatorial zone (cf. Fig. 50, southern hemisphere), but characteristic differences also are shown. A diurnal climatic fluctuation is characteristic of the altitudinal vegetation belts in the equatorial zone, whereas the temperature-conditioned annual seasons become more and more pronounced with increasing latitude. A gradual change occurs from the tropical evergreen rain forest to the subtropical, and finally to the temperate, evergreen rain forest.[1]

The following scheme summarises the changes described, in a horizontal direction, and those for the equatorial zone in the vertical direction (vertical arrows imply decreasing mean temperature):

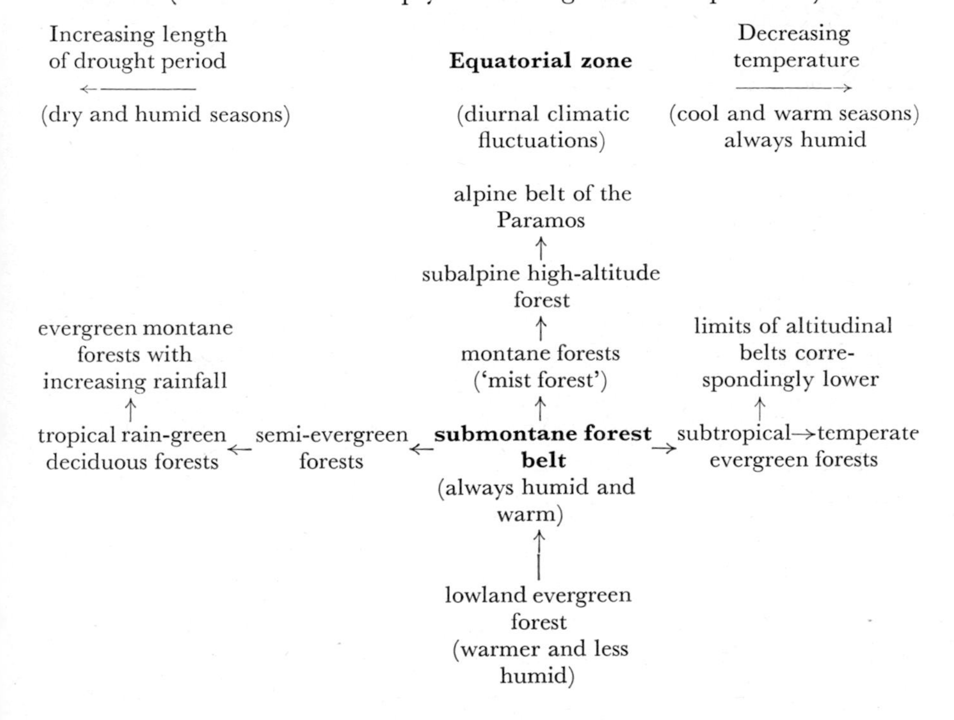

2. Evergreen subtropical rain forests

As an example of the changes which occur in evergreen rain forests with increasing latitude, I have selected the east coast of Australia with

[1] These two types occur also on the east coast of Africa, but lack any direct connection with the tropical evergreen rain forest, as the latter is formed only as islands on mountain slopes because of the dry monsoons.

Tasmania and New Zealand. The temperature relations change notably from north to south. In the tropical area (north Queensland) the mean annual temperature is 23·2-24·7°C, in the subtropical area (south Queensland, New South Wales), 17·4-19·3°C, and in the temperate area (south Victoria, Tasmania), 12·2-13·7°C. The average daily minima of the coldest month and the absolute minima (in brackets) are, in the tropical area 14·2-16·2°C (4·4-6·1°), in the subtropical area 6·9-8·4°C (−2·8 to +2·1°) and in the temperate area 4·7-6·8°C (−3·3 to −1·1°).

Physiognomically, the subtropical rain forest shows little difference from the tropical rain forest (Webb, 1959; Baur, 1957). In place of tree species with large leaves those with medium sized leaves occur and lianas and epiphytes are less frequent. The species present in the tropical and subtropical rain forest do not belong to the Australian floristic element but to the Indomalayan and, therefore, to the palaeotropical flora. A change in character occurs only with appearance of conifers (*Araucaria bidwillii*, *A. cunninghamii*) while the tree ferns become more frequent. The Kauri Pine (*Agathis*) which occurs only occasionally in north Queensland is dominant in the most northerly and subtropical part of New Zealand. In Borneo this tree is characteristic of the montane belt. *Agathis* produces in New Zealand the valuable Kauri resin. Since much resin has been found in subfossil form as large masses in the soil, the forests must have been largely destroyed for the easier harvesting of the resin, and they have been degraded to an unfertile heath through repeated burning. *Podocarpus* species, which are so typical of the tropical subalpine forests, are also well represented in New Zealand; in addition there are, among conifers, *Dacrydium* and *Phyllocladus* species together with many Melanesian elements (Fig. 104).

The subtropical character of these forests with many epiphytes and lianas (*Metrosideros*, *Freycinetia*) and, in the north, palms in addition is retained well, in spite of the fact that the climate is virtually temperate (Dawson, 1962).

However, the physiognomy of the forest changes entirely as soon as one enters the temperate evergreen forest, in which floristic elements of the antarctic are present. A dominant representative is, in particular, the evergreen but very small-leaved genus *Nothofagus*. An additional characteristic is that in these forests only one or few species predominate.

The distribution of the forests with subtropical character and that of the *Nothofagus* forests cannot be explained climatically. The latter are scattered as islands all over the volcanic plateau of the North Island and become more frequent to the south; however, they are entirely absent in certain areas on the South Island. The mosaic distribution of the two vegetation types in New Zealand seems to indicate that the present vegetation is not in an equilibrium with the climate. A strong glaciation on the South Island and an intensive volcanic activity on the North Island have prevented repeatedly the attainment of an equilibrium condition.

The pioneers upon reinvasion were Podocarps. These become slowly replaced by the subtropical broad-leaved forest, wherein an important role is played by *Metrosideros robusta* (*Myrtaceae*) as strangler on *Dacrydium*. Today the Podocarpaceae are often mixed with subtropical elements. The original Podocarp forest is preserved in relatively pure form only in the mountains. However, the *Nothofagus* species appear to compete more successfully so that they will probably gain in distribution in time (Robbins, 1962). The vegetation relations in New Zealand

FIG. 104. *Beilschmiedia tawa* forest with subtropical elements on the North Island of New Zealand, volcanic plateau at 600 m elevation (Phot. L. Cockayne).

are difficult to understand ecologically. It will always be necessary to consider their historical aspects.

The character and floristics of the New Zealand *Nothofagus* forests corresponds very closely to those of the West Patagonian forests in South Chile, which will not be discussed within the frame of this book. In the southern hemisphere, subtropical evergreen forests occur at the SE margin of Africa, at the east coast of South Brazil and on the east slope of the Andes at Tucuman. It is very surprising to suddenly find above the latter as the next higher belt (above 1200 m) deciduous forests (Hueck, 1954a, b; 1966), which are related to northern hemisphere taxa (*Alnus*, *Juglans*, *Sambucus*, *Prunus*, *Berberis*). These conditions are otherwise found only in the subtropical mountains of the northern hemisphere (Mexico, southern foot of the Himalayas, SE-Asia, South Japan.)

From all these areas there are good vegetation descriptions available, but no ecological investigations. So I only state this observation without further development.

3. Tropical montane rain forests of Java

The rain forests at low altitudes have been eliminated on Java, therefore, all the investigations done in this area refer to the montane rain forest.

The Primeval Forest Laboratory, Tjibodas is situated at an altitude of 1500 m on the slope of the Gede volcano above the Botanic Garden of Buitenzorg, now called Bogor.

Nearly untouched, primary virgin forest is only found at high altitudes in Java at Tjibodas, so this area has been more closely investigated. It is floristically well known through the works of Koorders (1918-23) (from Schimper-Faber, 1935). The forest investigated belongs to the lower montane belt which is very rich in epiphytic orchids and lianas but which also includes tree ferns, epiphytic lycopods, Hymenophyllaceae and many mosses.

Koorders (1914) cited a total of 575 species of flowering plants. Of these 165 (= 29%) were trees, 350 (= 61%) shrubs and herbs and 60 (= 10%) climbers.

The rain forest at this altitude, which is extremely humid (see p. 86, Fig. 62) is still very rich in trees that attain heights of up to 600 m, e.g. the Rasamala tree (*Altingia excelsa*). The utilisation of available space is so complete that Junghuhn even spoke of 'horror vacui'. Lianas occupy the trunks, epiphytes cover the stems and branches or occupy their branch forks (Fig. 81). Epiphytes with very pretty flowers are *Medinilla* species and *Rhododendron javanicum*. Above the forest limit the latter also grows on the ground. Among the tall herbaceous species wild bananas (*Musa* spp.) occur and near the ground several begonias.

Meijer (1959) has recently analysed a one-hectare area of entirely undisturbed forest on level terrain at 1500 m. He identified and counted all trees over 10 cm diameter at breast height. Within this sample plot he found 333 species of spermatophytes and ferns. Of these, 78 were trees, 40 shrubs, 30 lianas, 10 climbers (= creepers), about 100 epiphytes and 73 terrestrial herbs. Only 59 tree species had a diameter of more than 10 cm. Of these he measured 284 tree trunks.

TABLE 32

Distribution of 284 trunks by diameter classes (cm)

10–20	21–30	31–40	41–50	51–60	61–70	71–80	>81
121	44	39	21	16	20	20	3

Nineteen tree species failed to reach a diameter of 10 cm, of these 13 were still immature. Half of the species were represented by only one individual, the most abundant species by 33 individuals. The maximum height of the trees was 60 m. The upper tree layer was dominated by *Altingia excelsa*, *Castanea javanica* and *Schima wallachii*; less frequent species in this layer were *Quercus induta* and *Q. pseudomoluccana*; *Engelhardtia spicata* was rare. Seedlings were found of all the tree species. Regeneration can take place only if there is a gap in the crown canopy. Species of secondary forest occur wherever large canopy openings are formed. The maximum age was estimated as 200-250 years, the average age of the trees measured was estimated as 130 years. Annual mortality amounted to 0·77-0·83%.

The mean temperature in the area investigated is 17·7°C (limits 12-27°C), humidity is between 80-90%, and the annual rainfall amounts to 3400 mm with a maximal dry period of 3 weeks. The soils are volcanic and much enriched with humus in the upper 20-30 cm. Bedrock occurs at a depth of 70-80 cm. The richness of the soil is indicated by *Cyrtandra*, *Elatostema*, *Begonia* and *Strobilanthes*. These genera are absent or very sparsely represented on poor tropical forest soils.

It is characteristic of the montane forests of south-east Asia that there are a number of families (Fagaceae) or genera (*Rhododendron*, *Vaccinium*, *Pinus*), which have their origin in more northern areas. Some of these occur only at higher altitudes (above 2000-2500 m), for example *Androsace*, *Primula*, Ranunculaceae (van Steenis, 1962). Representatives of this group are also found on the highest mountains of Australia and New Zealand but *Rhododendron* is found only in one Australian locality in north Queensland.

Soil temperatures of −10°C were observed at 2000 m in grass-covered areas but not in forests. Once clear felled and converted to grassland, plateaus at high altitudes become pronounced frost pockets. Frost-sensitive seedlings cannot become established in these circumstances and as a result, such areas may remain treeless forever.

The other altitudinal belts are not typically developed because the mountains on Java are young volcanic domes, which are not yet high enough to extend beyond the climatic limit of the timber line. The peaks are, however, treeless but the reason for this is entirely edaphic. The volcanic materials form rankers with pioneer communities with representatives of certain alpine species. These include *Anaphalis javanica*, the growth of which has been observed by van Leeuwen over many years on the summit of Pangarango (3000 m). Seedlings required 13 years in order to attain a height of 20 cm. He estimated the age of individuals with a stem of 15 cm diameter to be at least 100 years (see Schimper-Faber, 1935).

The altitudinal belts were classified by van Steenis (after Troll, 1959)[1] (Table 33).

4. Tropical alpine vegetation of the Andes

The summits of the Andes in the vicinity of the equator reach far into the zone of perpetual snow. The alpine vegetation is here known as

[1] The author supplied me with a series of figures from his work, which is gratefully acknowledged.

'Paramos'. Unfortunately, there are no laboratory facilities for ecological work. However, temperature measurements have been recorded over many years in the Andes of south Peru (see Walter, 1960, Figs. 35-36). Daily night frosts are found throughout nearly the whole year at a little below 4000 m.

TABLE 33

Altitudinal belts in the mountains of Indonesia

Snow belt	above 4500 m snow limit	Perpetual snow
Alpine belt	4000–4500 m tree line	Stone desert with mosses and lichens, and only a few woody species, particularly grasses and sedges
Subalpine belt	3600–4000 m forest border	low shrubs, isolated or in groups, also conifers
	2400–3600 m	dense shrub forest with a few, scattered larger trees, often moss-covered, conifers
Montane belt	1500–2400 m	closed high-forest, decrease in mosses below 2000 m
Submontane belt	1000–1500 m	closed high-forest, poor in mosses

A tabular summary of the temperature relations in the alpine zone of Ecuador by R. Espinosa is shown in C. Troll's (1959) work (Table 34).

TABLE 34

Temperatures at different altitudes in the Paramos of Ecuador (after R. Espinosa)

Station	Height above sea level m	Period of observation	Highest monthly means	Lowest monthly means	Abs. max.	Abs. min.	Max. daily range	Min. daily range
Cotopaxi	3600	VI. 1930 -IX. 1931	7·5° (II)	5·4° (VII)	17·3° (XII)	−1·5° (VIII)	14·9° (VIII 1931)	2·2° (VIII 1930)
Cruz Loma	3950	V.-X. 1931	6·8° (V)	5·9° (IX)	14·0° (X)	1·5° (VIII)	10·2° (X)	2·6° (X)
Gomessiat I	4450	VI.-X. 1931	4·5° (VI)	2·1° (VIII)	12·7° (VI)	−1·8° (VIII)	9·2° (IX)	1·1° (X)
Gomessiat II	4720	VII.-X. 1931	0·9°	0·7°	3·5° (VII)	−2·0° (IX)	4·3° (IX)	0·6° (VII)

Paramo vegetation occurs in the high mountains of Central and South America from 10° N to about 8° (15°) S. The most northern Paramos of

FIG. 105. *Lobelia telekii*, a woolly-candle plant, and tree-like *Senecio keniodendron* on Mt. Kenya (East Africa) at 4100 m (after C. Troll, 1959).

FIG. 106. Paramo vegetation in the alpine zone at Bogota in Colombia at about 3500 m. *Espeletia grandiflora* in the *Calamagrostis effusa* grassland (Phot. C. Troll).

Costa Rica have been investigated by H. Weber (1958). He emphasised the continuously wet and cool climate, which leads to strong humus accumulation with a pH of 3·8-4·0. Tall bunch grasses are characteristic of the Paramos, as well as of the alpine belt in New Zealand, where they are known as 'tussock' grasses. These are about 0·5-1 m or more tall and have about the same diameter. They look much like sedge hummocks. The leaves are sclerophyllous and persist after turning yellow. Thus the

FIG. 107. *Lupinus alopecuroides* in the Paramo of the Chimborazo (Ecuador) at 4800 m (Phot. H. Weber) (after C. Troll, 1959).

tussock grassland never looks green and at times it appears completely yellow. The tussock grasses include species of *Festuca*, *Deschampsia*, *Danthonia*, *Stipa*, *Calamagrostis*, and *Andropogon*. Equally characteristic of the equatorial alpine zone are the unique tree forms, which in Africa are represented by the giant Senecios (Fig. 105), in South America by the Espeletias (Fig. 106), and in Indonesia by *Anaphalis*. These are all Compositae. In addition there are the stem-forming ferns and the woolly-candle plants (Figs. 105 and 107).

The character of the mountain vegetation changes somewhat farther from the equator. Seasons become noticeable, the temperature ranges increase and may reach 50°C as shown in Table 35 after Troll

Table 35

Temperatures of high-altitude stations of the Paramo-Puna-Andes (after Knoch, Stenz and Cabrera, cited from C. Troll)

	Station and latitude	Height above sea level (m)	Mean annual temp.	Annual temp. range	Absolute max.	Absolute min.	Absolute temp. range	Daily periodic range Max.	Daily periodic range Min.
Populated areas	Quito 0° 14′	2850	12·6°	0·4°	24°	3°	21°	14·8° (VIII)	11·4° (IV)
	Cuzco 13° 27′	3380	10·0°	3·6°	23°	−2°	25°	16·3° (VII)	10·9° (II)
	La Paz 16° 30′	3690	9·3°	4·6°	24·2°	−2·7°	26·9°	15·8° (VIII)	12·1° (I)
	Oruro 17° 38′	3706	8·3°	6·7°	23·5°	−10·1°	33·6°	17·2° (VIII)	9·6° (I)
	La Quiaca 22° 10′	3458	9·5°	9·6°	30·7°	−18·0°	48·7°	26·6° (VIII)	18·7° (I)
Puna	Vincocaya	4380	1·9°	7·1°	19°	−22°	41°	24·1° (VIII)	14·4° (II)
	Mina Aguilar / Tres Cruces	4600	3·1°	11·0°	16·3°	−14·2°	30·5°	—	—

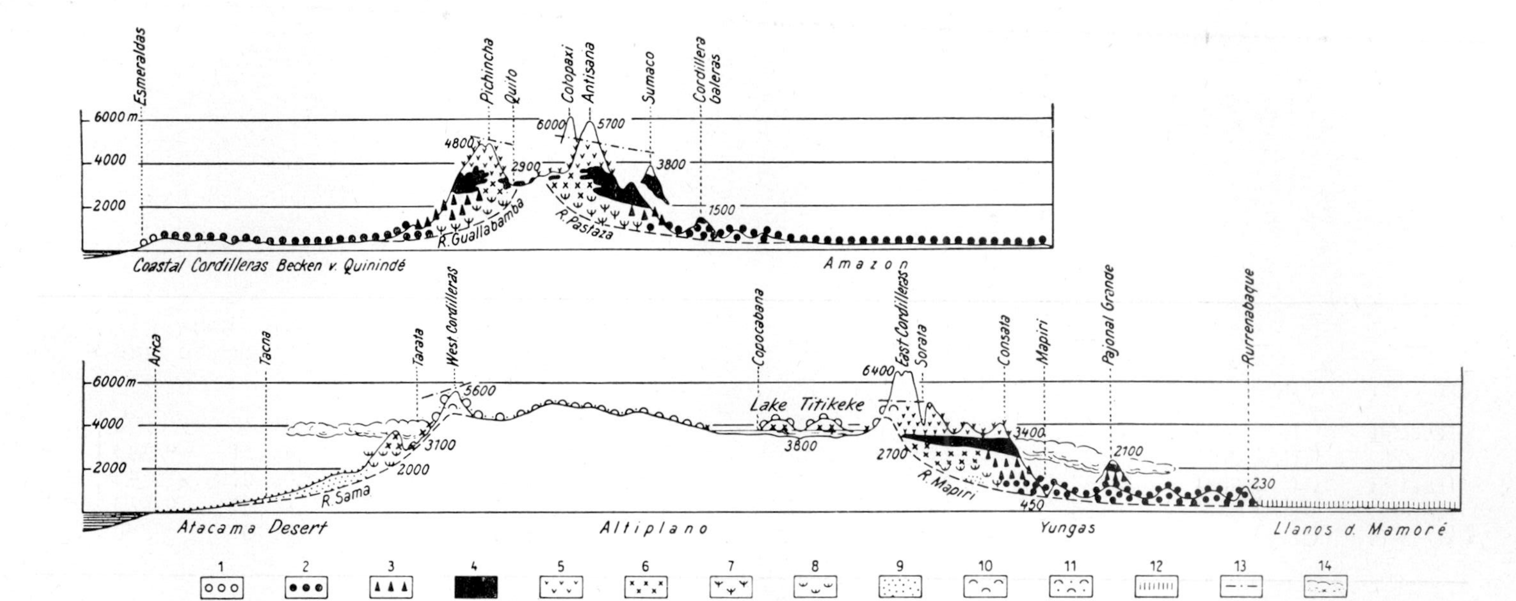

FIG. 108. Paramos (5, see key) above cloud forest and Puna (10 and 11) in the drier part of the High Andes. Upper vegetation profile on the equator, the lower one at 16-18°S at the height of Lake Titicaca (see Fig. 109). After C. Troll (1959).

1. Semi-evergreen rain forest; 2. Tropical rain forest; 3. Tropical evergreen mountain forest of the Tierra templada; 4. High and cloud forest of the Tierra fria (Ceja de la Montana); 5. Paramos; 6. Mesophytic shrubs and *Polylepis* woods; 7. Thorn and succulent shrubs of the Low-Valles; 8. Thorn and succulent shrubs of the High-Valles; 9. Desert; 10. Moist Puna; 11. Dry and thorn Puna; 12. Savannahs of the hot lowland; 13. Snow limit, 14. Fog and cloud belt.

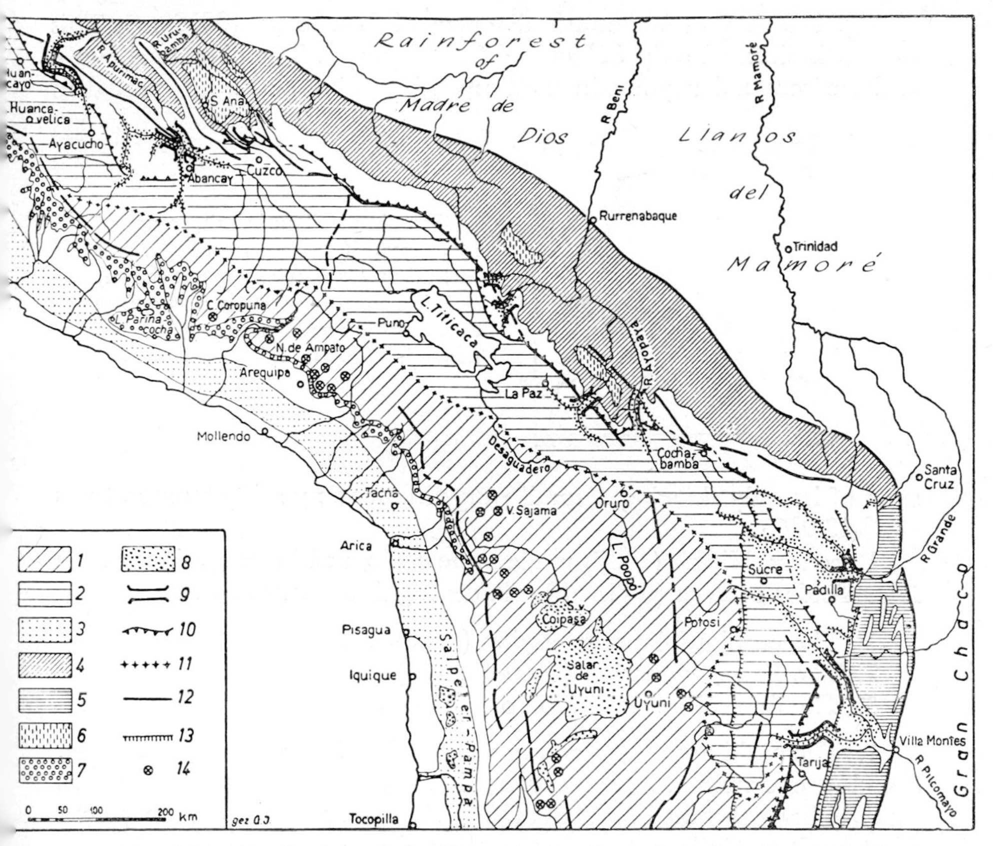

Fig. 109. Distribution of the Puna in the Central Andes around Lake Titicaca (after Troll, 1959).

1. Zone of the dry and salt Puna; 2. Zone of the moist Puna; 3. Desert plains (pampas) of Atacama; 4. Evergreen and high-forests (Montana) of the east Andian slopes; 5. Summergreen, changing upwards into evergreen, tropical/subtropical transition forests (Tucumano-Bolivian forest); 6. Savannah islands in the Montana (Medio Yungas); 7. Mesophytic shrub belt on the west slopes of the Puna-Andes; 8. Hot-dry valleys of the east Andian slopes (Low-Valles); 9. Channels of the valleys which cut through at their lowest points (wind 'gates' permitting passage of counterbalancing easterly winds); 10. Main climatic boundary of the Andes (western border of the ascending rains and fogs that spread from the continuously humid vegetation located to the east); 11. Eastern boundary of the *Lepidophyllum* heath vegetation; 12. Important mountain ranges and eastern foot hil's of the Andes; 13. Downward slope of the Puna block in Southeast Bolivia; 14. Tall cylindrical mountains.

(1959). In addition dry periods occur and the period of summer rain decreases. The vegetation becomes correspondingly, more and more xeromorphic in character and finally turns into thorn-succulent deserts and deserts with saline soils. This series can best be observed in the South American Andes. There, the drier alpine formations are known as 'Puna', in contrast to the 'Paramos' (Figs. 108 and 109).

Fig. 110 shows the climatic diagram of Huancayo (3353 m). This station is situated only 12° south of the equator. The climate shows a definite seasonal rhythm in the form of a rainy and a dry period in

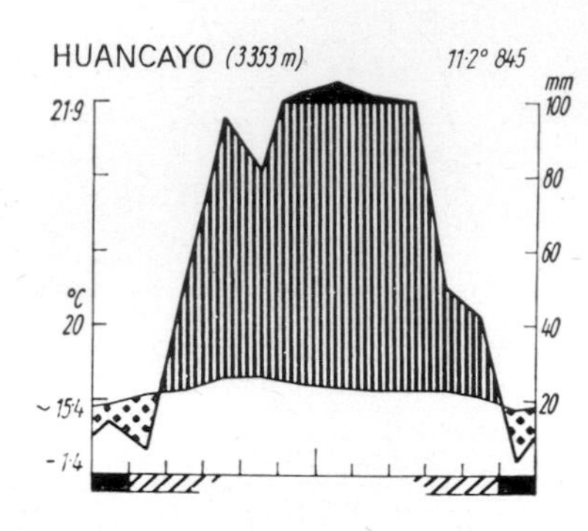

FIG. 110. Climatic diagram of Huancayo (3353 m); the lowest monthly temperatures are not shown, so it is uncertain whether frost occurs during the warm season (October through March).

spite of its only showing a small range in mean monthly temperatures (8·9°-13·0°C).

Recently, Hirsch (1957) made some ecological investigations in the High Andes of Peru (8°-15° S). He distinguished three zones:

1. The cushion-plant vegetation (4200-5100 m)
2. The Puna vegetation (4000-4500 m)
3. The dry-valley vegetation of the deep valleys

1. The cushion-plant vegetation is restricted to wet marginal areas. Stands are formed of interesting cushion-plants whose morphology has recently been classified by Rauh (1939). These are species of the genera *Plantago*, *Pycnophyllum*, *Azorella* and *Werneria*.

The climate is characterised by daily night frost but there is no season without growth. The soil surface temperature increases rapidly in the mornings and remains at this level during the day and only declines at 1600 hr. The air temperature around the cushion plants shows a similar pattern. Air humidity near the soil decreases in a corresponding manner to nearly 60%. However, the saturation deficits are never very high (not even 5 mm) because of the low temperatures.

Plantago rigida, which grows in the form of dense, low cushions of 50 cm height was investigated. Its leaves are not particularly tough and the stomata occur on the upper surface of the leaves. A second plant investigated was a *Pycnophyllum* sp. (Caryophyllaceae), which forms globose cushions up to 1 m diameter resembling those of certain mosses.

In the case of *Plantago*, maximal transpiration around noon amounted to 800 mg/dm^2/h, in the case of *Pycnophyllum* the amount was only half this. Hirsch believed the transpiration curve to indicate restricted transpiration during the forenoon. However, since he made only single measurements, it appears equally likely to be related to random variations. A restriction in the forenoon appears highly improbable since the noon values are nearly four times as high.

2. The Puna vegetation covers the entire uplands within the borders of the Andes. Precipitation in this area is less and the dry period causes an interruption of the growth period. The vegetation is open. Stands are formed by bunch grasses, such as *Calamagrostis*, *Festuca* and *Stipa* species. Additional small herbs occur during the rainy period. However, the succulent cacti are particularly characteristic. The soil consists of a humus layer up to 1 m deep, which is readily subject to surface desiccation. The surface is heated considerably, up to 38°C during the day, while the air humidity decreases to 28%, and the saturation deficit amounts to nearly 10 mm. Unfortunately, there are no evaporation records. At night, the conditions are similar to those in the cushion-plant zone.

Transpiration measurements were only carried out with cacti, i.e. the cushion-forming *Tephrocactus floccosus*, and *Oroya borchersii*, a globular cactus of diameter 20 cm. *Tephrocactus* is covered with a dense, long-woolly felt of hair and with thorns. *Oroya* is hairless, but it has thorns. The plants showed very high surface temperatures (up to 24°C) during the day, but *Tephrocactus* was heated more strongly than *Oroya*. In spite of this, the transpiration rate of *Tephrocactus* did not even attain 150 $mg/dm^2/h$, probably because of the hair-felt, while that of *Oroya* was twice as high. In general, it was very small as is always the case with cacti.

3. The dry-valley vegetation is restricted to the stony, deep valleys of desert-like character. At the lowest altitude the temperature always remains above 0°C. The air humidity near the ground decreases from 100% at night to 17% around noon. At that time the saturation deficit amounts to 39 mm.

Three entirely different plant types were used for transpiration measurements: a low-growing *Acacia*, *Tillandsia paleacea* and *Opuntia subulata*. The results showed correspondingly great differences.

Acacia showed a maximum transpiration rate of more than 700 $mg/dm^2/hr$, but the rate decreased to less than 100 $mg/dm^2/hr$ at noon, when conditions were extreme. *Tillandsia* transpired only half the amount and showed a less extreme decrease, while *Opuntia* only transpired a little, as one would expect. The shallow roots of the Tillandsias and Opuntias were unable to extract any water from the soil, but *Tillandsia* can probably absorb moisture with its 'suction scales' from the night dew. *Acacia* roots more deeply in the soil.

Ellenberg (1958) believes that the Paramos of Peru represent anthropogenically degraded stages of a natural forest belt.

Support for this assumption is shown in Fig. 50, where the upper boundary of *Polylepis* is indicated as between 10 and 30° S. The species of this genus of Rosaceae are either tree-like, or resemble the 'mountain pines' of the Alps. They differ from other tree species in that they are insensitive to daily night frost. Stands are found in scattered patches up to 4500 m, so that these are not usually related to the actual tree line (Hueck, 1962). In contrast to the settlement pattern in the European Alps, the lower alpine belt in the Andes has been covered, in part, with settlements for thousands of years. *Polylepis* provides the only source of firewood for the population. Even the mines in this part of the country depend on this wood. In addition, the copper-red bark of *Polylepis* is utilised for tanning purposes. It is, therefore, possible that the present-day stands represent remnants of a formerly more widely and uniformly distributed forest,[1] particularly since wood increments are very small at these elevations. The annual temperature of 6·9-2·8°C in combination with the occurrence of daily night frosts means there is relatively little warmth. Rainfall probably varies between 600-800 mm. If the upper boundary of *Polylepis* in Fig. 50 is considered to represent the upper tree limit, the asymmetrical pattern of the tree line on the northern and southern hemisphere would disappear and only the depression expected in the moist equatorial region of the actual Paramos would remain.

Troll (1959) suggested using the South American name 'Paramos' in general for the alpine belt in equatorial regions of the southern hemisphere as this belt neither corresponds climatically nor physiognomically to the alpine belt of the Alps. However, the term 'alpine' is nowadays understood as referring only to the belt between the tree- and snow-limit. This belt is different in all climatic zones, e.g. in the north African mountains and in the mountains of Asia Minor. The only difficulty exists in classifying the shrub zone above the forest belt, which is well marked in most regions, but which, in certain areas, is represented by a forest that decreases in height very gradually, as for example in New Zealand. Clearly defined tussock grassland only begins above this zone. In such cases one should not approach classification dogmatically for one cannot accommodate all altitudinal sequences of the world into one system. However, the scheme of C. Troll with its excellent comparative description of the tropical mountains gives a very good guide (Fig. 111). This scheme as well as Fig. 50 demonstrates the entirely different relations which exist in the tropical mountains in the southern hemisphere from those of the northern hemisphere. Here, the high forests are formed from boreal elements: in Central America and Asia these are represented by the genera *Pinus* and *Quercus*, in Asia there occur in addition numerous *Lithocarpus* species, etc., as well as the holarctic genera *Rhododendron* and *Vaccinium*. *Rhododendron* even extends across the south-east Asian mountains into Australia (north Queensland).

As mentioned by C. Troll, the upper tree limit formed by *Pinus hartwegii* in Mexico shows a unique feature that is otherwise found only at the dry-zone border between forest and prairie—the trees grow farther and farther apart with increasing altitude.

[1] This may be true for the slopes and stony soils, but the plains were probably never covered with trees.

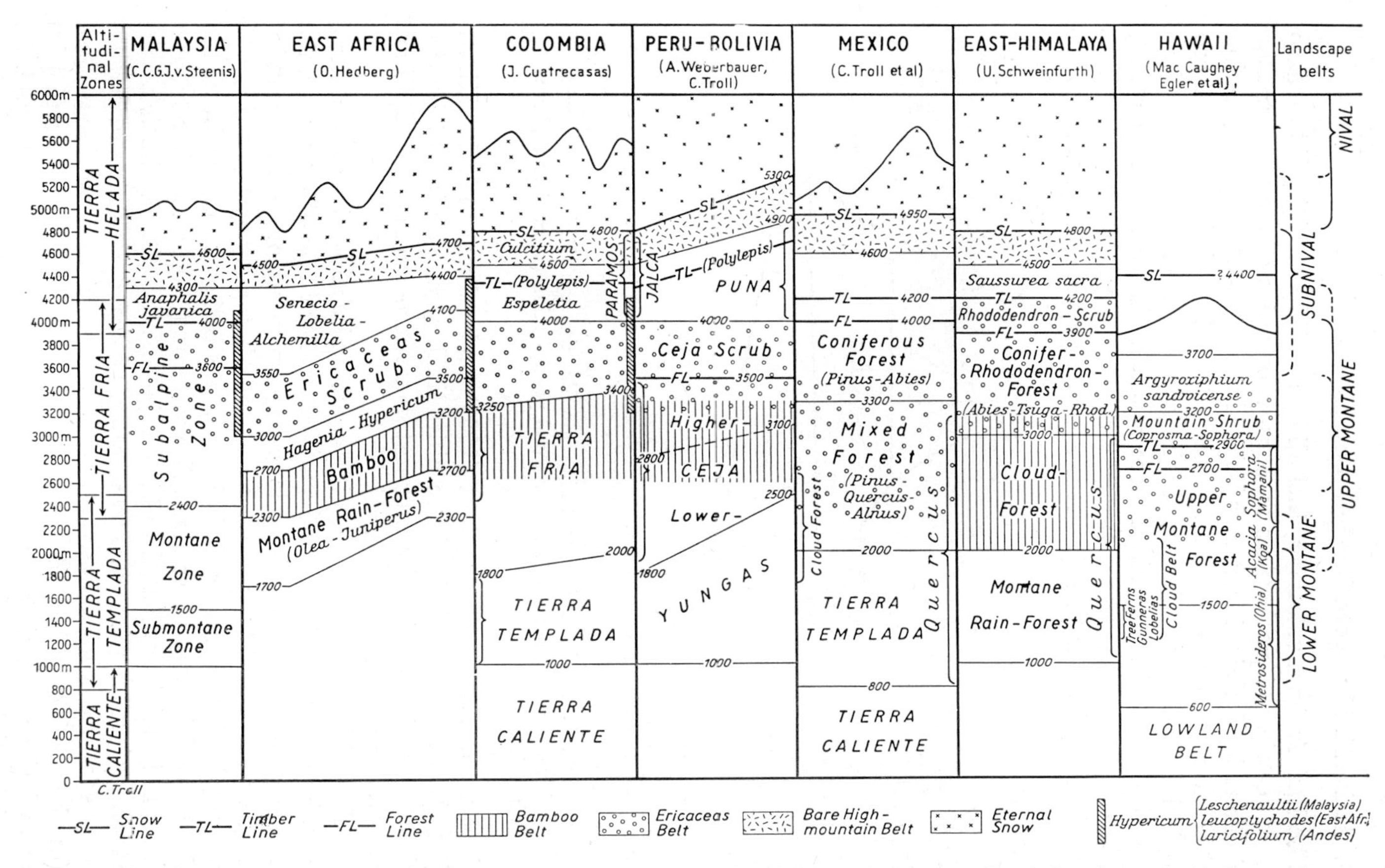

FIG. 111. A comparative scheme of the altitudinal zones of the tropical mountains (after C. Troll, 1959).

5. Altitudinal vegetation belts on Kilimanjaro

The East African volcanoes, including the 6000 m high Kilimanjaro, have certain unique features amongst mountains of the tropics. They rise from level plains in a rather isolated way without direct connections either to north or to south. In January 1935 I carried out several ecological investigations in the high-alpine belt of the Kilimanjaro. Meteorological investigations were made available by Klute (1920). The vegetation had already been described in detail by Volkens in 1897 and later by Tobler (1914). More detailed phytosociological investiga-

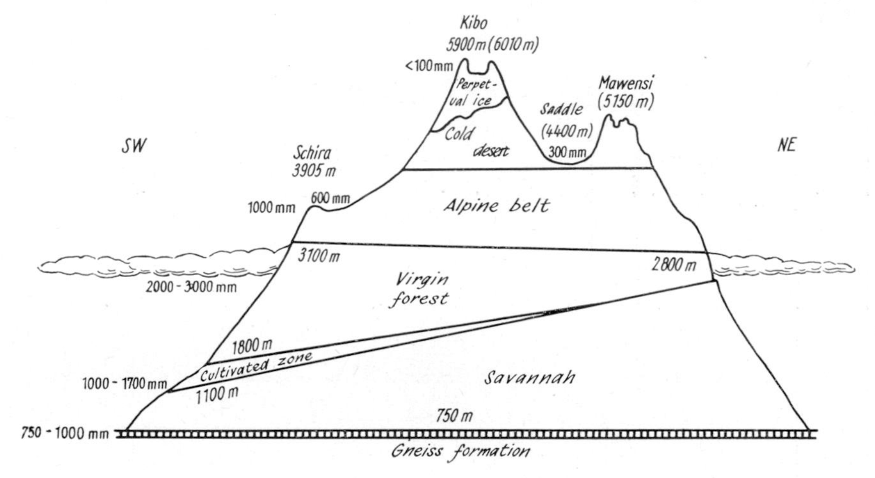

FIG. 112. Schematic vegetation profile through Kilimanjaro with information on the approximate annual precipitation (in mm) at different altitudes.

tions were contributed by Salt (1954) and subsequently by Klötzli (1958). The latter classified the altitudinal vegetation belts following the usual trail from the Bismarck rest house across to Peter's and Kibo's rest houses and so to the summit.

The Kilimanjaro (Fig. 112) is a combined strato-volcano with three eruption-centres and many adventitious craters. The oldest centre, the Schira plateau (3906 m), is no longer conspicuous, of the two summits, the Mawensi (5150 m) consists merely of a lava-plug on the main vent, surrounded by greatly eroded cinders. It is believed to have formed in the early Pliocene, while the Kibo-summit (5900 m, earlier records show 6010 m) is estimated to have originated in the middle Pliocene. Its crater is still preserved. The summit is covered with ice and glacier-tongues which reached down to 4000 m on the south-west side in 1898 (on the north-east side to 5700 m). In 1930 they only extended to 4500 m and since that time they have retreated even farther (Fig. 115). Glacial activity from Pluvial times can be traced down to 3600 m.

Fig. 112 shows a schematic representation of the altitudinal belts.

The south-west slope is very moist since it lies in the direction of the south-west monsoon; the leeward north-east slope is much drier (cf. a similarly dry forest type of the West Usambara Mts. in Fig. 119). The virgin forest zone, therefore, forms a wedge, whose narrow edge almost disappears on the east slope, while the glacier-tongues extend farther down the south-west slope.

The base of the volcano rises above a plain formed from gneiss at an altitude of 750 m. The gradually rising slopes are covered with open savannah up to 1100 m. They have been described as orchard-grassland, for the trees appear to be distributed in a manner reminiscent of orchards while their sizes are similar to those of orchard trees. This formation is followed by a zone of cultivated land consisting primarily

FIG. 113. Epiphytic ferns; *Begonia meyeri-johannis* in a branch fork in the foreground. Virgin forest on Kilimanjaro (Phot. E. Walter).

of coffee and, higher up, of banana plantations. The plantations are situated, in part, on originally virgin forest soil. Today virgin forest begins at about 1800 m. It covers the lower steep-slope zone and extends up to 3100 m on the south-west side. This is a typical tropical cloud forest of the upper montane belt as is found on nearly all tropical mountains at an appropriate altitude. Dominant representatives amongst epiphytes are *Lycopodium* and ferns, particularly the Hymenophyllaceae and epiphytic mosses (Fig. 113) (Biebl, 1964a). The ground is covered with a carpet of *Selaginella*.

The subalpine forest consists of *Olea chrysophylla*, *Ilex mitis*, *Juniperus procera*, *Podocarpus milanjianus* and *Hagenia abyssinica*. Candelabra-Euphorbias extend up to 2130 m in sunny situations, where *Cynodon dactylon* forms grass-mats.

Podocarpus and *Hagenia* (covered densely with *Usnea*) are particularly frequent at the upper forest border (Fig. 114). Admixed plants are the tree-like *Crotolaria* and a beautiful *Hypericum* species with very large flowers, which attains heights up to 6 m. Above this occurs an alpine transition belt into which tongues of forest extend along the valleys. The transition is marked by an ericaceous shrub zone. The shrubs,

Erica arborea and *Philippia excelsa*, are 6 m and more tall and scattered among them is the 2 m tall bracken fern (*Pteridium aquilinum*). The ground is densely covered with *Lycopodium clavatum*.

Above this occurs the alpine belt. Its classification will be discussed later. It does not quite extend to the saddle between Kibo and Mawensi.

Fig. 114. Forest border at the Bismarck rest house. Trees (*Hagenia abyssinica* and *Podocarpus*) densely covered with bearded lichens (*Usnea*); to the left in front of hut *Erica arborea* (Phot. E. Walter).

At this altitude, the vegetation becomes extremely sparse (Fig. 115). Klötzli considers the lower limit of the cold desert with a few isolated cushion plants to have been reached by 4000 m. He found solitary specimens of *Koeleria gracilis* and *Helichrysum newii* as the last plants at altitudes somewhat above 5000 m.

Fig. 115. The Kibo-summit with glacier-tongues as seen from the saddle between Kibo and Mawensi (Phot. E. Walter).

The lower alpine belt is typical tussock-grassland (Fig. 116). The soil is very peaty and was therefore described earlier by Volkens as an organic soil belt. Klötzli lists as dominant grasses *Exotheca abyssinica* and *Agrostis volkensii*. I noticed flowering *Kniphofia* in this belt and scattered shrubs of *Protea kilimandscharica*. Klötzli mentions also *Anemone thomsoni*, *Dierama pendulum* (Iridaceae) and the orchid *Disa stairsii* mixed with *Helichrysum* species. In drier places a community occurs of *Adenocarpus mannii* with *Artemisia afra.* Salt mentions the hummock-forming *Carex monostachya* with *Deschampsia caespitosa* as occurring in extremely peaty places. An associate of these is *Alchemilla argyrophylla.*

FIG. 116. Lower alpine belt (tussock-grassland) with a clump of trees of *Agauria salicifolia* at their limit. On the slope (darker) the upper alpine belt with Ericaceae. To the left the ice-covered Kibo-summit, to the right Mawensi (Phot. E. Walter).

The upper alpine belt is dominated by Ericaceae 0·5-1 m tall, i.e. *Philippia trimera* and *Blaeria johnstonii. Protea* and *Adenocarpus* also extend into this belt. A strange life form is presented by *Lobelia deckenii* (Fig. 117) which attains a height of more than 1 m.

FIG. 117. *Lobelia deckenii* growing among Ericaceae in the upper alpine belt of Kilimanjaro (Phot. E. Walter).

Tree-like Senecios (Fig. 118) occur together with *Carex monostachya* in swampy places. They were earlier known as *Senecio johnstonii* and are now divided into two groups, *Senecio kilimanjari* (distributed between

3100-4000 m) and *S. cottonii* (3700-4500 m). They attain heights of up to 7·5 m. A beautiful group that was mentioned by H. Meyer in 1898 was rediscovered by Salt in nearly unchanged form after 50 years. Thus, the trees must be able to achieve a considerable age (cf. *Anaphalis*, p. 183).

A region characterised by springs occurs above 4000 m. At such a site, the ericaceous plants disappear and the vegetation becomes less dense. The main community is here formed by Compositae, i.e. by *Euryops dacrydioides* and *Helichrysum newii*, etc. In addition the grasses *Festuca abyssinica* and *Pentachistis borussica* occur. In the swampy

FIG. 118. Gully site with giant *Senecio* at 3800 m (Phot. E. Walter).

spring flushes pure stands of *Euryops* and *Carex monostachya* grow with *Alchemilla johnstonii*. Here, giant Senecios are absent. At 4400 m elevation the whitish *Helichrysum* cushions are dominant together with *Festuca kilimanjarica* and *Koeleria gracilis* but these occur only as scattered, single plants protected by rocks.

Klötzli emphasised that the climax communities (altitudinal vegetation belts) merge into one another in a continuous manner, this was also my impression.

Very similar vegetation patterns occur in the alpine belt of the other East and Central African volcanic mountains (Hedberg, 1951). A very informative sketch by Haumann is shown in Fig. 120, which portrays the lay-out of the alpine belt of the Ruwenzori.

This non-volcanic mountain area is characterised by an extremely moist climate. At 2200 m the annual precipitation is around 4000 mm and at the summit area is still about 1600 mm. It rains practically every day, which explains the presence of *Sphagnum*.

A bamboo zone occurs on very wet soils (Virunga, Ruwenzori, Elgon, Aberdare, Kenya). It is only weakly represented on Meru and entirely absent on Kilimanjaro.

The Afroalpine flora has been studied by Hedberg (1957, 1961). It includes 278 species, of which 225 (or 81%) are endemic to Africa; 94

FIG. 119. Dry type of mountain forest with climbers and epiphytes, in a clearing tree-like Lobelias; Shume-forest on the West Usambara Mountains, East Africa (Phot. E. Walter).

species are known only from one mountain, vicarious species are common. Seven of the non-endemic species have a Boreal distribution, 7 a South African, 4 a Mediterranean and 2 a Himalayan. The remaining species occur in the temperate zones of both hemispheres and are nearly cosmopolitan.

I shall now discuss ecological investigations.

Since the tropical climate is characterised by diurnal rather than annual variations, the temperature fluctuations are not greatly reflected in the soil. If one measures soil temperatures downwards in a freshly dug soil pit, one soon obtains constant temperatures. This constant temperature corresponds to the average air temperature at the soil surface both because of the absence of snow in tropical regions and because of the constant heat conduction, characteristic of continuously moist soils. The average air temperature measured by meteorologists in a weather shelter is equal to the constant soil temperature in shaded

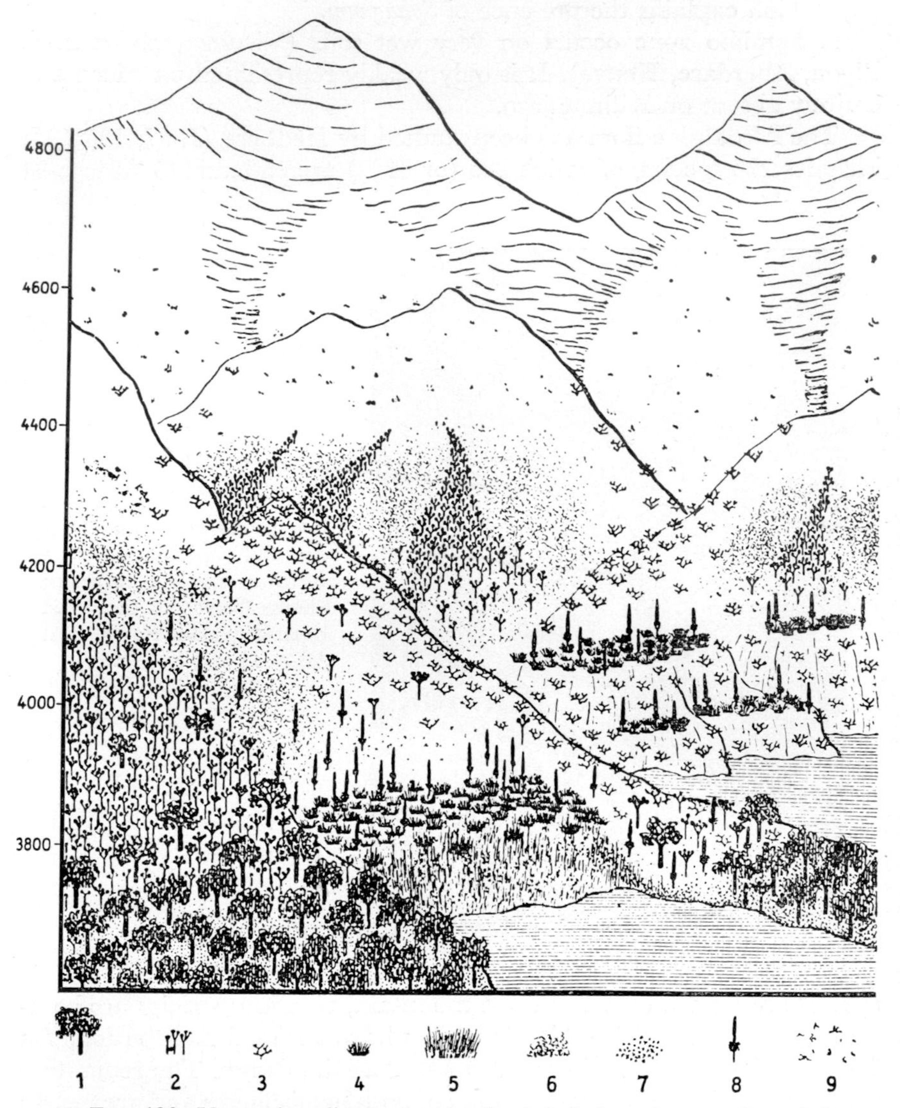

FIG. 120. Vegetation distribution in the belt below the snowline of the Ruwenzori, above 3600 m (after L. Haumann) (cited from C. Troll, 1959).

1. *Philippia*; 2. *Senecio* spp.; 3. *Helichrysum stuhlmannii*; 4. *Carex runssorensis*; 5. *Deschampsia & Subularia* et al.; 6. *Alchemilla* spp.; 7. *Sphagnum*; 8. *Lobelia wollastonii*; 9. *Mosses & lichens.*

localities. This relationship was supported by tests in Amani. The constant soil temperature is increased by 2-3°C in open, sunny localities. This is shown clearly in Table 36.

This tabulation shows that the soil temperature in the shade has already become constant by about 30 cm while in an exposed place it only becomes constant below 60 cm.

TABLE 36

Soil temperatures at different depths at the southwest foot of the Kilimanjaro, coffee plantation at 1180 m elevation, 2.I.1935 at 0800 hr. Deep, very homogeneous loose, volcanic, well weathered soil.

Open sunny places			Under a very shady tree		
Depth	t°C		Depth	t°C	
5 cm	19·8		6 cm	19·3	
16 cm	22·8		14 cm	19·8	
44 cm	24·4		23 cm	20·4	
52 cm	24·4		27 cm	21·4	
60 cm	24·5		39 cm	21·8	constant
68 cm	24·7	constant	52 cm	21·8	constant
70 cm	24·7	constant	70 cm	21·8	constant
87 cm	24·7	constant	94 cm	21·8	constant

On the basis of such measurements, the following constant soil temperatures were determined and compared with the average air temperatures of the stations (bold type) (Table 37).

TABLE 37

Meteorological stations or vegetation belts	Height above sea level (m)	Temperature (°C) habitat shaded	Temperature (°C) habitat sunny
Kibo-Mawensi saddle	4400	—	8·0
Gentle slope with scattered bunch grass	about 4300	—	8·0
Erosion-gully (cold-air drainage)	3800–4000	5·3	—
Heath-covered area above gully	3800–4000	—	8·9
Upper border of virgin forest	2600	11·2	—
Same area in grassland	2600	12·0	—
Same area, open place	2600	—	12·7
Open place in virgin forest	2200	—	16·5
Lower virgin forest, very shaded	1840	16·5	—
Mamba Station	**1550**	**17·8**	—
Coffee Plantation	1180	21·8	24·7
Moshi Station	**1150**	**20·8**	—
Usambara Mts.: **Amani Station**	**900**	**19·9**	—
Same area, in virgin forest	900	19–20°	—
At the coast: **Tanga Station**	**28**	**25·4**	—
Same area, exposed place	28	—	27·5–28

The climate in the East Usambara Mountains at Amani is much moister and thus cooler than in the savannah area at Moshi. The temperature of 5·3°C at 4000 m on Kilimanjaro is most likely caused by the cold air drainage (nightly minimum— 11°C). On the other hand, the temperature of 8°C measured at 4400 m in a sunny place appears too high for the mean annual temperature. One might estimate the mean annual temperature as about 5°C at 4400 m, i.e. at the upper limit of the vegetation. It is thus of the same magnitude as in the densely forested area of central Sweden. This gives an indication of the unfavourable conditions for plant growth produced by the daily fluctuating climate at these altitudes and may explain the broad cold desert belts that are found in tropical mountains below the snow line. Such cold desert belts are absent in mountains of the temperate zone.

Information on the temperature relations in the alpine belt on Kilimanjaro has been provided by Klute (1930), who made various measurements in radiation screens at 4160 m elevation in the period between August 19 and October 12, 1912 (Table 38).

TABLE 38

Mean values					
7 a.m. 6·1°C	minimum	average	−1·8°C	freely exposed	4030 m −7·3°C
2 p.m. 5·6°C		absolute	−4·0°C	minimum	4330 m −7·9°C
9 p.m. 0·5°C	maximum		11·5°C	thermometers	4340 m −8·2°C
	daily range average		10·0°C	showed:	

The rainfall situation is very peculiar. In the tropics the alpine belt extends far above the cloud cover and thus receives only a little rain. The amount decreases rapidly in the subalpine forest belt as well.

Salt mentions the results of measurements that were done over a period of 5 years on the south-west side of Kilimanjaro. The rainfall maximum occurred near 2400 m. At 3600 m, precipitation varied greatly with exposure. At three places rainfall averages were 1016 mm, 608 mm and 174 mm, respectively. At 4200 m annual precipitation was 174 mm and at 4800 m it was only 75 mm. According to more recent measurements of the East African Meteorological Department, the saddle at 4400 m receives about 300 mm per year and near the crater precipitation is less than 100 mm (Klötzli, 1958).

However, the non-measurable precipitation must play an important role in the alpine belt. In the mornings, around 10-11 a.m., the clouds often ascend to the alpine belt. At 4000 m I was often surrounded by fogs which suddenly appeared. These certainly have a considerable wetting effect, particularly on steep rock-faces. However, the distribution of precipitation is very irregular.

Insolation increases continuously with elevation. I measured the following radiation intensities as differences in °C (in the period between January 1-6, 1935) with a black bulb thermometer:

1180 m—13·4° 2180 m—17·2° 3800 m—18·2°

The absolute maximum was probably not measured since the area always became clouded at noon.[1]

The strong insolation of the alpine belt, which was also determined on Java (Fig. 121), causes much heating of the air near the soil surface. At an air temperature of only 7·2°C, the temperature in a grass-tussock

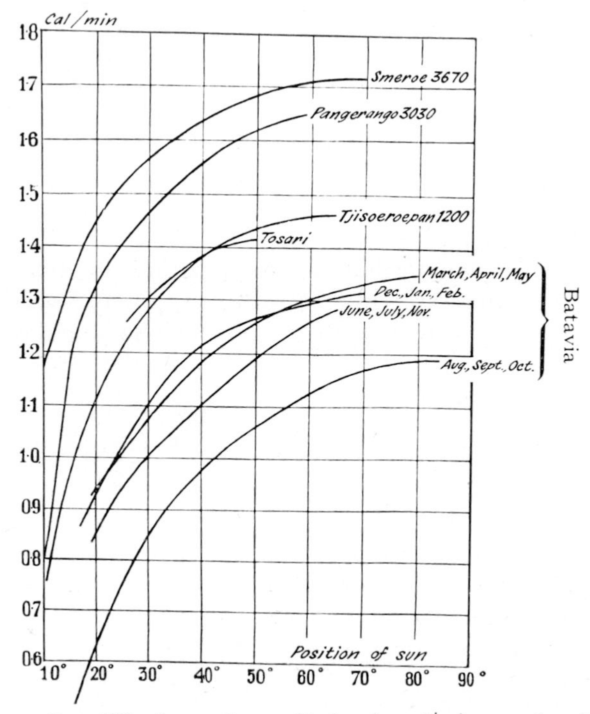

Fig. 121. Increasing radiation in cal/min correlated with increasing altitude in Java (after Boerema); Batavia (Djakarta) is located at the sea coast. The reduced radiation there in August and October is caused by the larger dust content of the air during this relatively dry period.

was 21°C and in cushions of *Helichrysum* it varied from 17-29°C. At 4000 m (air temperature 8·2°C) temperatures of 18·5°C were measured in *Alchemilla*-cushions and 41°C in a grass-tussock.

However, as soon as the sun is blanked out by fog, the temperature decreases rapidly. An unshaded minimum thermometer registered −3·5°C in a ravine; −11°C at the base of a giant *Senecio* (Fig. 118); 2°C in a *Helichrysum* cushion; and between rocks 2·1°C.

These extreme temperature fluctuations are repeated daily throughout the whole year and make for unique climatic conditions which are not found elsewhere in the world.

In spite of low air temperatures, evaporation is rather high due to

[1] At Amani (980 m) the maximum difference between a blackened and non-blackened thermometer was 17·2°C on a cloudless day (December 18, 1935).

the strong radiation. The evaporation values (Table 39) in ml/h were obtained before clouding on January 6 at 3800 m (standard Piche with green paper).

TABLE 39

Time of day	Height above ground 150 cm	30 cm	2 cm	Temperature relations
8:00– 9:00 a.m.	1·1	0·84	0·77	air temperature 11·4°
9:00–10:25 a.m.	1·15	1·10	1·00	soil surface 28°, grass-clump 36·5°
10:25–11:40 a.m.	1·12	1·05	1·05	soil surface 33·5°, grass-clump 36·5°

On January 7, evaporation during a period of sunshine between 0710 and 1040 hr amounted to 0·66 ml/h. In spite of fog and recurring light rain, daily evaporation on these and the following days amounted to the following values:

Distance above soil (cm)	Evaporation (ml)
150	7·7–10·9
30	6·6– 8·9
2	5·8– 8·1

Several transpiration experiments were done before the appearance of a fog at 1025 hr on January 6. The following transpiration values were obtained at an evaporation rate of 1·05 ml/h.

TABLE 40

Ericaceae		Compositae	
Erica arborea	2·6 mg/g/min	*Euryops dacrydioides*	1·4 mg/g/min
Philippia cf. *trimera*	6·1 mg/g/min	*Helichrysum* cf. *newii*	3·6 mg/g/min

These values correspond roughly to those of the temperate zone alpine *Loiseleuria procumbens*, thus, they are very low.

Under these extreme climatic conditions the vegetation shows a rather xeromorphic character. Most of the species are covered with a white hair-felt, others have ericaceous foliage. The giant *Senecio* species have large, broad leaves that are arranged in terminal rosettes with a loose anatomical structure reminiscent of *Tussilago farfara*.

The conditions on Kilimanjaro are identical with those found by Hedberg (1963) on Mt. Kenya. A thermograph in Teleki valley at 4200 m recorded the following maximum and minimum temperatures between August 2-9:

1. slope position, 0·5 m above the soil: +9·5° and −1·0°
2. valley bottom position, 1·0 m above the soil: +15·0° and −6·5°

At the second site there was a stand of *Lobelia keniensis.*

The daily night frost that occurs throughout the whole year requires special adaptations by the plants and it is of interest to know their nature. Hedberg pointed out that there are structural devices that protect the growing organs from extreme temperature fluctuations. Protection against night frost is afforded in *Lobelia* and *Senecio* through the closing of their rosettes. In *Lobelia,* the growing point of the shoot is, in effect, enclosed in a water tank 10 cm deep but the water only freezes at night on the surface. The stem has thick cork layers and is surrounded by old leaves. Dead leaf-bases are retained also by the tussock-grasses, such as *Festuca pilgeri* ssp. *supina.*

The alternating freezing and thawing also cause local solifluction which in turn results in characteristic vegetation patterns of *Helichrysum citrispinum* var. *hoehnelli, Carduus keniensis, Festuca* sp. and *Senecio majeri-johannis,* amongst others.

References

BAUR, G. N. 1957. Nature and distribution of rain forests in New South Wales. *Austr. J. Bot.*, **5**, 190-233.

BIEBL, R. 1964a. Austrocknungsresistenz tropischer Urwaldmoose auf Puerto Rico. *Protoplasma*, **59**, 277-97.

BIEBL, R. 1964b. Temperaturresistenz tropischer Pflanzen auf Puerto Rico. *Protoplasma*, **59**, 134-56.

DAWSON, J. W. 1962. The New Zealand lowland Podocarp forest. Is it subtropical? *Tuatara*, **9**, 98-116.

ELLENBERG, H. 1958. Wald oder Steppe? Die natürliche Pflanzendecke der Anden Perus, I und II. *Wiss. Techn.*, 645-48, 679-81.

HEDBERG, O. 1963. Ekologisk specialesering in den afroalpine floran. *Svensk. Naturv.*, 158-70.

HEDBERG, O. 1961. The phytogeographical position of the Afroalpine Flora. *Rec. Adv. Bot.*, Toronto, 914-19.

HEDBERG, O. 1957. Afroalpine vascular plants, a taxonomic revision. *Symb. Bot. Upsal.*, **15**, 1-411.

HEDBERG, O. 1951. Vegetation belts of the East African Mountains. *Svensk. Bot. Tidskr.*, **45**, 140-202.

HIRSCH, G. 1957. Zur Klimatologie und Transpiration an Vegetationsgrenzen. *Beitr. Biol. Pflanzen.*, **33**, 371-422.

HUECK, K. 1954a. *Waldbäume und Waldtypen aus Nordwest-Argentinien.* Berlin.

HUECK, K. 1954b. Pflanzengeographisch-forstwirtschaftliche Probleme aus dem Nordwesten Argentiniens. *Zschr. f. Weltforstwirtschaft*, **17**, 219-225.

HUECK, K. 1962. Der *Polylepis*-Wald in den venezolanischen Anden, eine Parallele zum mitteleuropäischen Latschenwald. *Angew. Pflanzensoz.*, **17**, 57-76.

HUECK, K. 1966. *Die Wälder Südamerikas. Vegetationsmonographien der einzelnen Grossräume.* Vol. 2, Stuttgart.

KLÖTZLI, F. 1958. Zur Pflanzensoziologie des Südhanges der alpinen Stufe des Kilimandscharo. *Ber. Geobot. Forschungs-Inst. Rübel.* 1957, 33-58.

KLUTE, F. 1920. *Ergebnisse der Forschungen am Kilimandscharo* 1912. Berlin.

KOORDERS, S. H. 1914. Floristischer Überblick über die Blütenpflanzen des Urwaldes von Tjibodas. *Englers. Bot. Jahrb.*, **50**, *Suppl.* 278-303.

KOORDERS, S. H. 1918-23. *Flora von Tjibodas*, Batavia.

LAMPRECHT, H. 1954. Estudios selviculturales en los bosques del valle de la Mucuy cerca de Medina. *Univ. Andes. Fac. Ingen. Forest. Merida* 1-130.

MEIJER, W. 1959. Plantsociological analysis of montane rain forest in Tjibodas, West Java. *Acta. Bot. Neerlandica*, **8**, 277-91.

RAUH, W. 1939. Über polsterförmigen Wuchs. *Nova Acta Leopoldina*, **7** (49) 268-508.

ROBBINS, R. G. 1962. The podocarp-broadleaf forest of New Zealand. *Trans. Roy. Soc. N.Z., Bot.*, **1**, 33-75.

SALT, G. 1954. A contribution to the ecology of upper Kilimanjaro. *J. Ecol.*, **42**, 375-423.

SCHIMPER, A. F. W. and FABER, F. C. VON 1935. *Pflanzengeographie auf physiologischer Grundlage*. 3rd ed. Jena.

VAN STEENIS, C. G. G. J. 1962. The mountain floras of the Malaysian Tropics. *Endeavour*, **21**, 183-94.

TOBLER-WOLFF, G., and TOBLER, F. 1914. Vegetationsbilder vom Kilimandscharo. *Veg. Bilder*, **12**, 1, Plates 7-81.

TROLL, C. 1959. Die tropischen Gebirge. *Bonner Geogr. Abh.*, **25**.

VOLKENS, G. 1897. *Der Kilimandscharo*. Berlin.

WALTER, H. 1960. *Einführung in die Phytologie*. Vol III. *Grundlagen der Pflanzenverbreitung*. Part 1, *Standortslehre*. 2nd edn. Stuttgart.

WEBB, L. J. 1959. A physiognomic classification of Australian rain forest. *J. Ecol.*, **47**, 551-70.

WEBER, H. 1958. Die Paramos von Costa Rica und ihre pflanzengeographische Verkettung mit den Hochanden Südamerikas. *Akad. Wiss., Mainz, Math.-Naturw. Kl., Jahrg.*, **3**.

V. *Tropical Semi-evergreen and Deciduous Forests*

1. The ecological significance of annual leaf-fall

In the tropical rain forest leaf-shedding does not occur simultaneously at a particular time of year, while in the semi-evergreen and tropical deciduous forests leaf-shedding is a definite seasonal phenomenon. Here, the impression is given that shedding is caused by the dry season,[1] just as the winter season is assumed to cause leaf-shedding in the temperate zone. However, in the latter case we know that the causative relationship is only indirect; the 'summer-green' trees lose their leaves in the winter even when they are kept in a heated greenhouse, or if they are transplanted into a climatic region with very mild, frost-free winters, e.g. to Cape Province. However, under such circumstances the leafless period is reduced and as soon as the temperature rises in the spring, the new leaves emerge, some a little sooner, others a little later.

In the tropical rain-green forests, relations are somewhat more complicated than might be assumed initially. Leaf-shedding occurs earlier in dry years and later after a good rainy season. However, leaf emergence is not correlated with the onset of the rainy season but rather precedes its onset, often coinciding with flowering. This is well documented in the literature. I observed this phenomenon in the semi-evergreen forests in Panama, also among the deciduous eucalypts of north Australia and the thorn-acacias of south-west Africa. So, the triggering effect cannot be attributed to the water regime but, once again, to the increase in temperature.

One might also expect to find deciduous trees in areas with a very dry and hot and a somewhat cooler, yet rainy season, where the trees would develop their foliage at the end of the dry season. However, this does not occur. Apparently, woody plants do not develop their foliage with a decrease in temperature and the triggering effect must come through an increased temperature after the rest period. In regions with winter rain we find evergreen woody species and these too develop their new foliage in the spring; that is, at the beginning of the dry season. Wherever deciduous species are present in such regions (e.g. flood plain trees in the Mediterranean region), they are in foliage during the summer; that is, during the drought period.

Only one shrub, *Fouquieria splendens*, in the Sonoran desert (north Mexico, Arizona), is known to develop new foliage after each heavy rain, presumably at any time of the year, frequently six times a year (see p. 327), and then to shed it after a week's drought. Thus, this shrub has

[1] Hence the more descriptive German term 'regengrüne' which can best be translated 'rain-green'.

no definite rest period. Periodic development of new foliage appears to be determined by temperature in all other cases.

In regions having an unfavourable season, the ecological adaptation of the plants takes two directions:

1. The leaves assume a xeromorphic structure so that they can endure the unfavourable season (sclerophyllous plants of the winter-rain areas, *Eucalyptus*, Acacias with phyllodes, winter-green hardwood species, such as *Ilex*, *Hedera*, etc., as well as winter-green conifers.

2. Leaf-organs become short-lived and are not retained during the unfavourable season, although they can be mesomorphic as in the case of all deciduous woody plants.

Which of the two types prevails in a given area is entirely a function of their competitive ability. The xeromorphic evergreen plants have the advantage that they can carry out photosynthesis and thus produce dry matter at any time with the onset of favourable conditions. However, their utilisation of dry matter for foliage is relatively high per unit leaf area. Deciduous trees have to renew their foliage every year but they use a relatively small quantity of matter per unit leaf area.

Deciduous woody species have attained dominance in tropical regions with a summer-rain period and a long drought period during the cool season. The only exception is Australia, where the corresponding life forms are absent and where besides the evergreen eucalypts there are *Acacia* species with phyllodes. Apparently, the genetic prerequisites were not available for the development of the deciduous habit. There are only a few deciduous forms among the north Australian elements that originated in south-east Asia such as *Gyrocarpus*, *Bauhinia*, *Brachychiton*, *Cochlospermum* (Bixaceae), etc.

Thus, my surprise was so much greater when I found an area of deciduous eucalypts south of Darwin. These were just developing their leaves at the end of October in the form of large, soft, fresh-green leaves. This type has not been mentioned in the literature. The most frequent form was *Eucalyptus alba*. Further search showed that similar responses have been noted in *E. grandifolia*, *E. bigalerita*, *E. confertifolia* and *E. clavigera*. The two latter species supposedly have harder and thicker leaves. Leaf-shedding may not be complete after a season with very heavy rainfall. Thus, these are facultative deciduous species, such as many others of the semi-evergreen forest. In a similar way to the semi-evergreen forests these species occur in summer-rain areas with very high amounts of rainfall and a short drought period or in areas transitional to the evergreen rain forest. However, here they seem to be competitive only in unfavourable habitats, as for example on stony ridges or in depressions with heavy textured soils, which dry up considerably during the dry season. The evergreen eucalypts of the moist forests cannot cope with such situations and the evergreen eucalypts of the dry areas with pronounced xeromorphic leaves probably lag behind the deciduous forms in growth capacity and dry-matter production.

The deciduous trees, on the other hand, have an advantage in that they form a large leaf area with a relatively small investment of dry organic matter. This allows for a greater production of dry matter and more rapid growth (see p. 21 *et seq.*).

Thus, it is seen that the deciduous habit may result in a greater competitive ability under certain climatic or edaphic conditions. However, the genetic prerequisites for this ability may be entirely lacking in certain floristic regions, e.g. in New Zealand where only evergreen species occur, forming the forest border to the arid region of Otago as well as at high altitudes in the mountains. Today, introduced deciduous *Rosa* species spread rapidly in the borders of grassland as an undesirable weed. Afforestation experiments in the southern part of South Island have also shown that *Acer pseudoplatanus* becomes a weed, and that *Betula* and *Larix* easily become established as well as such shrubs as *Sambucus nigra*, *Ribes grossularia*, *R. sanguineum* and several *Salix* species.

2. Semi-evergreen and rain-green tropical forests

The evergreen rain forest can maintain itself in areas with a drought period of 1-2·5 months, provided that it receives a very high rainfall of 2500-3000 mm.

However, even in these areas with such a short drought period an annual periodicity can be observed. Certain species of tree shed their leaves simultaneously, or come into leaf at the same time; epiphytic orchids can shrink up during the dry season, and the lushness of the forest is noticeably reduced. Beard (1944) called these the evergreen seasonal forests of the tropics. Ashton (1958) has made light measurements in such a forest in the Amazon basin near Monte Alegre, Para, which coincide with those made by Evans (1939, 1956) in south Nigeria. The transition from typical tropical rain forest to seasonal rain forest is a gradual one. A pronounced difference in character is only found where the dry period is longer or where the rainfall is less. In such forests, the upper tree layer is defoliated for some of the year, while the lower tree layer retains its foliage. In such cases we can speak of *semi-evergreen forests*.

Wherever the water relations become increasingly unfavourable the deciduous covering extends to successively lower layers. I distinguish, therefore, the humid rain-green forest and the dry rain-green forest respectively.

Woodland types in these categories occupy nearly the entire area south of the Sahara in Africa, and even across the Cunene and Zambezi to the south. Thus, they enclose the equatorial region of the evergreen rain forest on all sides (Manshard, 1960-61). In Asia, they occupy much of India and the dry areas of south-east Asia, and in South America by far the largest part of the continent is probably occupied by these formations. However, in Australia the deciduous tree species are almost entirely absent from this climatic zone and are replaced by evergreen *Eucalyptus* species.

The main differentiating feature of the evergreen rain forest is the pronounced periodicity of the climate, which is primarily determined by the unequal distribution of rainfall. However, the seasonal

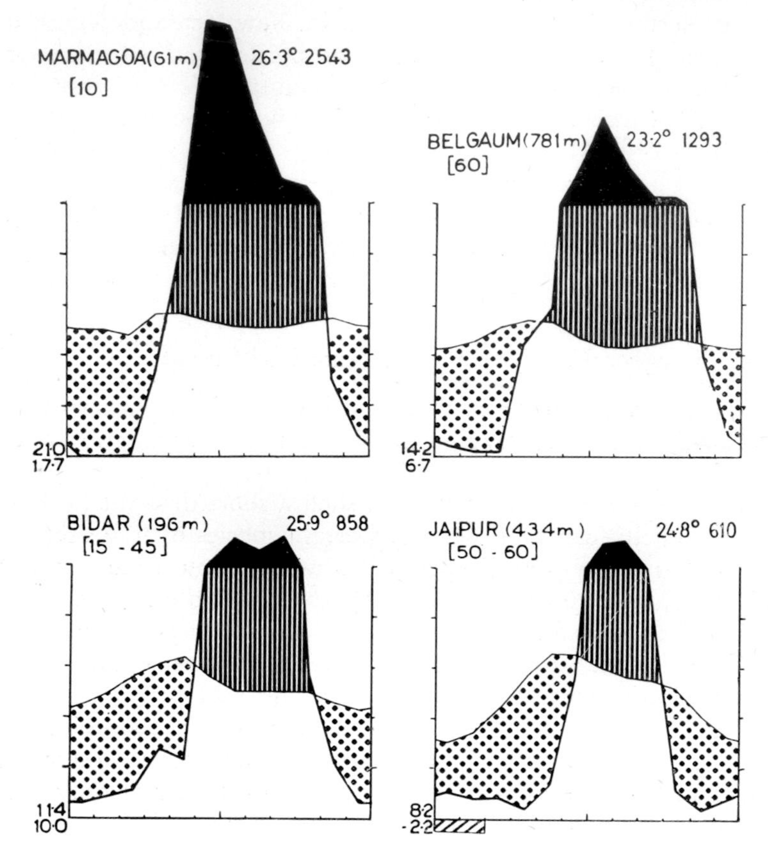

FIG. 122. Climatic diagrams of India for evergreen (Marmagoa), semi-evergreen (Belgaum), humid (Bidar) and dry (Jaipur) monsoon forests (from *Klimadiagramm-Weltatlas*).

temperature variations are also of importance, particularly with increasing distance from the equator and with increasing duration of the dry period. Nevertheless, this zone is still tropical, i.e. frosts do not occur in general, but the temperature is markedly decreased during the dry period so that one can now speak of a warm season with rain and a dry, cool season.

The partitioning of the year into a rainy and a dry period is particularly pronounced in monsoon areas, such as India, so one speaks frequently of dry and moist monsoon forests, respectively. However, it is of little consequence for the vegetation whether the rainfall is zenithal or monsoonal in origin. Lauer (1952) only uses the length of the dry period for the classification of the tropical vegetation zones. However, it has become quite evident that this is not enough. One cannot consider a single climatic parameter but one must view the climate in its entirety. I, therefore, present the relevant climatic diagrams (Fig. 122).

Neither the amount of precipitation nor the length of the dry period alone are decisive for delimiting the different forest types. A longer dry period can be compensated by higher precipitation.

So, rainfall and length of dry period have to be considered simultaneously. This has become clear through my investigations on India which were done at the request of UNESCO (1962). For classifying the vegetation I used Champion's (1936) map of natural forest regions which he had developed on the basis of many years of experience as a forester in India. The rainfall and length of the dry period could be extracted from the corresponding climatic diagrams in the *Weltatlas*. By entering the rainfall on the ordinate and the length of the dry period on the abscissa, the points representing a particular vegetation type fell into a well-defined pattern (Fig. 123). Vertical separation would be expected if the length of the dry season were the only factor responsible, but horizontal separation would be expected if rainfall was the only factor. In fact, however, the mean lines run at a steeper angle in the moist area and more horizontally in the dry area.

It may be concluded, therefore, that the change from tropical rain forest to semi-evergreen rain forest and from this to the rain-green dry forest is determined more by the length of the dry season than by the rainfall, while the rainfall is more critical than the length of the dry period for the change of this last type of forest into savannah and desert.

Similar relationships can be assumed to hold for the natural forest regions of the African and South American tropics. The values for rainfall and length of dry season may vary marginally and in detail with differences in total climate. An accurate check is difficult, for the present vegetation in these areas has been greatly modified by man and a detailed vegetation study comparable to that for India is not available.

Closed forest does not offer a secure basis for the maintenance of human populations and is today still a refuge for relict tribes (Pygmies). Pioneers always see the forest as an enemy when they begin the settlement of forest areas. This explains why the destruction of forest always exceeds the tolerable limit. The reverse movement, to re-afforest denuded areas, always begins when it is far too late. The notion still prevails today among the farmers of Australia, in the saying, that could lead toward a complete destruction of the forest, 'one blade of grass is more valuable than two trees'.

In the tropics, as elsewhere, the fight against the forest begins with a partial clearing. However, wherever forest is cleared a little, and wherever the denuded areas are invaded by grasses, the forest stands can easily be reduced by annual burning of the grass cover. The final result is an open savannah-like landscape. Since the rainy period shows a more regular pattern in the semi-evergreen and the humid rain-green forest zone than in the rain-green dry-forest, it is understandable that these two former types, in particular, have been destroyed (except for very small remnants), since their climate is more suitable for agriculture.

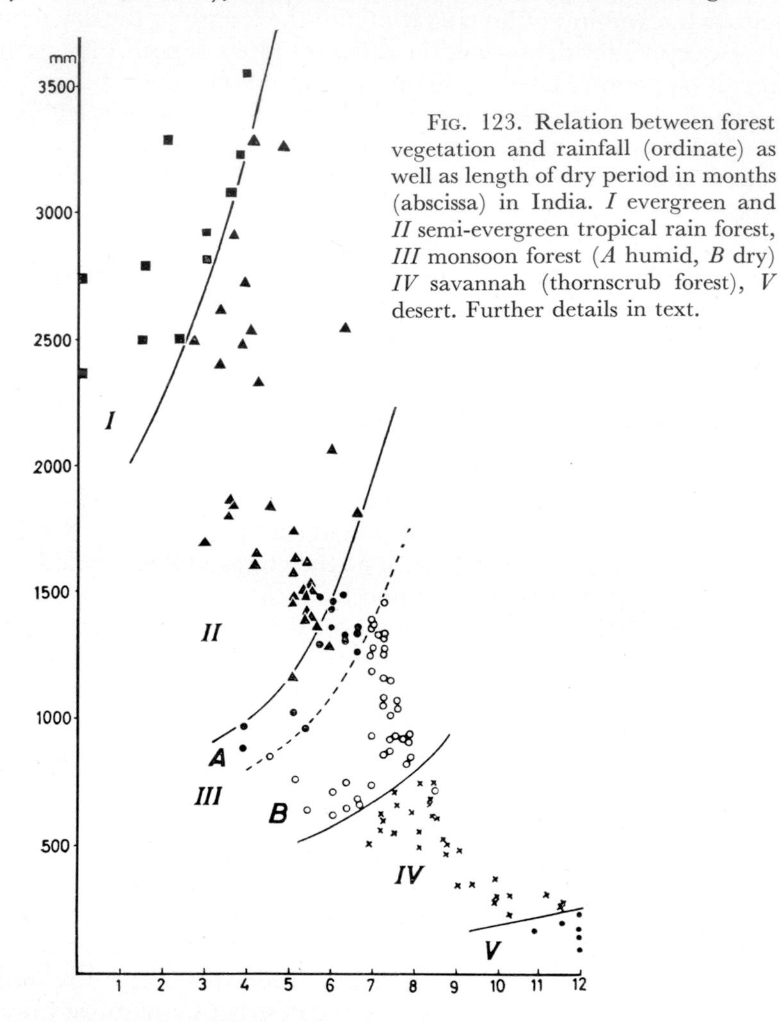

FIG. 123. Relation between forest vegetation and rainfall (ordinate) as well as length of dry period in months (abscissa) in India. *I* evergreen and *II* semi-evergreen tropical rain forest, *III* monsoon forest (*A* humid, *B* dry) *IV* savannah (thornscrub forest), *V* desert. Further details in text.

This may be the reason why geographers speak of moist savannahs. By contrast, much larger areas of rain-green dry-forest are still preserved, since it is less favourable for settlement.

More detailed ecological investigations are not available yet on this type of forest. Its floristics will not be discussed here but I will rather restrict my discussion to a short, general description. A schematic representation of the individual types as found in Peru, including the arid regions, is shown in Fig. 124. I intend to discuss the arid regions later and will treat the first two types only briefly here; the third type is dealt with in the succeeding section.

(*a*) Semi-evergreen tropical rain forests.
(*b*) Rain-green tropical forests or humid monsoon forests.
(*c*) Tropical dry-forests or dry monsoon forests.

(*a*) *Semi-evergreen tropical rain forests*

Their structure does not differ from that of the evergreen forests but the number of species is usually somewhat less in the semi-evergreen

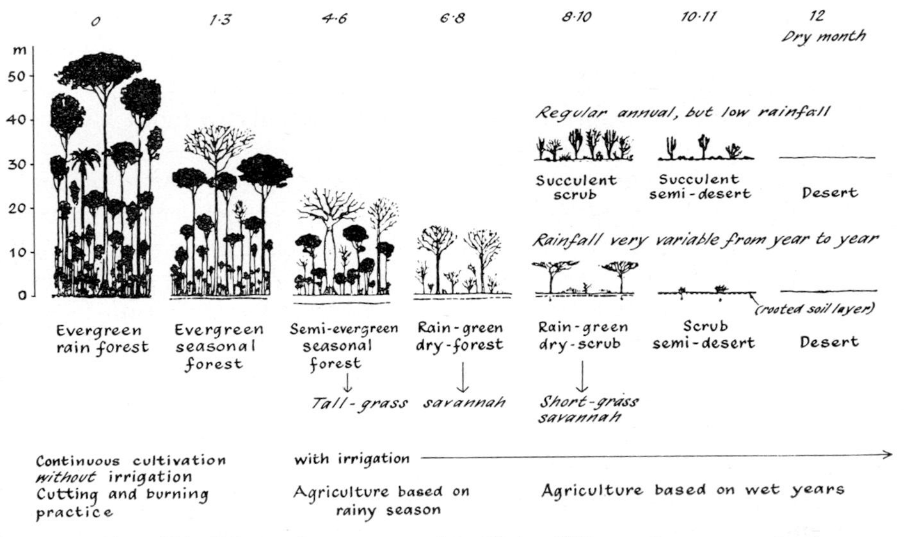

FIG. 124. Schematic representation of the different forest types in the tropical zone of Peru in relation to increasing length of dry period and decreasing rainfall (after Ellenberg). Savannah communities after burning, and comments on general suitability for agriculture have been added.

forests. Some tree species tend to occur in pure stands. Buttressed roots and cauliflory are less common.

The semi-evergreen forest is best developed in the monsoon area of India and Burma, Thailand and Malaya and occurs immediately adjacent to the evergreen forest (Champion, 1936; Puri, 1960; Ogawa *et al.*, 1961; Thai-van-Trung, 1962). The latter does not differ very much from the semi-evergreen forest during the rainy season. Their average heights too are similar but the semi-evergreen forest does not have the typical, isolated extremely high trees such as are found in the tropical rain forest. The semi-evergreen forests are still rich in lianas and epiphytes, differences only become noticeable during the drought season. Trees that are completely barren at times are also found in the rain forest, but their leafless period is not correlated with any particular season of the year. In the semi-evergreen forest up to 30% of the trees in the upper stratum are deciduous. The lower strata may still be composed largely of evergreen species whose average leaf size is smaller than in the evergreen rain forest. Leaf emergence usually occurs one month before the beginning of the monsoon rains and is triggered off by the increasing temperature. This shows that the soil still contains sufficient water at the end of the dry period to allow the emergence of new foliage.

The annual rainfall in these forest areas is usually still over 1500 mm. At higher altitudes in the mountains, where low temperatures and very high humidity (mist) prevails, the evergreen montane rain forest can maintain itself in spite of the rainless months. For example, near the coastal villages at the foot of the Western Ghats in India there are semi-evergreen forests and there are rain forests, still evergreen, near the mountain villages which, in part, show annual rainfall records of more than 5000 mm. Such forests also occur extrazonally in moist habitats along rivers at lower altitudes.

In Africa, the semi-evergreen and evergreen forests extend beyond 5° S in the Kwango area (Guinea region). However, between these and the rain-green dry-forests of the Zambesi region wide areas of grassland occur in spite of the forest climate. Devred (1957) believes that these areas were formerly covered with a semi-evergreen transition forest. However, through increasing dryness of the climate these areas have supposedly become sensitive to fire and cutting, so that the forests have eventually been entirely destroyed by human influences. Grassland has developed in their place.

This transition forest zone appears, in general, to be destroyed in Africa, because the extensive areas of tropical rain forest are immediately adjacent to open areas, the anthropogenically conditioned savannahs, which represent a degradation stage of the semi-evergreen forests.

This applies also to the Guinea-savannahs, with a rainfall of 1000-1500 mm, which extend across Nigeria where they join with the rain forest to the south. The area is densely populated and repeatedly burnt over. Originally, this area was occupied by a semi-evergreen forest with *Albizzia adianthifolia*, *A. zygia* and *Bombax buonopozense* in the upper stratum and an evergreen lower tree stratum. Today, one finds in the savannah *Daniella* with *Elaeis* or *Uapaca* (Clayton, 1961). Hopkins (1962) has also pointed out that secondary savannah has replaced the semi-evergreen forest in Nigeria after fire in an area receiving 1232 mm annual rainfall and having a dry period of 4-5 months.

However, degradation of the forests may occur even in protected areas. The inhabitants of the Murchison Falls National Park area in Uganda were deported in 1912 because of the danger of sleeping sickness. This caused a great increase in the number of elephants. These de-bark and girdle the trees which then die in the annual fires. As a consequence, these areas have been converted from tropical rain forests with *Chlorophora excelsa* and semi-evergreen forests with *Terminalia glaucescens* into treeless grasslands. The forested area decreased by 55-60% between 1932-56. Only a few species are spreading, such as *Hapoptelia grandis*, which is not touched by elephants, or fire-resistant species, such as *Lonchocarpus laxiflorus* (Buechner and Dawkins, 1961).

Similar semi-evergreen forests are also found next to the evergreen forests in South America (Beard, 1944; Hueck, 1961a, 1965).

(*b*) *Rain-green tropical forests*

The humid type of these forests is known in India as monsoon forest. Natural grasslands are entirely absent in India; there they are merely the result of human influences (Whyte, *et al.*, 1954). The teak forests (*Tectona grandis*) and the sal forests (*Shorea robusta*) with their associated tree species are amongst the most important. The teak- or djati-tree also plays an important role in economic forestry. *Tectona grandis* grows up to 40 m and forms the zonal vegetation together with *Terminalia tomentosa* and *Diospyros melanoxylon* in Central India, which, today, however, is present only in the form of small, disturbed relicts (Bhatia, 1958). The forest gives a hygrophilous impression during the rainy season. It also remains green during the dry season because the leaves are only dropped gradually. New foliage emerges a little before the beginning of the rainy season so that the forests are usually leafless for a short period only. The number of tree species is smaller and one species is often dominant. A closed herb layer is developed because of the favourable light relations beneath the trees and lianas and epiphytes are still common. Most plants flower during the dry season.

Pentacme suavis becomes more frequent in areas with less rain. In still dryer areas the *Terminalia tomentosa* forests become dominant. These are already becoming transitional to dry forests.

As an example of the three-dimensional classification of the vegetation in the monsoon area, I will discuss the conditions in Burma (Vidal, 1959).

A semi-evergreen rain forest (40-50 m tall) grows in the warm coastal area under relatively high rainfall conditions (above 1500 mm). It has two tree strata (Dipterocarpaceae, Leguminosae, Meliaceae, Sapindeae, etc.), a shrub layer (Anonaceae, Rubiaceae, Euphorbiaceae), a herb layer (Araceae, Zingiberaceae), lianas and epiphytes (Orchidaceae). *Eupatorium odoratum* spreads on clear-felled land. Rain-green forests with *Pentacme*, *Shorea*, *Terminalia tomentosa* and *Dipterocarpus* occur on shallow soils. Hence the tree canopy is interrupted, and lower and undergrowth vegetation is very dense. There is also a transitional type in which *Tectona* dominates. The montane forests (600-800 m with 3600 mm rain) are evergreen Dipterocarp forests with palms and bamboo species; Fagaceae (*Castanopsis*, *Lithocarpus*, *Pasania* and *Quercus*) with Lauraceae dominate between 1000 and 2000 m. Half the species are deciduous. Above 2000 m the trees are covered with *Usnea* and mosses and the ground with *Selaginella*. Here too conifers occur (*Pinus merkussii*, *P. khasya*; *Keteleeria*), whose spread is favoured by fire and grazing.

The annual phenological cycle in the Indian monsoon area has been described by Misra (1959), where he distinguished six periods in the upper Gangetic plain:

1. The temperatures increase in March and April, when the daily ranges attain 25°C. As a consequence the trees shoot rapidly. This is the main flowering period for herbaceous plants.
2. May and June are the driest and hottest months. Growth can only be observed in moist habitats.
3. The highest rainfall occurs during July and August. It becomes cooler again; in this period the smallest daily temperature ranges occur. Most seeds germinate at this time and the trees exhibit maximum growth.
4. Showers alternate with sunny days in September and October. The growing conditions are ideal and annuals flower and set seed.
5. November and December are characterised by cooler weather. The growing season comes to an end.
6. The lowest temperatures are observed during January and February. The growth period is interrupted during this, the coolest, period. The ground vegetation is rather open.

As a result of the high rainfall, the low-lying areas remain flooded for 4-8 months. Here, mud communities quickly become established after the recession of the flood waters. Forests are preserved only at a few places because the level areas are heavily grazed and covered with thorn plants. The higher lying areas are used for cultivation.

Ogawa and his collaborators (1961) have not only given an exact description of the vegetation of Thailand with a vegetation map and topographic profiles but they included also productivity investigations. They investigated wetter and drier plant communities as well as some secondary formations, viz.:

1. An open Dipterocarp dry-forest at 300 m elevation with solitary 20 m trees and grass cover 20-30 cm tall on the ground.
2. A somewhat moister, mixed rain-green forest, with trees 20-25 m tall and a sparse grass cover on the ground.
3. A wet, evergreen gallery forest to 30 m, very dense, with lianas but only few epiphytes.
4. A temperate evergreen forest at 1200 m above sea level with 20 m tall *Pinus khasia*, *Terminalia*, *Artocarpus*, etc. In the lower tree stratum: *Symplocos*, *Quercus*, *Castanopsis*, etc., with frequent epiphytic orchids and no lianas.
5. A grassland community with *Eragrostis*, *Andropogon*, *Agrostis*, *Setaria*, etc.

6. A vigorous tall-grass community (up to 2·3 m) with *Themeda* and *Arundinaria* growing on termite mounds.

7. An *Arundo donax* thicket at an altitude of 300 m which occurs along rivers, having 7 m tall, very densely growing bamboo-like shoots.

The authors calculated the dry weight in ton/ha, (*a*) of stems and branches, (*b*) of leaves, (*c*) of the total mass (including undergrowth and roots), (*d*) the leaf-area index (LAI) and (*e*) the net-production (NP) of the aerial parts but without allowing for incremental wood.

The results on the experimental areas of communities 1-7 are shown in Table 41.

TABLE 41

Dry weight (tons/hectare)

	(*a*) Branches	(*b*) Leaves	(*c*) Total	(*d*) LAI	(*e*) NP
1. Dipterocarp dry-forest	44·7	4·9	65·9	4·3	7·8
2. Mixed forest	52·6	4·9	77·0	4·2	8·0
3. Gallery forest	270·0	19·5	377·6	16·6	25·3
4. Temperate forest	163·4	14·5	231·4	12·6	18·9
5. Grassland	—	—	13·5	—	11·3
6. Grass on termite mounds	—	—	42·1	—	38·5
7. *Arundo donax* stand	—	—	162·5	23·3	51·9

The carbon and nitrogen content of the plant biomass and the soil were also investigated. The leaf-area index for the gallery forest (16·6) appears to be unexpectedly high. By comparison, I quote the leaf-area index for tropical rain forests as 8·5 (Ivory Coast) and 9·5 (Nigeria), for a temperate forest in north Japan as <7 and in south Japan >7. The LAI for the *Arundo* stand is unusually high. However, here the leaf position is more vertical, the stand was only 3-5 m wide and also received light at its sides. The enormous net-production is related to the high LAI.

The net-production in an *Abies sacchalinensis* forest in north Japan amounts to: needles 4·5, wood 16·8, total 21·4 ton/ha (gross 47·7). The net-production of an evergreen *Distylium racemosum* forest in south Japan: leaves 11·4, wood 9·2, total 21·6 ton/ha (gross 73·1).

The net-production of tropical forests was estimated as 40 ton/ha by the author, and upon the assumption that respiration losses amount to 60-70 ton/ha, the gross production was estimated to be above 100 ton/ha (see pp. 33-4, 111).

3. Tropical dry-woodland and thorn-scrub thickets

These formations are also known as dry monsoon or savannah woodland. I shall discuss these together with the driest formations of the equatorial region, the thorn-scrub thickets which contain many very interesting succulents. The transition from dry-woodland to savannahs with decreasing summer-rains at increasing distance from the equator will be discussed later. The dry type of tropical rain-green forest in South America and particularly in Africa has been designated by many different names. The name 'dry-woodland' is most fitting in respect of its condition during the dry season because, at that time, these woodlands look absolutely barren and brown.

The leafless branches give no protection against the intense sun. Remnants of yellow grass-shoot are seen on the ground. If a fire has previously swept through the stand, the surface is covered with a dusty

FIG. 125. Cross-section of a stem of *Sweetia dasycarpa* from the Brazilian dry-forest (after Warming).

ash and the smell of fire remains in the air for a long time. Before the start of the summer-rains, the trees become covered with light-green foliage that often has a varnished appearance. Many begin to flower but most of the flowers look rather unattractive. Only rarely can one see trees that are fully covered with flowering racemes, as for example *Burkea africana*, or the *Erythrina* species which unfold their large, soft, pink coloured flowers before leaf emergence. These are reminiscent of peach trees in the spring. The grasses begin to grow immediately after the start of the rains and soon herbs and shoots from perennials are visible everywhere.

The trees of this woodland have buds that are well protected against desiccation and also thick bark (Fig. 125), which may protect them against damage during the frequent forest fires. Epiphytes are absent and lianas are rare or, where they occur, they are thick-stemmed and woody (Figs 126 and 127).

The woodland of the summer-rain region shows much variation in floristic composition in relation both to climate and habitat. Many descriptions of the vegetation are available, but eco-physiological

FIG. 126. The liana *Fockea damarana* (Asclep.) in a *Sclerocarya schweinfurthiana-Gyrocarpus americana* dry-woodland on dolomite. Otavi mountain area (S.W. Africa) (Phot. E. Walter).

FIG. 127. As Fig. 126, *Fockea damarana*. Stem resembles a giant python (Phot. E. Walter).

investigations are entirely lacking. The most dominant 'Miombo'-types in Zambia and Tanzania are represented by stands of *Brachystegia* with *Isoberlinia* (all legumes) and many other tree species. These woodlands are usually no taller than 9-12 m and are reminiscent of the low oak-forests (Niederwald) of central Europe. The crown attains 60-80% cover. The leaves of the trees open and they usually bloom one month before the start of the summer-rain season. The duration of the leafless condition during the dry season—which is also the cooler season of the year—varies with the moisture relations of the habitats. Undergrowth is present, yet a grass layer is hardly formed. Poikilohydrous ferns, such as *Pellaea hastata* and *Actiniopteris radiata*, have been reported as being present. A more detailed description of the different types is given by Burtt (Burtt, 1942; Rattray and Wild, 1961-62).

Contrary to a widely held opinion, the tree species of the Miombo are not fire-resistant. The woodland can be destroyed by repeated burning, particularly if the fires occur shortly before the onset of the rainy season, when they are very hot (Trapnell, 1959). The Mopane (*Colophospermum mopane*) forests are more open types with a much denser, grassy undergrowth. The same applies to the more open forest types with Combretaceae and *Pterocarpus* or with *Acacia*, *Albizzia* and *Commiphora* species.

In areas near the equator with a very low rainfall concentrated in two rainy periods, as for example in East Africa, very dense, impenetrable thorn-scrub thickets often occur. They are composed of Acacias, which are conspicuous by their umbrella-shaped crowns, and other thorny scrubs. They are particularly well developed even in high rainfall areas on very permeable coral-rock at the coast. Baobab trees (*Adansonia digitata*) with their monstrous branches and twigs stand high above the thorn-scrub. They are leafless during the dry season when only their large, elongated fruits hang suspended from the branches. Today, these coastal strips are usually cleared but the Baobab trees have remained to form characteristic landmarks in these areas (Fig. 128). Because of this, the tree is mentioned in all travellers' accounts. Giant individuals, little more than 20 m high attain a girth of more than 20 m. The water stored in the soft wood was estimated, in one instance, as 120,000 litres! As a consequence, this species can easily endure long dry periods. As shown on the distribution map (Fig. 129), *Adansonia digitata* occurs both at the most arid limits of the woodland area and in the transition zone to climatically conditioned savannahs. Wherever this species is found in areas of higher rainfall, it is restricted to dry habitats, e.g. permeable, calcareous coral rock at the East African coast. Trees occurring in settlements are commonly planted, thus, they are not in their native habitats. This applies also to the specimens on the Canary and Cape Verde Islands on Madeira or the islands in the Indian Ocean. Several related species are found in Madagascar and in north-west Australia. The age of the largest individuals can easily be overestimated. A tree planted in

FIG. 128. A big Baobab (*Adansonia digitata*), Kruger National Park, South Africa (Phot. E. Walter).

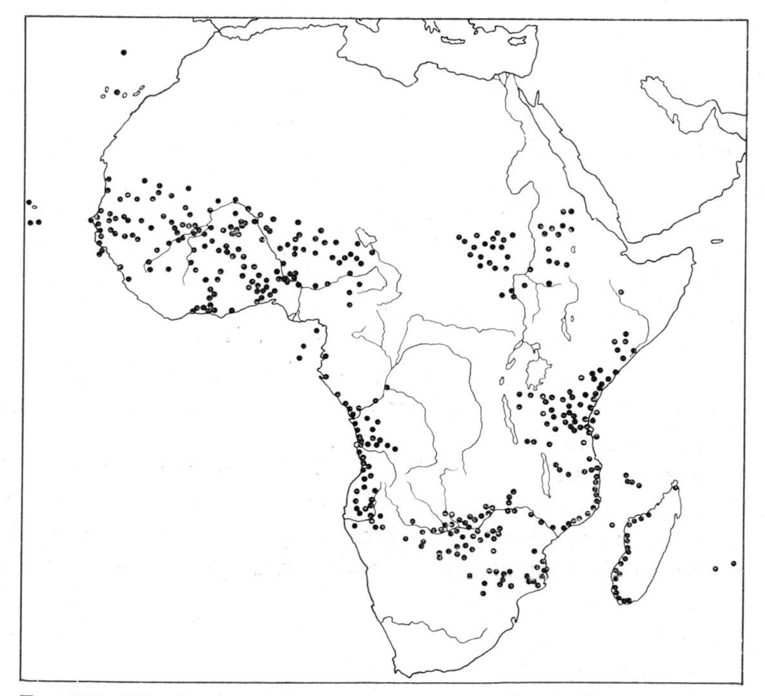

FIG. 129. Distribution of the Baobab, *Adansonia digitata* (after Springer).

Khartoum on the Blue Nile had attained a diameter of near 1 m in 30 years. According to E. R. Swart (Salisbury), an age of 1000 years was obtained by the Carbon-14 dating method in the case of a giant, but not hollow, tree.[1]

There are still drier semi-desert areas near the equator, situated on the sheltered side of the mountains. Windward and leeward sides can be very different climatically especially in the trade-wind areas, where the winds constantly blow from one direction (Wilhelmy, 1957).

Particularly strange vegetation formations are found, for example, in the small arid area between the Pare and Usambara Mountains in East Africa.

This is probably the driest area near to the equator (4-5° S). Exact climatic measurements are not available. The annual precipitation only amounts to 100-200 mm at a mean annual temperature of about 28°C (soil temperature 30·5°C at 40 cm depth). The rainfall is distributed between two seasons. This favours the development of succulents, which has resulted here in the formation of one of the strangest of succulent-semi-deserts. It contains very interesting species: the colony-forming Sansevierias and the stem-succulents, *Euphorbia tirucalli* and *Caralluma* (Stapeliaceae) (Fig. 130). Others include *Pyrenacantha* (Icacinaceae), which resembles grey pieces of rock, and *Adenia globosa*, whose stems form bulbous green mounds. Leafless, bow-shaped, but thorny, green branches grow in all directions from their apices and the stem only becomes visible when these branches are removed. *Cissus* species cling to the shrubs as climbers. They are related to the grape vine but their succulent stems are reminiscent of certain *Cereus* (Cactaceae) species. One can also find the very poisonous *Adenium* with its beautiful flowers (Fig. 131) amongst the thorn scrub; a species of this genus forms very peculiar misshapen trees in Socotra.

Succulents also occur in other tropical areas in especially dry habitats, e.g. on rocky slopes. In Africa they include the candelabra-euphorbias and *Aloë* species, in America the cacti.

Woodlands of the drier type are also widely distributed in South America, for example in the Chaco region. However, I shall restrict my discussion to the Brazilian 'Campos cerrados' and the 'Caatinga,' for which ecological studies by Rawitscher (1948) and Ferri and Coutinho (1959) are available.

The Campos cerrados occur in a typical summer-rain area with more than 1300 mm annual rainfall and an average temperature of 21·2°C. The soils are very deep (up to 20 m) and they hold the rainfall for 3 years, so that they remain moist throughout the whole year

[1] I extracted these facts from a thorough monographic work on the Baobab tree by A. Springer, *Der Affenbrotbaum* (*Adansonia digitata* L.), Ms. 139 pp., 1 map and many figures. This work includes a careful evaluation of the literature, a detailed distribution map and accounts of the importance of this tree and its mythology in the life of the natives. It is not yet published; however, a copy is available to interested readers through the Botanisches Institut, Stuttgart-Hohenheim.

FIG. 130. Arid area between Pare- and Usambara mountains. Among Sansevierias an almost black flowered *Caralluma* (Asclep.), which gives off a very bad odour (Phot. E. Walter).

FIG. 131. *Adenium obesum* (Apocynac.) flowering among thorn-scrub (Phot. W. Busse). The beautifully flowering but very poisonous *Adenium* spp. are used for poisoned arrows.

from a depth of 2 m downwards. Deep-rooting woody plants are all evergreen and transpire large amounts of water throughout the whole year. Shallow-rooting plants include both the grasses which dry up during the drought period, and deciduous woody species with very strange xylopodia, i.e. underground stems. This latter group includes the 'stemless' palm *Acanthococcos*. On germination its stem grows downwards and only the negatively geotropic leaves appear above the soil. When the stem apex has reached a depth of 50-60 cm it bends through 180° and grows upwards; the downward portion of the stem then dies and disintegrates. Such species are very resistant towards the annual fires, which occur during the dry season.

Amongst epiphytes there are only drought-resistant species, such as Bromeliaceae and cacti. A recent review has shown rather clearly that the concept of the Campos cerrados includes a number of very different vegetation types ranging from closed forests to open savannah.[1] These are in part natural and, in part, anthropogenic vegetation types. The savannah-like appearance is usually the result of annual burning, practised by the Indians even before the arrival of white settlers. The zonal vegetation appears to have been predominantly a moist rain-green forest, but its poor growth is often the result of very poor sandy soil which is probably deficient in micro-elements as well. Such sparse woodland may, however, be natural and, indeed, grow where there are nutrient or trace-element deficiencies. In the state of São Paulo, near Pirassununga I noted the remains of a semi-evergreen rain forest, present on a relict basalt-soil, and here it was obviously the zonal vegetation type. The Campos cerrados with the proteaceous *Roplea* are on poor, sandy soils (cf. with similar situations in Queensland, p. 226). The following table, by J. S. Veiga of the IZIP Research Station, shows that such sandy soils are not only poor in phosphorus and potassium but also in trace elements, using cotton, maize and soya as indicators. The yield in field experiments for different kinds of fertilisation of the soil is given in kg/ha:

Kinds of fertiliser	Cotton	Maize	Soya
Complete	2460	4860	1200
—N	1440	3600	—
—P	270	860	240
—K	70	1300	450
—S	1570	4170	680
—Zn & B	870	4080	750
—Ca	980	3900	1100
Unfertilised	—	—	240

From this it appears that cotton suffers more from Zn + B deficiency than from nitrogen deficiency, the yield falling to about one-third of

[1] *Simposo sôbre do Cerrado*, Univ. São Paulo 1963.

that obtained with a complete nutrient solution. Maize is scarcely affected by the lack of Zn + B, while soya reaches about three-fifths of its potential maximum yield when these elements are missing. An experimental *Panicum maximum* sward also indicated a deficiency of Zn and B very clearly.

Similar considerations also seem to apply to the 'Chaparrales' in Venezuela, which represent an open tree savannah. In contrast to Vareschi (1960), Hueck (1961b) does not attribute the savannah-like character of this formation to burning. Yet, semi-evergreen forest should be able to exist in an area with an annual rainfall of 1700 mm, where the dry period is less than three months despite the uniform mean temperature of 28·2°C.

The Caatinga in the dry north-east of Brazil can be considered as the true extreme dry-woodland or semi-desert. Hueck (1960-61) describes such woodland as a 'deciduous, rain-green, open woodland composed of low and medium-sized trees and thorny shrubs. The type

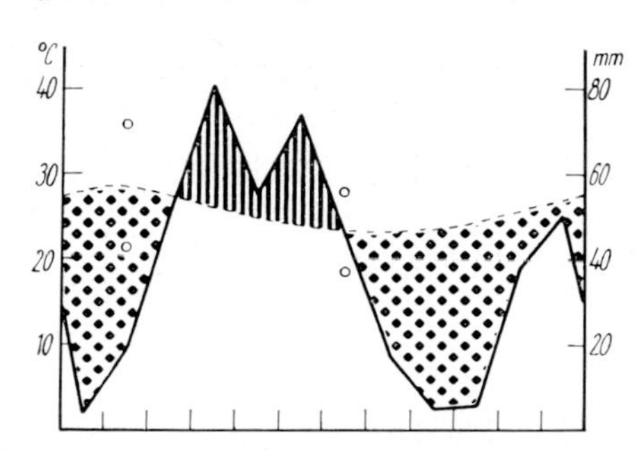

FIG. 132. Climatic-diagram from the Caatinga area from data of Ferri. The humid season is only brief, the two rainy seasons favour the growth of succulents. Abscissa: months from July to June. Temperature curve interpolated.

occurs on any, even the poorest, soil and, indeed, corresponds to the thorn scrub of other hot countries. The particularly characteristic ecological conditions are lack of rain during many months, great irregularity of precipitation, relatively low air-humidity and strong evaporating conditions. In the centre of this arid area, the 'Poligono da Sêca', a drought lasting for several years becomes a 'scourge to the country".' The Caatinga is characterised by water-storing 'bottle-trees' and by numerous succulents, such as cacti, non-epiphytic Bromeliaceae and succulent Euphorbiaceae. The abundance of succulents is related to the distribution of the rainfall. In this area, which lies close to the equator, the annual rainfall of 500 mm is spread over a larger wet period from October to December and a smaller one from May to June (Fig. 132). The conditions are similar, therefore, to those in Africa in the rainshadow of the Pare mountains (see p. 222).

Ferri (1955), who investigated the transpiration of the succulent plants of the Caatinga, found that they keep their stomata closed during the dry season (those that did not shed their leaves). The plants are characterised by very low cuticular transpiration, although their leaves do not exhibit any xeromorphic features. Transpiration is supposedly reduced even during the rainy season. Thus, their response is exactly opposite to that of the plants in the Campos cerrados.

Very peculiar ecological relations exist in the summer-rain areas of Australia. The tropical and subtropical rain forests, composed of south-east Asian elements, and occurring on the high-rainfall east coast of Australia, have already been discussed (p. 180). Semi-evergreen forests floristically related to them are only found as relicts in a very few localities. I visited such a site with Dr Webb, 60 km south-west of Mt. Garnet. The forest showed the following structure:

Tree layer (20-25 m) open, consisting of *Brachychiton australe* (bottle-tree), *Gyrocarpus americana*, *Pleiogynium cerasiferum*, *Polyscias elegans* and a *Ficus* species, the only evergreen tree.

Shrub layer (2·5-3 m) very dense and rich in species, in part evergreen, including *Celastrus balocularis* (leaves like *Ilex*), *Strychnos arborea*, *Carissa ovata*, *Melia dubia*, *Alstonia constricta*, *Maba humilis*, *Trema aspera*, *Ehretia membranifolia*, *Canthium lucidum*, *Geijera salicifolia*, *Erythroxylon australe*, *Turraca brownii* and many others.

Ground layer virtually absent, scattered ferns (*Cheilanthes tenuifolia*) and small moss cushions.

Lianas were often as thick as an arm singly or intertwined, e.g. *Parsonia*, *Cissus*, *Cynanchum*, *Celastrus*.

Epiphytes included some such as *Bolbophyllum* and *Dendrobium* spp. as well as poikilohydrous Polypodiaceae.

Forests with south-east Asian elements are found on volcanic soils (basalt), particularly in the marginal areas.

Wherever there is a change in bedrock and a reduction in rainfall, the dense rain forest is immediately replaced by open *Eucalyptus* forest. The boundary is very abrupt in terms of physiognomic and floristic criteria. Only a few south-east Asian elements extend into the moist *Eucalyptus* forest. In all other respects, the character of the *Eucalyptus* forest is entirely different from that of either the semi-evergreen, the humid rain-green, or the dry-forests of other tropical regions. Except for the facultatively deciduous eucalypts (see p. 208), which occur only in the more humid, warmer areas, all *Eucalyptus* species, including those of the arid regions, are evergreen with leathery, although not really sclerophyllous, foliage. In addition, they are characterised by the vertical position of their leaves. Thus, all *Eucalyptus* forests have an open appearance, even those of the humid areas, and all are associated with a relatively tall, closed grassy undergrowth, at least in the summer-rain areas. *Eucalyptus* species are particularly suitable for growing in combination with grasses (see p. 263). This is also the reason why physiognomically there are only very gradual differences between the humid and dry *Eucalyptus* forests and the open *Eucalyptus* savannahs (Fig. 133). Evergreen *Eucalyptus* species also extend up to the timberline in the Australian Alps, where they may be in snow for up to 6 months.

Eco-physiological investigations of the *Eucalyptus* forests in the summer-rain regions are not yet available.

FIG. 133. *Eucalyptus marginata*—forest with *Kingia australis*, *Xanthorrhoea preissii* and the author. Porongorup Mountains, West Australia (Phot. E. Walter).

4. Tropical parkland areas

In the more humid areas of the tropical summer-rain region, a great deal of rain (i.e. 1000-2000 mm) commonly falls in a relatively short rainy season. As a consequence, on level terrain vast areas with clayey soils remain flooded for several months. However, the same areas dry up very markedly during the dry season. Thus such habitats are extreme with respect to the dynamic alternations in soil moisture. Forest cannot develop in such habitats. However, grasses can cope with such situations and certain woody plants. The latter include palms such as *Borassus* and *Hyphaene* in Africa, which here occur scattered in the grassland. Occasionally they are mixed with some Acacias. Thus, here we have edaphically conditioned grassland (occasionally a palm savannah) associated with a particular topography. The grassland is commonly interrupted by islands of forest resulting in a parkland formation. Unfortunately no ecological investigations of the habitat relations such as duration of flooding, aeration of the soil, its water content during the dry season, rooting relations, seasonality of the vegetation, etc., are as yet available. Normally, these areas are only visited during the dry season since they are impassable during the rainy season.

As an example, I will discuss here the parkland area in northern south-west Africa at the Omuramba u Ovambo within the zone of dry rain-green woodland. This is an area that is perfectly flat. It is very accessible during the dry season. Smaller or larger groups of trees or small spinneys are scattered in a picturesque manner over the dry, grass-covered plain, which also boasts a few scattered palms here and there. The ecological factors that control the distribution of the tree colonies and the grassland are not readily apparent. However, when I had the opportunity to get into this area in a four-wheel-drive under considerable difficulties after several very heavy rains, their relationship became immediately obvious. The entire grassland was covered with water whereas the tree-islands stood out of the flooded landscape just like real islands with dry surfaces. The microrelief is hardly detectable but difference in level of 20-50 cm is sufficient to be significant. Continuing rain does not result in the flooding of the tree-islands, instead the water masses slowly begin to move in broad flood-sheets, although it is still uncertain which direction they take, whether to the Cunene in the north or to the Etosha Pan in the south. Soil under the grass may be saturated for a long time during the rainy period, which can last 4-5 months, but such conditions cannot be tolerated by the trees of this area. The flood waters stagnate for an even longer period in the lowest depressions, from which they evaporate only slowly during the dry season. These are the clay pans, the Vleys, which support the development of a swamp vegetation including beautiful Amaryllidaceae. Swamp forests cannot develop in such areas because the soil dries out completely during the drought season. The grasses turn yellow and the swamp plants in the Vleys dry up (map of the area, Fig. 198, p. 339, cf. also Fig. 134). Clearly, an important problem is the origin of the microrelief on these plains.

In northern south-west Africa it was possible, in many cases, to detect that the tree colonies were growing on very flat, spread-out, old termite mounds. If a group of trees originates in this manner, the raised patch can become enlarged subsequently through wind-blown deposits that become lodged around them during the dry season.

C. Troll has described a similar parkland area in East Africa, which he has called 'termite savannah' or 'flood savannah'. The wooded areas are much higher than the grassland and are readily recognised as termite mounds. C. Troll thought that these mounds were richer in nutrients and would thus favour tree growth. This question has been investigated more thoroughly by Hesse (1955). He could not substantiate Troll's opinion. The termites use subsoil exclusively for their termite mounds and they do not modify the soil. So long as the mounds are occupied, the termites maintain them but unoccupied mounds readily degenerate. Termites prevent the establishment of vegetation although lush vegetation develops on old termite mounds. Sisal-agaves also grew much better in plantations on such old termite mounds. However, the better growth is not related to a higher nutrient content but to more favourable physical conditions. The soil is deeper and is better drained and aerated. Soil from termite mounds was found to be

FIG. 134. Termite savannah in the north-west of Kenya; on the left is a shrubby 'Candlestick-*Euphorbia*' (Phot. E. Walter).

unfavourable for plant growth when spread on land, preparatory to cultivation, because it is derived from sterile subsoil.

It has also been observed that dry woodland areas and the lusher vegetation on the termite mounds may be surrounded by fires, so that the development of grass was encouraged further in such burned-over areas. This explains the origin of the 'termite savannah' (see Fig. 135, p. 240), which resembles flood savannah, inasmuch as it originates through a difference in topography and is not a homogeneous community but a mosaic and, therefore, more a parkland than a 'true savannah'.

Similar conditions have been described for the central Sudan by Bunting and Lea (1962). Flooded areas on clay soils are grassland yet minor height differences of 10 cm can cause a change in the vegetation. Rain only penetrates to a depth of 30-40 cm and the water stored in the soil corresponds to a rainfall of about 100 mm, which is used up by the vegetation in 3 weeks, provided there is no rain. The annuals die rapidly under these conditions. The presence of annual, and the absence of perennial grasses is probably the result of the extremely fierce fires.

Edaphically conditioned savannahs are also found in north Surinam with an annual precipitation of 2000-2500 mm. They are probably caused through the presence of shallow, impermeable kaolin-horizons and the formation of hardpans during soil development, or, also, through natural surface erosion near deeply cut rivers which greatly decrease the water reserves (Bakker, 1954).

Even in tropical landscapes little modified by man, the traveller is often confronted with a complex pattern that is not very easily clarified. However, wherever there is an opportunity for a longer stay combined with thorough ecological investigations, it is possible to detect the significant relationships between vegetation and habitat. They are determined by microclimatic differences, by differences in topography, exposure or soil. But firmly substantiated analyses can only be made after ecological experimental stations have been established in these as yet little known areas, where they could provide a basis for prolonged observations.

5. Tropical grassland

Grasslands determined by climate are found in tropical forest areas, just as in the forest areas of the temperate zone, but only in the alpine areas above the timber line. In spite of this, natural grassland is widely distributed in the tropics (Michelmore, 1939).

Here we distinguish:

I. Primary or natural grassland, that is:

A. Edaphically conditioned grassland

1. On alternating wet and dry habitats

(*a*). grassland in the area of the plain water divides or plains, known as 'Dambo', in East Africa,
(*b*). periodically flooded grassland in river valleys,
(*c*). grassland in depressions, which are very wet during the rainy season, but which dry up during the drought season, known as 'Mbuga'.

2. On shallow, dry soils.

B. Grassland in the mountains caused through natural fires.

II. Secondary or anthropogenic grassland, originating from the intervention of man (logging, fire, grazing).

The entire region from south Sudan to north Transvaal is climatically a potential forest region. Viewed from above these vast populated and, in Zambia and Tanzania, largely unoccupied regions give an idea of the principal distribution of the different vegetation types, which is here related to topography. In particular, one can gain an impression of the size of the areas not occupied by forest, but by natural grassland and swamp instead.

Consider first the conditions at the watershed on this very plain (at 1100-1400 m above sea level) between the Atlantic Ocean (at the origin of the Congo River system) and the Indian Ocean. This area receives heavy summer-rains that are followed by extreme drought during the somewhat cooler season of the year.

This marked seasonal contrast between humid and very arid conditions is further enhanced by the topography. Water moves during the rainy season as flood-sheets from the higher parts to the lower, even where topographic differences are slight, bringing about the transfer of soil particles.

The hummocks show eluviated soils from which particles have been removed, the hollows illuviated soils where particles have been deposited. Transition areas commonly show colluvial soils, in which both processes have occurred, removal and deposition. The sequence, eluvial—colluvial—illuvial soil is correlated with a sieving of the soil particles according to their size. The more coarse particles are deposited first, on the slope, while the finer clay particles are deposited in the hollows.

This results in a topographically dependent sequence of soils—the catena (= chain) of Milne—which is repeated on every slope. It is correlated with a similarly recurring sequence of vegetation (Morrison, *et al.*, 1948).

The eluvial soils are relatively dry, the colluvial soils are moister but still well aerated whereas the illuvial, clayey soils are impermeable to water and periodically have standing water at the surface. During the rainy period these last swell when they are sticky and poorly aerated but, when the water has evaporated and been lost, they dry out completely and become as hard as rock. These habitats show, therefore, a notable alteration of water content. Such biotopes are unsuitable for tree growth and are occupied instead by grasses.

When flying above the watershed in Zambia, one can see the origins of the river systems very clearly in the form of broad, flat strips of grass. The angle of slope is so slight that rain wash forms floods in these shallow, branched hollows, from where it drains away only very slowly. One only notices the formation of a river bed in the centre further down. However, broad strips of grass still continue to be associated with the rivers for a long way. Where the slope increases, so that the river cuts a deeper bed, the strips of grassland disappear entirely and the narrow river valleys continue through a closed region of woodland. However, when any slope is lacking, the rivers spread into extensive swamp areas.

These grasslands have recently been investigated more closely by Vesey-Fitzgerald (1963). Characteristic grasses are *Loudetia simplex*, several *Hyparrhenia* species and several other Andropogoneae and Paniceae. The height of the leaf layer is 50 cm, the stems reach up to 1 m and 50% of the soil remains exposed between the basal parts of the bunch grasses. In addition, one finds scattered geophytes and some herbs. These 'Dambos' are separated by a distinct boundary from *Brachystegia* woodland; they are less sharply separated from the *Combretum* stands, while *Acacia* stands often grade very transiently into grassland.

Grassland can also occur on slopes, wherever there is constant seepage during the rainy season; that is, where the soil remains waterlogged for a long period.

Very large areas are occupied by grassland in broad hollows or in rift-valleys, where rivers have been dammed by tectonic movements or where basins have been formed that lack outlets. In such habitats one can find the black cotton soils that are completely waterlogged during the rainy season. These are called 'Mbuga' in East Africa. They occur in Kenya and also in the Serengeti as well as in the Massai-steppe, and are often impassable for several months during the rainy season. They contract very greatly during the dry season, when they become hard, like rock and furrowed with deep cracks so that they can only be traversed by car with extreme difficulty. The clay content is very high. The dark colour can only be attributed to dark humus-clay complexes in the absence of high organic content. In spite of their wetness and stickiness during the rainy season, the water only penetrates a very little into the soil but exact measurements of the water relations are not available. Such grassland is invaded only by very few woody plants comprising several Flötenakazie, of which *Acacia drepanolobium* is most widely distributed. This species reaches about 3 m. At the base of the two lateral leaf-thorns are, regularly, two globular galls, whose walls are penetrated by ants when soft and immature. Their larvae are found inside the hollow galls, which serve for nesting, and as soon as one touches a twig the ants appear outside, ready to attack an enemy. When the wind blows across the openings of the galls, a special sound is produced (hence called 'Whistling Thorns'). After the start of the rainy season in February, the white inflorescences are formed together with the delicate, feathery leaves but, usually, these Acacias are only seen in the defoliated condition.

Other gall-acacias are *A. formicarum*, *A. seyal* var. *fistula* and the arabic gum-producing *A. malacocephala*, etc.

In addition to these Acacias there occurs the very tolerant *Balanites aegyptiaca*, which has long, green thorns. These trees occur south-east of Nairobi, widely dispersed but at regular distances from one another. Perhaps they have very long lateral roots. Their crown cover was hardly over 1%, the grass layer was nearly closed and composed mainly of *Pennisetum mezianum* with much *Themeda triandra*, some *Digitaria milanjana* and scattered *Pennisetum stramineum*.

Frequently, a solid, impervious clay hard-pan is formed in the hollows. After rain, such soils become slippery at their surface but the water does not penetrate to any depth. Occasionally, a layer of sedimented sand is found on top. These soils too dry out during the dry season and are covered with a sparse grass cover. Such pans are frequently found interspersed in woodland areas.

Grass cover is also found on the flood plains of large river valleys and on flats surrounding swamps and lakes. This is where one often finds the 'termite savannah', i.e. the previously mentioned vegetation mosaic where tree growth is associated with termite mounds. Tree growth commonly accompanies the rivers along the levées also, while the adjacent flood plains are covered with grass. The designation of such landscapes as 'savannahs', as has become common practice among geographers, is not satisfactory from an ecological viewpoint.

Only a few tree species occur as components of the grassland. These are especially certain palms, species of *Borassus* and *Hyphaene*, which can occur in scattered patches. By contrast, *Syzygium cordatum* prefers terraced slopes or termite mounds. *Acacia nilotica* is a typical flood-plain tree, which can withstand 2-3 months of flooding. However, it can survive only in moving water, which provides better aeration than stagnant water.

It is more difficult to find an ecological explanation for grassland in mountainous areas where one can find it over 1700 m. In tropical Africa this is still below, or within, the *Podocarpus* forest belt, which extends from 1800-2200 m. This grassland has only scattered woody plants, such as *Protea*, *Erythrina*, *Dombeya*, *Cussonia*, *Faurea*, *Myrica* and *Upaca* (Euphorbiaceae). The dominant grass is *Themeda triandra*. Michelmore explains the absence of forest in this case as due to poor soil aeration, even on moderate slopes. He refers to the higher rainfall and the lower evaporation rate in the mountains. Woodland occurs only on stony, steep slopes and in deeply cut, active erosion valleys which are better drained. However, other authorities consider the grassland to be a secondary formation determined by repeated burning with grazing or cultivation; only the moist ravines have, supposedly, been spared fires. Here also it is desirable to initiate detailed ecological investigations. The question of the naturalness of these grasslands will be discussed later.

Much smaller grassland areas occur on shallow, dry soils. They are found on rocky outcrops or knolls or above hardened iron-oxide pans. These pans may only have been formed, however, beneath secondary grassland on former woodland soil, after the soil had been more strongly exposed to direct sunlight upon the removal of the tree cover. Such iron pans then interfere with tree seedling establishment.

Secondary grassland is widely distributed in all populated areas, which, as a replacement community, has resulted from the destruction of forest through cutting, burning or grazing.

Wherever single trees remain as remnants, a savannah-like plant cover results. However, these are anthropogenic savannahs. Once a grass cover has become established, repeated fires are responsible for its continued spread. Even humid forests become scorched each year at their margins and are thus gradually pushed back. The sharp boundary between forest and secondary grassland is usually the result of such fires. Commonly, a cover of pure grass with but a single dominant species becomes established. In place of the rain forests one finds the elephant-grass, *Pennisetum purpureum*, while drier forest types are replaced by *Imperata cylindrica*. Annual fires alone do not damage the grasses. Instead they often show better development, because of the removal of the dead grass. It is only when grazing animals are driven on to the burned areas just when they start to become green after the rain that the better grass species are so heavily exploited that the poorer species become dominant.

These secondary savannahs and grassy parklands have become so widespread that today they commonly dominate the general picture of the landscape. Since they are often hard to identify as replacement-communities, they are commonly considered by geographers to represent the natural vegetation cover. This accounts for the separation of a 'moist savannah' with tall grass cover from a 'dry savannah' with short grass cover as separate vegetation zones that occur next to the rain forest zone. In reality, however, these grasslands are merely replacement communities of certain forest or woodland types. As truly zonal vegetation, savannahs and grassland only occur with a very much lower annual rainfall (Chap. VI, 3).

However, the question arises, whether fire should not be considered a natural factor. Complete protection of woodland against fire is practically impossible. It may even be dangerous, because dry litter may accumulate in such amounts that a fire caused through lightning or negligence may become particularly hot and devastating.

With this question, we return once more to the grasslands of mountainous areas. The problem has received particular attention in Natal. There a *Themeda triandra* dominated grassland covers all the slopes of the Drakensberg mountains in the montane *Podocarpus latifolius* forest belt, while the forest is restricted only to steep slopes and ravines. In the grassland only *Protea* species occur to a height of 3-4 m, with bark 2 cm thick that is very resistant to fire. However, even these trees may succumb to a very hot grass fire. The fire is less severe only in those places where the grass cover is subjected to annual burning, which allows for little accumulation of litter. *Protea* seedlings can survive such fires and after a certain time a dense *Protea* thicket can become established. Grass fires that run down-slope are similarly not so intense.

Today, grass fires are usually caused by man, either intentionally or by accident. However, Killick (1963) emphasises that grass fires caused by lightning are frequently observed in the Drakensberg mountains. In 1957, lightning caused some 17% of all fires, in other years 6-13%. These fires are usually caused by the first thunderstorms in the spring, when the dead felted grass is still dry. However, grass fires can also start during the dry period, for example through the breakage and fall of hard, quartzite rocks, producing a shower of sparks when they strike other rocks on the ground. Thus, grass fires occurred before man's intervention. They were not then as frequent, yet frequent enough to be reflected as a factor of the natural environment in the plant cover. The same is also probably true for the mountainous areas of tropical Africa. Grass fires are now, in part, used as a management tool to reduce the danger of natural fires.

This shows that the grassland in mountainous areas can be considered as the natural primary vegetation cover, conditioned primarily neither through climate nor through soil, but through natural grass fires. This view is also held by Bayer (1955). If such grassland is artificially protected from fires for many years, the characteristic *Themeda triandra* disappears entirely. It is soon eliminated through competition with other grass species.

However, one should not overstress the importance of such natural grass fires. According to Talbot, fires in Uganda are practically always man-induced. Without these, the forest would spread. Besides climate, soil and fire, the equilibrium between grasses and woody plants is also influenced by big-game animals (Talbot, 1964). Strong grazing pressure eliminates inflammable material, which favours woody plants. Low grazing pressure, on the other hand, favours the taller grasses, which then gain dominance. However, while cattle specialise, primarily, in their feeding habits on grasses, the grazing of big-game animals shows a less degradative, complementary pattern.

Giraffes feed on tree foliage, rhinoceros' on shrubs, gnus on fresh, green grass shoots only, zebras on somewhat older, but still green leaves, other wildlife species on still drier leaves and stalks. In addition, one distinguishes between wild animals that roam from place to place and those that remain in one area. For these reasons, the feeding capacity of natural pastures in East Africa is about 2-10 times greater for wild animals than for cattle. The biomass, in living animals, can amount to about 12,000-25,000 kg/ha in the case of wild animals, whereas in the case of cattle on farms it can amount to about 5,000-7,500 kg/ha and for native cattle only to 2,500-3,500 kg/ha.

However, it would be wrong to overestimate the role of big-game animals with regard to causal relations in tropical grassland. The big-game herds are unique to Africa. They are entirely absent in South America. In spite of this, grassland is widely distributed in tropical South America as an edaphic formation together with annual rainfalls of 500-2500 mm. The extensive Llanos of the Orinoco region may be mentioned as an example. The term 'Savannah' was originally applied to these South American grassland areas. For this reason, Beard (1953, 1964) defines savannah as the natural vegetation on mature soils of ancient land formations with poor drainage, i.e. those with temporary water logging at one time and extreme drying out at another. Grass fires are frequent in savannah, which according to Beard represents an 'edaphic climax'. However, fires are not a prerequisite for either the initiation or maintenance of the savannah formation. We see, therefore, that they belong to the topogenic grasslands series which have been discussed in some detail. Climatically, these areas are potential forest or woodland.

References

ASHTON, P. S. 1958. Light intensity measurements in rain forest near Santarem, Brazil. *J. Ecol.*, **46**, 65-70.

BAKKER, J. P. 1954. Über den Einfluss von Klima, jüngerer Sedimentation und Bodenprofilentwicklung auf die Savannen Nord-Surinams. *Erdkunde*, **8**, 89-112.

BAYER, A. W. 1955. *The ecology of grasslands. The grasses and pastures of South Africa.* Johannesburg.

BEARD, J. S. 1944. Climax vegetation in tropical America. *Ecology*, **25**, 127-58.

BEARD, J. S. 1953. The savannah vegetation of northern tropical America. *Ecol. Monogr.*, **33**, 149-215.

BEARD, J. S. 1964. Savanna. *I.U.C.N. publ., N.S.*, **4**, 98-103.

BHATIA, K. K. 1958. A mixed teak forest in central India. *J. Ecol.*, **46**, 43-63.

BUECHNER, H. K., and DAWKINS, H. C. 1961. Vegetation change induced by elephants and fire. *J. Ecol.*, **42**, 752-66.

BUNTING, A. H., and LEA, J. D. 1962. The soils and vegetation of the Fung, East Central Sudan. *J. Ecol.*, **50**, 528-58.

BURTT, B. D. 1942. Some East African vegetation communities. *J. Ecol.*, **30**, 65-146.

CHAMPION, H. G. 1936. A preliminary survey of forest types of India and Burma. *Indian Forest Records, N.S. Silvicult.*, **1**, 1-287.

CLAYTON, W. D. 1961. Derived savanna in Kabba-Province, Nigeria. *J. Ecol.*, **49**, 595-604.

DEVRED, R. 1957. Limite phytogéographique occidentoméridionale de la region guinéenne au Kwango. *Bull. Jard. Bot. Bruxelles*, **27**, 417-31.

EVANS, G. C. 1939. Ecological studies in the rain forest of southern Nigeria. *J. Ecol.*, **27**, 436-83.

EVANS, G. C. 1956. An area survey method of investigating the distribution of light intensity in woodlands, with particular reference to sun flecks. *J. Ecol.*, **44**, 391-428.

FERRI, M. G. 1955. Contribuição ao conhicimento da ecologia do Cerrado e da Caatinga. *Univ. São Paulo, Bot.*, **12**, 195.

FERRI, M. G., and COUTINHO, L. M. 1959. Contribuçâo ao conhicimento da ecologia do Cerrado. *Univ. São Paulo Bol.*, **224**, 103-50.

HESSE, P. R. 1955. A chemical and physical study of the soils of termite mounds in East Africa. *J. Ecol.*, **43**, 449-61.

HOPKINS, B. 1962. Vegetation of Olokemeji Forest Reserve, Nigeria, I. *J. Ecol.*, **50**, 559-98.

HUECK, K. 1960-61. Die waldgeographischen Regionen und Unterregionen von Südamerika. *Geogr. Taschenbuch*, 224-34.

HUECK, K. 1961a. Die Wälder Venezuelas. *Beih. Forstw. Centbl.*, **14**.

HUECK, K. 1961b. Verbreitung. Ökologie und wirtschaftliche Bedeutung der 'Chaparrales' in Venezuela. *Ber. Geobot. Inst. Rübel.* **32**, 192-203.

HUECK, K. 1965. *Die Wälder Südamerikas. Vegetationsmonographien der einzelnen Grossräume.* Stuttgart.

KILLICK, D. J. B. 1963. An account of the plant ecology of the Cathedral Peak Area of the Natal Drakensberg. *Bot. Surv. S. Afr.*, **34**.

LAUER, W. 1952. Humide und aride Jahreszeiten in Afrika und Südamerika und ihre Beziehungen zu den Vegetationsgürteln. *Bonner Geogr. Abh.*, **9**.

MANSHARD, W. 1960-61. Ein Vorschlag zur Gliederung und Benennung von Vegetationsformationen in Afrika südlich der Sahara. *Geogr. Taschenb.*, 454-63.

MICHELMORE, A. D. P. 1939. Observations on tropical grasslands. *J. Ecol.*, **27**, 292-312.

MISRA, R. 1959. The status of the plant communities on the upper Gangetic plain. *J. Ind. Bot. Soc.*, **38**, 1-7.

MORRISON, C. G. T., HOYLE, A. C., and HOPE-SIMPSON, J. A. 1948. Tropical soil-vegetation catenas and mosaics. *J. Ecol.*, **36**, 1-84.

OGAWA, H., YODA, K., and KIRA, T. 1961. A preliminary survey on the vegetation of Thailand. *Nature & Life in SE Asia*, **1**, 21-157.

PURI, G. S. 1960. *Indian Forest ecology*. 2 vols. New Delhi and Calcutta.

RATTRAY, J. M., and WILD, H. 1961-62. Vegetation map of the Federation of Rhodesia and Nyasaland. *Kirkia*, **2**, 94-104.

RAWITSCHER, F. 1948. The water economy of the vegetation of the 'Campos cerrados' in southern Brazil. *J. Ecol.*, **36**, 237-68.

TALBOT, L. M. 1964. Savanna. *I.U.C.N. publ., N.S.*, **4**, 88-97.

THAI-VAN-TRUNG. 1962. Ecology and classification of the forest vegetation of Vietnam. *Akad. Wiss. Leningrad* (in Russian).

TRAPNELL, C. G. 1959. Ecological results of woodland burning experiments in Northern Rhodesia. *J. Ecol.*, **47**, 129-68.

UNESCO. 1962. *Humid Tropics Res. Progr.*, **106 B**.

VARESCHI, V. 1960. La estacion biologica de los llanos de la sociedad Venezolana de Ciencias Naturales y su tarea. *Bol. Soc. Venez. Cienc. Natur.*, **96**, 107-34.

VESEY-FITZGERALD, D. F. 1963. Central African grasslands. *J. Ecol.*, **51**, 243-74.

VIDAL, J. 1959. Conditions écologiques, groupments végétaux et flore de Laos. *Bull. Soc. Bot. Fr.*, 3-41.

WHYTE, R. V., VENKATURAMANAN, S. V., and DABADGHAE, P. M. 1954. The grassland of India. *8th Congr. int. Bot., Compt. Revd. Sect.*, **7** & **8**, 46-53.

WILHELMY, H. 1954. Die klimamorphologie und pflanzengeographie Entwicklung des Trockengebietes am Nordrand Südamerikas seit dem Pleistozän. *Die Erde*, **6**, 244-73.

VI. *Natural Savannahs as a Transition to the Arid Zone*

1. The savannah concept

In spite of the original usage of the terms 'savannah' and 'steppe' for very definite vegetation types, it has recently become common practice among geographers to use these terms for designating certain climatic zones. C. Troll (1935) at first suggested designating all hygrophilous plant communities in the tropics as 'savannahs' and all xerophytic communities as 'steppes', regardless of the kind of vegetation present. However, since the terms hygrophilous and xerophilous are rather difficult to define, Jaeger (1945) interpreted 'savannah' in a purely climatic sense. Thus, his 'savannah' concept has little to do with any specific zonal vegetation.

The 'humid-savannah' concept embraces the climatic zone of tropical summer-rain with a dry period of 2·5-5 months, that of dry-savannah the zone with a dry period of 5-7·5 months and thorn-scrub savannah is understood as coinciding with a dry period of 7·5-10 months. C. Troll (1952) has accepted these definitions and remarked that 'savannah' always reflects a combination of edaphic, biotic and anthropogenic factors and, therefore, includes all forms, from treeless grassland to forest. When used in this way, the concept loses its original plant-ecological meaning and becomes a climatic zone, encompassing various plant formations. With this type of usage it would appear more appropriate to use a climatic designation such as 'summer-rain area of the tropics', or something similar. If designations descriptive of vegetation are employed, however, they should match reality. It appears absolutely essential for a functional analysis to distinguish natural formations from purely anthropogenic ones, even if both have physiognomic similarities; otherwise, an ecological viewpoint cannot be maintained.

Therefore, I will use the term 'steppe'—which has been misused so commonly, particularly by investigators concerned with African vegetation—only for the grasslands of the non-tropical zone. The Russian term 'stepj' was originally applied only to grassland in the southern part of eastern Europe, corresponding to the American prairie.

The term savannah as here interpreted—in accordance with Schimper, Drude and investigators from other countries—refers to the natural, homogeneous zonal vegetation of the tropical summer-rain zone showing a closed grass cover and scattered, individual woody plants, either shrubs or trees. This climatically conditioned savannah probably occurs in Africa only in areas with rainfall below 600 mm and corresponds most closely to the 'thorn-scrub savannah' of Jaeger. Where grass occurs without trees on sandy soils in areas marginal to desert, it is best simply to speak of 'grassland'.

Physiognomically very similar 'savannahs' are also widely distributed in the more humid parts of Africa and South America but there is little doubt now that these are secondary plant communities which have replaced the various woodland types discussed in the preceding chapter. Here, their origin is due to burning, carried out annually by man to destroy the dry grass. In this way vegetation can be changed over wide areas, even in almost unpopulated regions. Where the grass fires are repeated regularly, a characteristic equilibrium is eventually achieved; that is, vegetation with a closed cover of grass and scattered, fire-resistant trees. Whenever burning is discontinued, the original forest vegetation re-invades provided there are relict stands remaining in moist ravines, on ungrazed termite mounds or on rocky hills. Such a landscape presents a rather natural appearance to the traveller, just as with the European meadows or heaths. Only on closer investigation is the anthropogenic character of such vegetation revealed. However, the decision is not usually easy to make in any particular area, if one has only a limited time for investigation (Schmithüsen, 1959; for S. America, see Goodland, 1966). There is no objection to designating these as 'secondary savannahs' but one must differentiate them clearly from the climatically conditioned, natural savannahs, which cover a much smaller area.

The secondary savannahs also show characteristic differences in relation to the climate. Much taller grasses develop in the more humid than in the drier areas and the fire-resistant tree species are also different ones. These features cannot be accepted, however, as evidence against their secondary character. One should examine the concept of natural vegetation critically even in the least populated areas of Africa and South America. Natural vegetation is usually only found after a long and careful search.

It is often argued that fires can originate naturally from lightning. This is absolutely correct. However, lightning-caused fires are much less frequent and they do not usually spread very far in natural woodland. Only when forests are opened up by repeated burning in combination with grazing and after broad grassland areas have become established does the effect of fire become so pronounced that the natural vegetation is eventually completely eliminated.

Another natural factor is often emphasised: this is the presence of the many herds of antelope and zebra, which particularly impressed the earlier African explorers. These animals certainly are of especial importance to the ecosystems of these regions. However, it is doubtful whether they play a decisive role in the development of vegetation. Commonly, these herds are forced into unpopulated areas by man, who acquires ever increasing areas for his own use. Moreover, the abundant wildlife is only conspicuous where herds gather in areas with water reserves and sustained fresh grazing during dry years. The semi-arid areas of the natural tall-grass savannahs are especially rich in wildlife.

One thing, however, is striking. All Acacias in Africa have thorns, where the large herds of big game occur, while Acacias in the Australian savannahs are thornless. These latter are inhabited by kangaroos, which never occur in large numbers (Walter, 1961 see p. 254).

From an ecological viewpoint, parkland formations should not be called 'savannahs', either, as was done by C. Troll (1950) (Fig. 135). They are not uniform vegetation types, but macromosaics comprising many different communities. In one instance, they are made up of natural extrazonal woodland communities along rivers, in damp ravines or on raised land that is not flooded during the rainy season; in

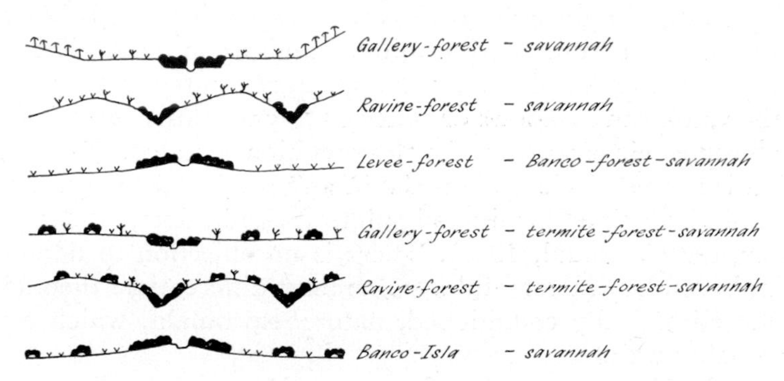

FIG. 135. Landscape types, after C. Troll, called by him 'savannahs'. However, from an ecological viewpoint these represent macromosaics of very different formations, such as pure grassland, savannahs and larger or smaller forest stands.

another, they are of natural grasslands and palm savannahs on flood plains or possibly also of secondary savannahs. A savannah concept which is too broad devalues the term from a scientific viewpoint.

The soils of the secondary savannahs are the same as those under the corresponding woodland types from which they have originated. However, their organic content is reduced by fires so that the soils become short of humus; their nitrogen and phosphorus contents also decrease. The very wet condition during the rainy season and the great desiccation during periods of drought bring about the formation of a shallow, hard iron-pan, which may become somewhat exposed through erosion. The soils are poor and produce good yields only after appropriate treatment with fertilisers.

The soil catena (*sensu* Milne) is particularly well shown here. The catena concept relates to a pedogenic series of soil types from top to the bottom on slopes. While the tops of the hill show red, non-swelling, kaolin-containing soils, the hollows show black, swelling, montmorillonite-containing soils that crack during the dry season. Although the latter are more fertile they require much more cultivation and are thus not used in more primitive agriculture. They represent a big obstacle to the traffic across the land during the rainy season.

In the following sections I will concern myself with the climatically conditioned savannah which forms the zonal vegetation in the dry tropics wherever woodland cannot develop on plateau-habitats.

2. Grass and trees as competitors

'Natural savannah' is understood to mean a homogeneous plant community in which the individual woody plants grow more or less widely apart on a grassy background (Fig. 136). If the woody plants are trees,

FIG. 136. View across a tree savannah, Graslaagte Farm near Otjiwarongo, S.W. Africa (Phot. E. Walter).

whose distance apart is about 5-10 times their height, one is subject to a peculiar illusion by the landscape. This is commonly the case, for example, in the central Kalahari. It appears as if one is standing in a forest glade with a few scattered trees, which is surrounded on all sides by dense forest. However, if one walks towards this forest, one never reaches it, the glade moves with one and the forest border retreats constantly (Fig. 137).

Two plant life-forms occur together in savannah that are usually antagonists in other situations. In the temperate zone, for example, we find either pure grasslands (steppe, prairie) or woodlands (forests, shrublands). An admixture of the two does not occur. The 'forest-steppe' is a mosaic-like macrocomplex, in which forest and grassland are usually well separated.

In natural savannah there are ecologically very different types among the woody plants and the grasses, but in spite of this we can recognise certain characteristic properties, which apply to most of the woody species as well as to the typical grasses and which are decisive for the development of a particular kind of vegetation type (Walter,

1939). Here I consider perennial grasses, exclusive of the spiny and sclerophyllous grasses. The latter two are foreign to the savannah and their physiological-ecological behaviour is quite different.

The ecological properties of the typical grasses are primarily related to moisture conditions; with regard to temperature we find representatives among both the woody plants and the grasses, which can flourish from the tropics to the arctic zones. As far as moisture is concerned, differences of both loss and absorption are apparent.

In general, the grasses are characterised by very active transpiration which is not reduced even under conditions of soil moisture stress. The consequence is a very sudden decline in the moisture content of the leaf cells and hence a rapid increase in osmotic pressure resulting in the death of the cells. Necrosis starts at the leaf tips, and continues to the

FIG. 137. Tree savannah with *Acacia giraffae* and a grass-cover of *Schmidtia quinqueseta*. Dornfontain Farm, S.W. Africa (Phot. E. Walter).

base. The grassy areas get 'burned' very rapidly and fresh, green areas can become dry and yellowish brown within a short time.

Since the meristematic tissues of most grasses are located at the leaf base, they are usually protected from complete desiccation by the dry leaf sheaths which totally enclose these growing points. Thus the grasses remain alive for many months during a drought and develop new shoots upon resumption of favourable temperature and water relations.

Amongst frost-sensitive grasses of the subtropics a sudden die-back of the aerial parts may, occasionally, be caused by frost. The vegetative apices remain alive in this case also, just as does the root system.

During desiccation, a strange phenomenon can be observed in the roots of these grasses (*Aristida, Eragrostis*, etc.). The root epidermis and cortex die and their cells shrink. The dead outer layer then persists, with the remaining root hairs and sand grains, as a loose 'sleeve' around the living central cylinder (Henrici, 1929; Goossens, 1935; and see Fig. 3 of Walter, 1939).

In the arid areas bunch-grasses also occur. These not only branch at the base but also somewhat higher up. The branches, which support the growing points, survive the dry season. As an example, I cite *Aristida uniplumis* (Fig. 141, p. 249), which is widely distributed in south-west Africa. The shoots of this grass do not dry up completely but turn yellow. Before re-sprouting, chlorophyll is regenerated and the shoots again become green. This phenomenon has not yet been investigated in detail (Henrici, 1927).

The grasses only use a very little water in the dormant condition, i.e. after the death of the transpiring organs. Therefore, they can survive long dry spells without requiring large water reserves in the soil.

I carried out measurements in south and south-west Africa and showed that the residual water in the soil even near the end of the dry season is sufficient to replace the plant's water losses. The taller the bunch grasses, the greater will be their water consumption during the period of drought, therefore tall species only occur in those savannah areas where the dry season is not too long.

Grasses can regenerate even before the onset of the rainy season, where there is a reasonable amount of water left in the soil. This is the case, for example, on the borders of small erosion channels. However, initially the leaves remain small but they enlarge suddenly after the first good rain. On sand dunes that hold much water under a sparse vegetation cover, individual bunches of *Aristida ciliata* started to flower even before the rain began. Tests on soil samples showed that the water content at 1 m amounted to 1·25% and at 1·25 m even to 2%. The soil could be moulded manually into a ball so it was clearly moist.

However, the fact that the grasses in such habitats regenerate only near the end of the dry season is related to the low temperatures. Tropical grasses remain dormant so long as there are night frosts and a daily maximum of less than 20°C. Therefore, my experiments to induce the growth of grasses during the cooler season by irrigation did not succeed. In contrast to this, *Eragrostis lehmannia* after transfer into a warm, damp room formed green leaves within 11 days. In the case of *Eustachys paspaloides* young roots develop after 2 weeks at the nodes of the rhizomes and also soon after the leaves emerged. It is quite possible that tropical grasses undergo a winter rest period during the cooler part of the year which can only be broken at a characteristic temperature threshold.

When grasses transpire actively, water absorption has to be equally effective. Since grasses always grow in dense stands, there is only a limited soil volume available to the root system of each individual. Thus, a great absorption of water is only possible if the soil volume is exploited very considerably by the roots—which is certainly the case for grassland —and also, if the soil contains sufficient water. Thus, it is understandable that the development of grasses is favoured on not too coarse textured soils with a relatively large water-holding capacity and that they develop best in a climate in which there is sufficient rain during the growing season; that is, in the tropical summer-rain region (Walter, 1960, p. 284).

The behaviour of the woody plants is different. They differ generally from other ecological types by a much more precise regulation of transpiration, whereby they begin to close their stomata at very small water deficits. Evergreen species with sclerophyllous foliage can endure long periods of drought in this condition. Other species shed their foliage during such periods either in part, or completely.

However, evergreen and also deciduous woody plants still lose so much water by transpiration during the dry season that some, even though small, absorption from the soil is necessary.

The transpiration of shrubs in a resting condition was determined during the dry period in November in the shrub savannah of south-west Africa. This shrub savannah is dominated by *Aristida uniplumis* in the grass layer and by shrubs of *Boscia foetida* (Capparidaceae) *Catophractes alexandri* and *Rhigozum trichotomum* (both Bignoniaceae). *Boscia* transpired an average of 7·8% of its fresh weight per day, *Catophractes* 6·4% and *Rhigozum* 5%. Whole branches and small shrubs were weighed. After cutting the former off, there is no decrease in rate at this small transpiration rate. Half-hourly weighings showed that the transpiration curve, in general, follows the evaporation curve.

I was unable to determine the water loss of *Acacia* shrubs in the defoliated condition during the dry period. In the area investigated they only grew in habitats with more favourable water balances, as, for example, in erosion channels or on rocky habitats. They were already in flower in November. Transpiration of flowering twigs amounted to 2·0-3·5% of the fresh weight per hour. Since the flowers lose only a very little water and since the leaflets were still very small, one may conclude from the relatively high transpiration values that the transpiration of the twigs alone was considerable. It was probably higher than in the shrubs already mentioned, which transpired 5-8% per day. Acacias occur as components of the zonal vegetation only in areas with somewhat higher rainfall. In the drier areas they only occur as extrazonal vegetation in particularly favourable habitats (Figs 138 and 139).

If one calculates the water loss of shrubs on a 100 m^2 area in the *Catophractes-Rhigozum* shrub savannah, one obtains a figure of about 350 g per day. This would correspond to a water loss of 84 litres in 8 months, i.e. to a rate of precipitation of less than 1 mm. The rainfall in the preceding year was 265 mm. Even if the water losses calculated represent only a rough estimate, they indicate how little storage water apparently needs to be present for survival of the plants through the dormant period during the entire dry season.

However, some water absorption from the soil is necessary for the shrubs, because their stored water would hardly last more than a week.

In contrast to the grasses, the shrubs have a more extensive root system, i.e. they extend laterally very far[1] and, to a certain extent, they also reach very deep vertically, but the soil is only very loosely exploited by roots. Stony soils in which only small amounts of fine soil are held among the rocks are very favourable, therefore. Areas with winter rains are likewise favourable, since in these areas water percolates to greater depths.

By contrast, on fine-textured soils an extensive root system is at a disadvantage compared with a compact root system in competition for small quantities of residual water. In such situations, therefore, shrubs can only maintain a foothold among grasses where more water is left in the soil during the dry period than is used up by the grasses.

The requirements of woody plants and grasses for water can be summarised as follows:

Woody plants:

1. Require more rain.
2. Rainfall can be in winter.
3. The soil must hold enough water during the dry period to allow some, although a very small, water uptake.
4. The water-holding capacity of the soil does not need to be high, i.e. the soil can be coarse textured, or even stony, or rocky since the extensive root system reaches a long way both laterally and vertically.

Grasses (perennial, not sclerophyllous):

1. Can grow with a much lower annual rainfall.
2. Rain, however, must fall during the growing season—in the summer.
3. Practically no water is absorbed during the dormant period.
4. The soils must have a relatively high water-holding capacity—they should not be too coarse textured, so that the compact root system can absorb sufficient water from the small volume of soil it permeates during the growing season. This is essential because grasses transpire very actively so long as there is sufficient water in the soil; with lack of water, the leaves dry out very rapidly.

When these aspects are considered, the distribution of woody plants and grasses in drier areas becomes quite understandable. It depends upon climate and soil.

The following scheme can be suggested:

I. Winter-rain areas: sclerophyllous woody plants, grasses less dominant.

[1] For *Zizyphus lotus* 20 m were recorded, for *Tamarix* and *Acacia tortilis* up to 50 m (after Kausch, 1959).

Figs. 138 and 139. Vegetation profiles and maps drawn to scale (without grasses) near Voigtsgrund in S.W. Africa (drawn by W. Huss).

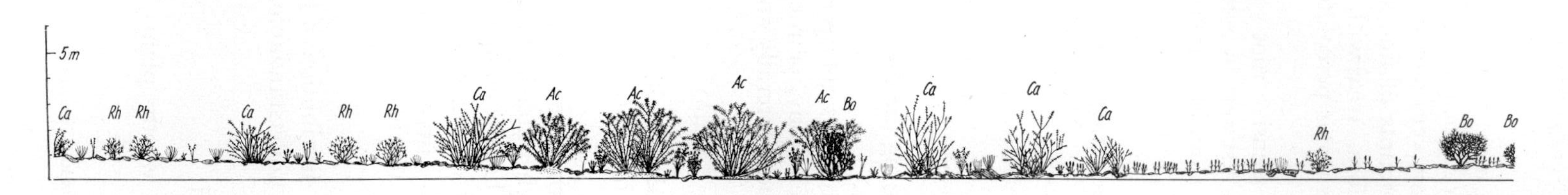

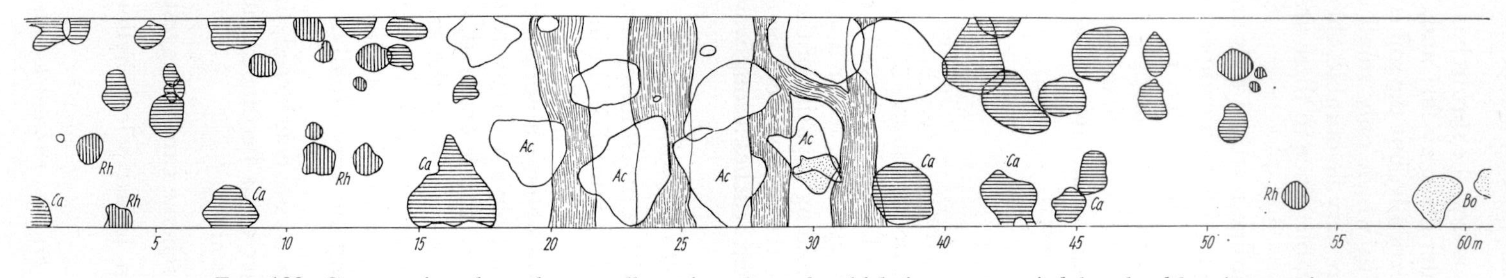

Fig. 138. Cross-section through a small erosion channel, which is accompanied by gland-bearing acacias (*Acacia nebrownii* = *glandulifera*).

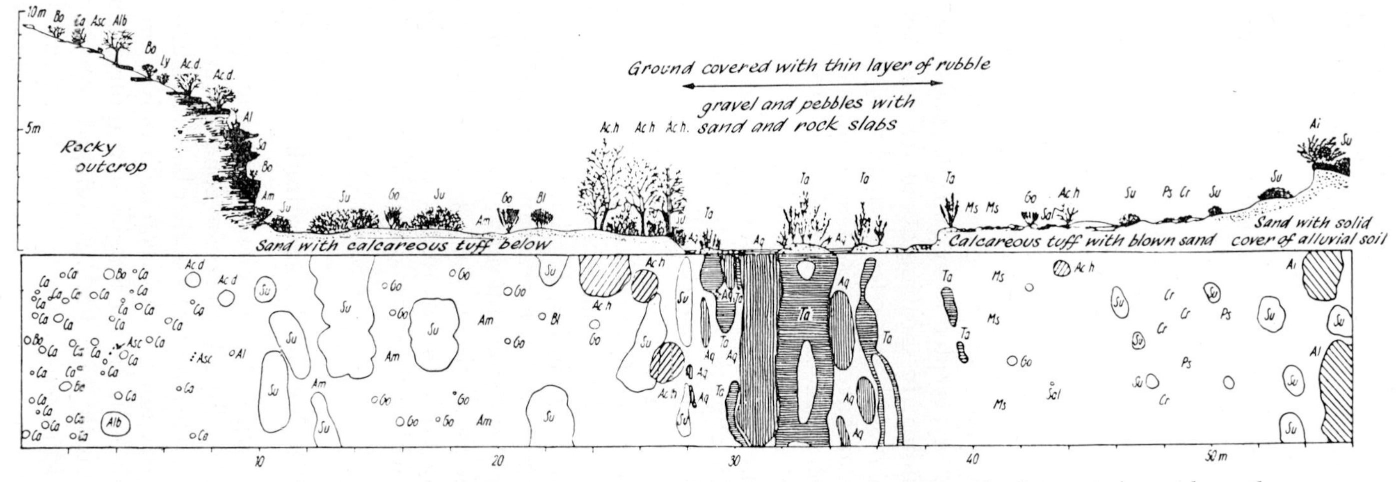

FIG. 139. Cross-section through the Tsub-Rivier with halophytic and psammophytic vegetation and a rocky slope (left). The *Acacia* spp. and *Albizzia*, which require more water, root in the rock fissures, or close to the erosion channels or in the ground water (non-brackish) of the Riviers. Explanation of the symbols:

Ac	*Acacia nebrownii* (glandulifera)	Aq	Slightly brackish surface water	Su	*Suaeda monoica*
Ac.d.	*Acacia detinens*	Asc	Asclepiadaceae	Go	*Gomphocarpus fruticosus*
Ac.h.	*Acacia horrida*	Bo	*Boscia foetida*	Sal	*Salsola aphylla*
Ai	*Tetragonia schenckii*	Bl	*Blumea gariepina*	Ms	*Mesembryanthemum*
Al	*Aloë*	Ca	*Catophractes alexandri*	Cr	*Cryophytum*
Alb	*Albizzia anthelmintica*	Ly	*Lycium*	Ps	*Psilocaulon*
Am	*Argemone mexicana*	Sa	*Sansevieria*	Rh	*Rhigozum trichetomum*

II. Summer-rain areas:

A. Soils stony: woody plants dominate over grasses.

B. Soils of finer texture:

(*a*) Low precipitation (about 100-250 mm): pure grassland, woody plants absent.

(*b*) Precipitation higher (about 250-500 mm): savannah, grasses dominate but woody plants present also.

(*c*) Precipitation high (over 500 mm): woodland, grasses less dominant.

Schimper had earlier used the terms grassland- and forest-climate. C. Troll's statement that, 'Schimper's opinion that forest and grassland are related to contrasting climates was a basic error in his plant geographical concepts', cannot be accepted from an ecological viewpoint. It is merely a misunderstanding. In speaking about relations between vegetation and climate one should only refer to zonal vegetation, that is, vegetation on level areas and average soils. The occurrence of grassland as zonal vegetation—on average soils—is, in fact, related to a very definite climate: the rainy season must coincide with the warm season, and the rainfall should not be so high that forest can develop. Wherever grassland occurs outside this climatic zone, it is either extrazonal or azonal vegetation. In these cases, it is conditioned either by topography, extreme soils or by man's interference (fire, grazing, mowing). One should not merely draw conclusions from the appearance of the vegetation but one should always look for causal relationships as well. These are capable of being very different in any particular situation.[1] Thus, the ecologist should not simply equate one grassland with another. This is the privilege of the geographer. Ecologists must, therefore, pursue the problem of the nature of the association of the two competitors, woody plants and grasses, as presented in the savannah. In general, the two life-forms compete strongly with each other, so that one is present to the exclusion of the other.

3. The competition-equilibrium in the savannah

In the temperate zones there are no tree species able to maintain themselves in the midst of grassland for all temperate zone tree species are species of closed forest stands. Therefore, the border between steppe and forest communities is relatively abrupt. An ecotone is only commonly formed by a narrow shrub belt. In the tropics it is quite different. Here are woody species, usually with umbrella-shaped crowns, particularly *Acacia* spp. (Fig. 140), which are not forced to form closed stands. They can also tolerate grass-fires and they may even regenerate

[1] In the Serengeti grassland of East Africa it is true that the annual precipitation can amount to 530-570 mm but rainfall is quite unreliable: there are often months without rain. 'Hard-pan' is also discussed by Anderson and Talbot (1965).

after the loss of the aerial parts. So, a broad transition zone is formed between the closed woodland zone and the pure grassland, in which single trees and shrubs persist in the midst of a grassy plain. These are the savannahs of the summer-rain regions (Fig. 141).

As an example, I will discuss the relationships which I investigated in south-west Africa. For the purpose of understanding the competitive

FIG. 140. *Acacia detinens—Eragrostis curvula*—savannah in S.W. Africa (Phot. E. Walter).

relationships, I shall proceed in my discussion from arid to humid areas. Since I am here dealing with zonal vegetation, I shall assume a relatively level topography and a soil that absorbs and stores all the water received by precipitation (Fig. 142*a-d*).

In (*a*) precipitation will amount to about 100 mm. The rain water

FIG. 141. View across a tree savannah seen from a dune ridge (western Kalahari), Donnersberg farm, S.W. Africa. In the distant background another dune ridge. The trees are *Acacia giraffae*; the flowering grass is *Aristida* (recently named Stipagrostis) *uniplumis* (Phot. E. Walter).

does not penetrate very deeply into the soil. It is absorbed by the upper layers. The small bunch grasses root in these layers. All the water supply is absorbed during the growing season by their compact root systems and transpired from their leaves. When most of the water supply is exhausted, the aerial parts of the grasses die down to the basal meristems. The small amounts of water remaining are used up by the grasses during the dry spell. For this reason woody plants cannot become established. Where the summer rainfall increases to 200 mm (see *b*), water penetrates a little deeper into the soil and here larger grass species become established. However, these also take up nearly all the soil water during the summer. Only with a further increase in

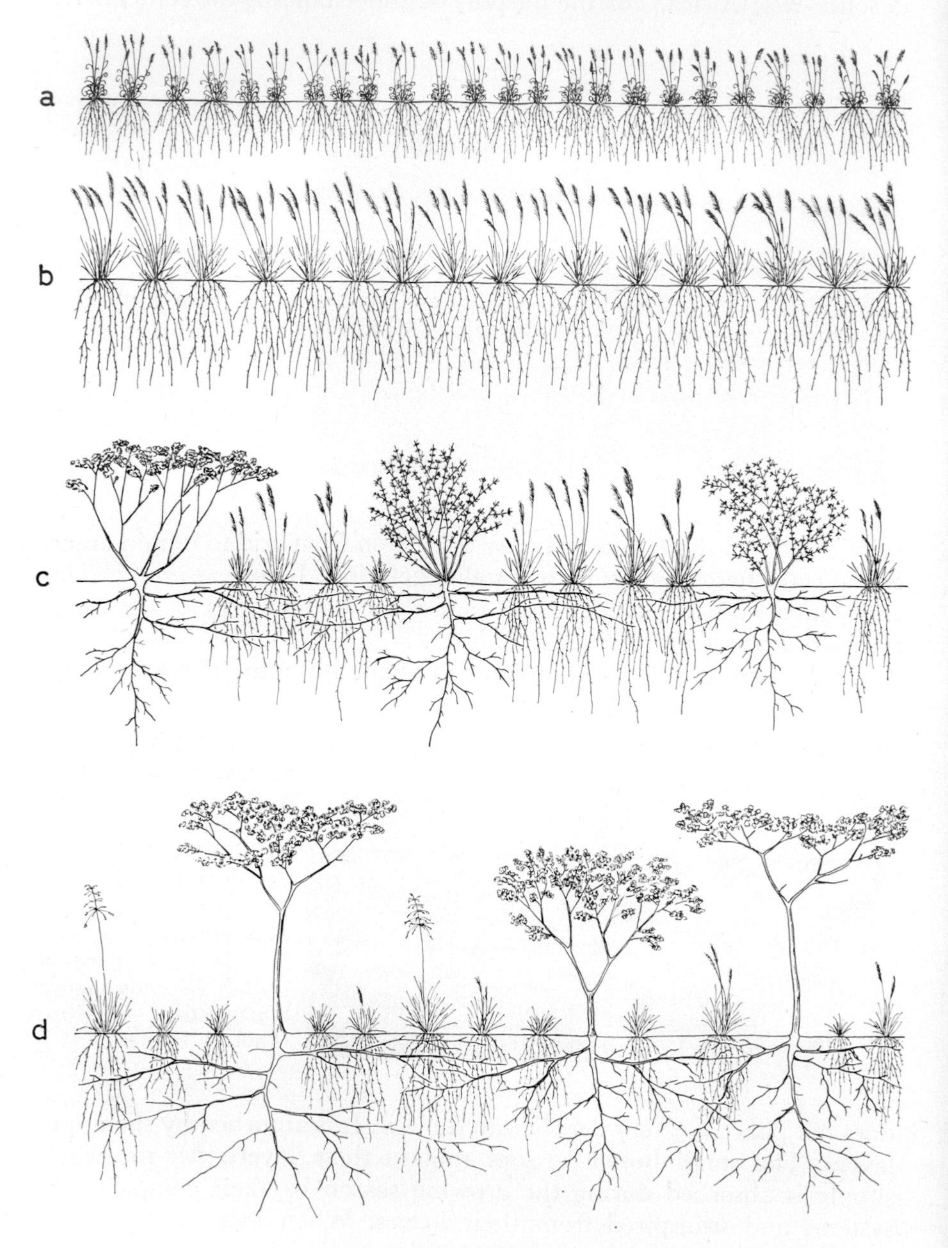

FIG. 142. Schematic representation of the transition from grassland (*a* and *b*) to savannah (*c*) and to the dry woodland (*d*).

precipitation (see *c*) will the grasses not use all the water. Some residual water remains even during the dry season and only this provides conditions sufficiently favourable for the establishment of solitary, woody species, initially only shrubs. The damper the climate the more residual water is left by the grasses and the more, and larger, woody plants which can maintain themselves with the grasses (see *d*). The shrub savannah becomes a tree savannah. The controlling element in all these cases is the grasses. The woody plants are merely tolerated. In dry years, many of them die because of insufficient residual water while in moist years more residual water is available and then either woody plants develop more vigorously or more seedlings may become established.

This interdependent relationship only changes where the woody plants close up to form a dry woodland. Then the grasses become less vigorous through shading and only maintain their position in the more open places beneath the trees. Now, they have assumed the role of the tolerated component (see *d*).

Another life form which plays a role in the dry areas of the tropics are the succulents (leaf or stem succulents). They are characterised by water-storage organs, which become 'recharged' after rain and which can then serve as a water source for a long period. Succulents are plants that transpire only very little with but a small rate of production of dry matter. For this reason they grow very slowly and their competitive ability is inferior to the other forms. Succulents are only found, therefore, where the other life forms die out; that is, in some deserts (see pp. 316 and 393).

Succulents can also grow in humid climates on very shallow soils, which dry up rapidly after rain. However, this provides sufficient opportunity for temporary water storage and the stored water is then used slowly between successive rain spells.

This demonstrates that the dominance of a single life form determining the facies of the vegetation is always related to the soil in addition to climate. Wherever different soils occur side by side within a climatic zone there too are different vegetation types side by side; climate and soil together determining the landscape. A very clear example will be discussed later in connection with the upper Karroo (South Africa) (see p. 392 *et seq*. and Fig. 244).

An interesting phenomenon in arid areas—which can be seen particularly well on air photos—is the strip-like arrangement of vegetation, e.g. in Somaliland (Hodge, 1962), or in central Australia, where I observed a similar pattern of strips with *Acacia* shrubs alternating with strips that are almost barren of vegetation or which are occupied only by grasses. In all these cases some differences could be noted in soil structure or texture, but such an origin is not always clear. In most cases the strips occur perpendicular to a very slight slope on which some deposits may have resulted from the action of water or wind.

4. Invasive scrub—a threat to farm management in the savannah zone

As mentioned earlier, the transition from woodland to grassland occurs in the drier part of the tropics with a decrease of annual precipitation from 600 mm to about 250 mm. Here, the competitive ability of the grasses increases more and more over that of woody plants, resulting in a sequence from tree savannah to shrub savannah to bush savannah. However, recurrent fires seem to play a role in addition to the climate. Fires must be considered as a causal factor even under natural conditions but their frequency has been considerably increased by the practice of the local inhabitants of burning the grass cover annually. The lighting of fires has only recently been prohibited in farmed areas.

Fire has a selective effect particularly on the woody plants. Only those species develop in savannah that can either survive a grass fire or that regenerate from their below-ground parts. Many Acacias have dormant buds at the root collar to a depth of about 20 cm beneath the soil surface, where they are not affected by fire. Temperatures during grass fires have been measured at different heights above the ground giving the following results (Pitot and Masson, 1951; Beadle, 1940): the highest temperatures, amounting to 280-560°C, were recorded at 50 cm above the soil: at the soil surface the temperature may increase to 70-100°C for 2-3 minutes, while at 2 cm depth fire has no effect.

Trees with thick bark endure fires without any damage. Only 'pyrophytes' remain in areas with frequent fires. These are plants that are, in part, directly adapted to fires. Among the Proteaceae and to a certain extent also among the Myrtaceae in Australia there are species whose woody fruits remain hanging in the branches in closed condition (*Hakea*, *Xylomelum*, *Banksia*, *Callistemon*, etc.). These fruits only open after a fire has gone through the stand, but there is still some delay before the seeds fall on the cool ash, where they find a good seed bed.

The reason for prohibiting fires today in farmed areas is, first, that rains cause severe soil erosion after the burning of dead grass and the mineral nutrients are washed away with the ash to a considerable extent. In addition, too-frequent fires cause negative selection among the grasses. Since cattle prefer burned-over areas where the new grass shoots appear, the better fodder grasses in particular are very much weakened and the poor species eventually become dominant. Grass fires would probably be much less detrimental without grazing.

Farmers blame the discontinuation of fires for the invasion by shrub thickets that today poses a threat, approaching epidemic proportions, in the drier tropics of all continents. By this is meant that the thorn-shrubs of the savannah increase greatly in density resulting, in time, in a thorn-scrub vegetation which is useless for grazing. Thus, the take-over by thickets poses a threat to the livelihood of the farmers (Walter, 1954).

FIG. 143. Numerous seedlings of *Acacia detinens* becoming established on a barren, overgrazed area. Osombosatjuru farm.

FIG. 144. Later stage. *Acacia detinens* and *Dichrostochys nutans* now form a dense stand. In the background trees of *Acacia giraffae* at the Rivier. Omambonde-Ost farm.

FIG. 145. Here *Acacia maras* has formed an impenetrable thorn-scrub thicket. Hazeldene farm.

Closer investigation has shown that invasive scrub is not generally caused by discontinuation of fires, but that it occurs only on old farms with large numbers of grazing animals or near watering places, i.e. always where the perennial grasses have been much suppressed or destroyed by over grazing. As soon as the leaf area of the grasses is much reduced, the utilisation of water by the grasses through transpiration is decreased, thus, more unused water is left in the soil after the rainy season than before. This favours the woody plants. They become more vigorous and reproduce more easily. Under normal conditions tree seedlings cannot compete successfully with the grass roots and only very few survive under these conditions. Therefore, the woody plants always occur scattered. But, whenever the competitors, i.e. the grasses, become destroyed by grazing, practically all the woody seedlings survive and an impenetrable thorn-scrub thicket develops within a few years (Figs. 143-145).

The spread of thorn-scrub is favoured also by the fact that animals feed on the pods of the Acacias when the grass shoots become scarce. The seeds pass unharmed through their intestines and then find a good seed bed in the deposited dung (Leistner, 1961). I investigated seeds of *Acacia giraffae* from ripe pods, from elephant droppings and from cow dung. The germination percentages after 5 weeks were 92%, 78% and 59%, respectively. The first seeds germinated after 11, 9 and 5 days, respectively. The rate of germination is, therefore, much increased, which is of great importance in arid areas.

The fact that nearly all the woody plants of the African savannah are equipped with thorns is probably related to the selective action of big game over thousands of years. In this connection, it is remarkable that in Australia, where only the kangaroo can be considered as a grazing animal, and in New Zealand, where no mammals have occurred (with exception of some bats), thorny species are nearly absent. Only species with thorny leaves are present. However, these occur more in woodland than in ecologically open areas. The 'spinifex' grasses (*Triodia*, *Plectrachne*) are also thorny or, better, xeromorphic.

However, if grazing by big game has had such a selective effect on woody plants, the question arises, why did this not lead to the spread of invasive scrub by the weakening of the grass component. Here, the very different pattern of grazing by big-game animals as compared with cattle appears to be decisive. Big-game animals change their grazing areas continuously, also they do not spend much time at watering places, where there is danger from predators. Thus, a pattern of grazing results, comparable with what one tries to achieve by rotation of grazing or alternate grazing. In this way the perennial grasses are not systematically clipped down to the ground one by one.

By contrast, the exploitation of the grass cover is very different on the stationary grazing land of the farms in the tropics which is generally fenced in. Cattle are lazy and move about very little. They feed, initially, only on the grasses near the watering places and move farther away only when the grass is cropped short. The perennial grasses become destroyed entirely around the watering places but they are not even touched in the more distant places on the farm. Therefore, scrub invasion is confined at first to near the watering places and only later does it extend farther away from them. The areas never reached by the animals do not develop thickets although burning may also have been discontinued in these areas. By introducing the practice of rotation grazing, the grass cover would be less exploited and this would also prevent invasive scrub (Walter and Volk, 1954).[1]

The zonal vegetation is apparently changed by invasive scrub into a type characteristic of a more humid climate and the question arises, whether or not this change can be considered permanent. Experience

FIG. 146. Overmature thicket with *Acacia detinens*. The approximately 50-year-old bushes die out. The openings are invaded by *Aristida uniplumis*. Upon protection from grazing a shrub savannah will become re-established. Otjitambi Farm.

over a long enough period is not yet available. However, some observations indicate that with the further growth of the shrubs water use will, in time, no longer be balanced by rainfall. Shaded-out shrubs die, particularly in drought years, the thickets open up and grasses can become re-established in the openings. Regeneration of the woody plants does not take place in the dense thickets, probably because of the high light requirements of the seedlings (Fig. 146). So when the shrubs attain their age limit, after a few decades, and begin to go over, a grass cover has already been established which can then close up readily. Only a few younger bushes remain in between. Of course, a farmer does not want to wait so long and wishes to eliminate the thorn-scrub rapidly. How this can be accomplished is mainly a question of economics. Simply to burn it is ineffective (Fig. 147).

Thicket formation is more a threat to the thinly populated, farmed

[1] Invasive scrubs of the flood-savannah consequent upon grazing depends upon the soil becoming compressed through a brief period of retention of flood-water and its rapid discharge. See, for example, Wilhelmy (1957).

areas rather than to the areas occupied by the natives. A farm is occupied by one family only with a few farm-hands. Utilisation of wood by the household is very limited. It is quite different among the natives, who live in large numbers around the water-places and who need much wood for cooking purposes. For this they use thorn-shrubs.

FIG. 147. A thicket area after burning. The dormant buds regenerate from the root collar and the result is an increased density of the thicket of *Acacia giraffae*. Omambonde-Ost farm.

Invasive scrub is absent in such areas, instead therefore, a bare, desert-like area develops with over-grazing which is subject to severe soil erosion. From an economic point of view this development is even worse.

5. Edaphically determined vegetation in the savannah-grassland zone

In dry tropical areas, the zonal vegetation only becomes established on relatively deep, sandy soils in level regions. Loamy and clayey soils are only formed in hollows because of the very slow rate of chemical weathering. In the damper parts of the savannah zone, these periodically wet habitats are still occupied by pure grassland but in part, also, by grassland with scattered palms. The soils in these hollows are black and with a high clay content. When dry they develop very deep cracks and a coal-like appearance. They become sticky and slippery after heavy rain and present a dangerous obstacle to driving. These are not chernozem soils. Their black colour is due to strongly swelling, stable clay-humus complexes. In the dry savannah zone a weak degree of salinity is usually noticeable in the periodically wet soils. They are base-saturated and are occupied by *Salsola* species in south-west Africa where succulent members of the *Mesembryanthemum* family are also found.

More commonly a true soil is lacking in the savannah areas, instead the surface is weathered rock, especially on slopes. Grasses are less dominant in such situations and are replaced by an open shrub formation whose degree of cover decreases proportionately with rainfall, gradually changing into desert. However, in transition areas succulents or specialised woody species with bottle-like, water-storing stems become common (see p. 358). Savannah-like stands can also occur on sandy soils under climatic conditions where one would expect pure grassland. This is the case where the sandy soil is only shallow (Walter, 1939).

As an example I shall discuss conditions in the southern part of south-west Africa near Mariental, where the mean rainfall is 185 mm. An impression of the vegetation is gained from Fig. 148. The ground

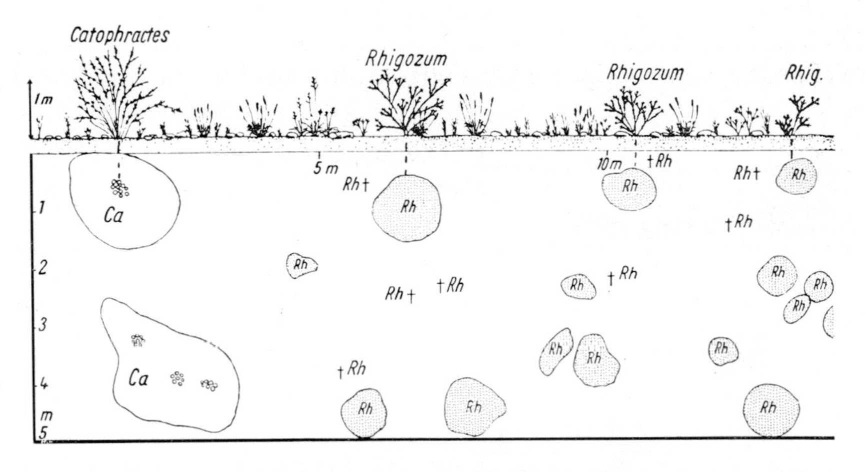

FIG. 148. Belt transect and profile (1 m wide) through typical vegetation near Voigtsgrund (S.W. Africa). Grasses poorly developed during the dry season. Below, plan of transect showing the vegetation cover without grasses. Ca *Catophractes*. Rh *Rhigozum* (†dead). From Walter and Volk (1954).

cover amounts to 40% and the yield of dry matter of grasses and annual herbs is 400 kg/ha; the dominant is *Aristida uniplumis*, which attains some 60 cm. The soil profile shows that the bedrock of sandstone occurs at a shallow depth, where it is either finely layered and nearly free of crevices or coarsely layered and broken up, resulting in larger cracks. The sandy soil stores only part of the rainfall, the rest penetrates the rock. The rainwater stored in the sand is completely used up by the grasses but that held in the sandstone cracks cannot be reached by their fine rootlets. However, the longer roots of the shrubs penetrate into these crevices.

In this case there is no direct competition between the grasses and shrubs, their relationship is complementary with regard to the exploitation of the soil water. By contrast, there is great competition between the three shrub species which occur together here: *Boscia foetida*, *Catophractes alexandri* and *Rhigozum trichotomum*. Of these, *Boscia* is the most vigorous but also uses the highest amount of water, *Catophractes* is intermediate, and *Rhigozum* is the smallest and least exacting species. Suitable habitats for these shrubs are only found where there is sufficient water held in the cracks in the underlying sandstone. Their seeds are distributed uniformly across the area, but seedlings can only get established in wet years when the roots can reach a moisture-filled crevice, not occupied by roots of other plants, before the start of the dry season. Commonly, seedlings of all three species are found together at the same site. However, as the plants age, the stored ground water is no longer sufficient for their combined requirements and then *Boscia* succeeds over the other two species. These die out and their remains can be seen frequently still beneath the bushes of *Boscia*. Less favourable habitats that do not provide enough water for *Boscia* can be colonised by *Catophractes*, while *Rhigozum* has to be satisfied with the most unfavour-

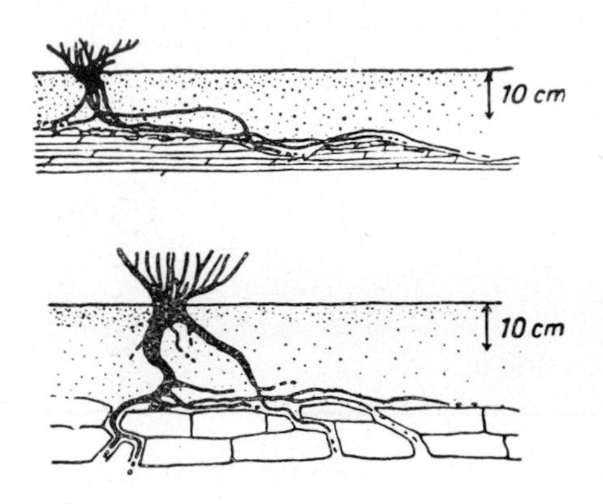

Fig. 149. Sandstone layering and rooting relations of *Rhigozum* (above) and *Catophractes* (below). The relatively finer-textured soil is shown by dots; torn-off roots indicated by dashes. After Walter and Volk (1954).

able habitats. These are usually those places where the finely layered sandstone underlies the sand and wherever this type of sandstone occurs over wide areas, *Rhigozum* is found as the only woody plant (Fig. 149). More favourable habitats are near the erosion channels, since the sandstone here receives more water through surface seepage. Here, at the margin, one finds vigorous *Catophractes* or *Boscia* shrubs, while in the channel itself a characteristic species of *Acacia* is found to form dense stands (Fig. 138).

Wherever the soil conditions change, e.g. in places where the sandy soil is deeper, the grasses push out the shrubs. Conversely, where the sandy layer is less deep or altogether absent, grasses are less abundant. Without investigation of the soil relations, the distribution pattern of the grasses and shrubs cannot be understood.

The vegetation also changes on steeply terraced slopes. Where the strata dip toward the lower terrace permitting the lateral transfer of some seepage water, dense stands of trees occur. Otherwise, the habitat is very dry and only suitable for succulents, such as large species of *Alöe* (Fig. 139).

6. Vegetation zones in relation to decreasing amounts of rainfall in the subtropical region

The following vegetation zones can be distinguished in relation to decreasing summer rain in south west Africa:

Rain-green dry-woodland	→	savannah	→	grassland	→	desert
(Fig. 150)		(Fig. 141)				

the different climatic savannah types occurring in the range of 500-250 mm annual rainfall. A similar sequence would also be expected in the summer rain area south of the Sahara. However, the very detailed

Fig. 150. Mopane-forest (*Colophospermum mopane*) with monkey family, near the Victoria Falls, Rhodesia (Phot. E. Walter).

investigations of Smith (1949) have shown that this is not so. The grassland zone is absent and the savannah vegetation extends into areas with 900 mm annual rainfall. The annual rainfall varies in the Sudan, near Wadi Halfa on the Egyptian border from zero to more than 1400 mm in the transition area to the Congo. The region is flat and the soils are sandy in the north-west and clayey in the south-east. The following vegetation zones can be distinguished from north to south with increasing rainfall (Fig. 151):

1. Desert, less than 50 mm rain per year
2. *Acacia* desert-scrub
3. *Acacia* short-grass country
4. *Acacia* tall-grass country
5. Mixed deciduous fireswept forest

1. The desert zone only develops a grassy growth (known as 'Gezzu') after periodic rains and this then attracts the nomads with their camel herds. Otherwise, the clay soils are completely free of vegetation. *Acacia flava* (*A. ehrenbergiana*) occurs on sands south of the 50 mm isohyet. Along erosion channels this tree extends still further

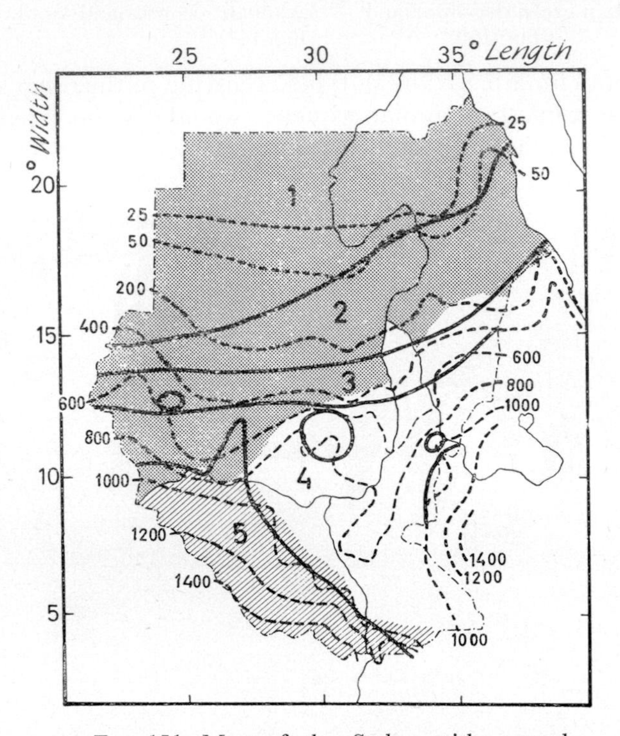

FIG. 151. Map of the Sudan with annual rainfall (dashed lines) from 25-1400 mm, soil types (dotted—sandy soils, cross-hatched—iron-sandstone, plain—clay soil) and vegetation zones, viz. 1. Desert; 2. Acacia—desert scrub; 3. Acacia—short grass country; 4. Acacia—Tall grass country; 5. Mixed deciduous forest. Circles: mountain massif with forest in the Acacia zone (after Smith, modified).

north, where it is associated with *Capparis decidua*, *Maerua crassifolia* and *Leptadenia spartium*.

2. The *Acacia* desert-scrub occupies the region between 50-250 mm on sand and up to 400 mm on clay soils. The typical species is *Acacia tortilis*, other associates are—in addition to those named above—*Acacia raddiana* and *Boscia senegalensis*. Along erosion channels there is also *Acacia seyal* and, in broader valleys, *A. mellifera*. On sand dunes overlying clay *Salvadora persica* grows and in the sandy region *Panicum turgidum* and *Cyperus* sp. Along the shores of the Nile one finds *Acacia*

albida, *A. seyal*, *Zizyphus spina-cristi*, *Balanites aegyptiaca* and also the doum-palm, *Hyphaene thebaica*.

3. In the *Acacia* short-grass country with *Acacia mellifera*, *Boscia* and *Cadaba rotundifolia* dominate on clay soils and *Acacia senegal* (which produces the genuine gum arabic) together with *Albizzia* spp. on sands. This vegetation type extends on to sandy soils with 250-400 mm rainfall and on clay soils with 400-600 mm. On soils subject to flooding during the rainy period one finds pure stands of *Hyphaene*. The grass layer is comprised of annuals and is subject to grazing and annual burning. Commonly, even-aged stands are formed by *Acacia mellifera*, which die simultaneously and are then destroyed by fire. Subsequently, a shrub-grassland with *Cadaba* develops.

4. Even in the *Acacia* tall-grass country, the 1·5-2·5 m tall grasses are annuals. Huge grass fires sweep across the entire area and are stopped only by water courses or by occasional grass-free areas ('Mahal'), on which seeds have germinated after an early rain and subsequently succumbed to drought. Pure grassland occupies those areas flooded during the rainy season and tree growth is restricted here only to termite mounds (see p. 229). The dominant tree species is *Acacia seyal*, which either grows in pure stands or admixed with *Balanites aegyptiaca*. The spread of this vegetation type into areas with an annual precipitation of 900 mm is conditioned by fire. On stony hills with sparse grass cover which are not overrun by the fire, one finds species of the dry forest, such as *Lonchocarpus*, *Stereospermum*, *Sterculia*, *Anogeissus*, *Boswellia*, *Ficus* and even bamboo (*Oxytenanthera*).[1]

The change from grass-free areas ('Mahal', caused by irregularities in the rainfall) to *Acacia* tall-grass stands is characteristic of this zone. The 'Mahal' areas can readily develop into even-aged Acacia thickets (see pp. 251, 253) and on the death of the shrubs an open grassland with *Balanites* develops temporarily. These different successional stages can occur side by side.

5. Only where the annual rainfall exceeds 900 mm are the grasses so strongly suppressed by the trees that the effectiveness of the grass fires decreases. Here, a series of broad-leaved tree species is found forming dry woodlands. These are usually mixed woodlands. Pure stands are formed only by *Isoberlinia doka* and *Anogeissus schimperi*. Grasses of the undergrowth are perennials which spring up after the first rains. In addition, there are plants with bulbs and corms. Many of the tree species also occur in South Africa (*Sclerocarya*, *Burkea*, *Pterocarpus*, *Strychnos*, etc.), others are related to Indian species (*Tamarindus*, *Mimusops*, *Dalbergia*).

Relict stands of rain forest occur at some sites with favourable water relations having such species as *Chlorophora excelsa*, *Ceiba pentandra*, etc., with much *Coffea robusta* in the shrub layer.

In reviewing these vegetation zones in the arid parts, it is remarkable that the Acacias play a greater role than the grasses even on sandy soils

[1] The vast Sudd swamps also occur in this zone of the Upper Nile (p. 145).

in areas bordering desert and that perennial grasses are absent from the savannah region. This would suggest that they are not natural vegetation zones but anthropogenically conditioned degradation stages. In south-west Africa, also, the perennial grasses disappear entirely after a prolonged period of over-grazing and this causes a change in the competition-equilibrium of the savannah. Annual grasses are weaker competitors with shrubs because they only germinate after the first rains and have to re-establish both the shoot system and the root system annually. By contrast, the perennials come into leaf immediately and can extract more water from the rooted soil layer.

The regions south of the Sahara have been subjected for centuries to over-grazing by the herds of camel, goat and sheep of the nomads. Even where thinly populated, these areas have not been protected from animal browsing and grazing. An additional factor is the practice of annual burning, which damages the perennial grasses in particular, while seeds of the annual grasses survive on the soil surface where they are exposed to temperatures hardly exceeding 100°C (see p. 252). However, not only is the grass layer changed but the woody plants are also affected. They are browsed by camels and goats and an additional factor is the high consumption of wood by the settlements and the charcoal manufacturers. Even in remote areas the shrubs are cut so that protective measures against over-exploitation have had to be passed (Halwagy, 1962: Rosetti, 1962). As a consequence, a completely over-grazed, secondary vegetation composed of thorn-shrubs and annuals has become established which appears to be greatly modified from the original vegetation type. The only perennial grass in the desert area, *Panicum turgidum*, is very rare. It is thus not surprising that on protected areas after a few years the shrubs show better development, but otherwise there is no significant upgrading of the vegetation.

The spread of tall-grass savannah deep into the humid region appears likewise to be a consequence of fierce annual burning. Smith called this vegetation type a fire-climax community, which is probably correct. This, therefore, is not a climatic zonal vegetation.

In the examples described so far the soils are more or less fine-textured (sandy-clayey). However, where the soils are stony the grasses are less dominant and the sequence on such soils in south-west Africa is as follows:

Rain-green dry-woodland → Rain-green dry-scrub → Dwarf-scrub semi-desert → Desert

Areas with winter rain favour the development of woody plants, as previously discussed, while grassland is practically absent. With decreasing rainfall, the woody plants decrease in size finally forming dwarf-scrub semi-desert, which gradually changes into desert. Thus, we here have a series in which grasses play no part:

Sclerophyllous forest → Sclerophyllous scrub → Dwarf-scrub semi-desert → Desert

Where rainfall is distributed in two rainy seasons, or where it occurs less regularly, succulents become more important. The series is:

Thorn-scrub (with succulents) → Succulent semi-desert → Desert

However, a prerequisite for these different sequences is that the various life forms must already occur as floristic elements of the region. For example, in Australia succulents and rain-green savannah shrubs are completely absent. The Eucalypts have leathery evergreen leaves, whereas Acacias have sclerophyllous evergreen, flattened or cylindrical phyllodes. Instead, one finds a grass form which is represented nowhere else. These are the spinifex grasses, also called Porcupine grasses. They form semi-globose cushions with a diameter of 1 m or more. In one species the diameter increases constantly with age, while in another the cushions disintegrate with age, since they grow only centrifugally on the north side. The Porcupine grasses belong to the genera *Triodia* and *Plectrachne*. Moreover, in Australia the winter- and summer-rain regions are not separated by a rain-free desert, instead there is a region in which the mean annual rainfall is over 100 mm and appears, on the average, to be very uniformly distributed over the entire year. However, there may be no rainfall for more than a year. In rainy years, rainfall can occur at any season. Semi-succulent dwarf-shrubs of the family Chenopodiaceae are characteristic of these areas. Porcupine grasses occur in dry areas, in which summer rains are clearly more common, but some winter rain may occur as well. Thus, very complicated series of plant communities develop (*a*) in winter-rain areas and (*b*) in summer-rain areas, and may be schematically represented as follows:

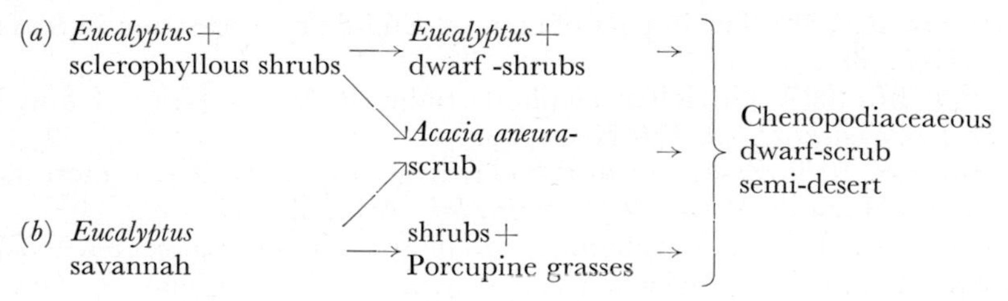

When a vegetation profile is taken to avoid the central part of Australia, but through the west instead, the chenopodiaceaeous dwarf-shrubs (salt-bush) are absent and, in the driest summer-rain areas, pure Porcupine-grasslands dominate. In the driest winter-rain areas pure *Acacia aneura* (Mulga) shrubland dominates. These shrubs have phyllodes like pine needles. Both formations form a mosaic pattern in the transition zone.

This shows how varied the vegetation patterns can be in relation to rainfall distribution and floristic conditions. Each region has certain unique features. The Eucalypts of the savannah areas probably do not show quite the same ecological relations as the rain-green shrubs of the African and South American savannahs. They are even better suited to an association with grasses in the rain-rich summer-rain zones since they produce little shade. Savannah-like formations are found, therefore, even with a very high annual rainfall (see p. 226 *et seq.*).

Ellenberg gives the following sequences for Peru (Fig. 124, p. 213):

(*a*) Annual uniform but low rainfall:

Rain-green dry-forest → Succulent scrub → Succulent semi-desert → Desert

(*b*) Rainfall highly variable from year to year:

Rain-green dry-forest → Rain-green dry-scrub → Scrub semi-desert → Desert

These considerations have led me to touch upon the main relationships of the vegetation in the arid regions. These will be discussed in greater detail in the next chapters.

References

ANDERSON, G. D., and TALBOT, L. M. 1965. Soil factors affecting the distribution of the grassland types on the Serengeti plains. *J. Ecol.*, **53**, 33-56.

BEADLE, N. C. W. 1940. Soil temperatures during forest fires and their effect on the survival of vegetation. *J. Ecol.*, **28**, 180-92.

GOODLAND, R. 1966. The savannah vegetation of Calabozo, Venezuela, and Rupununi, British Guiana. *Bol. Soc. Venez. Cienc. Natur.*, **26**, 341-49.

GOOSSENS, A. P. 1935. Notes on the anatomy of grass roots. *Trans. Roy. Soc. S. Afr.*, **23**, 1-21.

HALWAGY, R. 1962. The impact of man on semi-desert vegetation in Sudan. *J. Ecol.*, **50**, 263-73.

HENRICI, M. 1927. Physiological plant studies in South Africa, I and II. 11*th*, 12*th Reps Direct. Vet. Educn Res.*

HENRICI, M. 1929. Structure of the cortex of grass roots in the more arid regions of South Africa. *Dept. Agric., Sci., Bull.*, **85**.

HODGE, C. A. H. 1962. Vegetation strips in Somaliland. *J. Ecol.*, **50**, 465-74.

JAEGER, F. 1945. Zur Gliederung und Benennung des tropischen Graslandgürtels. *Verh. Naturf. Ges. Basel*, **56**, 509-20.

KAUSCH, W. 1959. *Der Einfluss von edaphischen und klimatischen Faktoren auf die Ausbildung des Wurzelwerkes der Pflanzen unter besonderer Berücksichtigung einiger algerischer Wüstenpflanzen.* Dissertation. Darmstadt.

LEISTNER, O. A. 1961. On the dispersal of *Acacia giraffae* by game. *Koedoe*, **4**, 101-04.

PITOT, A., and MASSON, H. 1951. Quelques données sur la temperature au cours des feux de brousse aux environs de Dakar. *Bull. Inst. Fr. Afr. Noire*, **13**, 711-32.

ROSETTI, C. 1962. Observations sur la vegetation au Mali Oriental. *FAO No. UNSF/DL/ES/4.*

SCHMITHÜSEN, J. 1959. *Allgemeine Vegetationsgeographie*. Berlin.

SMITH, J. 1949. *Distribution of tree species in the Sudan in relation to rainfall and soil texture*. Khartoum.

TROLL, C. 1935. Gedanken und Bemerkungen zur ökologischen Pflanzengeographie. *Geogr. Zsch.*, **41**, 380-8.

TROLL, C. 1950. Savannentypen und das Problem der Primärsavannen. *Proc. 7th Int. Bot. Congr. Stockholm*, 670-4.

TROLL, C. 1952. Das Pflanzenkleid der Tropen in seiner Abhängigkeit von Klima, Boden und Mensch. *Dtsch. Geogr.*, **28**, 35-66.

WALTER, H. 1939. Grasland, Savanne und Busch der ariden Teile Afrikas in ihrer ökologischen Bedingtheit. *Jb. Wiss. Bot.*, **87**, 750-860.

WALTER, H. 1954. Die Verbuschung, eine Erscheinang der subtropischen Savannengebiete, und ihre ökologischen Ursachen. *Vegetatio*, **5**/**6**, 6-10.

WALTER, H. 1960. *Einführung in die Phytologie*. Vol. III. *Grundlagen der Pflanzenverbreitung*. Part 1, *Standortslehre*. 2nd edn. Stuttgart.

WALTER, H. 1961. Die Bedeutung des Grosswilds für die Ausbildung der Pflanzendecke. *Stuttgart Beitr. Naturk.*, **69**.

WALTER, H., and VOLK, O. H. 1954. *Die Grundlagen der Weidewirtschaft in Südwestafrika*. Stuttgart.

WILHELMY, H. 1957. Das grosse Pantanal in Mato Grosso. *Verh. dt. Geogrtags*, **31**, 45-71.

VII. *General Features of Vegetation of Subtropical Arid Regions*

1. The concept of arid regions

An arid region is one in which the plants suffer from lack of water as a result of low rainfall and high evaporation throughout the longer part of the year. Thus, plant cover is only sparsely developed and shows various adaptations to the unfavourable water conditions.

Arid regions occupy about 35% of the surface of the earth. They show a symmetrical distribution on either side of the tropical zone (Fig. 152). They belong to the sub-tropical zone of the northern and southern hemispheres and also extend deep into the temperate zone of Asia and North America. An accurate criterion for separating arid from wet areas is not easily established.

In most cases, rainfall cannot be used for this purpose. For example, with equal precipitation, we find a desert climate in the sub-tropics and a very wet climate in the northern regions.

Usually, the separation is made when the annual evaporation from an open water surface exceeds the annual rainfall. However, since there is no absolutely satisfactory method for measuring evaporation, one is forced to use indirect indications viz.:

1. Lakes without outlets can only exist in areas where evaporation exceeds rainfall. In a wet area, where the evaporation from a water surface is less than the rainfall falling on it, all hollows without subterranean outlets become filled with water which, in time, finally overflows. Thus, they must always have a surface outlet. However, one should not assume that lakes with surface outlets are never encountered in arid regions. For example, if the catchment area of a lake is very large, it will receive additional water through numerous inflow channels. Such lakes may show a temporary overflow at least in spite of the high evaporation, characteristic of the arid climate.

2. A second feature of arid regions is the presence of intermittently flowing streams. Exceptions occur where the stream-heads are located outside the arid region, as for example in case of the Nile, the Orange, the Colorado and the Volga rivers. Conversely, intermittently active streams can also occur in high rainfall areas, for example in the Mediterranean region, where the summers are so dry that they result in a temporary drying-up of the headwater.

3. Finally, it should be pointed out that saline soils are frequently found in arid regions. These result from the low rainfall that is insufficient to transport the salts formed during weathering to the sea.

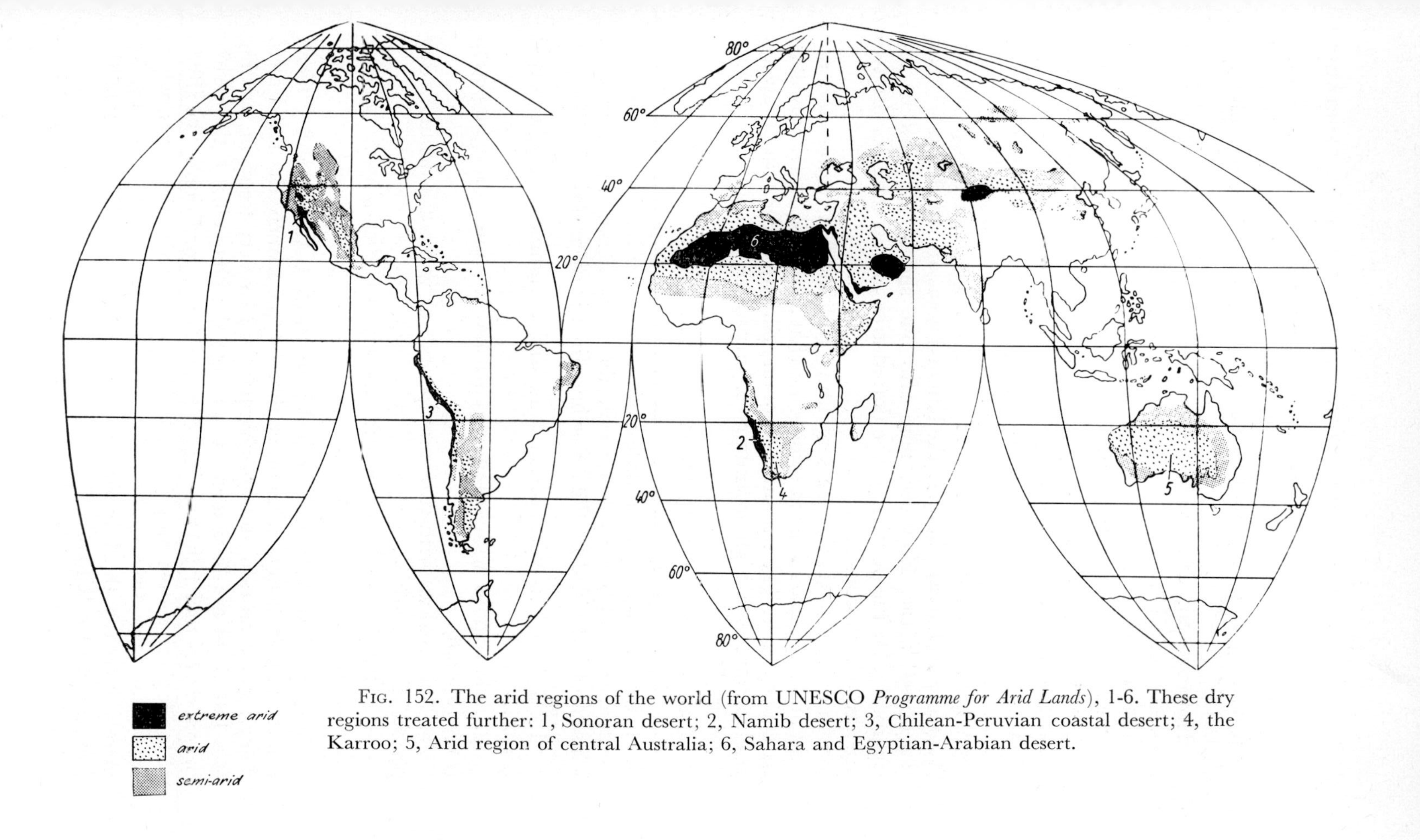

FIG. 152. The arid regions of the world (from UNESCO *Programme for Arid Lands*), 1-6. These dry regions treated further: 1, Sonoran desert; 2, Namib desert; 3, Chilean-Peruvian coastal desert; 4, the Karroo; 5, Arid region of central Australia; 6, Sahara and Egyptian-Arabian desert.

Saline soils occur only locally in wet regions e.g. around salt springs, which bring readily soluble salts to the surface from particular geological strata, or on the coast in those places where sea water influences them.

However, all these pointers are insufficient for an unequivocal separation of arid and wet regions. There are only a few arid areas that do not also exhibit wet conditions during some times of the year. Conversely, there are regions with wet climates that have more-or-less long-lasting dry periods, e.g. the tropical regions of the dry forest or savannah, which I have already discussed.

In arid regions we must also consider the annual march of the climate. Therefore, I shall once again employ the method of graphically representing climate by climatic diagrams. Through these, one can recognise arid areas wherever the hatched areas are much smaller than the dotted areas. However, here also one should not proceed too rigorously, for the position of the dotted area and its division into two are also of importance.

An objective criterion for separating arid and wet regions does not exist and any separation has to be somewhat arbitrary; it is simply done for convenience. The transition from one climate to the other is always gradual, unless a high mountain ridge provides an abrupt climatic division.

There is thus not one uniform arid climate, instead we must distinguish a series of different types, which are similarly more or less arbitrarily delimited. One may say that almost every arid area has its own climatic peculiarities and thereby differs from the others in one way or another. This can be clarified by the following climatic diagrams.

1. Arid regions with winter rains and a summer drought period (Figs 153-155).
2. Arid regions with summer rains and a winter drought period (Figs 156-158).
3. Arid regions with two rainy periods or without a definite rainy period (Figs 159-161).
4. Extremely arid regions with merely episodic rains or rain-free (Figs 162-164).

The two last diagrams, from the South-west African and Chilean coasts actually belong to another desert type, namely the fog-deserts. They receive no rain and are characterised by very frequent, wetting fogs but these are insufficient to change the character of the nearly barren desert under the existing, extreme desert conditions. The effects of the fog are shown only on rock-ridges and mountain slopes, where condensation occurs together with air-movements. Otherwise, the fog merely increases the humidity of the air, causing a lowering of the temperature (see Chaps IX and X).

In addition to the sub-tropical arid areas that are discussed here there are also arid areas in the temperate and cold climates. These are subject to different temperature regimes.

Extremely arid regions will be designated deserts. However, deserts also cannot be defined absolutely (Schiffers, 1950; Gabriel, 1961). Only very few deserts have such an extremely arid climate that they can be

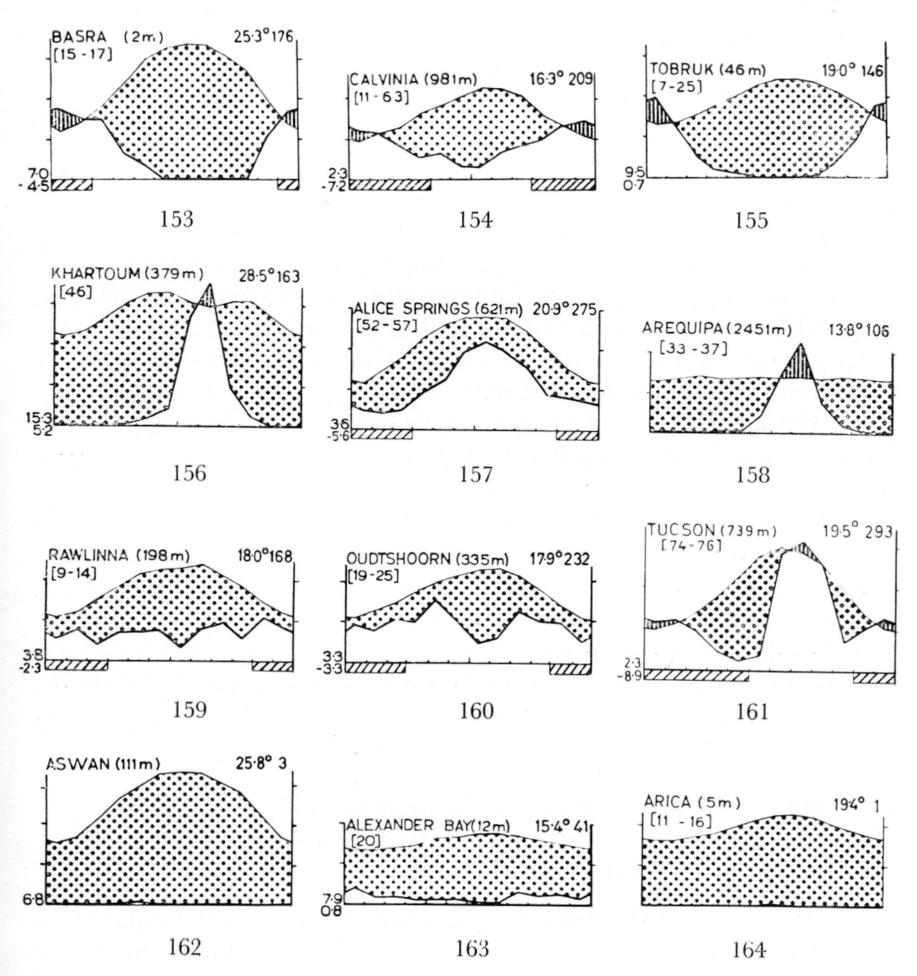

FIGS. 153-164. Climatic diagrams of desert regions. Upper row, with winter rains; second row, with summer rains; third row, with rains throughout the year or with two rainy seasons; bottom row, almost rain-free (from *Klimadiagramm-Weltatlas*).

considered 'absolute deserts'. This applies to regions where the vegetation is completely absent—except in oases—or where the vegetation only develops for a short period after a rare rain (ephemeral vegetation).

Very commonly economic reasons are decisive in designating an area as 'desert'. By this, one understands an area which cannot be utilised agriculturally by a settled population unless the area is irrigated.

Extensive ranching gains ground in the desert whilst rain-dependent agriculture, better suited to a less extremely arid climate, gives way. Often, the completely unpopulated 'deserts'—for example in central Australia or in the Kalahari—are by no means the driest parts of the continents. In such cases, factors other than climate are decisive. For example, location relative to transport facilities may play a role, or sand dunes may be an obstacle to accessibility. This factor may render an area unsuitable for range management. Moreover, difficulties in obtaining good drinking water may be decisive, or permeable limestone soils (karst formation) may prevent the establishment of a cover of vegetation.

The designation 'desert' depends very much upon a comparison with an observer's background. For the North American coming from a wet region, the south-western part bordering Mexico is already the Sonoran 'desert'. The Egyptian, who is familiar with the most extreme desert in the world, does not consider the coastal strip of his country a part of the desert, in spite of the fact that this area is much drier and more poorly vegetated than the Sonoran desert. The religious desert-dweller considers the desert the garden of God, from which He has removed all human and animal life so that it is a place where man can move about in peace.

Thus, it is not surprising that the name 'desert' is used for a variety of very different landscapes. However, in this chapter, I shall consider only climatically determined deserts; arid areas characterised by a relatively sparse plant cover. Even among these there are so many different types that it seems necessary to treat them separately. The few features that are common to all deserts will be discussed in this chapter, while their individual characteristics will be treated in Chaps VIII-XIV.

2. Sources of water for plants in arid areas: decreasing vegetation density and unequal water distribution in the soil

When investigating the reasons responsible for the great similarity in landscapes of all arid regions of the earth, I must highlight as the most important feature the decreasing dominance of the plant cover and the increasing dominance of surface rocks. Whereas in other areas vegetation covers the earth like a blanket, thereby protecting it from erosion, this is not so in arid areas. Thus, weathering and erosion bring about much more strongly varied relief in the desert, often reminiscent, even at low altitudes, of high mountainous terrain. Only the density of vegetation in arid areas makes it conceivable that sufficient water for plant growth is available at all with the very low rainfall.[1]

[1] This has been emphasised by Lavrenko (1962); also by Sweschnikowa and Salenski (1956) which was not available for consultation.

One has often wondered how plants grow at all in the very dry deserts, and tended to relate this to some sort of special physiological drought resistance of the desert plants. However, more detailed investigations have shown, in all cases, that such is not the case and that there are only characteristic differences of degree. Whenever a growing plant occurs in the desert, water is present within range of its roots if it has water-absorbing roots and not water-storing organs.

It is important to remember that rainfall cited in mm refers to the amount of water in litres falling on a square metre of soil. Therefore, one should always consider this quantity of water in relation to the plant mass per square metre of soil. So far, however, this has rarely been done.

To determine the dependency of vegetation density on rainfall it is necessary to keep the other factors constant, such as the temperature or soil-type. One should also use similar vegetation types only for comparison.

If these conditions are met, it can be shown that the plant mass—and thus also the transpiring surface—decreases proportionately with rainfall. I first obtained evidence of this in South-west Africa by comparing *ungrazed* grasslands on sandy soil (Walter, 1939). In this area the rainfall occurs as summer rain and its amount decreases from the moister north-east part of the country to the coast; from more than 500 mm to practically zero (Fig. 165). At the same time, the grasses become smaller and the surface cover decreases, until the actual desert begins at a rainfall of 100 mm per year. The quantitative relationship between production of aerial dry matter and rainfall is shown in Fig. 166. This shows that the annual productivity, in fact, decreases proportionately with the amount of rainfall.

Since only grasses were subjected to this comparison, we may assume that this principle holds not only for dry matter but also for the transpiring surface of the grasses, i.e. that the transpiring surface also decreases proportionately with the amount of rain.

This, then, would imply that in a wet and an arid area the same quantity of rain would be available per unit transpiring surface. Thus, the water supply to the individual plant is not much worse in the arid than in the wet areas.[1]

Furthermore, I was able to show that this principle also applied to other vegetation, e.g. to the *Eucalyptus* forests in Australia. In this case I determined both the annual leaf-mass (leaf litter) and leaf area directly. Here, one must compare the virgin forests with their undergrowth

[1] A worsening of the water conditions can occur in so far as the potential evaporation increases with decreasing rainfall at an equal level of water supply. However, this is true only for the annual amount. During the rainy period as such, which coincides with the growth season, the differences in potential evaporation are not very great. Instead, the average length of the growth season as a whole decreases with decreasing rainfall. Thus, we find that different grass species complete their development at faster rates in the drier areas. Here they replace the larger species of the wetter regions.

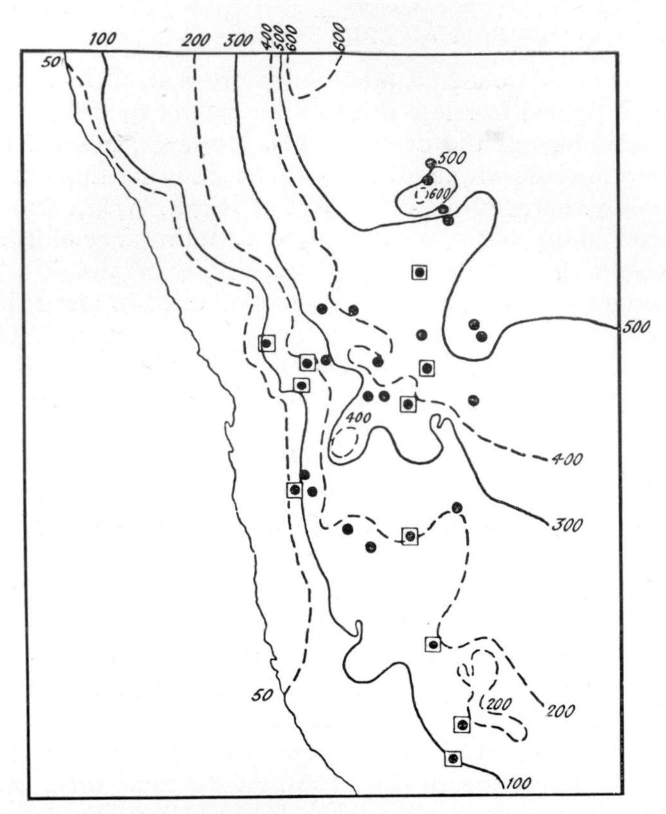

Fig. 165. Rainfall distribution in S.W. Africa (annual rainfall in mm). ⊡Sample quadrats for determining the productivity of the grassland; ● sample locations for forage-plant analyses (after H. Walter).

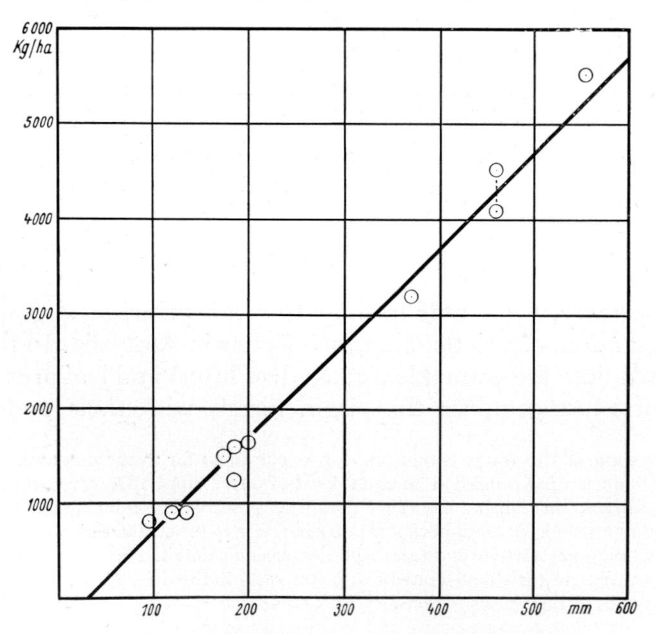

Fig. 166. Productivity of grassland in S.W. Africa in relation to amount of rainfall. Ordinate: Dry matter in kg/ha; abscissa: average annual rainfall (after H. Walter).

separately from artificially established, second-growth stands lacking undergrowth, which show greater leaf-masses.

Fig. 167 shows rainfall on the abscissa and relative quantities of litter and leaf areas, respectively, on the ordinate (highest values, in each case, were taken as 100). Apart from some irregularities, which cannot be avoided in such determinations, the proportionality is shown rather clearly, in this case, with much higher rainfall, from about 500 mm to more than 1500 mm, and in a winter-rain region.[1]

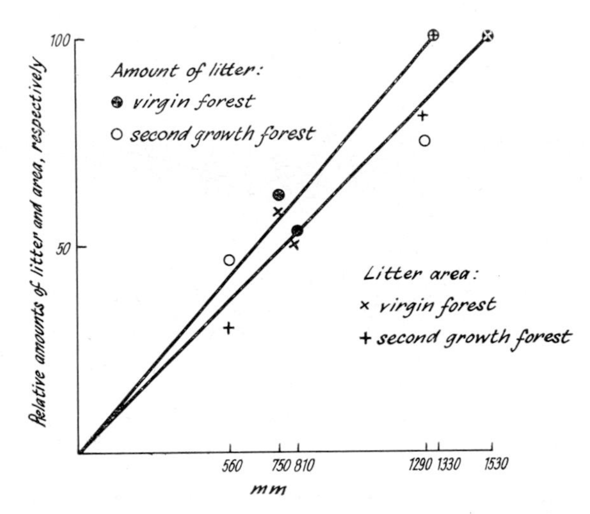

FIG. 167. The relationship between quantity of leaf litter and leaf litter area to rainfall in Australian *Eucalyptus* forests (data from O. W. Loneragan, Forest Dept. of W. Australia).

From this principle, which has been demonstrated so clearly here, one can conclude that *the water supply per unit leaf area remains almost unchanged in wet and arid regions*. The only prerequisite is that the plant must always possess the ability to extract the necessary water by its root system. One observes, in fact, that root system development in arid areas is correlated more strongly with a reduction in leaf area; the roots extend less deeply, but are always more developed laterally. The relation between above- and below-ground parts alters with increasing aridity more and more in favour of the latter. An exception is found only in the succulents, which are not comparable with the others.

A second factor must be considered in extremely arid areas where rainfall is less than 100 mm. This is the unequal distribution of the water in the soil, which is reflected in the plant cover. The absence of a

[1] A similar proportionality has been determined also by Le Houerou in Tunisia. He found that the yields/ha of olive plants decreased in proportion with the rainfall. Although the yield per *individual tree* remained unchanged, the *number of trees* per hectare, i.e. the density of the plantation, decreased with decreasing rainfall (Le Houerou, 1959).

closed plant cover results in a very pronounced surface run-off. Therefore, on higher ground less penetration occurs than corresponds to the rainfall, while much more penetration occurs in hollows. The distribution of vegetation is correlated with this water distribution and thus is concentrated wherever the soil water storage is greatest, and here the water supply is commonly no worse than in wetter areas (Fig. 168). If

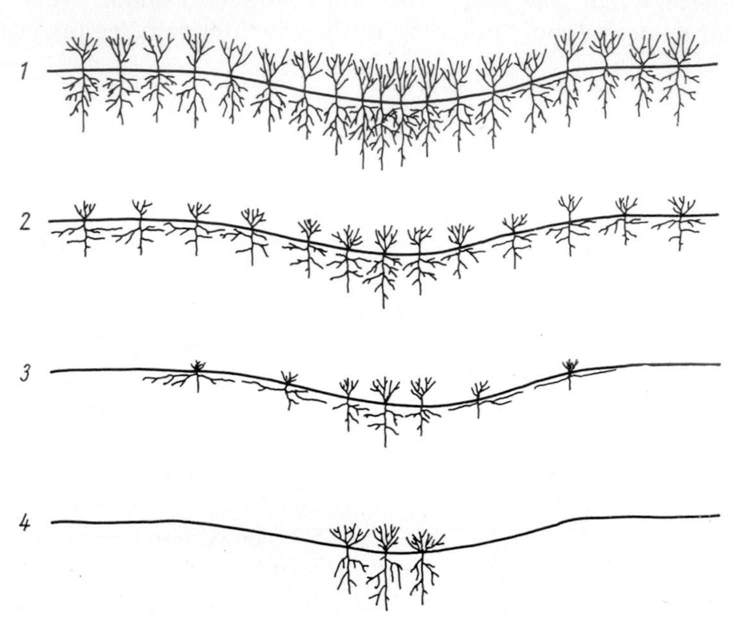

Fig. 168. Schematic representation of the change from a 'randomly' distributed vegetation (1 and 2) to a 'contagiously' distributed one (3 and 4) with decreasing rainfall in extreme arid areas.

one assumes, for example, that with an annual rainfall of 25 mm a hollow may collect 40% of the rainwater from an area 50 times as large, the plants in this hollow would have water available corresponding to a rainfall of 500 mm. This change in the character of vegetation can be observed particularly well from an aeroplane. While the vegetation is 'randomly'[1] distributed over the whole surface in rain-rich regions showing greater density only in hollows, it changes into a more and more 'contagious'[1] distribution in extremely arid areas (Monod, 1954).[2] By this we imply that the largest part of the area is completely barren of

[1] Random and contagious are used here for Monod's 'diffus' and 'contractés', respectively.

[2] Contagiously distributed vegetation may be arranged in strips through sheet-erosion in nearly flat plains. In areas in Somaliland with a rainfall of 150 mm many parallel strips, at right angles to the slope, were covered by *Andropogon* with wider strips lacking vegetation between them. The *Andropogon* strips receive by run-off twice as much water as the others. The water content of the soil under *Andropogon* was 10%, that of the vegetation-free areas only 6-7% (Hemming, 1965).

plants and that the vegetation becomes more restricted to the valleys.[1] In hollows, the roots of the plants grow much deeper into the soil since the water penetrates relatively deeply into the soil at these places, where it is also protected from evaporation (see Figs 271, 272 and 283). This demonstrates that no general principle can be established concerning rooting depth in relation to decreasing rainfall.

Contagious distribution of vegetation has been used for agricultural purposes also. Old ditch and dike systems have been discovered in the Negev desert which, in pre-Arabian times, served to collect the run-off from higher areas after rain in order to transport it to small crop fields in lower areas. Recently these systems have been restored and it then became possible to produce a crop with a very low annual rainfall (Evenari *et al.*, 1961; see also Evenari *et al.*, 1963). The run-off is estimated as 20-40% of the total rainfall.

Similar systems, from the Roman period, are also known on the Egyptian coast, which then served for viniculture. Even today, one can cultivate olives with an annual rainfall of 100 mm in this region, if the run-off from stony ridges is led into the olive orchard. The catchment area must be at least four times that of the cultivated area, and all rainwater must collect as run-off. Some cultivation is possible, therefore, where such features are provided.

Unequal water distribution is noticeable not only in deserts with pronounced relief, but even in nearly flat terrain. However, here the water does not collect in special channels but rather it moves as a broad flood-sheet across large areas; even a hardly detectable slope of 1:2000 is sufficient. The flood-sheet moves very slowly so that the water has a chance to penetrate the soil. The longer the surface is covered by the flood, the more water penetrates the soil and so the density of the vegetation increases more and more in the direction along which the flood-sheet moves. These conditions only alter again at the lowest part of the area, because here the water may sometimes have moved farther, before it has all penetrated, and sometimes not so far, depending on the heaviness of the rain. At the same time undissolved salts, which are carried with the water, accumulate at the lower end of the flood-layer and this is shown by some soil salinity at these places. The largest flood-sheets fill the hollows lacking outlets, where the undissolved salts are deposited after evaporation. This results in the formation of salt-pans.

Where, instead, the bedrock consists of permeable limestone, surface run-off is almost entirely absent; such areas are almost devoid of vegetation, as for example, the large Nullarbor Plain between west and south Australia. Only a few single woody plants can maintain themselves in small 'dolines' (i.e. erosion hollows in the 'karst' land-

[1] A similar feature can be also noted in the transition zone between a forest and a steppe area: the drier the region the more interrupted the forest and the more concentrated its growth in the valleys. Thus here we have a 'contracted' distribution of forest vegetation.

scape) that are filled with fine soil. Run-off is also small on sandy flats. However, by contrast, sandy soil has favourable water storage characteristics in arid—as opposed to wet—areas. This will be discussed later.

The characteristics emphasised here refer primarily to perennial species of desert vegetation which alone can adjust to average climatic conditions. In unfavourable years they can lose large parts of their shoot system or they may die off completely in some places. In good rain-years, they regenerate more vigorously or they invade new areas by means of seedlings. In addition, there are also numerous species in all arid areas which comprise the 'ephemeral' vegetation. These plants depend upon the rainfall in any particular year. They form, so to speak, a 'buffering' vegetation, which serves as a balancing element to variations in rainfall. The perennials cannot use all the available water in favourable years. This is used much more efficiently by the ephemerals. They germinate by the thousands and cover the entire desert as a green, or coloured carpet during the flowering period. They are much less tied to a particular habitat and grow anywhere, provided that the ground remains damp for a long enough period. If, in poor rain-years, this is the case only in hollows, then they only occur there. Where soil moisture is slight, the osmotic values of the annuals increase sharply and many dwarfed individuals develop, bloom early and produce only few fruits and seeds. If, instead, the rainy period lasts longer, the plants develop very vigorously under the favourable water conditions and produce a large number of flowers and fruits at the end of the growth period. Thus, the seed content of the desert soil becomes replenished. The density of ephemerals always depends, therefore, on the current rainfall. Moreover, the current rainfall also determines the floristic composition, for the following reasons.

1. The individual annual species only germinate within very narrow ranges of temperature, which may be higher or lower. Therefore, the temperature relations after the rain determine which species germinate and which remain dormant (Went, 1948, 1949).

2. Among ephemerals we find a range of rates of growth and water requirements. There are hygromorphic plants with delicate leaves that grow rapidly but require good aeration of the soil. These cannot endure extreme conditions of high evaporation. Others are of a more xeromorphic structure, often hairy or succulent. These can endure a short dry period within the rainy season much better. They also develop a stronger root system but have a slower growth rate.

3. Competition amongst ephemerals occurs during the favourable season just as noticeably as in any other plant community. If much rain falls during the rainy season keeping the soil and air damp for a longer period, the hygromorphic and fast-growing species are favoured and the more xeromorphic species are suppressed. If, by contrast, rainfall occurs irregularly so that the upper soil layers dry up in between

showers, the xeromorphic species become dominant since they can endure a greater decline of soil water.

This makes it understandable how ephemeral vegetation, in contrast to the perennial vegetation, shows a very variable picture from year to year. In 1934, a unique, extremely rainy season occurred in South-west Africa. It lasted for several months and was followed by a period of three extreme drought years. Local farmers claimed to have observed entirely new species that they had never seen before. Investigations showed, however, that these were species that prefer damp habitats. In normal years, they are very restricted in their distribution and flower only in protected locations between rock bluffs or in erosion channels remaining unnoticed by farmers. An example is the composite, *Nidorella*. The extreme rainfall made the soil so damp everywhere that *Nidorella* became dominant within a short time, thereby rapidly colouring wide areas with its yellow flowers. This, of course, could be noticed by anyone.

The geophytes also show a close dependency in their development on the current rainfall. These can, at least in part, be considered ephemerals, as they usually have a very short growing period and reduce their shoots by necrosis each time with the increasing drought.

They are mostly plants with bulbs or corms whose storage organs are often deeply buried in the ground. According to observations of local people, these geophytes do not develop shoots for many years if the rainy seasons are unfavourable. Instead, they grow vigorously in some years after very strong rainfall. I could not check these observations as this would have required several years. However, this behaviour appears quite possible. The soil, in which the bulbs or corms are buried, dries up completely during the dry season. Conditions for shoot and leaf emergence are provided only when the rains penetrate sufficiently deeply so that roots, newly formed from the storage organs, can absorb sufficient water. Where this is not the case the bulbs and corms remain dormant. They are so well protected from desiccation in the soil that they do not dry out even after many years of drought.

Geophytes may have already formed their shoots before the beginning of the rainy season in certain years. This suggests the presence of some small quantity of residual water stored at lower soil depths from the preceding rain of good years. This is frequently so in heavily overgrazed areas. Perhaps some geophytes may also store enough water in their bulbs and corms to enable them to produce a flowering shoot without absorbing water from the soil.

There is hardly any competition between the perennials and the ephemerals. As already pointed out, the perennials cannot utilise all the water which becomes available themselves during a good rainy season. They rarely form a continuous root system through the upper soil layers. Moreover, the perennials occur at such low densities that the ephemerals are hardly affected by shading. On the contrary, one can observe that shading by the crowns of perennials often has a favourable effect, because evaporation is somewhat reduced in these positions and delicate plants are protected from direct irradiation. A similar situation can often be observed after a rainy period, when the ephemerals are already dried up in the open sites but still remain active for some time in the shade of woody plants.

The behaviour of ephemerals in deserts with regularly occurring rainy seasons is the same as in deserts with rare episodic rainfalls. However, in the latter they do not recur annually but only in those years and areas with a particularly heavy rainfall. For example, the

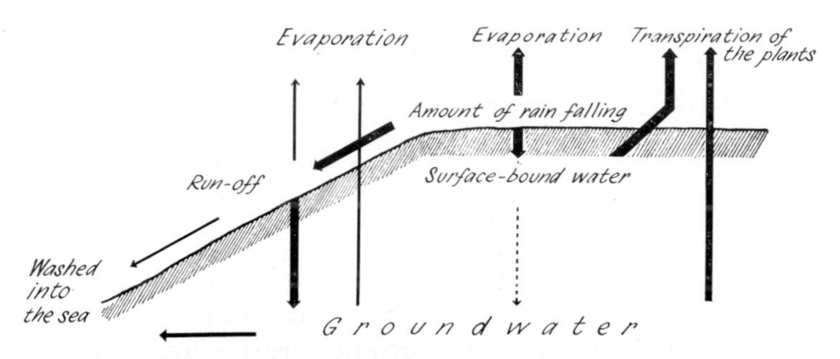

FIG. 169. Schematic representation of the water economy of arid regions. Further explanation in the text (after Walter and Volk, 1954).

outer Namib desert in South-west Africa is normally completely barren of vegetation. However, in 1934 more than 100 mm rain fell in this area and the desert became covered with a rich carpet of Mesembryanthemums. I saw this *Mesembryanthemum* cover still persisting in 1935, although it was already somewhat broken and weakened. These succulent species can store so much water that they can remain alive without any more rainfall for over a year and they may even continue to bloom then.

3. The importance of soil texture to the water relations of arid regions

The amount of water which penetrates the soil does not only depend on the amount of rainfall but, as we have seen, also on the amount of run-off or seepage. The total amount is only partly available for plant growth. A schematic illustration is provided to explain these rather complex relationships (Fig. 169).

Rain falling on a soil surface will—added to by surface-flow—eventually penetrate partly into the soil and, in part, it will evaporate directly or will disappear by surface run-off. Water that penetrates more or less deeply into the soil will be held by the soil particles as capillary water. Any excess water, which, as gravitational water, would make up the ground water, is usually absent in arid regions. The soil is only wetted periodically to a limited depth; beneath this occurs the 'dead soil', which contains no water available for plant growth.

The surface run-off reaches the valleys through erosion channels. Streams are only formed after very heavy rains, which may reach the sea in coastal deserts, otherwise they only flow into hollows lacking outlets. After light rain, the water seeps rapidly through the river bed, from whence it contributes to the ground water. In places where ground water is near the surface it can be lost by evaporation from the soil surface.

Apart from the river valleys, ground water is supplied mainly through rock crevices and dry, periodically-flooded valleys in arid regions. But ground water may be entirely absent. The water held on the surfaces of soil-particles is of importance to vegetation. This provides a water reserve available for plant growth, thus the water stored in the soil is dependent on the quantity and frequency of rainfall, and to a great measure on the nature of the soil.

The proportions of surface adsorption and run-off are dependent, both on the slope and, in addition, on the soil drainage.

In general, run-off is greatest on clay soils, least on sandy soils and non-existent on loose, stony shingle. The greater the run-off, the less water penetrates the soil. This factor, therefore, increases the dryness of clay soils. Run-off from solid rock surfaces is also great, but it is decreased by the presence of cracks. Smith (1949) pointed out that any bare soil in arid regions becomes superficially compressed through trampling by grazing animals, pressure from car wheels, or through the fall of rain drops. This increases the run-off markedly and the same also applies to periodically-flooded soils, particularly where these are somewhat clayey. Flood-terraces, therefore, only become covered with vegetation with some difficulty, because the flood-wash removes any seeds after each rain.

However, run-off is dependent upon soil texture and the proportion of water which penetrates the soil. This may be either evaporated or stored as capillary water. Here it must be emphasised particularly that in arid regions soil texture has the opposite effect on water-plant relationships to that in wet areas (see Walter, 1932, 1960, pp. 191f). In wet areas, clay soils with their high water-holding capacity are considered to be the wettest soils; sandy soils and especially rocky soils are considered to be the driest. The opposite occurs in arid areas. The reasons are as follows. Soils in arid regions are never completely wetted; great quantities of storage water are found in the upper layers, only as deep as the penetration of rain water during the rainy season. Let us assume a rainfall of 50 mm. It will only wet the upper 10 cm of a clay soil; in a sandy soil it will penetrate to a depth of 50 cm, and even deeper in a stony soil, where the water is retained only by the soil present in crevices in the rock.

The rain is followed by evaporation. In the clay-soil, which soon cracks, the upper 5 cm rapidly dries out, i.e. 50% of the rain is lost in this way. In a sandy soil, the surface dries in the same way but strands of capillary water are quickly broken and the rest, more than 90% of the rain, remains in the soil. The stony soil shows virtually no loss by evaporation. Thus, its storage capacity is maximal (Fig. 170). In arid areas, therefore, one finds, under uniform climatic conditions, trees on stony-soils, grass on sandy soils, while clay-soils remain nearly barren of vegetation. This explanation, which I gave for arid areas in the west of North America, were supported by Hillel and Tadmor (1962) in the

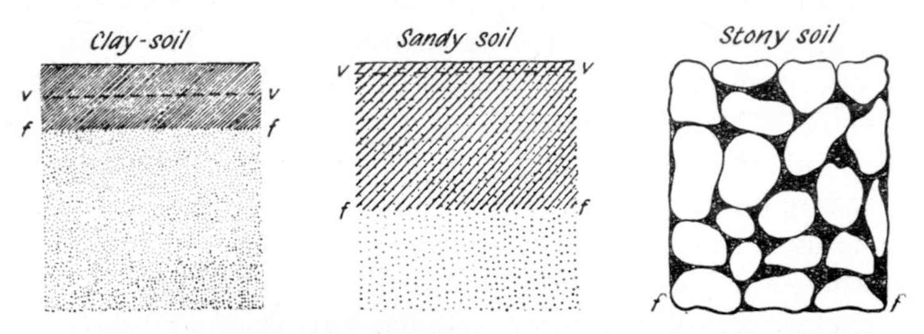

FIG. 170. Schematic representation of the degree of penetration of equal quantities of rainfall in different soil types (after H. Walter, 1960). *f–f* lower limit of wet soil after a rainfall; *v–v* lower limit of water which evaporates; note the clay-soil loses half the water, the sandy soil about 10% and the stony soil practically nothing.

Negev desert. These authors found, in an area of uniform rainfall, 35 mm of available water on loess soils, about 50 mm in stony soils (from which there was much surface run-off), 90 mm in sandy soils, and from 250-500 mm in dry valleys (fed by water seepage).

If the soil of a rock outcrop has only a few cracks, water can percolate downward in these to the ground water. Thus, the fine soil in these cracks is damp from the soil surface to the ground water table. A plant root has the chance of growing right down to the ground water in these crevices, so ensuring its water supply throughout the year. Such habitats are occupied by species with very deep-penetrating taproots. This type of root system, however, in contrast to information commonly given in textbooks, is rather exceptional. Rooting depth is entirely dependent on the depth water penetrates and this is generally shallow in arid regions (see p. 273).

I saw an interesting example near Basrah in Mesopotamia of how deep-lying ground water can be used for the afforestation of desert areas. An inexhaustible reservoir of ground water occurs at 15 m, supplied by the Euphrates and Tigris rivers through gravel strata.

However, with an annual rainfall of 120 mm, only the upper soil layers become wetted and the dry soil below prevents roots penetrating to the ground water. Only ephemeral vegetation develops after winter rains. Nevertheless, ground water is made available from wells, often in a very primitive manner, and is then used for watering vegetable crops (onions, tomatoes, squash) on a small scale. With the very high summer temperatures (up to 50°C), water has to be poured into the ditches 8 times a day. Because of the high evaporation, the soil rapidly becomes salty and vegetable crops can only be cultivated for one year. However, cuttings of *Tamarix articulata* are planted between the vegetables and root rapidly. Since the soil is damp from the surface down to the ground water table, because of the abundant watering, the *Tamarix* roots can extend down after the cessation of surface watering until they reach the water table. The persistence of the trees is then secure. They are cut every 25 years, and regenerate from the stumps. In this way, a forest becomes established on the abandoned irrigation-fields instead of a desert.

If one considers, in addition to climatic influences, the influence of topography, soil texture and soil structure on the soil-water relations, one cannot but be impressed by the multitude of habitat variations. Hardly any two habitats can be considered alike. Therefore, one cannot support Stocker's (1935) opinion, without some reservation, that there are more numerous adaptive forms among plants than the number of habitat variations. Such an opinion is only justified if one classifies the habitats very broadly into a few types, and, on the other hand, recognises every individual reaction of the plants, e.g. differences in transpiration rate, as a special feature. It is perhaps correct in so far as plants which belong to different ecological types can occur together in a restricted, apparently uniform habitat. However, we are not usually in a position to determine whether or not the soil really provides uniform conditions with respect to any differences in the rooting habit of these plants. Moreover, every plant species has a characteristic ecological range, so that similar ecological types may grow in different habitats. Since the number of plant species and, even more, the number of ecological types, is limited, whereas habitats are characterised by innumerable fine and gradual variations, one could even suggest the opposite view to that of Stocker.

4. The principle of relative constancy of habitat with changing ecological niche (or biotope)

In South-west Africa the very wide ecological amplitude of many tree species is rather striking (Walter, 1953). There are species that range from the borders of desert, with 100 mm rainfall, into the savannah. However, the ecological niches (biotopes) throughout this distribution range are very unequal. Whereas the species are found in erosion channels and between rocks in dry regions they occur in the wetter parts of their range on sand flats. In regions with even higher rainfall, they become restricted to dry slopes, because they are unable to compete with faster growing species on the plains. Thus, there is a pronounced change in ecological niche, which enables one to recognise a definite principle: the drier the climate the more the species becomes restricted to the damper ecological niche; that is, decreased rainfall at the drier end of a climatic gradient is compensated by greater soil moisture. Thus, the overall water status of the different habitats remains more or less constant.

With respect to differences in water relations due to differences in soil texture, topography and run-off, the following sequence of niches may be recognised. These are, from the driest to the wettest:

1. Slopes with impermeable soils and high rates of run-off.
2. Flood-terraces with surface-compacted soils.
3. Clay soils on level flats with high evaporation rates.
4. Sand flats easily penetrated by rain.
5. Stable sand dunes.
6. Talus slopes and stony plains with low evaporation rates.
7. Fissured rocky outcrops, which can hold water easily.
8. Bottoms of slopes with inflow of seepage.
9. Erosion channels, carrying water temporarily and storing water in the underlying soil.
10. Dry valleys containing a continuously flowing stream of shallow ground water below the surface.

If one traces the distribution of a particular species along a climatic gradient from its wetter to its drier end, one observes that the species changes its ecological niches along this gradient. Of course, differences among species concerning water requirements and rooting habits are also to be seen.

Two species seldom show a completely parallel distribution pattern. Some prefer rocky, others sandy habitats. Particular species of tree in wetter areas are suddenly found to grow in fissures of rocky outcrops in drier areas or else they occur at the foot of granitic mountain slopes. By contrast, others are found only in valleys with ground water, where they are able to withstand periodic flooding and the flashes of temporary, torrential streams. The general tendency, however, is clear. It enables one to formulate the principle of relative constancy of the overall environment with changing ecological niche (see Walter, 1954, pp. 41ff) namely:

Wherever climate changes in a particular direction across the geographical range of a species, there is a correlated sequence of ecological niches, which compensates for the change in climate, i.e. the overall environmental conditions remain more or less the same in each different ecological niche. This principle applies to water relations in arid regions and relates to all factors in general which are determined by climate (see p. 12).

Particularly fine examples illustrating the truth of this principle are given by Smith (1949). He distinguished 14 different niches related to increasingly favourable water relations in the Sudan. These enable various species each to grow under very different climatic conditions. For example, the range in annual rainfall of the following species is:

Acacia seyal	370– 800 mm
Khaya senegalensis	400–1050 mm
Prosopis africana	480–1100 mm
Tamarindus indica	500–1200 mm
Dalbergia melanoxylon	360–1200 mm

Through this rainfall range they occupy different niches. For example, *Acacia tortilis* grows:

in the desert in erosion channels in 50 mm,
near Khartoum on sandy soils in 150 mm,
in the Kassala district on clayey soils in 300 mm and
on the dry slopes of the Butana hills in 500 mm.

Hyphaene thebaica occurs:

in erosion channels and along streams in 100 mm,
on red loamy soils in 600 mm,
on loamy ridges in 750 mm and
on heavy clay soils in 900 mm.

Sterculia setigera, which grows in 1200 mm on plains, grows in only 300 mm on rocky outcropping knolls. *Albizzia aylmeri* grows in 800 mm on clay soils, while on sand dunes it even survives 300 mm. Smith cites very many examples especially for more favourable conditions on sandy soils as compared with the less favourable ones on clay soils. The Sudan is particularly suitable for such comparisons since its western part consists of vast sand flats, while its eastern lowland is characterised by clay soils. One generally observes that only two-thirds of the rainfall is necessary for the presence of the same species on sandy soils as compared with clay soils. For example, *Acacia senegal* occurs most abundantly on sandy soils with 400 mm rainfall, but on clay soils at 600 mm.

The water requirements of different species can only be assessed by comparing their presence on uniformly flat soils without in-flow or run-off. On this basis, the following sequence of *Acacia* species dominate in relation to increasing rainfall in the Sudan: *Acacia flava*, *A. orfota*, *A. tortilis*, *A. raddiana*, *A. mellifera*, *A. fistula*, *A. senegal*, *A. seyal*, *A. drepanolobium*, *A. campylacantha*, *A. sieberiana*, *A. albida*, *A. hebecladoides*, *A. abyssinica*. Of course, several species may occur side by side at any particular level of annual rainfall, but if so they occur in different niches. The less demanding species are restricted to drier habitats, the more exacting ones in the wetter habitats. Since the clay content exerts an adverse effect on the soil-water balance in arid regions, there are obvious relationships between the presence of particular species, the rainfall and the clay content of the soils (Fig. 171).

According to Le Houerou (1959) the optimum yield of olives in Tunis lies on sandy soils with an annual rainfall of 230 mm and on those heavy soils with over 500 mm. In very dry climates plantations

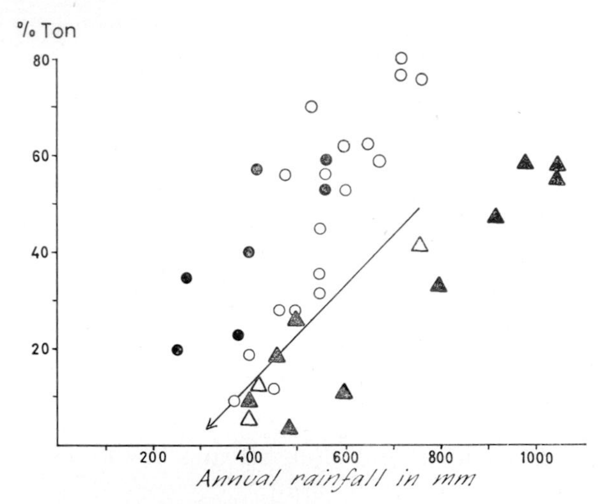

FIG. 171. Distribution of different woody plants in the Sudan in relation to the clay content of the soil (ordinate) and annual rainfall in mm (abscissa). Unshaded circles—*Acacia seyal*, black circles—*Aracia mellifera*, unshaded triangles—*Prosopis africana*, black triangles—*Khaya senegalensis*. The long arrow separates the woody species of the *Acacia* zone from those of the tropical dry-forest (after Smith, modified).

are only possible on light soils but in wet climates they are leached rapidly from sandy soils.

The higher the clay content, the more rainfall is necessary. Conditions on clay soils can be improved for woody plants when blown sand is deposited as dunes around the plants, so gradually increasing their height until the knolls of larger sand dunes are covered with shrubs (Fig. 267).

In stationary dunes, the dune ridges are especially favourable habitats. This can be seen, for example, in the southern Kalahari, where the dunes—they range from 5-51 m high and up to 3 km long—are oriented in parallel from north-west to south-east at distances of about 230 m from each other, covering in all an area of 117,000 km². The sand on the dune ridges is coarser than in the hollows, grasses as competitors are less abundant and the soil is also more exposed. Therefore, tree species such as the following can grow: *Albizzia anthelminthica*, *Terminalia sericea*, *Acacia giraffae*, *Boscia albitrunca* and *Acacia haematoxylon*. Among these, *Albizzia* and *Terminalia* have their main centres of distribution much farther to the north in an area with higher rainfall, because they have greater water requirements. The endemic *Acacia haematoxylon* is a weak competitor and restricted exclusively to this favourable habitat (Leistner, 1964).

5. Water relations and the main ecological plant types of the non-saline soils of arid regions

The water factor is of decisive importance for plants of the arid regions. In ecological investigations it is insufficient merely to obtain an insight into the water relations of different habitats. One has to study in addition the response of the plants to these conditions. This can be done by determining the *hydrature* (see below) *of the protoplasm* in the plants (Chap. I, 8).

The relationship between hydrature (which may be termed relative water activity) as measured in per cent of relative vapour pressure (% hy) and the potential imbibition pressure, or potential osmotic pressure (referred to more briefly as osmotic value) can be expressed by the following mathematical relationship:

$$\Pi_T = -\frac{1000R \cdot T \cdot s}{M} \ln \frac{p}{p_0}.$$

Here Π_T stands for the atmospheric value of the potential imbibition pressure or the osmotic value, respectively, at the absolute temperature T; R is the gas constant (= 0·08207), T the absolute temperature (at 20°C = 293), s the specific weight of water (at 20°C = 0·9982), M the molecular weight of water (= 18) and p/p_0 the relative vapour pressure, which when expressed as a percentage, I term the *hydrature*.

Inserting the constants in the formula, it may be written, for a temperature of 20°C as:

$$\Pi_{20} = -1331 \ln \frac{p}{p_0} = -3067 \log \frac{p}{p_0}.$$

One can calculate from this relationship between hydrature and osmotic value at 20°C the values shown in Table 42.

This treatment is necessary to understand the hydrature relationship of living cells.

As is well known, the protoplasm of a mature plant cell is bounded by a semi-permeable membrane to the outside and inside, at the vacuole. The cell as a whole is surrounded by an elastic cellulose-wall. The osmotic relations are expressed by the following equation (all values in atm.):

$$S = \Pi - P$$

where S stands for the suction tension (= Diffusion Pressure Deficit, DPD) of the cell, which indicates the negative water potential of the cell; Π stands for the potential osmotic pressure (or negative osmotic potential) of cell sap and P for the hydrostatic pressure existing in the cell, or turgor pressure. Since the living protoplasm is only separated

TABLE 42

hy (%)	Π (atm)	hy (%)	Π (atm)	hy (%)	Π (atm)
100	0	95	68·4	85	217
99·5	6·7	94	82·5	80	298
99	13·4	93	96·7	75	384
98·5	20·1	92	111	70	476
98	26·9	91	125	60	681
97·5	33·8	90	140	50	924
97	40·6	89	165	40	1221
96·5	47·5	88	171	30	1604
96	54·5	87	186	20	2144
95·5	61·4	86	201	10	3067

from the cell sap by a semi-permeable membrane and since it is under the same pressure, it must have, in equilibrium conditions, the same hydrature (i.e. the same relative vapour pressure) as the cell sap.

We know that the living protoplasm—if not subjected to physico-chemical changes—shows the same hydration curve in response to changes in hydrature as do dead colloids (Walter, 1923, 1963). Therefore, any increase in osmotic value (decrease in hydrature) is correlated with dehydration of the protoplasm. Such a water loss is not without influence on the physiological processes in the protoplasm. The osmotic value can, therefore, be used as an indicator of such changes. For this reason, I consider the determination of the potential osmotic pressure to be of particular value. It serves as a measure of the hydrature of the plant's protoplasm and, therefore, for its degree of hydration or state of imbibition.[1] The determination of the osmotic value by the cryoscopic method is very simple. Criticisms raised against this method were investigated and found not to apply (Walter and Weissman, 1935).

The water content of a plant decreases when the plant loses more water by transpiration than it can absorb through the roots. This may only be a temporary disturbance of its water balance.

Since most of the water—with the exception of young seedlings—is in the vacuoles, a decrease in water content will result in a reduction

[1] Of course, the water taken up by protoplasm also depends on other factors, e.g. the electrostatic status, which is influenced by ion absorption. Such influences on the water relations will be reflected physiologically. However, whether these influences are of any significance in ecological investigations of the water regime in nature, is not yet known. This is probably so in the case of halophytes, which are ignored in this discussion (see next section).

of cell wall tension through a decrease in turgor pressure and in an increased cell sap concentration, i.e. an increase in the osmotic value. Correlated with this is a water reduction of the protoplasm.[1]

An increase in potential osmotic pressure and, therefore, a decrease of the water content of the protoplasm is *the first measurable and physiologically significant disturbance of the water balance.* When the disturbance has been eliminated, the cells regain turgor and the original condition can be considered to be restored. If the plant suffers very great water losses, the increase in osmotic value and the decrease in the hydrature of the protoplasm may induce damage and the cell dies. I call the threshold value at which damage has been observed in many cells the *maximum osmotic value.*

If a non-lethal disturbance of the water balance recurs frequently or if it lasts for a long period, regulatory processes occur in the protoplasm to effect an *active increase in the osmotic value through the production of new osmotically active substances.*

Consequently the protoplasm assumes changed physiological properties. This may be reflected morphologically in newly formed organs and their physiological behaviour (xeromorphic structures, increased resistance to water loss). Set out below is the chain reaction resulting from lack of water, in so far as we can review it in the present state of knowledge without resorting to unproven hypotheses:

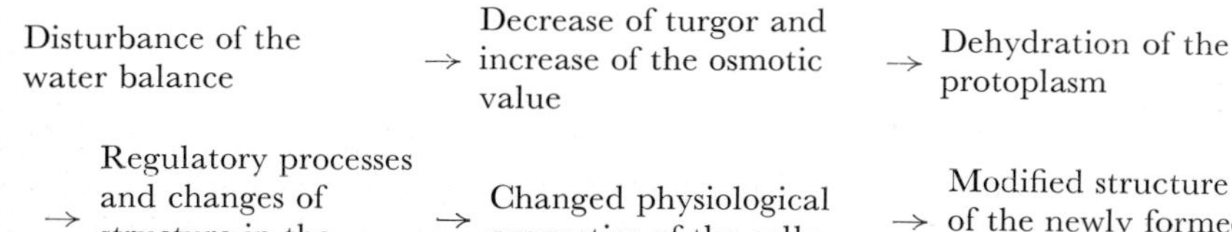

If we compare on the same tree, for example, a shade-leaf with a sun-leaf which has developed under different conditions of hydrature, we can see that the osmotic value of the sun-leaf will always be higher than that of the shade-leaf even in a water-saturated and fully turgid condition. Accordingly, the sun-leaf has a xeromorphic structure and is physiologically better adapted to water deficiency. These differences can neither be detected by measuring the degree of water saturation nor by testing the suction tension (water potential) of the cells. Yet they are indicated by the potential osmotic pressure (= osmotic value).

The same applies to comparative investigations of the same plant, first in the spring under favourable water relations and then later, during the dry summer, when the plant has frequently suffered disturbances to its water balance. The leaves formed in the summer will be quite different morphologically and physiologically from those

[1] Whether a reduction in cell wall tension alone may have some influence on physiological processes in the protoplasm is not yet known. However, evidence has been obtained in many cases that water loss from protoplasm influences all life-processes.

formed in the spring. The potential osmotic pressure of the summer leaves will also be higher when fully turgid, i.e. when completely water-saturated. By contrast, the suction tension in such a case will equal zero. The water content per gram dry-weight alone is less in the case of summer leaves, even when fully saturated. These notions will be applied to examples in the following paragraphs.

These explanations should make it explicable *why I always refer to the osmotic value (= potential osmotic pressure) especially in ecological investigations in arid regions, where water is the limiting factor*. The osmotic value serves as a measure of the hydrature of the plant protoplasm, which it is necessary to know, not to explain water uptake by the plant, but to understand the *overall ecological response of the plant*.

I also use hydrature as a criterion in classifying ecological types. On the basis of the water regime of plants we distinguish first of all two main types (Walter and Kreeb, 1970):

1. The *'non-water regulating' or poikilohydrous* lower plants, which do not differ notably in their water activity from that of their environment. They become dormant as soon as the surrounding humidity decreases only slightly below 100%.
2. The *'water-regulating' or homoiohydrous* higher plants which, even under the most extreme conditions in the desert, maintain their own, very high internal hydrature and which cannot endure the drying out of their organs.

Both types are represented in arid regions. Among the former are the lower algae, particularly Cyanophyceae, and lichens. Much less common are mosses. One finds Hepaticae, such as xerophilous Ricciaceae to some extent, a number of ferns, some species of *Selaginella* and only a very few angiosperms, e.g. *Myrothamnus*, shown in Figs 220-223 (see also Walter, 1960, pp. 209f).[1]

The presence of these plants indicates the existence of a wet season in that particular desert region. Even though the wet season may be extremely brief, it permits these plants to grow and reproduce. Although these completely dried-up plants become fresh and green in a very short time, after each rain, and then resume growth at once (see Fig. 185) (Oppenheimer and Halevy, 1962), their photosynthetic abilities have not yet been investigated.

The most favourable period for the development of poikilohydrous species is the cool season, since this period is characterised by high air humidity even when there is little rain; fog formation also occurs during this period in certain coastal deserts. This favours the lichens particu-

[1] Recently a sedge, *Trilepis pilosa*, was discovered on the granite mountain islands of West Africa, which endures complete desiccation and turns yellow upon drying. If such plants are put in water, their water content rises from 8 to 56% and their leaves become green within 24 hours in the light. Rooted plants green from their base upwards if irrigated; the greening does not occur in the dark. The plant can remain alive in its dried-up condition for at least a year. In this connection see Hambler, D. J. (1961).

larly. This group of plants does not occur in the most extreme deserts, in which a high air humidity does not persist for any length of time.

During the greater part of the year, the poikilohydrous desert plants are in a dormant condition. They dry up completely and are exposed at the soil surface to high temperatures of probably more than 60°C. However, some are found in shaded places in the shelter of rock bluffs, where temperatures do not reach as high and where moisture is retained longer after a rain. Some algae become established beneath translucent stones of quartz. No investigations are available from arid areas which deal with the drought and heat resistance of these plants. However, we know that lichens on rock substrates are extremely resistant in Europe (see Walter, 1960, p. 50). Their presence in desert regions proves that they are adapted to that environment.

The water relations of the homoiohydrous higher plants are much more complicated. The maintenance of their protoplasmic hydrature at little below 100% under the extremely dry conditions of the desert is very remarkable and is a phenomenon that has long attracted the attention of ecologists.

Both the literature and the many hypotheses have been discussed in detail in the section on hydrature and the xerophyte problem in Walter (1960). Therefore, here I will only summarise once more the important points leaving out the details.

First, it should be emphasised that in the present case also the water economy cannot be investigated independently of the dry-matter production of the plants. A reduction in transpiration through closure of the stomata also causes an interruption in the absorption of CO_2 and thus a reduction in photosynthesis.

Plants of arid regions must be able to prevent a dangerous reduction of their water content or hydrature and, at the same time, must maintain an active gas exchange with the atmosphere. However, in order to survive during periods of reduced water uptake, they must be able to prevent loss of water almost entirely, even at the expense of dry-matter production.

These two requirements are fulfilled in very different ways by the different types of homoiohydrous plants:

1. *The ephemerals* complete their development within a very short period. In arid regions they develop only in regularly, or in episodically-occurring, favourable periods when the soil contains sufficient water to maintain a high plant-water content even with a high rate of gas exchange.

The much longer unfavourable periods they survive in the form of:

(*a*) seeds, in the case of *annuals* (therophytes),
(*b*) subsurface storage organs, in the case of *geophytes*.

These ephemerals only differ a very little ecologically, and with regard to their hydrature, from the therophytes and geophytes of sunny habitats in the temperate zone. They are characterised by very rapid growth rates.

2. *The succulents* (non-halophytic) have water storage cells in their leaves, stems or below-ground organs, which are always recharged during the favourable period, when the soil is wet.[1] They seal themselves off almost entirely from their surroundings (both atmosphere and soil) during the long periods of drought and they use up their stored water extremely slowly. Their dry-matter production and rate of growth is slow because of the inhibited gas exchange. Their CO_2 metabolism may show deviations from the usual behaviour (i.e. accumulation of organic acids during the night by intake of CO_2, that is used for photosynthesis during the day).[2] Succulents are the principal dominants in arid areas with two short rainy periods per year. They are absent in Australia except for the rare *Sarcostemma australis*.

3. *The xerophytes* survive the drought periods in a more-or-less active condition. They maintain their gas exchange during the dry periods for as long as possible and merely decrease it temporarily. Thus, they depend upon continuous soil water absorption to replenish their transpiration losses and maintain their water balance. As previously discussed (p. 270ff), this is quite possible since the water supply in arid areas, when calculated per transpiring surface, is generally not as unfavourable as normally assumed. However, when considering their plight in more detail there are still many problems.

Since the amount of rain falling per unit area is relatively small in arid regions, the plant roots must extend a long way laterally. The high root/shoot ratio mentioned earlier only applies to this group of plants. The hydrature and therefore also the water content of the protoplasm of these plants is somewhat less than that of mesophytes. A lower hydrature inhibits shoot growth more than root growth, so resulting in greater root development.

Water deficits occur in the plant, upon the depletion of the available soil water to a dangerous level, which are correlated with a decrease of the hydrature and an increase in the osmotic concentration. At this point, the transpiring surface is often reduced also, particularly in the *malacophyllous xerophytes* (i.e. those having soft hairy leaves).

Initially, in this process the larger, mesomorphic leaves are generally replaced by smaller xeromorphic leaves. If this is not sufficient to reduce transpiration losses, nearly all leaves are shed to maintain a satisfactory water balance in the buds (pp. 333-4).

Orshan and Zand (1962) investigated the reduction of the trans-

[1] Succulents with underground water storage (roots, rhizomes, bulbs or tubers) are much more numerous in arid areas than might be imagined (Walter, 1960, pp. 264-265). For a small region in South Africa can be quoted, many Leguminosae, *Euphorbia* 7 spp., *Fockea edulis*, etc., *Pachypodium succulentum*, *P. bispinosum*, *Pelargonium* spp., *Mesembryanthemum* spp., *Raphionacme leyhreri*, *Ceropegia ampeiata*, *Ipomoea argyrioides*, *Kedrostis* spp., *Chamarea* sp., *Othonna auricuzifolia*, *Senecio* spp., *Doria carnosa*, *D. eriocarpa* (cf. Dyer, 1937).

[2] The so called 'de Saussure effect' (cf. p. 325).

piring surfaces of typical desert plants of the Negev desert during the dry period (*Anabasis articulata, Zygophyllum dumosum, Noea macrantha, Haloxylon articulatum, Artemisia herba-alba, A. monosperma*). They repeated the same investigation on malacophyllous species of the Mediterranean zone of Israel (*Poterium spinosum, Phlomis viscosa, Teucrium polium, Thymus capitatus*). Among desert plants, the reduction ranged from 96-72%, among Mediterranean plants from 49-25%. *Artemisia monosperma*, a desert plant of sand dune habitats, showed an intermediate value of 55%. Thus, the more extreme the drought, the greater the reduction in transpiring surface.

In *sclerophyllous xerophytes* with tough leaves, or in leafless switch-plants, or in species with phyllodes, cladophylls, or cladodes, a reduction in transpiring surface only occurs with very extreme danger. Then, whole parts of the plant die off. Sclerophyllous species can reduce transpiration losses during dangerous periods much more effectively than others and when they discontinue their gas exchange almost entirely, by closing the stomata, they can reduce their water losses to a very small fraction. It has not yet been shown conclusively whether high suction tensions resulting from negative turgor pressure have any significance under such conditions.[1] Such a response appears quite possible, if one considers the mechanically rigid structure of the leaves, which could present a high resistance to cohesion tensions.

The two groups of xerophytes here distinguished are not sharply separated. Transitional forms are also found between succulents and xerophytes. The individual species often show very different physiological behaviour, for example, with regard to transpiration and photosynthesis.

4. A type also occurs that is frequently found in the dry regions of the temperate zone, namely the *stenohydric xerophytes* which close their stomata immediately with the onset of dry conditions, so that photosynthesis is brought to a standstill. With long continuous periods of drought they become starved, their leaves do not dry out but turn yellow and their osmotic value does not rise but, on the contrary, often even falls. Many monocotyledonous geophytes and herbaceous Euphorbias etc., belong to this group. A very typical example is found in the Sonoran Desert, *Fouquieria splendens* (cf. Chap. VIII, 3).

Any arid region has certain niches in which the plants receive a sufficiency of water throughout the whole year. Here, I am not referring only to open water or spring flushes but to habitats that often appear rather dry where, however, ground water or soil water occurs a little below the surface.

These habitats may be occupied by species that are actually alien to the desert. They are usually characterised by mesomorphic leaf-structure and also by high transpiration rates, which are either not, or

[1] Investigations concerning this are now available, see Kreeb (1961, 1963).

only a little, reduced even during the dry period. Lange (1959) has shown that their high rate of transpiration is often a prerequisite for maintaining themselves in the arid climate. No selection for heat tolerance has taken place among these species, since they belong to taxa that have their main centre of distribution in tropical summer-rain regions, where the higher annual temperatures do not coincide with the dry period. For example, the heat tolerance of the Cucurbitaceae of the desert regions or of *Abutilon* spp. is so small (only a little above 40°C) that only the cooling effect of transpiration prevents their leaves from reaching lethal temperatures. A decrease in transpiration caused by inadequate water absorption is followed by a rapid increase in leaf temperature and death from overheating. In typical desert plants, which show a considerable lowering of the transpiration rate during the dry season, leaf temperatures even increase above 50°C, yet this is not lethal. Natural selection has operated in this way, otherwise, these species would not be able to grow in the extreme desert.

With regard to their water relations I distinguish *stenohydrous* species, which have a small range between their optimum and maximum osmotic value, and *euryhydrous* species that have a larger range. However, whether it is *hydrostable* or *hydrolabile* is more important for the drought resistance of a plant i.e. whether its osmotic value increases rapidly at the beginning of the drought season or whether it can maintain its hydrature unchanged for a long period. Several morphological-physiological properties are decisive in determining the latter, for example, the way in which the transpiring organs and the root systems have developed, their anatomy and, furthermore, the physiological regulatory mechanisms controlling their water loss and water uptake and also the efficiency of their water-conducting system. The extent of their root systems, which it is difficult to determine in arid areas where the soils are usually very hard, is especially important.

Ephemeral species are generally stenohydrous, hydrolabile species with only a low resistance to drought. Succulents are stenohydrous but extremely hydrostable species and, therefore, often especially resistant to drought. The greater proportion of xerophytes are probably strongly euryhydrous species, so that malacophyllous species are relatively hydrolabile, in contrast to the sclerophyllous species which are hydrostable. Thus, the former only possess limited drought resistance, the latter, however, are in part very highly resistant.[1]

This has been a rather schematic review. Each desert region has its peculiarities and harbours its own unique plant types. These will be discussed in the following chapters concerning the eco-physiologically

[1] From an ecological viewpoint, it is only of limited value to investigate drought resistance (lethal water deficits) with cut-off shoots, leaves or in potted plants, as has often been done, so that the root system has no chance to spread out laterally. Only the behaviour of whole plants in their natural habitats can give information about their ability to survive drought periods. Water loss and uptake are correlated and dependent on one another. The plant is a whole organism, i.e. a living unit.

better-investigated regions. However, before I proceed to the individual regions, I must draw attention to a further important factor, besides water, characteristic of arid regions, that is the accumulation of readily soluble salts.

6. Halophytes and the salt factor

In addition to water relations, the salt factor plays an important role in the vegetation of arid regions. Salinisation of the soil is in some respects climatically conditioned. Readily soluble salts cannot be washed into the sea in regions lacking outlets to the sea. Thus, they are bound to accumulate locally since water is always evaporated. Salt accumulation, however, is not a purely climatic phenomenon because, under the same climatic conditions, we can find strong salinisation in one situation, while it is entirely absent in others.

It is usually held that soluble salts originate by the weathering of rocks. But this is not correct; the problem is much more complicated.

Among the salts of the arid regions, sodium chloride (NaCl) plays a major role, in addition $MgCl_2$. To some extent easily soluble sulphates also occur, such as Na_2SO_4 and $MgSO_4$. Most sulphates are deposited in the form of gypsum ($CaSO_4$), which, however, plays no role in determining salinity of the soil as it is much less soluble than the others.

Sodium is present in large quantities in nearly all rocks exposed to weathering, but a very different situation exists with regard to chlorine. It is only found as a component of very few and rare minerals. Thus, no significant quantity of NaCl can originate from the weathering of crystalline rocks. In fact, it can be shown that saline soils are absent in arid regions with predominantly crystalline rock-outcrops. By contrast, soil salinity is very pronounced in regions with marine-sedimentary rocks (Jurassic, Cretaceous, Tertiary). Here one can often observe directly the salt blooms from specific strata of the weathered rock.

In these cases the sedimentary rocks are always of marine origin. During deposition in the sea, salt water becomes incorporated into the rock and thus the soluble salts also, i.e. mainly NaCl. Upon the weathering of these rocks, the marine salts are again exposed at the surface and become dissolved in rain water and carried to pans without outlets. Here, salt accumulation occurs in time through repeated evaporation of the water.

Thus, in arid areas easily soluble salts migrate from higher to lower sites during rain. As a consequence, the higher altitudes are rarely saline, while most depressions are. Most salt occurs in places where all the residual water evaporates and in those that are only reached by capillary water (Fig. 172). Before final evaporation of the water and precipitation of the salts, there is merely an increase in the concentration of the solution but salts are still moved along with the water. Salt deposition occurs only at the place where the water movement ceases provided the solution has not become over-saturated at an earlier point, which then also results in salt precipitation. Thus, the salts occur along the path of the water in a sequence determined by their solubility. The first to precipitate is $CaCO_3$, then $CaSO_4$ and finally NaCl together with other easily soluble salts. Where the ground-water table in an arid region is so close to the surface that it is kept moist by capillary movement, salt accumulation is particularly strong through constant evaporation and the soil becomes covered with a crust of salt. Without exception, such damp habitats are always brackish in arid regions, even if the ground water or spring water contains only a relatively low concentration of soluble salts.

Raised habitats are only saline under very dry conditions, where the low rainfall is insufficient to dissolve and transport the salts (e.g. in the Egyptian desert).

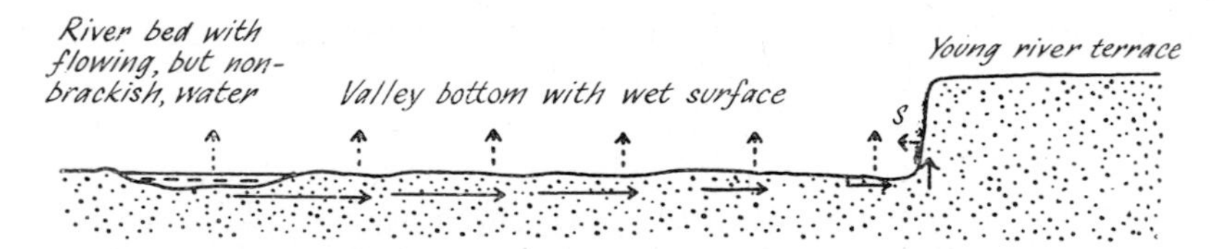

FIG. 172. Salt accumulation in Swakop-valley (Namib desert, S.W. Africa). The arrows indicate the direction and strength of the water flow; the dashed arrows indicate evaporation. The salt concentration increases at the edge of the valley, and the salt forms blooms at S, at the foot of the terrace, where the capillary water stream ceases (after Walter, 1960).

Since any significant salt accumulation can only take place where water occurs as transporting agent, soil salinity is particularly pronounced on the borders of deserts, where rainfall is still appreciable. By contrast, soil salinity is less apparent in rainless deserts, since the water necessary for salt transport is not available and since the quantity of salt in the rock itself is only very small.

On my route through the rainless, southern part of the Libyan desert, from Asyut at the Nile to the Oasis El Kharga, I did not see a single plant or any signs of salt-crusts over a stretch of more than 200 km. Yet salt-crusts developed immediately in the Oasis area itself, where ground water was present.

However, salinisation can also occur by other means, i.e. not only by salt transport in water. Commonly, salt can be of aolic origin. Sea water can form spray from strong surf and the water can evaporate from the small droplets. This type of salt accumulation may often be observed on sub-tropical sea coasts. The residual salt dust is blown inland where it is washed into the soil by the rain. In the Namib desert (South-west Africa) the coastal fog contains salt and the desert soil is salty as far inland as the fog reaches.

This problem was investigated very precisely in west Australia. Here the sodium chloride content of the rain water was determined. According to Teakle (1937), it amounts to 15-50 p.p.m. at the coast, inland to 4-20 p.p.m. In rainy coastal areas the salt is washed back into the sea by the rivers. Thus the soil does not become saline, and one speaks of a 'cyclical salt' movement. However, this does not apply to the arid inland areas without outflow. From 1933 to 1936 all the salt contained in rain water was determined at two stations in Australia; near Merredin, 250 km east of Perth on the west coast of Australia and at Salmon Gums, 100 km north of Esperance on the south coast. The annual quantity of salt added to the soil was, on average:

in Merredin	18 kg/ha
in Salmon Gums	30 kg/ha

Correlated with this, the salt content of the soils is higher at Salmon Gums. Further data on the salt content of rain water in 1936 was obtained at 4 coastal stations and 5 inland stations and the amount of salt deposited was estimated:

TABLE 43

Coastal stations	Average salt content of rain water (p.p.m.)	Amount of salt deposited (kg/ha)	Inland stations	Average salt content of rain water (p.p.m.)	Amount of salt deposited (kg/ha)
Perth	27·2	340	Coolgardie	19·8	40
Geraldton	47·3	194	Cue	15·8	41
Esperance	34·1	290	Mundiwindi	5·8	11
Cordon	13·5	43	Wiluna	6·6	11
			Rawlina	10·4	10

As one could have assumed, salt deposition from rain is less in inland stations that are at least 600 km away from the coast, and the added salt appears to be greater from the west than from the south coast as a result of the prevailing west winds. However, one should not assume that all the salt contained in rain has originated from the sea. Within a short distance from the west coast, small salt pans are formed (because of the aridity) which remain dry throughout the whole summer. From their salt crust surfaces the wind can blow the salt out as dust during the dry part of the year. According to Russian authors, this 'salt dusting' also plays a role in the formation of the solonetz soils in the Russian steppes. Here the Siwash sea, north of the Crimea, which dries considerably during the summer serves as a supply area for the salt.

Finally, saline soils are found in areas where the sea bed dries up as a result of a retreating coast line, e.g. in the Caspian lowlands, or where formerly large water basins have nearly dried up, e.g. around the

Great Salt Lake in Utah or near the Tuz Gölü in Anatolia (Turkey). Salt accumulation can occur also on irrigated, cultivated fields that are poorly drained. The widespread soil salinity in Mesopotamia (Iraq) has probably resulted from irrigation. Today, a salt-desert occurs around Ur, where a flourishing agriculture existed several thousand years ago. Evidence for this is given by the ancient irrigation system (Kreeb, 1964).

Thus, it can be seen that salt accumulation in the soils of arid areas may occur in different ways and its causes must be investigated in each individual case.

Plants on salt-soils are called halophytes. Most of the salt-soils are wet because they occur in hollows where the run-off water accumulates.

For the plants of these wet habitats, the *hygrohalophytes*, the water balance is less of a problem than the salt balance. The osmotic concentration of the salt-containing soil solution is compensated by a corresponding absorption and accumulation of salts in the cell sap. This, however, requires sufficient salt resistance by the protoplasm, which is characteristic of typical halophytes in contrast to the non-halophytic plants (see Walter, 1960, pp. 477-91). Still more complicated are the relationships of the *xerohalophytes*. They occur on less salty soils but these dry up temporarily so that here high salt concentration is correlated with lack of water.

From the point of view of soil science, soil suction tensions produced at low water contents by the physical properties of the soils, such as capillary forces, imbibition forces, etc. (= matric potential), are distinguished from those due to the soluble salt content of the soil water (= osmotic potential). The sum of both parameters (= total water stress or water potential) is considered to be important for the water uptake of the plant. However, this assumption appears to be wrong. Since xerohalophytes can absorb readily soluble salts from the soil solution and accumulate them in their cell sap (see p. 368), the osmotically conditioned suction tensions must be balanced and only the physical forces can be active. It has also been shown in the germination of different seeds, for example, that only the physically-conditioned suction tension has any inhibiting effect. The osmotic suction tension was without effect, even when 100 times as strong. The necessary prerequisite is, however, that the absorbed salts should not cause damage to the protoplasm (Collis-George and Sands, 1962).

References

Collis-George N., and Sands, J. E. 1962. Comparison of the effects of the physical and chemical components of soil water energy on seed germination. *Austr. J. Agric. Res.*, **13**, 575-84.

Dyer, R. A. 1937. The vegetation of the Albany and Bathurst divisions. *Bot. Surv. S. Afr.*, **17**.

EVENARI, M., SCHANA, L., TADMOR, M., and AHARONI, Y. 1961. Ancient agriculture in the Negev. *Science*, **133**, 979-96.

EVENARI, M., SHANA, L., and TADMOR, N. H. 1963. Run-off farming in the Negev Desert of Israel. *Natn. and Univ. Inst. Agric.*, Rehovot Spec. Publ. 393.

GABRIEL, A. 1961. Die Wüsten der Erde und ihre Erforschung. *Verständl. Wiss.*, *Naturw. Abt.*, **76**.

HAMBLER, D. J. 1961. A poikilohydrous and poikilochlorophyllous angiosperm from Africa. *Nature, Lond.*, **191**, 1415-16.

HEMMING, C. F. 1965. Vegetation areas in Somaliland. *J. Ecol.*, **53**, 57-67.

HILLEL, D., and TADMOR, M. 1962. Water regime and vegetation in the Central Negev Highlands of Israel. *Ecology*, **43**, 33-41.

KREEB, K. 1961. Zur Frage des negativen Turgors bei mediterranen Hartlaubpflanzen unter natürlichen Bedingungen. *Planta*, **56**, 479-89.

KREEB, K. 1963. Untersuchungen zum Wasserhaushalt der Pflanzen unter extrem ariden Bedingungen. *Planta*, **59**, 442-58.

KREEB, K. 1964. *Ökologische Grundlagen der Bewässerungskulturen in den Subtropen.* Stuttgart.

LANGE, O. L. 1959. Untersuchungen über Wärmehaushalt und Hitzeresistenz mauritanischer Wüsten- und Savannenpflanzen. *Flora*, **147**, 595-651.

LAVRENKO, E. M. 1962. The characteristic features of the plant geography of the Euro-Asiatic and north African deserts (in Russian). *Akad. Wiss. Leningr.*, **15**.

LE HOUEROU, H. N. 1959. Écologie, phytosociologie et productivité de l'olivier en Tunisie méridionale. *Bull. Serv. Carte Phytogeogr. Sér. B*, **4**, 7-22.

LEISTNER, O. A. 1964. *The plant ecology of the southern Kalahari.* Diss. Stellenbosch.

MONOD, TH. 1954. Modes 'contractés' et 'diffus' de la végétation saharienne. In *Biology of deserts*, 35-44. London.

OPPENHEIMER, H. R., and HALEVY, A. H. 1962. Anabiosis von *Ceterach officinarum* Lam. et D. *Bull. Res. Coun. Israel, D*, **11**, 127-47.

ORSHAN, G., and ZAND, G. 1962. Seasonal body reduction of certain desert halfshrubs. *Bull. Res. Coun. Israel. D*, **11**, 35-42.

SCHIFFERS, H. 1950. *Die Sahara.* Stuttgart.

SMITH, J. 1949. *Distribution of tree species in the Sudan in relation to rainfall and soil texture.* Khartoum.

STOCKER, O. 1935. Transpiration und Wasserhaushalt in verschiedenen Klimazonen. *Jb. wiss. Bot.*, **81**, 464-96.

SWESCHNIKOWA, V., and SALENSKI, O. 1956. *Water economy of arid areas of central Asia and Kazakhstan* (in Russian).

TEAKLE, L. J. H. 1937. The salt (sodium chloride) content of rain water. *J. Dept. Agric. W. Austr.*, **14**, 115-33.

WALTER, H. 1923. Protoplasma- und Membranquellung bei Plasmolyse. *Jb. wiss. Bot.*, **62**, 145-213.

WALTER, H. 1932. Die Wasserverhältnisse an verschiedenen Standorten in humiden und ariden Gebieten. *Beih. Bot. Zbl.*, **49**, 495-519.

WALTER, H. 1939. Grasland, Savanna und Busch der ariden Teile Afrikas in ihrer ökologischen Bedingtheit. *Jb. Wiss. Bot.*, **87**, 750-860.

WALTER, H. 1954. *Einführung in die Phytologie.* Vol. III, *Grundlagen der Pflanzenverbreitung.* Part 2, *Arealkunde.* Stuttgart.

WALTER, H. 1960. *Einführung in die Phytologie.* Vol. III, *Grundlagen der Pflanzenverbreitung.* Part 1, *Standortslehre.* 2nd ed. Stuttgart.

WALTER, H. 1963. Zur Klärung des spezifischen Wasserzustandes im Plasma und in der Zellwand bei den höheren Pflanzen und seine Bestimmung. *Ber. Dtsch. Bot. Ges.*, **76**, 40-70.

WALTER, H. and E., 1953. Einige allgemeine Ergebnisse unserer Forschungsreise nach Südwestafrika 1952/53: Das Gesetz der relativen Standortskonstanz; das Wesen der Pflanzengemeinschaften. *Ber. Dtsch. Bot. Ges.*, **66**, 228-36.

WALTER, H., and KREEB, K, 1970. Die Hydration und Hydratur des Protoplasmas. *Protoplasmatologia,* **2**, C/6, Vienna.

WALTER, H., and STADELMANN, E. 1968. The physiological prerequisites for the transition of autotrophic plants from water to terrestrial life. *Bio-Science,* **18**, 694-701.

WALTER, H., and WEISSMAN, O. 1935. Über die Gefrierpunkte und osmotischen Werte lebender und toter pflanzlicher Gewebe. *Jb. Wiss. Bot.*, **82**, 273-310.

WENT, F. W. 1948. Ecology of desert plants, I. Observations on germination in the Joshua Tree National Monument, California. *Ecology,* **29**, 242-253.

WENT, F. W. 1949. Ecology of desert plants, II. The effect of rain and temperature on germination and growth. *Ecology,* **30**, 1-13.

WENT, F. W., and WESTERGAARD, W. 1949. Ecology of desert plants, III. Development of plants in the Death Valley National Monument, California. *Ecology,* **30**, 26-38.

VIII. *The Sonoran Desert*

1. Climate and habitat conditions

ECO-PHYSIOLOGICAL desert research had its beginning with the establishment in 1903 of the 'Desert Laboratory of the Carnegie Institution of Washington' in Tucson, Arizona. It is located in the northern part of the region which extends from the North Mexican state, Sonora, descending to the Colorado plateau in the U.S.A. The ecological relations of its vegetation were studied for nearly half a century in the desert laboratory, first under the direction of D. T. MacDougal and then under that of F. Shreve. I had the opportunity of working there as Rockefeller Fellow from 1929 to 1930, and the experience gained has shaped all my subsequent work.[1]

The compendious work, *Vegetation of the Sonoran Desert* by F. Shreve, was only published in 1950, after his death. With its publication, the research activities in Tucson were terminated and the desert laboratory dissolved. For this reason it seems appropriate to begin my discussion of the vegetation of arid regions with a summary of the more significant basic investigations that were done at Tucson.

The Sonoran desert extends from 35° to 23° N. The area around Tucson lies in the north-eastern part (32° N) at 728 m above sea level (Fig. 173). Its climatic relations are shown in the climatic diagrams (Figs 174-177). The mean annual temperature in Tucson is 19·5°C. The average daily minimum temperature for all months is above 0°C. However, night frosts occur in the winter months, particularly as the result of cold air drainage.

Rainfall varies in this desert region from 55 mm in the west at the lower Colorado to more than 400 mm in the east at its mountainous border. The average rainfall around Tucson is 293 mm. Thus, the area is not a real desert, but rather a semi-desert. Yet, it represents the driest zone in the U.S.A. and, therefore, is referred to as 'desert'.

The annual distribution of rainfall is of particular interest. While rain falls almost exclusively in the winter in the west, it occurs in Tucson in the winter, from December to March and again in the summer, from July to September. Further to the south, in Mexico, summer rains are more predominant. Thus, one can distinguish in Tucson four seasons: (*i*) the winter-rain season, which coincides with cool temperatures, (*ii*) the pre-summer dry season, which shows the highest temperature in June (av. maximum 37·1°C), (*iii*) the summer-rain season, mostly with isolated, strong thunderstorm showers, and (*iv*) the post-summer dry

[1] All data (osmotic values) relating to water conditions shown in this section are extracted from Walter (1931) which is not referred to further in the following discussion.

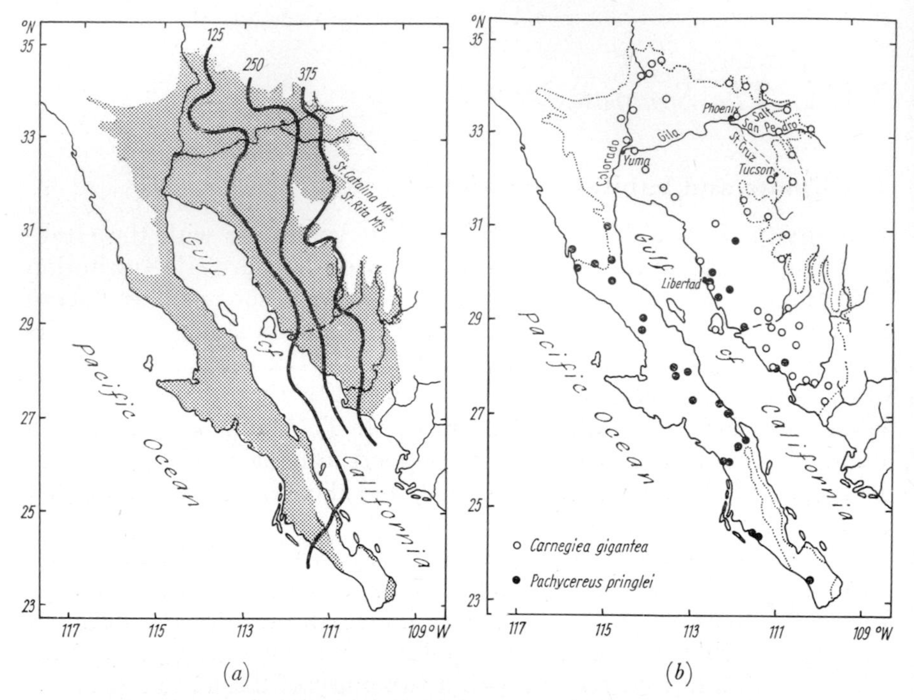

FIG. 173. (*a*) The Sonoran desert (dotted) with annual rainfall isohyets (in mm).

(*b*) The distribution of *Carnegiea* and *Pachycereus* in the Sonoran desert (after Shreve, modified); these are vicarious species, whose distributions only overlap in the Gulf of California.

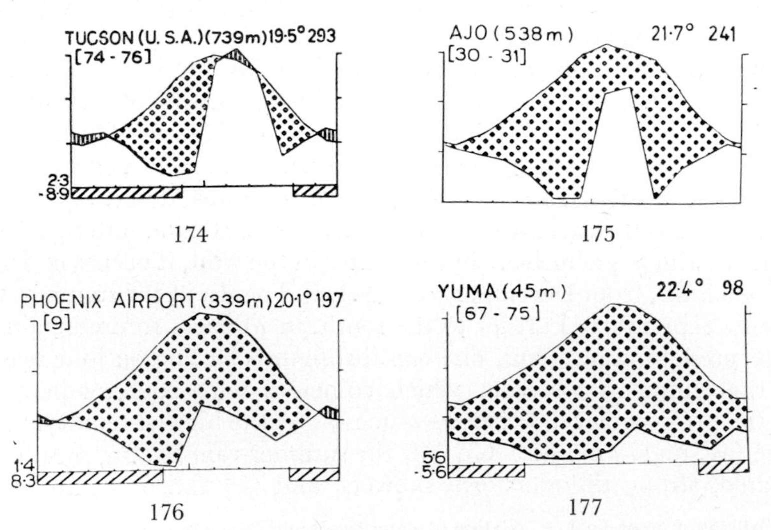

FIGS. 174-177. Climatic-diagrams of stations in the northern Sonoran desert.

season. The following table gives the distribution of rainfall in these four seasons:

TABLE 44

	Rainfall mm	per cent of annual total
Winter-rain season (December-March)	92	31
Pre-summer dry season (April-June)	22	7
Summer-rain reason (July-September)	144	49
Post-summer dry season (October-November)	37	13

However, these are mean values. In particular years the distribution can vary greatly. Shreve summarised the distribution of dry periods by individual years. He defined as a drought a sequence of at least 30 days without rain above 3·5 mm; because any rain below this amount has no effect on the vegetation. The result is shown in Fig. 178 (after Schratz, 1931). Fig. 179 indicates that the distribution and rainfall was almost normal during my stay in Arizona in 1929, while the year 1930 had no real pre-summer dry season and had also too short a post-summer dry season. Values for evaporation during the cooler season of the year are shown in Fig. 180. They are usually 1·0-1·2 ml/h, comparable to those in mid-summer in Central Europe.

Rock weathering in Arizona is predominantly physical as in all desert areas. Rocks, which in sunny locations may become heated to 70°C, can be rapidly cooled to 20°C by a sudden shower. The unequal contraction of individual minerals causes disintegration into finer and finer fragments. Sandy material which has originated in this way is swept up by the wind and so causes further erosion by a sand-blasting effect. However, sand dune formation only occurs at the north end of the Gulf of California, west of Yuma, which is well beyond the Sonoran desert (Rempel, 1936).

The cactus desert around Tucson consists of 25% hills and lower mountain ranges, formed from volcanic rock or granite. However these are covered in masses of their own detritus. The weathering products are transported downhill by flood-sheets which develop during heavy rain showers, so that the particles become sorted out according to their size. By this process, coarse stones remain at the top of the slope while the finer ones are transported farther towards the valley. Thus, one recognises down each slope a continuous sequence of texturally sorted desert soils, beginning with the rocky outcrop areas on the upper slopes, then stone rubble and gravel fields on the main slopes to fine silty soils of the valley floors. Particle sorting is also correlated with the steepness of the slope. Ridge tops and steep slopes are areas of rocky outcrops. At their bottoms the slopes decrease rapidly, levelling out,

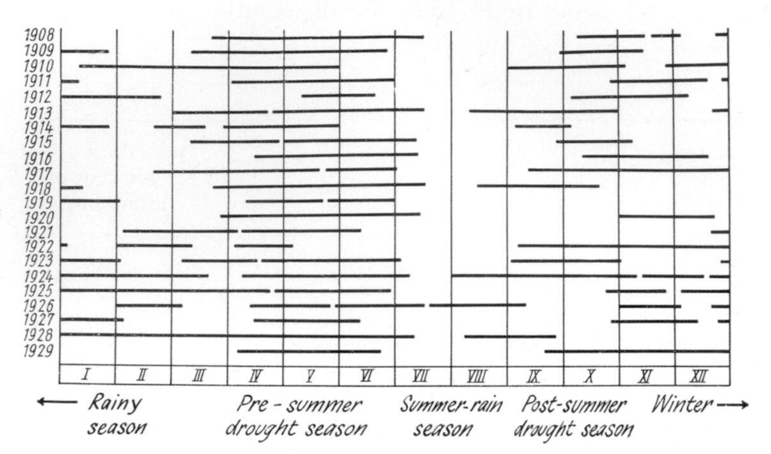

FIG. 178. Distribution and length of droughts in Tucson for the years 1908-29 (after F. Shreve).

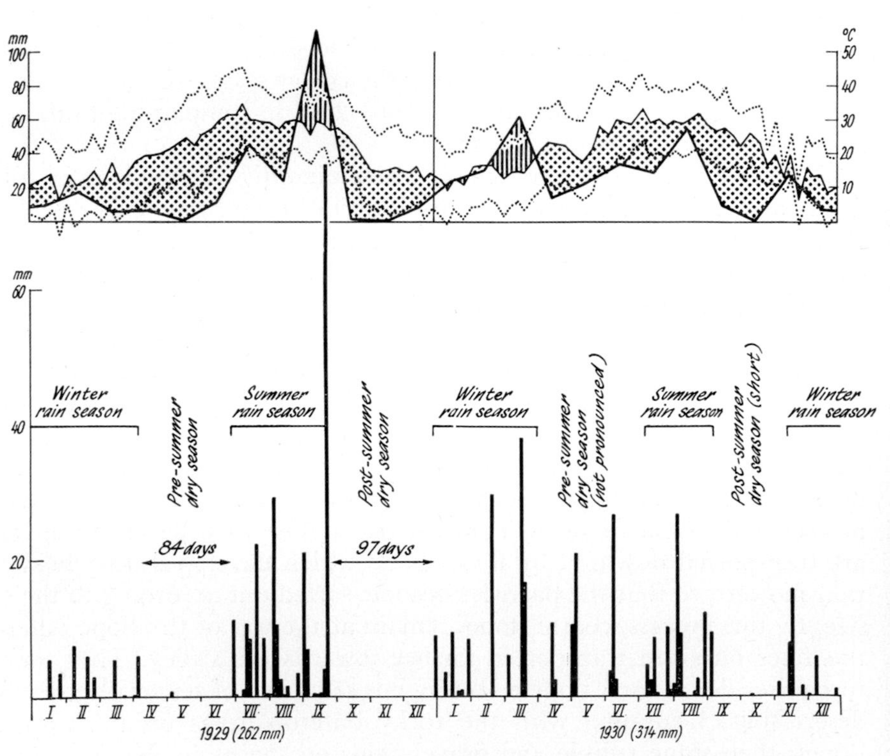

FIG. 179. Weather at the Desert Laboratory (Tucson) 1929-30. Below: rainfall (every 3 days). Above: extreme temperatures (weekly) dotted; mean temperatures plotted together with monthly rainfall amounts (as in the climatogram).

and grade into nearly level gravel flats (Bajada, wrongly called Mesa), which are smooth, like a table top, and are made up of a gravelly sandy soil-like material. They dip about 4-12 m per kilometre, i.e. in average 8/1000. Their distinctions from alluvial flood plains can only be clearly recognised where there is a strong erosion margin. In turn, the river bed itself cuts into the flood plain exposing steep-sided slopes and the river bottom is filled with sedimentary sand. The St Cruz river, a tributary of the Gila river, which flows near Yuma into the Colorado (see Fig. 173), is supplied by ground water throughout the year.

The almost level gravel flats (Bajadas) make up nearly half the total area. They are cut by numerous, more or less deep erosion channels

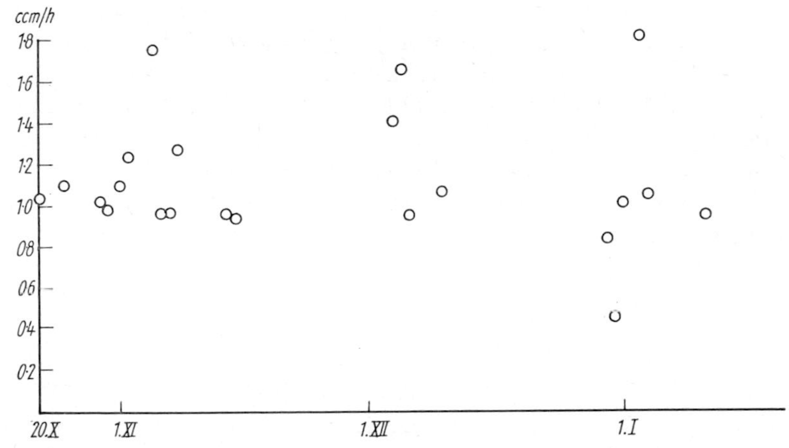

FIG. 180. Highest hourly evaporation rate (standard Piche with green paper) on the field-work days at the Desert Laboratory in Tucson (Arizona) in the winter of 1929/30.

(washes, arroyos). After torrential rain, commonly more than 50 mm in 24 hours, these erosion channels become like foaming mountain torrents for a short period, usually for a few hours only. Huge masses of water fill the river bed and cover the flood plains for several days. Here, the finest silt particles are deposited. The rain also washes all plant debris and litter into the valleys, so the soils are very low in humus content and no soil profile develops. Generally, the soils on the Bajadas are not as coarsely textured as one would assume from looking at the surface material. More or less large stones accumulate on the surface in the form of desert pavement, under which one finds much fine soil.

The soils are well aerated because of their low water content. However, after extremely heavy rains, the valleys and especially the pans, can remain submerged for several days. Woody plants cannot withstand such poor soil aeration and it was noted that 30-50-year-old trees died after such extreme flooding.

Salt accumulates in hollows without outflows, where stagnant water slowly evaporates. However, saline soils are only of local significance in the Sonoran desert since most of the parent rock is crystalline. Thus little NaCl can be dissolved by weathering. Sodium carbonate formation occurs only in valleys, with local accumulations of humus (black alkali soils) but this too is not of great significance area-wise.

However, lime crusts which occur a little below the surface (10-15 cm) are widely distributed. These are known as hardpans or caliche (Shreve and Mallery, 1933). The crust is from a few millimetres to more than 10 cm thick and cements the stones into a solid layer. Wherever the upper soil layers are removed by erosion, hardpan appears at the surface, where it gives the impression of a layer of breccia. The caliche is usually impermeable to water and prevents root penetration to lower layers. A prerequisite for its formation is some lime in the parent rock. It does not occur in soils derived from granite. Such lime concretions are found in other semi-arid regions, e.g. in chernozemic soils (see Walter, 1960, p. 338). They are formed at the lower limit of the mean penetration of rain water in the soils. The drier the climate, the closer they are to the surface and the harder they seem to be cemented. The hard lime crusts also probably form at shallow depths beneath the surface in places where upward moving capillary water or seepage evaporates.[1]

2. Classification of the vegetation

The vegetation shows a pattern corresponding to the habitat features described (Shreve, 1951; Spalding, 1909). A series of different communities can be distinguished.

(*a*) The *Carnegiea-Encelia* community (Fig. 181) of raised areas and steeper slopes.

The water relations in these habitats are favourable. Rainfall penetrates easily into the soil. Fine soil occurs between the stones and blocks of rock which stores the water, protecting it against evaporation and making it available to the plants. Deep penetration of water is slow and continues for several weeks after rain. Shreve has shown that less than 2 mm of rain does not increase the water content of the soil of these habitats. Three millimetres of rain is of importance only when the soil is still damp from a previous wetting (Shreve, F., 1914). The water content decreases during the periods of drought but it does not fall below 9% at 30 cm in May, and in November not below 10%. The amount of non-available water was not determined. However, that sufficient water is always available to the plants is shown by the osmotic

[1] In Spain lime crusts are developed under an annual rainfall of 100-500 mm (mostly 150-250 mm). Important features are the high temperatures and the long period of drought for, with an even distribution of rainfall, only lime concretions (loess-fragments, etc.) are formed. The crusts only form either at the surface or underground by water oozing out from bedding-planes (Rutte, 1958).

value of *Larrea* hardly increasing during the dry period in these habitats (see Fig. 196, p. 332).

The rocky and stony soil is favourable to shrubs. The main species of this community are, in addition to the two mentioned:

Cercidium microphyllum
Fouquieria splendens (Fig. 194, p. 328)
Acacia constricta
Celtis pallida
Opuntia versicolor etc.
Ferocactus wislizenii
Jatropha cardiophylla
Prosopis juliflora var. *velutina*
Ephedra trifurca
Heteropogon contortus
Hibiscus coulteri
Lippia wrightii
Selaginella arizonica
and five different fern species.

Larrea occurs rarely and only singly in these habitats, similarly *Prosopis*, which is more characteristic of the flood plains. The total

Fig. 181. Typical vegetation with *Carnegiea gigantea* and *Opuntia fulgida*. In the foreground, light-coloured dwarf shrubs of *Encelia farinosa*. Santa Catalina Mts., near Tucson, Arizona (Phot. Desert Lab.).

number of species in these habitats in Arizona is 300. *Carnegiea* only grows where the frost does not last longer than 24 hours (Shreve, 1911a; Turnage and Hinckley, 1938). This is always the case on south-facing slopes where the temperature during the day regularly rises above 0°C. during cloudless weather with night frosts. In these optimal conditions, *Carnegiea* always shows the greatest density and the lowest osmotic values during the winter months (Fig. 182).

(*b*) The *Larrea-Franseria* community on nearly level gravel flats.

The soil at the foot of the talus slopes is at first still stony and on such sites *Larrea* occurs in addition to the previously named community. Towards the valley bottom, the soil becomes gravelly to sandy and the habitat increasingly drier. At quite a shallow depth one often finds a solid lime hardpan (caliche). Ultimately, the only remaining shrub is *Larrea tridentata*, which is often poorly developed and which sheds nearly all its leaves during the dry period. A few *Franseria* species (soft-leaved Compositae) are associated and also a large number of different *Opuntia* species with flat or cylindrical shoots.

(*c*) The *Prosopis velutina* community on the alluvial flood plains with silt-loam soils.

Periodic flooding of these areas wets the soils as far down as the ground water. *Prosopis* has a root system which extends 20 m laterally and, at the same time, a considerable way downwards. On this substrate, an open forest can develop, whose umbrella-shaped crowns may

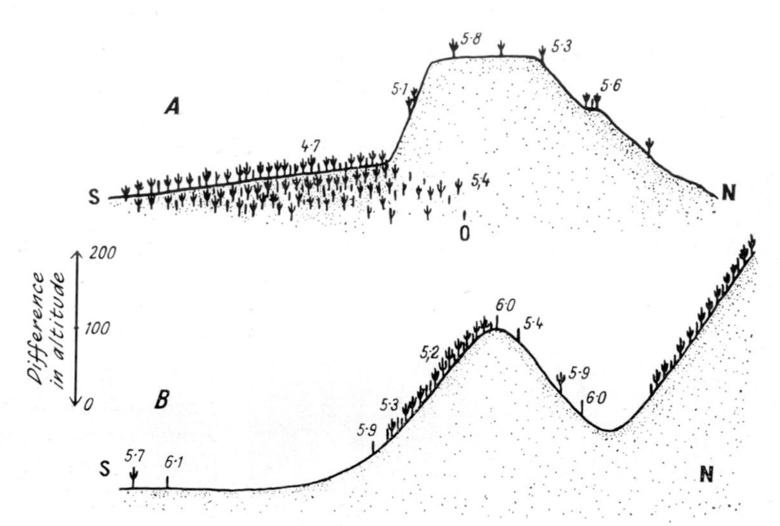

FIG. 182. Osmotic values of *Carnegiea gigantea* (comparable samples) related to its distribution (after H. Walter, 1931).

almost touch each other. Tree heights reach to 15-18 m and stem diameters to 1·2 to 1·5 m.

These forests were originally almost the only source of fuel and, therefore, are today nearly extinct. In some places one can still find dense *Prosopis* stands from regenerating stumps with *Acacia greggii*, *Celtis pallida*, *Condelia lycioides* and *Lycium* spp. and with lianas such as *Clematis drummondii*, *Vitis arizonica*, mostly species that can reach the ground water with their tap roots. However, in general, these areas are today covered with cotton, alfalfa and barley. The fields are easily irrigated if the water is hauled from wells with turbine pumps. The soil is very deep and fertile. Salt accumulation can occur in depressions where the water stagnates and slowly evaporates. At such places, one finds halophytes, such as *Atriplex* and even *Suaeda moquini*. The temperature relations are relatively unfavourable in the winter. Cold air drainage occurs on clear nights. Shreve compared the minimum night temperatures at the Desert Laboratory, which was situated on the Tumamoc hill, with those at the foot in the flood plain. The latter could be as much as 10°C cooler, but on rainy days the difference was only 0·2°C (Shreve, 1912). Thus, the number of night frosts is very high in the winter. All non-frost-resistant summer-ephemerals, which could otherwise persist longer in the moist soil, have already died from frost by the end of October and *Prosopis* sheds its leaves regularly in the winter, while those on the slopes remain green.

(*d*) The *Populus-Salix* bottomland of the river bed.

The dry rivers are usually deeply incised (1-2 m, occasionally sometimes up to 12-15 m) into the flood plains. This deep erosion probably only took place during the last 75 years when the plant cover was opened up more and more by increased grazing pressure. This greatly aggravated surface run-off and soil erosion.

Grazing, in general, has modified the whole vegetation markedly. In particular, certain species have become nearly extinct, while cacti have spread considerably since they are not eaten by grazing animals. In the case of some cylindrical *Opuntia* species with several long thorns, individual fragments come off at the slightest touch, as for example in *Opuntia fulgida* (Cholla or jumping cactus). These catch on to fur, are so distributed and then root easily after falling to the ground.

The river bed is usually barren of vegetation since the soil is shifted every time it is flooded; depending on the speed of the current, it is either stony, gravelly or sandy. In protected places at the river side trees grow, whose roots reach ground water a little below the surface, on soft bottomland. The vegetation here corresponds to the summer-green broad-leaved forests of the temperate zone, such as *Populus* and *Salix* species, *Fraxinus velutina* var. *toumeyi*, or *Sambucus mexicana*, which is very similar to *S. nigra*.

The roots of these trees are often washed out by turbulent floods. Occasionally, whole trees are torn loose and deposited farther down in the river bed. The sandy soil, frequently re-distributed by the water, adjacent to the river bed provides a habitat for fast growing shrubs, such as *Baccharis* and *Hymenoclea* species, which can withstand a partial burying by sand.

(*e*) The vegetation of erosion channels (washes, arroyos).

These channels begin as little furrows on the upper part of the slope which join to form ditches and then become deeper and broader. Commonly, they form deep canyons or gradually widen to form shallow dry-river beds. In their upper part where, after rain, the water gushes down with great force, the soil is covered with coarse rubble. As the stream velocity decreases in the lower part, gravel and sand are deposited.

At the borders of the erosion channels, water relations are much more favourable than away from them since the soil is better and more deeply wetted. On parts of the lower slopes, ground-water lakes can often develop.

Thus, species of the slope vegetation grow with greater vigour in such places. They form larger leaves which they often retain during the dry period. A few species, such as *Celtis pallida* and *Lycium*, are more abundant here and certain new ones become associated. *Cercidium torreyanum* occurs on the lower slope, and species that are otherwise found on alluvial plains only occur here also, such as *Prosopis* and *Acacia*.

The deep canyons are also affected by cold air drainage, so that sensitive species are absent. According to Shreve, the temperature by 8 p.m. was already nearly 5°C lower than on the slope, 31 m above. However, the depth of the cold air stream is not believed to be thicker than 18 m. The unfavourable temperature relations and, at the same time, more favourable water relations make it understandable that we can find representatives of higher mountain floras in the canyons, e.g. evergreen oak of the encinal belt. *Platanus wrightii* is also commonly found here.

(*f*) Several altitudinal belts of the higher mountains.

Not far from Tucson the 3000 m tall mountain ranges of the St Catalina and St Rita mountains rise up. These extend far above the desert zone into a damper climate and they are covered with snow for a long time in the winter.

Upon ascending the mountains the character of the vegetation has already changed very markedly by 1000 m, although certain representatives of the desert flora can be seen in favourable sites up to 1500 m, or even 2000 m. The large cacti disappear completely and the *Prosopis* savannah begins, showing scattered *Prosopis* shrubs intermixed with tussock grasses. Other associates are the leaf-succulents species of *Agave* and *Yucca* and the strange *Dasylirion* and *Nolina* species. Above 1300 m these first occur singly, then above 1500 m with increasing density, evergreen oaks of the encinal belt. This is a closed, *c.* 15-20 m tall, sclerophyllous forest with the species:

Quercus oblongifolia, *Q. emoryi*
Q. hypoleuca, rarely *Q. reticulata*
and *Arbutus arizonica*, *Arctostaphylos pungens*, etc.

Here, in these nearly untouched forests, one can get an impression approximating to the sclerophyllous Mediterranean forest as it must have looked in virgin condition.

At about 2000 m the evergreen oak forest is replaced by a pine forest consisting of *Pinus ponderosa* ssp. *scopulorum*, which is associated further up with *Pinus strobiformis*.

At 2400 m fir forests occur on the northern slope, with *Abies concolor* and *Pseudotsuga menziesii*. On southern slopes, pine can extend up to the highest tops.

A summer-green hardwood belt is absent under these climatic conditions. I noticed winter-deciduous oak scrub at only one location on a steep north slope of Mt. Baldy.

The Colorado plateau in northern Arizona, which is about 2100 m high, belongs to the *Pinus ponderosa* belt. Along its lower margin, one often finds savannah-like stands of *Juniperus utahensis* and *J. monosperma*. Here too occurs the strange *Pinus edulis* var. *monophylla*, which has rounded needles.

The Colorado plateau is divided by the most magnificent canyon in the world, the Grand Canyon of the Colorado. It is 350 km long, 800 to 1800 m deep. At its top it is 10 km broad, whence it drops in spectacular terraces—chiselled from the entire sequence of sedimentary rocks—to the lowest, narrow gulch cut into the crystalline, basal foot.

FIG. 183. Grand Canyon, Colorado, viewed from near Cape Royal. Note the scattered vegetation on scree slopes in the foreground (Phot. D. C. Smith).

The opaque waters of the Colorado River flow through this narrow gulch, forming many whirlpools.

A descending sequence of vegetation belts can also be recognised in the Colorado Canyon. However, this does not correspond to the zonation in South Arizona.

On the continuously moving scree slopes the plant cover is open. In its upper part there still occur a few, isolated *Pinus* and *Juniperus* trees (Fig. 183). *Artemisia tridentata* grows at 1200 m, where it forms a transition to the dwarf-shrub semi-desert with *Coleogyne ramosissima* (Rosacaceae), *Atriplex canescens*, *Ephedra viridis*, *Opuntia polyacantha*, etc.

In Arizona, the upper tree limit is only reached at San Francisco Peak, which rises from the Colorado plateau to nearly 4000 m. Above the *Abies-Pseudotsuga* belt one still finds a *Picea engelmannii* belt. Thus, the timber line is here formed by spruce. On the southern slopes near the upper tree limit yet another, low-growing *Abies* species occurs.

This discussion has shown how varied the vegetation is in Arizona. The most marked contrasts are found immediately adjacent.

However, with this brief review of the forest belts, we have left the arid region. I will not, therefore, go into further detail here for the excellent paper by F. Shreve on the Catalina Mts. goes into considerable detail (Shreve, F., 1915, 1922).

3. Ecological plant types of the Sonoran desert

As mentioned earlier, the rainfall varies greatly from year to year in arid regions. This is true also for Arizona (Humphrey, 1933). The excess water is utilised in wet years by ephemerals which do not develop in dry years (see p. 289).

Therefore, the occurrence of ephemerals (annuals and geophytes) in addition to the perennials would also be expected in the Sonoran desert. In fact, they are very numerous.

(*a*) *Ephemerals*

Arizona has two rainy seasons as mentioned already:

1. The winter-rain season with mean daily temperatures from 10-15°C. At night, the minimum temperatures may well drop below 0°C. Precipitation occurs usually in the form of light but long-lasting rain;

2. The summer-rain season with mean daily temperatures around 30°C and maximum temperatures of more than 40°C. Rain falls mostly in the form of heavy thunderstorm showers.

This division into two rainy seasons affects the occurrence of summer and winter ephemerals. Hardly a species can be found common to both rainy seasons.

The reason for this sharp distinction is the different temperature optima for germination. According to Shreve's (Shreve, F. 1915; see also Went, 1949) investigations, these are between 15-18°C for winter annuals and between 27-32°C for summer annuals. Nearly all species germinate rapidly at temperatures that do not vary more than 3-5°C from these optima. However, rapid germination is decisive, since the desert soil can quickly dry off at the surface after rain. Seedlings must therefore penetrate the soil quickly with their radicles.

Ephemerals respond very differently to the particular, current conditions for growth. If rain does not follow germination, the seedlings either die or develop into dwarf plants 1 cm high, which may, however, develop flower and fruit. Under favourable conditions, the same species may grow 20-30 cm tall and then flower abundantly. Some of the summer ephemerals may even reach 2-3 m in height.

Summer annuals do not germinate in winter, likewise, the seed dormancy of winter annuals is not interrupted by the summer-rains.

In the western part of the region, where rainfall occurs rarely in the summer, summer ephemerals hardly occur. However, they extend far into Texas, where rainfall occurs only in the summer. The life-span of the annuals rarely exceeds 6-8 weeks yet, in favourable habitats such as at the margin of irrigation ditches, they can even survive the period of drought and become perennial herbs. This applies especially to certain introduced, normally annual, weeds. The winter ephemerals that are adapted to low temperatures belong to the holarctic flora. Among these are representatives of central European genera and families and weeds including, *Erodium*, *Sonchus asper*, *Hordeum murinum*, *Malva silvestris*. In contrast to these, the summer ephemerals are neotropical elements and among the weeds some species occur that are distributed throughout the entire subtropics, as for example *Tribulus terrestris*.

When I began my investigations in October 1929, I had the good fortune that, just before, a 200 mm rain had fallen from June 28 to September 24 and the final shower alone had deposited 60 mm. Therefore, the summer ephemerals were particularly well developed. Nearly tropical conditions prevailed during the summer-rain reason, i.e. high humidities and temperatures. However, the wet soil did not develop any extreme temperatures at its surface. Since the rooting depth of summer ephemerals does not exceed 15 cm as a rule, their vegetative period is limited by the drying-off properties of the upper soil horizons. Hill-slopes with coarse-grained soil dry off to this depth in about 3 weeks after rain, the fine-grained soils of the valley flood plains in about 6 weeks. Since no rain fell between September 21 and December 31, the disappearance of the ephemeral vegetation could be observed in detail.

Species with the most delicate leaves (*Boerhavia*, *Mollugo*, *Datura*) had mostly already died by October 15 and the grasses (*Bouteloa*, Paniceae) had dried up. Other species began to yellow (*Amaranthus palmeri*, *Euphorbia heterophylla*) and remained green only in the shade of shrubs. However, species with succulent leaves (*Trianthema portulacastrum*, *Allionia incarnata*), or those with somewhat pubescent leaves (*Kallstroemia*, *Tribulus*, *Cladothrix lanuginosa*, *Solanum elaeagnifolium*), lasted through to the middle of November. Cannon (1911) reported a rooting depth of 22 cm for *Kallstroemia*. The very sappy *Martynia* (osm. value, 9·4 atm) was killed by a night frost on October 31, as were several other species in the flood-plain area of the valley.

The osmotic values of these species are characteristic of mesomorphic, herbaceous species (Table 45).

Daily variations in osmotic values were insignificant. However, the values increased continuously during the dry season until the maximum (lethal) value was reached at death.

The relationship between the development of the plant and its water economy as conditioned by habitat is clearly illustrated for

Solanum elaeagnifolium in Fig. 184. This is a prickly, xeromorphic species with white-pubescent leaves, which grows in the flood-plain area in a

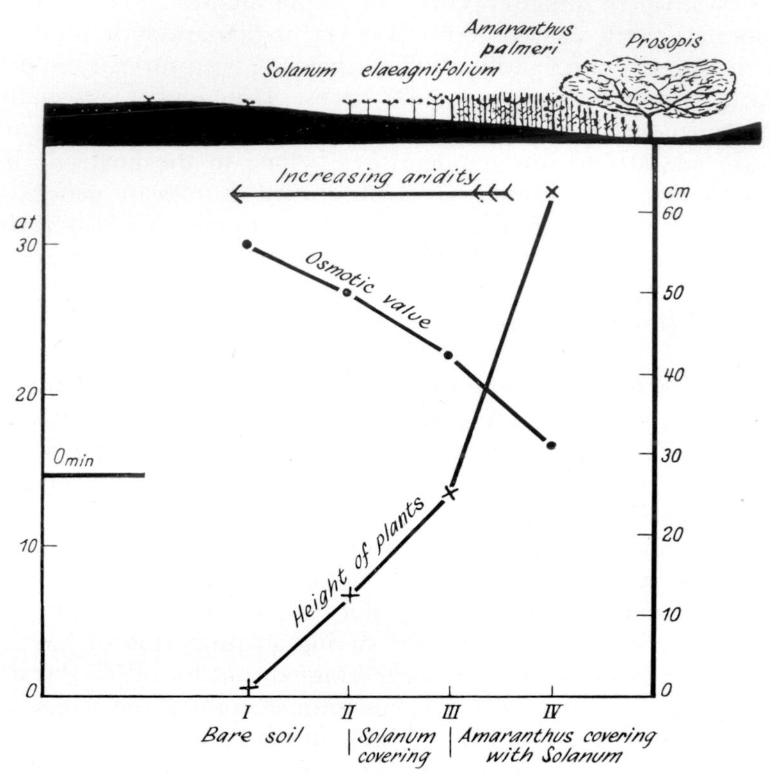

FIG. 184. Relationship between osmotic value and growth in *Solanum elaeagnifolium* (after H. Walter, 1931).

small depression. In the centre of this depression, the soil was more thoroughly wetted (Shreve, 1934) and the plants were vigorously developed. At the margin the soil was dry and they remained dwarfed.

TABLE 45

Osmotic values of summer annuals

	Optimum value (atm)	Maximum value (atm)
Euphorbia heterophylla	13–14	17
Kallstroemia grandiflora	15·4	18·4
Amaranthus palmeri	14·5	17·2
Tribulus terrestris	13·2	21·7
Cucurbita digitata	13·4	—
Cladothrix lanuginosa	18·9	34·8
Allionia incarnata	<20	25
Trianthema portulacastrum	18·4	30·9

The curve of the heights of the plants and that for the osmotic values are inversely related.

The winter-rain season began with rain on December 31 after a completely rainless drought of 97 days. A total of 127 mm of rain fell during the months January, February and March and, as a consequence, the winter ephemerals had ideal conditions for their development.

The ratio of species of summer to winter annuals at Tucson is

TABLE 46

Osmotic values of winter ephemerals

	Optimum value (atm)	Maximum value (atm)
Erodium cicutarium	9–10	> 14·7
Sonchus asper	10·6	
Hordeum murinum	12·6	
Malva silvestris	15·2	
Marrubium vulgare	15·4	
Brodiaea sp.	7·4	
Sicyos sp.	8–9	
Rumex hymenosepala	9·2	
Delphinium scaposum	10·5	
Anemone sphenophylla	14·3	
Verbena ciliata	14·4	
Penstemon wrightii	14·2	
Sphaeralcea pedata	17·1	
Erodium texanum	13·4	
Senecio douglasii	10·8	
Plantago ignota	11·5	
Streptanthus arizonicus	14·8	> 20·2
Parietaria obtusa	12·5	> 14·2
Phacelia tanacetifolia	16·2	

44:109. Further to the west in the Mohave desert, which lies in a winter-rain climate, the number of winter ephemerals increases to nearly 500. Seventy-three of these are common to both areas.

The winter ephemerals have a more delicate, hygromorphic structure since the saturation deficits of the air are much lower during the cooler winter months than during the summer. In addition, many of them, such as *Delphinium scaposum* and *Anemone sphenophylla*, prefer sun-shaded habitats on north slopes. The occurrence of several European weeds indicates that the climatic conditions during the winter-rain season correspond to those in central Europe during late spring to early summer. This is shown also by the osmotic values (Table 46).

It is characteristic of ephemerals that they do not show a strong correlation with any particular habitat and that they are indifferent to the type of soil. They occur wherever they can find enough water.

All soils are damp after heavy rain showers; however, after lighter rain only the erosion channels and depressions are moist. Often, individual species occur in dense stands where the rain water has washed the seeds together in certain places. On a moist clay soil, Shreve counted 1200 *Plantago* seedlings/square metre. Commonly the ephemerals are more dense in the shade of shrubs, since there the soil remains damp for a longer time. They are uniformly distributed on sandy soils. After the heavy rain in March, I saw extensive areas coloured yellow from *Eschscholtzia* in the western part of the desert in April and, in the Mohave desert, I saw a large area with white, short-stemmed Oenotheraceae and violet *Abronia* species.

Around this time, when the temperatures increase, the perennial species begin to spring up. Usually, the flowers emerge first and later the foliage.

(*b*) *Poikilohydrous pteridophytes*

This small group of ferns, which is restricted to rocky habitats, resemble the ephemerals in their behaviour. However, their leaves do not die during the dry season. They dry up, remain alive but dormant and become green again shortly after being wetted. In moist air, the leaves only unfold after a long time (Spalding, 1906a, 1906b). They behave similarly to mosses, however, the leaves of all these ferns appear to be xeromorphic in structure. They are either hard (*Pellaea mucronata*) or strongly hairy or scaly (*Notholaena* and *Cheilanthes* species, *Selaginella arizonica*). Very delicately leaved species, such as *Adiantum capillus-veneris*, are only found in shade by springs. The delicately leaved *Cheilanthes wrightii* also only grows in moist places which are never exposed to the full sun. Yet, even the xeromorphic species always grow in the shelter of the north side of boulders. This is related, perhaps, to the requirements of the prothalli, which may only develop fully in such a protected site. However, such a site is advantageous to the fern plant also since it can remain fresh for a longer period after a rain and can thus be photosynthetically active for a longer period. The ferns are so inconspicuous during the dry season that they are hard to detect. Their osmotic values can be determined only in the fresh condition (Table 47).

This plant life-form is also represented in the Karroo (South Africa) where I also used it for investigations (cf. p. 398 ff.).

(*c*) *General features of the perennial species*

The Sonoran desert is more truly described as semi-desert. The plant cover, therefore, is still relatively dense even if one only considers the perennial species. However, the degree of cover is less than 30%; often decreasing to less than 10%. Most species attain heights of 0·5-1·5 m, yet the giant cacti often reach 11 m and some individuals even 16-18 m. Tall trees are found only as gallery forests along the margins of dry-rivers with ground-water streams.

TABLE 47

Osmotic values (atm) of poikilohydrous species in fresh condition

Gymnopteris hispida	13·4
Pellaea mucronata	14·1–15·4
Notholaena hookeri	14·6–15·8
Notholaena sinuata var. *integerrima*	15·4
Notholaena aschenborniana	13·4–16·1
Cheilanthes wrightii	15·9–16·0
Cheilanthes fendleri	17·7–18·3
Cheilanthes lindheimeri (Fig. 185)	24·1
Selaginella arizonica (Fig. 185)	13·6–16·7

In spite of the relative openness of the vegetation cover the plants compete strongly and no free space is available for other perennials to invade. For example, the far-reaching lateral root systems of the very

FIG. 185. Poikilohydrous species (*Cheilanthes lindheimeri* and *Selaginella arizonica*) in the Sonoran desert near Tucson (Phot. E. Walter).

openly distributed creosote shrubs (*Larrea tridentata*) are in contact with one another on the dry, level, sandy, gravelly plains. On rocky soil, every favourable niche with deep fissures is likewise occupied by perennial plants. There is no room for seedlings, which are almost completely absent. The arborescent shrub *Cercidium microphyllum* attains an age of 300-400 years and seedlings only find space upon the death of an old individual. Shreve (1911b, 1917) studied the progress of seedlings of this species in a permanent quadrat for several years. The natural seed crop is very dense. In 8 years of observation, 923 seedlings were observed on an area of 557 m². However, only 4 remained alive for over 5 years. Most had already died by the first year (see Walter, 1960, p. 271). The persistence of the plant is assured only after attaining a certain size, when it can survive the loss of part of its shoot system during periods of water shortage, and when it is capable of regenerating the lost part in favourable years. So, one can only see very rarely a shrub in the desert which does not have many dead or, occasionally, only a few living branches. These are the scars of earlier, critical drought-years.

(*d*) *Succulents*

These are the most conspicuous life-forms over most of the Sonoran desert which, therefore, has also been called, 'the cactus-desert'. Indeed, most succulents are members of the family Cactaceae, which is here represented by a large number of species. The family Euphor-

FIG. 186. *Lemaireocereus thurberi* (centre) and *Fouquieria splendens* (with leaves and flowers) at Picu Pass, Sonora, Mexico (Phot. E. Walter).

biaceae which provides a large number of succulent forms in Africa, is here represented by only one member. This is *Pedilanthus macrocarpus*, which is found in Baja California.

Several shapes occur among the cacti:

(*i*) the pillar form, which branches at the centre (*Carnegiea*, *Pachycereus*) or even at the base (*Lophocereus*, *Lemaireocereus*) (Fig. 186),
(*ii*) the globular form (*Ferocactus*),

(*iii*) the cushion form, which originates from especially frequent branching at the base (*Echinocereus*, *Mammillaria*),

(*iv*) the cylindrical Opuntias (*Opuntia fulgida*, *O. acanthocarpa* and the smaller *O. bigelovii*, *O. arbuscula*, etc.),

(*v*) the flat Opuntias (very many species).

The ecology of the cacti has been investigated thoroughly in the Desert Laboratory but no comprehensive work has yet been published.

With the exception of *Penicereus*, which grows in erosion channels and which has a tap root, all cacti root only a few centimetres down. However, the roots are widely spread laterally (Fig. 187). Thus, cacti

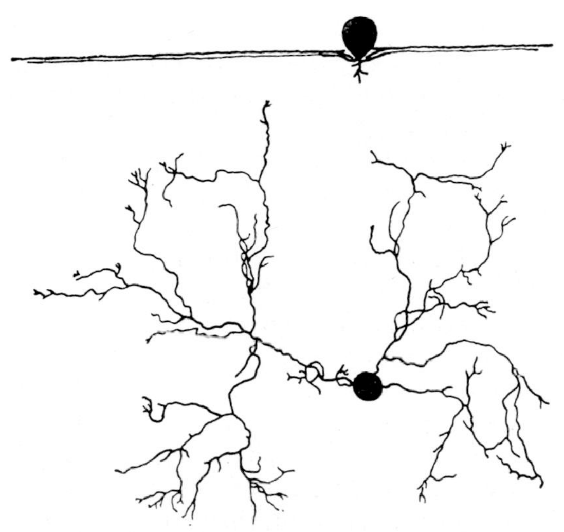

FIG. 187. Root system of *Ferocactus wislizenii* (after Cannon, modified). Above: Roots run horizontally at only 2 cm depth. Below: Map of lateral root distribution of the same plant (Walter, 1960).

can absorb water only after rain so long as the upper soil layers are still moist. The water is stored in the stem. The absorbing root tips die off during a drought and the plant depends entirely on its stored water.

Kausch (1959) investigated the formation of new roots by potted cacti. The plants were first kept dry for 6 months then the soil was moistened. New absorbing rootlets were formed within 8 hours in the case of an *Opuntia*, after 24 hours in 50% of the cacti and after 72 hours in 80%.

The large *Carnegiea* plants store 2000-3000 litres and they can survive for a year without water absorption (MacDougal, 1910, 1912; MacDougal and Spalding, 1910; MacDougal *et al.*, 1915). One can occasionally see living side-branches from a dead stem and these branches may even bear flowers in the following year. However, they never produce roots, even when buried in soil. *Ferocactus wislizenii* is used as a source of drinking water in the desert by Indians. They cut off the apex and push the central tissue down with a piece of wood. As a result, enough water collects in the cavity to scoop out by hand. My experience is that the taste is horrible. However, this is not important if one is nearly dying from thirst!

Transpiration experiments in the field have been done mainly by MacDougal.

(*i*) *Carnegiea gigantea* (Sahuaro), see Fig. 181.

This giant cactus attains heights of 10-12 m, occasionally even more. The diameter of the main stem varies from 30-40 cm. In plants 5-8 m tall some 5-6 candelabra-like side branches are formed, whose apices grow upwards as well. Secondary branching is exceptional in very old individuals.

The roots of *Carnegiea* hardly penetrate more than 1 m, but they spread laterally at this shallow depth up to 30 m. A solid, woody, cylindrical skeleton extends from the base to the apex. It consists of individual woody ribs, which anastomose. Commonly, woodpeckers build their nests in the pith of the stem, which then become surrounded by a corky callus growth. White flowers develop in May near the top, mostly on the south-west side; bees act as pollinators. The juicy fruits are eaten by birds and are also collected by the Papago Indians. An old giant cactus can bear up to 200 fruits, each of which may contain up to 1000 seeds. These are distributed in animal droppings (endozoochorously) and by ants (myrmechorously). Since the fruits ripen 2-4 weeks before the summer-rain season, they are able to germinate during it. However, only a few seedlings become established. They grow only a few millimetres in 2-3 years; after 10 years they are from 1·5-2 cm tall. A 1 m tall plant is about 30-50 years old. Only when the plants have reached a height of 2-3 m does their rate of increase change to about 10 cm per year. The oldest individuals are estimated to be between 150-200 years old.

In the Sonoran desert *Carnegiea* is restricted to the area east of the Colorado river and the east coast of the Gulf (Fig. 173b). In the mountains the species extends up to 1200 m (1500 m). It is frequent around Tucson. Old plants weigh 2000-4000 kg, in extreme cases 6-7 tons; their water content accounts for 85-91%. It is lowest towards the base because here the central woody cylinder is at its largest.

According to MacDougal, several plants lost 2-3 mg/day/g fresh weight in the laboratory, immediately after cutting in November. They were then kept outside. Until spring (February-May), transpiration decreased to 1·5 mg/day/g fresh weight, after a year (Dec.-May) to 1 mg/day/g fresh weight. The daily transpiration in an old plant amounted to between 1/1700 to 1/9300 of the fresh weight. In contrast, seedlings lose 1/12 of their fresh weight per day and somewhat older plants 1/46. Old plants cannot survive water losses in excess of 23% but, in seedlings, the critical point is at about 54%, which is reached in 100-120 days.

The volume change resulting from loss of water is made possible through the bellows-like mechanism of the ribs (Spalding, 1905). The ribs move closer together without altering the mechanical tissue of the outer cortex. Once water has been absorbed by the plant after rain, the ribs move apart again, the cross-section enlarges and so also the volume. Expansion is greater on the south side but the expanded condition persists longer after rain on the north side. The maximum increase between 2 ribs amounts to 35% or about 2·5 cm. MacDougal and Spalding (1910) recorded an annual curve of change in diameter. Minima occur during the pre-summer and post-summer drought periods. Any rain is reflected by an increase in diameter so long as the maximum swelling has not been attained. The rate and amount of swelling varies between individuals. Heating speeds up swelling, cooling has an inhibiting effect. Growth occurs mainly in the summer and the flowering period commences after the winter-rain season. Transpiration from the flowers is negligible.

The osmotic value of the cell sap was determined by MacDougal as 6-8 atm in fresh condition and as 9·8 atm after a water loss of 23·3%. However, one must make some allowance for variations in osmotic values of different tissues (Walter, 1931). The chlorophyll-containing layer of the ribs shows values 1-2 atm higher than the rib-parenchyma, while the cortical and pith parenchyma show values some 1-1·5 atm lower. In turn, the apical meristem has somewhat higher values, since water-storing tissue is absent there. In addition, values on the south side, where the ribs are closer together, are about 2 atm higher than those on the north side. It is only after rain has occurred that an equilibrium, or even a slight reversal of the osmotic relations, is achieved since the water-storing cells are filled more rapidly on the south side, where too the water-conducting tissue is better developed, viz.:

TABLE 48

Osmotic values (atm) of *Carnegiea* during the post-summer drought and winter-rain period (rain fell on 24.I and 19.III). (After Walter.)

Rib parenchyma from 1·5 m height	24.X	10.XI	5.I	24.I (Samples from 4 plants)				14.II	19.III	27.III
South side	—	6·0	7·2	5·9	4·7	5·7	5·9	5·3	5·2	4·1
West side	5·5	5·6	—	—	—	—	—	—	—	—
North side	—	5·1	5·3	5·9	5·2	6·0	6·0	4·7	5·8	—

In a broken-off side-branch that had been lying on the ground for 4 months, the osmotic value of the rib parenchyma had increased to 9·3-10·5 atm; in another one that had dried up for 9 months, the corresponding values were 10·7 atm (north side) and 13·6 atm (south side). However, osmotic values of cacti do not necessarily always increase upon drying.

When drying-out goes on at a very slow rate accompanied by yellowing, a decrease may even result through the respiration of osmotically active substances.

The distribution of osmotic values in a vertical direction were investigated on a longitudinal section of a 1·5 m tall individual:

TABLE 49

Osmotic values in relation to sampling height on *Carnegiea*. (After Walter.)

Distance from apex (cm)	Diameter of the stem (cm)	Osmotic values (atm)		
		Cortex (south side)	Pith	Cortex (north side)
9	23	6·3	5·6	6·1
30	29	6·4	6·2	5·8
60	30	—	6·0	—
90	25	—	6·3	—
120	23	6·1	—	5·7

Apical meristem = 10·0 atm

(*ii*) *Ferocactus* (*Echinocactus*) *wislizenii*

This cactus species, having a more globular form, rarely grows taller than 1 m (Fig. 188). Its ecology is very similar to that of *Carnegiea*.

Transpiration is somewhat less than in *Carnegiea*. It varies between 0·7-5 mg/day/g fresh weight depending on the degree of water saturation.

A plant with a surface area of 15,600 cm^2 weighed 49·39 kg and contained 45 kg water. A weight-change experiment was carried out by MacDougal over several years (Table 50).

TABLE 50

	fresh weight g
Initial weight, November 5, 1908	17,010
Without water till May 1909	11,370
On rooting in soil till October 1909	20,370
Without water in the lab. till May 1910	13,335

MacDougal gives as osmotic values 3-5 atm, after great water loss, 7 atm; for individuals that turn yellowish after a water loss of 36·2%, only 2·2 atm.

This shows that the drying-out process in cacti is very complicated (MacDougal, *et al.*, 1915). One can distinguish three phases:

1. Initial water loss; causing increase in the concentration of sugar and acid content.
2. Then, sugar loss by respiration, decrease in acid content. The water content (as fresh weight) remains constant in spite of continued water loss.
3. Beginning of rapid hydrolysis and yellowing.

FIG. 188. *Ferocactus wislizenii*, large individual on the left (Phot. E. Walter).

Another individual of *Ferocactus*, weighing 37·6 kg, was kept for 6 years in a shaded room. The average daily loss in weight for each year was as follows:

Experimental year	I	II	III	IV	V	VI	Av/annum
Weight loss in g	10·5	5·4	4	4·4	3·2	2·9	5

The water content was unchanged after 6 years (at 94%), but the content of hydrolysable carbohydrates had decreased from 26% to 11% in the storage tissue and from 32·5% to 23·3% in the outer cortex. The cell sap concentration was less than at the beginning of the experiment. Since the plant had lost 30% of its weight with an unchanged water content, 30% of the dry matter must have been respired.

I investigated the distribution of cell sap concentration in this species in much detail so that isosmotic lines could be superimposed on a cross-section (Fig. 189).

The asymmetrical arrangement of the vascular bundles and rib-width coincides precisely with the asymmetrical distribution of the osmotic values and also with the different growth rates on south and north sides. The slower growth rate on the south-west side, compared with that on the north-east side, is responsible for the asymmetrical shape of the cactus. Very old individuals can fall over eventually due to an imbalance in their weight resulting in their death.

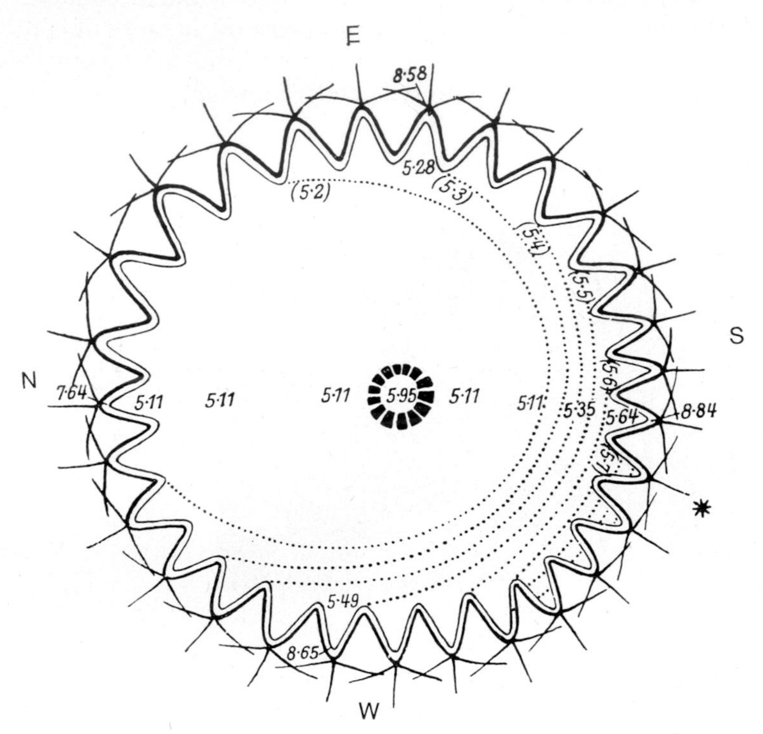

Fig. 189. Distribution of osmotic values on a cross-section of *Ferocactus wislizenii*. Isosmotic lines with osmotic value (atm) (after Walter, 1931).

The highest osmotic value in *Ferocactus* was found in the apical meristem with 6·9 atm. This contrasts with 4·4 atm for the pith and 5·3-5·6 atm for the cortical parenchyma. An individual that was uprooted for 9 months and laid on its side gave the values shown in Fig. 190. The position of the flower buds (asterisk in Fig. 189) shows a definite relation with the osmotic value distribution in the cross-section. This was shown particularly well in *Carnegiea* and *Pachycereus*. No flower buds are formed on the north side, where the lowest osmotic values occur. The flowers open first on the south-west side, which has the highest osmotic values.

(*iii*) *Opuntia* species

This group has the widest distribution, particularly the flat Opuntias. Cold-resistant species extend as far north as the Canadian border. In the Sonoran desert they are represented by many species with rather large cladodes and the species are difficult to distinguish from one another.

Maximum transpiration in the open has been determined as 5 mg/day/g fresh weight. However, Opuntias can reduce transpiration

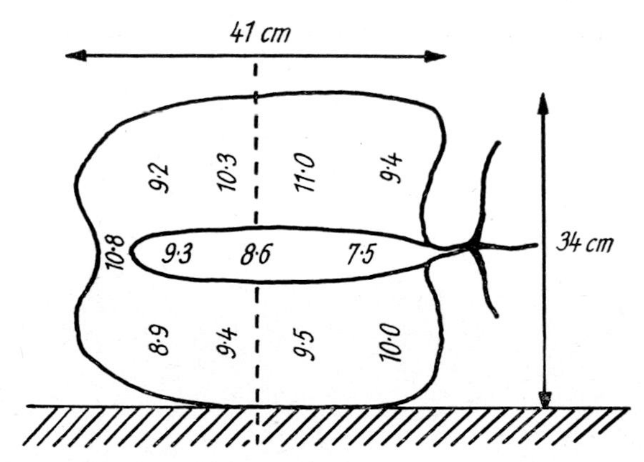

FIG. 190. Distribution of osmotic values (atm) in a *Ferocactus wislezenii* (longitudinal section) 9 months after uprooting.

considerably so that they may be able to survive without water uptake for 2-3 years. In this process, water is transferred from the old shoots to the young ones, while the old ones wilt and die. The young shoots can even produce roots. Fig. 191 shows the weight loss during a 3-month period in the sun compared with the laboratory.

Also the orientation of the flat shoot members has an influence on the transpiration rates (Spoehr):

TABLE 51

Transpiration losses (%) of *Opuntia blakeana* with different aspects

Period of Exposure	Orientation S-N	E-W	In the shade
Weight loss from 28.II to 5.IV (34 days)	18·5	16·3	5·8
Weight loss from 15.V to 28.VI (44 days)	24·7	26·7	23·3

Direct insolation of the shoot members results in high temperatures. Maximum heating occurs at 2 p.m. with an east-west aspect and at 11 a.m. with a north-south orientation, so that a minimum temperature is reached at 12.30 p.m. and the main maximum at 4 p.m. Under natural conditions, 53°C was the measured maximum in June, and 55°C in July.

Spoehr (1913, 1919) investigated the carbohydrate balance as well as the water economy. Cacti, characteristically, have a high slime content. The slime consists of polysaccharides which can be hydrolysed by 1% HCl to 34·1% D-glucose and 65·9% L-xylose. The evaporation rate of these slimes does not differ, in the first 4 hours, from that of pure water. So, it is difficult to imagine that they can function as a protection against transpiration. The low transpiration rate of the cacti is related to their small number of stomata and to the impermeability of their epidermis and outer cortical layers, particularly when these have shrunk through desiccation.

Young shoots are relatively rich in slime. Pentosans increase upon

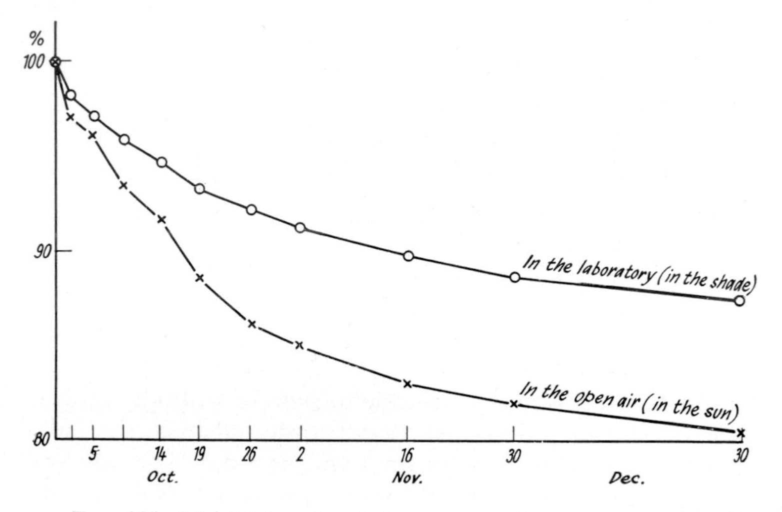

Fig. 191. Weight loss by drying expressed as a percentage of *Opuntia discata* from Sept. 30 to Dec. 30 (Tucson, Arizona) after the diagrams by MacDougal, Long and Brown.

loss of water, while the monosaccharides decrease thus resulting in a seasonal fluctuation.

Water is formed by respiration of the carbohydrates during desiccation. For this reason, the loss in weight is much greater than the reduction in water content.

An individual of *Opuntia phaeacantha* lost 60% of its initial weight in 189 days while the water content only decreased from 84·75% to 72·68%. Water produced by respiration also plays an important role in animals of arid ecological niches. In this way, many can survive for a long time without having to absorb water.

Livingston (1907) pointed out that the relative transpiration rate in cacti reaches a maximum at night and a minimum during the day while the absolute transpiration is never lower at night than during the day. This is a very strange fact.

E. B. Shreve (1915, 1916) supported this observation in a different way. Her investigation showed that the shoot-members, particularly in cylindrical Opuntias, carried out periodic movements. The shoot-members in *Opuntia versicolor* are slowly lowered during the dry season and rise again rapidly after rain. In addition, there are also daily movements, a raising during the day and a lowering at night. These movements are caused by changes in turgidity. Measurements of water loss and water balance showed that the water loss is greater at night when turgidity also decreases. The relation is reversed during the day. Changes in diameter and height, measured by auxanographs, also gave the same results with all cacti: a decrease at night and an increase during the day. This implies that the cacti open their stomata at night so enabling gas exchange to occur mainly at night. Therefore, CO_2 absorption must also take place at night. Thus, CO_2 is initially merely trapped in the plant, later to be reduced to carbohydrates in the photosynthetic process, i.e. the following day.

Spoehr (1913) has shown that the acid content of the cell sap in *Opuntia versicolor* is greatest at sunrise and lowest at 5 p.m. To neutralise 1 ml of cell sap, he needed 2·45 ml of 0·1 N KOH in the morning, and only 0·31 ml in the afternoon. The CO_2 absorbed during the night appears to be stored as organic acid.[1] This behaviour of the cacti is reminiscent of that of the desert rodents, who only leave their burrows at night in search for food and sleep in them during the day.

Thus, the cacti present a good example of an essential requirement in ecological investigations: never to consider any process in isolation below the level of the whole organism. Here, the water balance is intimately connected with the total metabolism.

I also made a detailed investigation of the osmotic values of Opuntias. In the flat Opuntias, the aspect of the shoot-members plays a role:

Opuntia engelmanii	0700 hr	1400 hr
Exposed to N-S	11·0 atm	11·7 atm
Exposed to E-W	10·6 atm	10·3 atm

Opuntia phaeacantha (E-W exposed, shoot-member was halved)

Sun-exposed side	15·5 atm
Shaded side	14·1 atm

Opuntia versicolor (cylindrical shoot-member was cut into two halves, apical and basal ends)

Apical end	15·7 atm
Basal end	15·2 atm

Daily variations in osmotic values could not be determined accurately. Shoot-members connected to one another showed only small differences in osmotic values. Drying-out resulted in an increase initially and then in a slight decrease and yellowing.

Continuing investigations in the field showed an increase until January 8 and then a decrease with beginning of the rainy season (Table 52).

[1] This so-called 'De Saussure effect' has been discussed in detail by Nuernbergk (1961), who includes a list of the succulents that show this effect. See also Epiphytes, p. 135.

TABLE 52

Osmotic values (atm) of Opuntias during the post-summer dry season and the winter-rain season

	October	Early January	March
Opuntia toumeyi	8·8	15·4–15·8	8·4
Opuntia blakeana	8·8	15·1	10·0
Opuntia versicolor	7·4	16·6–17·8	11·9

The lowest value, 6·7 atm, was obtained in young shoots. The highest value, 19·4 atm, was obtained in the hardly cylindrical succulent, *Opuntia arbuscula.*

(*iv*) Other cactus species

No precise ecological investigations are available for other cacti. *Pachycereus pringlei*, the largest form occurring in North Mexico, shows

FIG. 192. *Pachycereus pringlei* near Libertad, N. Mexico (Phot. E. Walter).

similar behaviour to *Carnegiea*. This also applies to the other pillar-cacti (Fig. 192).

The following osmotic values were determined in March after considerable rain in North Mexico near the Gulf of California:

Pachycereus pringlei:	I. Rib parenchyma 4·2 atm and chlorophyll-containing tissue 8·8 atm
	II. Rib parenchyma 5·4–6·2 atm
Carnegiea gigantea:	Rib parenchyma 4·1 atm
Lemaireocereus thurberi:	Cortical parenchyma 4·4 atm
Lophocereus schottii:	Cortical parenchyma 4·7–4·9 atm

The following are some values for small pillar and globular cacti:

Echinocereus fendleri:	8·1 atm (24.I.)
Echinocereus polyacanthus:	3·7 atm (7.III.)
Echinocereus rigidissimus:	5·4 atm (7.III.)
Mammillaria grahami:	5·8 atm (outer layers)
(24.I.)	4·8 atm (inner layers)

(*e*) *Characteristic shrubs*

Cacti are the best investigated group in the Sonoran desert but there are a few especially characteristic species among the shrubs which I will now discuss. These are *Fouquieria splendens*, which belongs to the small family Fouquieriaceae, *Cercidium* (*Parkinsonia*) *microphyllum*, a member of the Caesalpiniaceae, and *Larrea tridentata* (= *Covillea glutinosa*; Zygophyllaceae).

(*i*) *Fouquieria splendens*, Ocotillo

In the southern part of the Sonoran desert, *Fouquieria splendens* is replaced by other species. The genus is endemic to the Sonoran desert (Fig. 193a). *Fouquieria* is a unique ecological life-form. It consists of thin, 3-4 m (10 m) long, simple branches, which arise from the root collar, numbering up to 25, at an angle of more than 45° in all directions (Fig. 194). The long shoots only develop during the rainy season. They have delicate leaves, with deciduous lamina, the petioles being transformed into thorns. In their axils occur short shoots with delicate rosettes of leaves, each 1·5-4 cm long, and 0·5-1·5 cm wide. They emerge at any time of the year after heavy rain and turn yellow as soon as the soil has dried off. It may happen, therefore, that the plants produce new leaves 5-6 times a year. Stomata occur equally abundantly on both sides of the leaves, in number, 160/mm^2. Their movement has been studied by Lotfield (1921).

The flowers are formed at the end of the winter-rain season, when they droop in terminal clusters from the branches. They are bright red in colour and resemble torches.

In respect of its water economy, this plant behaves like a mesophytic plant, since it bears leaves only when the soil is wet and its

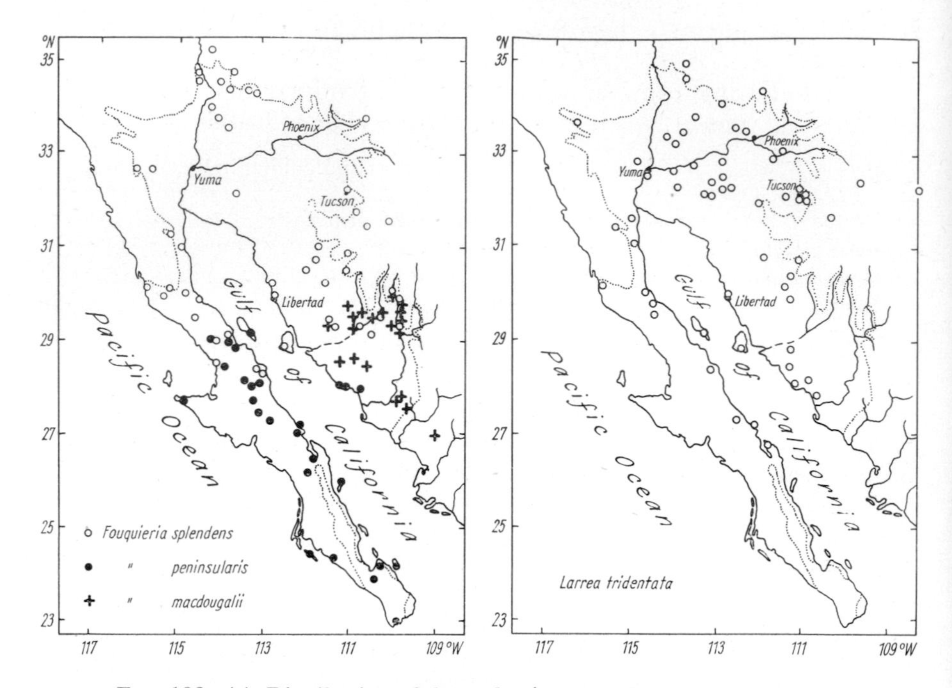

FIG. 193. (*a*) Distribution of the endemic genus *Fouquieria*.
(*b*) Distribution of the Creosote shrub (*Larrea tridentata*) in the Sonoran desert (both after F. Shreve, modified).

FIG. 194. Three large individuals of *Fouquieria splendens* near Tucson, Arizona (Phot. Edna M. Lind).

branches are protected against water loss by a layer of cork. Cannon (1905) and Lloyd (1906, 1912) have studied its behaviour.

The following observations show how rapidly the leaves are formed:

After rain on July 11, the buds were markedly enlarged on July 12. By July 13 they had reached a length of 7-8 mm and by July 14 the leaves had already unfolded to 1·5 cm. The shrub became green within 5 days also when watered. Lloyd tied a cloth around a twig which was kept wet. After 8 days, this twig had formed leaves 2 cm long. Water absorption probably took place through the buds and leaf scars.

Transpiration rates are very high, with a maximum of 8·25 mg/min/dm^2. At night, water absorption is greater than water loss, while the reverse is true during the day. The water content of the leaves reaches a maximum at sunrise. However, the water content always represents between 70 and 74% of the fresh weight. I could not demonstrate any daily fluctuations in osmotic values. At emergence, very young leaves have an osmotic value of 14·4 atm; mature leaves one of 17·2-19·6 atm. The leaves do not dry up with shortage of water, but turn yellow. During this process, the osmotic value increases initially to 21·4 atm then, however, the value decreases subsequently to 16 atm and finally drops to 11·7 atm. Yellowing enables the plant to transfer all important metabolites from the leaves to the axillary organs. This would appear to be of significance in relation to the frequent shedding of the leaves. Yellowing probably occurs as soon as the photosynthetic mechanism is disturbed by sufficient water loss to result in an adverse water balance, or to the closure of the stomata. A similar response to that of *Fouquieria* is also shown by a leafy shrub *Jatropha cordifolia* (Euphorbiaceae), which has leaves similar to birch. In this shrub, the osmotic value decreased upon yellowing from 17·6 to 11·8 atm.

Another member of the Fouquieriaceae is *Idria columnaris*, which occurs in but one locality in Baja California east of the Gulf near Libertad. This tree, which can attain heights of 6-8 m, belongs to the group of 'bottle trees', because its basal trunk is swollen and serves for water storage. Its diameter attains 50-60 (75) cm. The inner tissue has a spongy structure so that volume changes consequent upon water storage and loss are much facilitated. The branches are thin, attached horizontally and with somewhat xeromorphic leaves. Their cell sap showed osmotic values of 17·8-19·1 atm, while that of the swollen, lower trunk was 12·6 atm. Water storage in the stem occurs also in *Bursera* (*Elaphrium*) species (Burseraceae) which occur in association with *Idria*. In these, the osmotic value of the leaves was 18·5 atm, that of the stem tissue, 11 atm.

(*ii*) *Cercidium microphyllum*, Palo verde

This species belongs to the type of woody plants that have green stems and twigs, which act as the main photosynthetic organs. The leaves consist of 3-5 feathery, compound leaves, whose leaflets are rigid and round with a diameter of less than a millimetre. Young plants have leaflets with lamina 4 times as large which often redden in the winter and are less drought-resistant than mature leaves.

The leaves only appear after good winter- or summer-rains and they rarely remain functional for longer than 6-10 weeks. They then turn yellow and are shed, although the rachis remains attached to the branch for several more weeks. Even in full leaf, the total leaf area remains below that of the axial organs. This aborescent shrub becomes 5-6 (8-10) m high. It flowers after a good winter-rain in April or May. The seeds are hard-shelled and germinate only in the summer-rain season of the following year.

This species is restricted to rocky habitats and scree slopes. The roots penetrate vertically and laterally, wherever they find moisture in the rock crevices. A large proportion of the shoot system dies in drought-years but the tree can regenerate branches from its stem base. The anatomical relations, transpiration, water content and temperature relations were investigated by E. B. Shreve (1914).

The osmotic values of the xeromorphic leaves are relatively high and vary only little from November to February, i.e. between 26·6 and 31·5 atm. They can even attain 34·7-36·8 atm in dry habitats. Thus, they show the same values as sclerophyllous species. Similar values are also found among its relatives *Cercidium* (*Parkinsonia*) *torreyanum* (23·6-27·1 atm) and *Parkinsonia aculeata* (young 15·5 atm, mature 26·5 atm). This species is shown in Fig. 195.

FIG. 195. Arborescent *Parkinsonia aculeata* and further back *Olneya tesota*, in the foreground white shrubs of *Encelia farinosa*. Pitiquito, N. Mexico (Phot. E. Walter).

(*iii*) *Larrea* (*Covillea*) *tridentata*, Creosote bush

The Creosote bush gets its name from the lacquer that covers its leaves. This smells strongly of creosote after rain. The species is typical of many desert areas beyond the limits of the Sonoran desert. It also occurs in the Mohave desert and in part of the Great Basin desert, and related species play an important role in the arid areas of Argentina

(Morello, 1955, 1956). Its distribution in the Sonoran desert is shown in Fig. 193b, p. 328.

The Creosote bush is extremely adaptable and has low water requirements. It can endure a year without rain and yet it also occurs in regions with 500 mm rainfall. The habit of the plant is correspondingly different (Spalding, 1904; Runyon, 1934).

The shrub branches direct from the root collar and branches grow, with further branching, in all directions, some are horizontal and a little above the ground. Under favourable conditions, *Larrea* reaches a height of several metres, forms up to 100 branches and has many bipartite leaves up to 28 mm long. In unfavourable habitats, the shrub remains dwarfed when the leaves are often only 4 mm long. With increasing dryness it shed most of its leaves. However, even nearly leafless twigs remain alive for a long time and sprout again after rain. The Creosote bush can maintain itself for so long under extremely dry conditions because of its reduction in twig number and leaf area and the resulting great reduction in loss of water. In many areas it is almost the only plant that survives. However, favourable water conditions are immediately utilised by growth and dry-matter production. When rain falls after an extensive period of drought, the shrub too blooms by unfolding its many, small, yellow flowers.

Its osmotic values do not decrease below 28-30 atm even under favourable water conditions. A cell sap concentration of 22-24 atm was only found in young leaves of just-emerging shoots. However, as soon as water becomes limiting and the water balance cannot be maintained, the osmotic value increases beyond 55 atm without resulting in the death of the leaves. They only turn brownish and somewhat shrivelled. This increase is, of course, not an adaptation but a consequence of the water deficit. Its drought resistance lies in the fact that the cells can survive such a decrease in water content. Fig. 196 shows the annual fluctuations in osmotic values in relation to the habitat conditions.

(*iv*) Other ecological life-forms

Sclerophyllous plant types are by no means exhausted by the preceding three examples. Indeed, one may claim that each shrub species represents an individually distinct type.

Simmondsia chinensis (*californica*) is a low shrub with slender branches, yet with very thick, succulent leaves, reminiscent of *Cotyledon*, but they are tough. In contrast to the succulents, the normal osmotic values are very high, between 34 and 40 atm. After a dry period of 3 months 49·1 atm were measured. As usual, young leaves have low values at emergence but even these are already 20·8 atm. Fresh twigs lose 8·5-11% of their fresh weight in the first hours of drying, as much as 3·2-3·6% in the second hour and, in the following hours, only 0·6-0·9%. Thus, the plant maintains its water balance by its greatly reduced water loss.

However, around Tucson, it does not occur in the driest habitats, although it covers all the slopes near the northern border of the Sonoran desert at the Roosevelt dam, where it even occurs on southern exposures.

Mortonia scabrella is a small shrub that shows quite different behaviour. It occurs in a restricted, limestone area south-east of Tucson. The whole plant appears grey and is densely coated with almost scale-like, small, tough leaves, which can be retained for 6-7 years. The epidermal cells are large but they have very thick outer walls, the stomata are only a little set in and the palisade parenchyma consists of several layers.

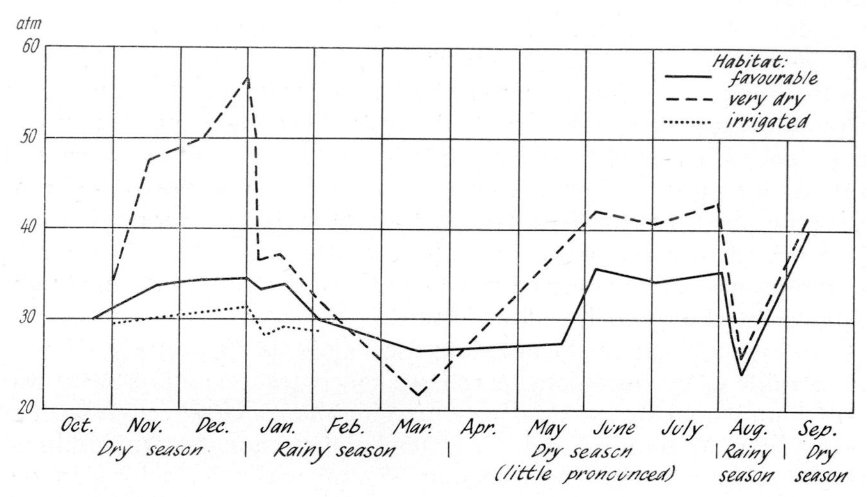

FIG. 196. Annual fluctuations of osmotic values in *Larrea tridentata* measured at the Desert Laboratory, Tucson, Arizona. The favourable habitat was the *Carnegiea-Encelia* community.

Twigs, which were kept in the laboratory, were air-dry within 5 days for the water content only amounts to about 20%. An osmotic value of 23·2-25·8 atm, determined after rain is, therefore, not at all high. Its anatomical structure hardly permits any great fluctuation in water content. Thus, the plant can probably only grow in limestone habitats, where there is always a sufficient supply of water available. Investigations in these conditions are not available.

Acacia is a genus that occurs more especially in tropical summer-rain regions but some species are also found in the Sonoran desert. They shed their fine, feathery leaves in the winter months or else retain them in part. They develop new foliage only when there is enough water in the soil, which they can absorb with their far-reaching root systems.

The osmotic values of the leaves under normal conditions are about 26 atm. However, with shortage of water the values can increase to more than 40 atm but the leaves could hardly remain alive for any length of time without a continuous water supply.

Prosopis juliflora ssp. *velutina*, Mesquite, is a characteristic species of the next higher altitudinal belt. It occurs however, extrazonally, in the Sonoran desert as a tree of bottom lands in the flood plains of the broader valleys. The tree has an extremely well developed root system so that it is well supplied with water even during a period of drought. Moreover, the shrubs of *Prosopis*, which occur at the edge of erosion channels or, less commonly, between large blocks of rock, are probably all in favourable situations. So it is not surprising that the osmotic values do not show pronounced increases during the dry season. The normal values are around 25-28 atm. These increase after a drought of one month to 30-32 atm and after a further month to 34-35 atm.

Prosopis sheds its leaves after the first frosts, which are particularly noticeable in valley sites. The leaves are not shed when *Prosopis* grows on slopes. The cell sap concentration was increased to 41·1 atm in dry habitats and, under these conditions of increased water tension, it was found that the leaves became smaller and more xeromorphic, a phenomenon that will be discussed further in connection with *Encelia farinosa*. In North Mexico, young leaves emerging at the end of March on some branches showed a value of 19·9 atm.

Deciduous trees and shrubs belonging to holarctic taxa also extend into the desert region. However, they are then restricted to very damp habitats, such as to the margins of river beds where a ground water supply is present throughout the year or to the lower parts of mountain valleys where springs occur. This group includes *Salix* and *Populus* species and others mentioned on p. 307. None of these species differs with respect to its habit and water relations from those in the temperate zone. The osmotic values of the trees were around 19-20 atm.

Vitis showed a cell sap concentration of 9·6 atm (young leaves). Our cultivated vine (*Vitis vinifera*) shows a value of 10·5 atm.

(*f*) *Dwarf-shrubs with soft, hairy leaves* (*malacophyllous plants*)

Encelia farinosa is a semi-shrub of the Compositae, which generally does not exceed 60-80 cm (Fig. 197). It has brittle twigs 5-10 mm thick and petioled leaves 3-7 cm long. The leaves and twigs are densely pubescent giving the plant a white appearance. The yellow flower heads rise on long stalks 10-15 cm above the leafy twigs and are formed twice a year during the two rainy seasons. In Arizona the species grows mainly on slopes. The transpiration rate of this shrub is relatively high when well supplied with water (Shreve, 1923, 1924). The water relations are of particular interest in relation to the morphology of the leaves. The daily fluctuations in osmotic values are very pronounced (averaging 4·5 atm) and correlated with the marked transpiration.

In favourable conditions of water supply on north slopes the osmotic value is 22-23 atm. Here, the leaves are large and less pubescent, they appear green and may be described as hygromorphic. The osmotic value is 28 atm in dry habitats and the leaves are mesomorphic, i.e. they are more densely pubescent, whitish and somewhat smaller. Such leaves are also formed on north slopes as soon as the cell sap concentration increases to 28 atm during a period of drought. When, with a longer-lasting drought, the osmotic values increase in all habitats, more reduced, xeromorphic leaves are always formed. At the same time the hygromorphic leaves die and are shed when their osmotic values reach 36-38 atm. The xeromorphic leaves are only shed with an osmotic value of more than 40 atm. Values of 55 atm were attained during the long

Fig. 197. *Encelia farinosa* a semi-shrub on Tumamoc Hill near Tucson, Arizona (Phot. E. Walter).

dry season at the end of December. Under these conditions, the plant is leafless except for small, highly pubescent leaves which occur only around the buds. The plants regenerated greatly after heavy rains in March, the young leaves again showed values of 22-23 atm and their structure was hygromorphic. Thus, *Encelia* shows the typical behaviour of a malacophyllous xerophyte, i.e. very marked fluctuations in osmotic values as the result of a relatively tardy reduction in transpiration. This results in large water deficits and in a reduction of the transpiring surface, initially by the production of successively smaller and more xeromorphic leaves and finally the successive and increased shedding of leaves.

Two species of *Franseria* also belong to the same category. They show a very similar response with respect to leaf formation and osmotic value. *Franseria dumosa* is more widely distributed and probably also more drought-resistant than *Encelia*.

By contrast, the labiate, *Lippia wrightii*, is less drought-resistant and restricted to protected habitats between rocks on north slopes. Its leaves are only sparsely covered with hairs and its osmotic values are lower (15-31 atm).

4. Conclusion

The discussion has shown that the responses of typical representatives of the Sonoran desert flora to the dry season are very different. They cannot be characterised by any one feature and they are difficult to classify into 'Schimper-type' or 'Maximow-type' xerophytes. This means that they cannot simply be classified into groups which survive a drought because of their low rate of transpiration as is most clearly applicable to the succulents, and also to *Cercidium* and others which can endure large water deficits. This can be said of *Larrea* and *Encelia* also.

The maintenance of some degree of hydration of the protoplasm and a low concentration of cell sap (high hydrature) is of importance to the plant. In this respect, stenohydrous species cannot endure any great increase in osmotic concentrations, while others, the euryhydrous plants, are less sensitive to this factor. However, between these two extreme types transitory forms occur. Moreover, those mechanisms which enable plants to avoid great changes in their water content during the dry season are themselves so varied that each species has to be studied separately. This, of course, means that any results obtained cannot be generalised very readily. Here the behaviour of the roots and the transpiration of the leaves are of importance.

Many species endure far longer dry spells than usually occur, even in extreme years. For example, several cacti can survive without taking up water for 2-3 years yet they are not found in the most extreme deserts.

It must never be forgotten that the critical phase of development which limits the distribution of plants in arid areas is the first few years of their lives, especially just after germination. The young plants are so sensitive that they only become fully established after several favourable successive years with short dry seasons. Such favourable situations are only found in the less extreme desert areas so that nothing is gained, from the point of view of species survival, if the plant can endure much more extreme conditions when mature, so long as its progeny have no chance of surviving the first critical years. Such a species will not maintain itself naturally in an area like this for any length of time. On the other hand, such a species can be cultivated, provided that sufficient care is given to it during early stages of development, for example, by watering it from time to time. Thus, the natural distribution of a species is always much more limited than its distribution in cultivation.

References

CANNON, W. A. 1905. On the transpiration of *Fouquieria splendens*. *Bull. Torrey. Bot. Club*, **32**, 397-414.

CANNON, W. A 1911. The root habits of desert plants. *Carnegie Inst. Wash. Publ.*, **131**, 1-96.

HUMPHREY, R. R. 1933. A detailed study of desert rainfall. *Ecology*, **14**, 31-34.

KAUSCH, W. 1959. *Der Einfluss von edaphischen und klimatischen Faktoren auf die Ausbildung des Wurzelwerkes der Pflanzen unter besonderer Berücksichtigung einiger algerischer Wüstenpflanzen*. Dissertation. Darmstadt.

LIVINGSTON, B. E. 1907. Relative transpiration in cacti. *Plant World*, **10**, 110-114.

LLOYD, F. E. 1906. The artificial induction of leaf formation in Octotillo. *Plant World*, **9**, 56.

LLOYD, F. E. 1912. The relation of transpiration and stomatal movements to the water-content of the leaves of *Fouquieria splendens*. *Plant World*, **15**, 1-4.

LOFTFIELD, J. V. 1921. The behaviour of stomata. *Carnegie Inst. Wash. Pub.*, **314**, 1-104.

MACDOUGAL, D. T. 1910. Variations of the water balance, 45-71 in MacDougal, D. T. and Spalding, E. S. 1910.

MACDOUGAL, D. T. 1912. The water balance of desert plants. *Ann. Bot.*, **26**, 71-93.

MACDOUGAL, D. T., LONG, E. R., and BROWN, I. G. 1915. End results of desiccation and respiration in succulent plants. *Physiol. Res.*, **1**, 289.

MACDOUGAL, D. T., and SPALDING, E. S. 1910. The water balance of succulent plants. *Carnegie Inst. Wash. Publ.*, **141**, 1-77.

MORELLO, J. 1955. Estudios botanicos en las regiones aridas de la Argentina, I, II. *Revista Agr. NW-Argentina*, **1**, 301-70, 385-524.

MORELLO, J. 1956. Estudios botanicos en las regiones aridas de la Argentina, III. *Revista. Agr. NW-Argentina*, **2**, 79-152.

NUERNBERGK, E. L. 1961. Endogener Rhythmus und CO_2-Stoffwechsel bei Pflanzen mit diurnalem Säurerythmus. *Planta*, **56**, 28-70.

REMPEL, P. J. 1936. The crescentic dunes of the Salton Sea and their relation to the vegetation. *Ecology*, **17**, 347-58.

RUNYON, E, H. 1934. The organisation of the creosote bush with respect to drought. *Ecology*, **15** 128-38.

RUTTE, E. 1958. Kalkkrusten in Spanien. *Neues Jb. Geol. Paläontol.*, **106**, 52-138.

SCHRATZ, E. 1931. Vergleichende Untersuchungen über den Wasserhaushalt von Pflanzen im Trockengebiet des südlichen Arizona. *Jb. Wiss. Bot.*, **74**, 153-290.

SHREVE, E. B. 1914. The daily march of transpiration in a desert perennial. *Carnegie Inst. Wash. Publ.*, **194**,1-64.

SHREVE, E. B. 1915. An investigation of the causes of autonomic movements in succulent plants. *Plant World*, **18**, 297, 331.

SHREVE, E. B. 1916. An analysis of the causes of variations in the transpiring power of cacti. *Physiol. Res.*, **2**, 73-127.

SHREVE, E. B. 1923. Seasonal changes in the water relations of desert plants. *Ecology*, **4**, 266-92.

SHREVE, E. B. 1924. Factors governing seasonal changes in transpiration of *Encelia farinosa*. *Bot. Gaz.*, **77**, 432-9.
SHREVE, F. 1911a. The influence of low temperatures on the distribution of giant cactus. *Plant World*, **14**, 136.
SHREVE, F. 1911b. Establishment behaviour of the Palo Verde. *Plant World*, **14**, 289.
SHREVE, F. 1912. Cold air drainage. *Plant World*, **15**, 110-15.
SHREVE, F. 1914. Rainfall as a determinant of soil moisture. *Plant World*, **17**, 8.
SHREVE, F. 1915. The vegetation of a desert mountain range as conditioned by climatic factors. *Carnegie Inst. Wash. Publ.*, **217**, 1-112.
SHREVE, F. 1917. The establishment of desert perennials. *J. Ecol.*, **5**, 210-16.
SHREVE, F. 1922. Conditions indirectly affecting vertical distribution on desert mountains. *Ecology*, **3**, 269-74.
SHREVE, F. 1934. Rainfall, runoff and soil moisture under desert conditions. *Ann. Assn Amer. Geogr.*, **24**, 131-56.
SHREVE, F. 1951. Vegetation of the Sonoran Desert, I. *Carnegie Inst. Wash. Publ.*, **591**,1-178.
SHREVE, F., and MALLERY, T. D. 1933. The relation of caliche to desert plants. *Soil Sci.*, **35**, 99-112.
SPALDING, E. S. 1905. Mechanical adjustment of the Suaharo (*Cereus giganteus*) to varying quantities of stored water. *Bull. Torrey Bot. Club*, **32**, 57-68.
SPALDING, V. M. 1904. Biological relations of certain desert shrubs, I. The Creosote Bush (*Covillea tridentata*) in its relation to water supply. *Bot. Gaz.*, **38**, 122-38.
SPALDING, V. M. 1906a. Absorption of atmospheric moisture by desert shrubs. *Bull. Torrey Bot. Club*, **33**, 367-75.
SPALDING, V. M. 1906b. Biological relations of certain desert shrubs, II. Absorption of water by leaves. *Bot. Gaz.*, **41**, 262-82.
SPALDING, V. M. 1909. Distribution and movement of desert plants. *Carnegie Inst. Wash. Publ.*, **113**, 1-144.
SPOEHR, H. A. 1913. Photochemische Vorgänge bei der diurnalen Entsäuerung der Sukkulenten. *Biochem. Zschr.*, **57**, 95-111.
SPOEHR, H. A. 1919. The carbohydrate economy of cacti. *Carnegie Inst. Wash. Publ.*, **287**, 1-79.
TURNAGE, W. V., and HINCKLEY, A. L. 1938. Freezing weather in relation to plant distribution in the Sonoran Desert. *Ecol. Monogr.*, **8**, 539-50.
WALTER, H. 1931. *Die Hydratur der Pflanze*, Jena.
WALTER, H. 1960. *Einführung in die Phytologie*. Vol. III, *Grundlagen der Pflanzenverbreitung*. Part 1. *Standortslehre*. 2nd ed. Stuttgart.
WENT, F. W. 1949. Ecology of desert plants, II. The effect of rain and temperature on germination and growth. *Ecology*, **30**, 1-13.
WENT, E. W., and WESTERGAARD, M. 1949. Ecology of desert plants, III. Development of plants in the Death Valley National Monument, California. *Ecology*, **30**, 26-38.

IX. *The Namib Fog-desert*

1. Precipitation

As a second desert type for detailed discussion I have chosen an extremely arid region on the west coast of Africa. The area receives rainfall only rarely but is characterised many days by fog. This desert region, the Namib, extends in a strip, about 100 km wide, southwards along the Atlantic coast from Angola in the north (south of Mossâmedes, about 15°S), across the Orange river and Little-Namaqualand into western Cape Province (at about 33°S). Along this stretch the Atlantic coast is washed by the cold Benguela current. Winter-rains are still present in the south part of the Lüderitz Bay, while tropical summer-rains are experienced in the middle and northern parts, provided rainfall occurs at all.

The centre region of the Namib, near Swakopmund, is the best investigated ecologically (Walter, 1936).[1]

The mean annual rainfall is less than 10 mm and rain occurs, on average, only once in two years. In certain years, therefore, rainfall is hardly measurable. Very rarely the rainfall amounts to 30 mm. However, in the year 1934 there was a rainfall of nearly 150 mm, when 113 mm fell in March alone and a further 27 mm in April. Such a rainfall has, so far, only been observed this once. I had the opportunity of seeing the effect of this rain the following year, in the rich plant cover that was still present in 1935. However, on my next two expeditions in 1938 and 1953 the normal, almost vegetation-free, desert aspect was restored. The only permanent rivers in this region are the Orange to the south and the Cunene to the north. All others are 'riviers' (dry-valleys), which only rarely function as streams.

The Namib can be divided into several sections (Fig. 198):

1. The southern Namib includes parts south of the Orange River and extends north up to Lüderitz Bay. 2. The middle Namib extends from here to Cape Cross. 3. The northern Namib, north of Cape Cross, is little known. Within the middle section, we are interested only in the area north of Kuiseb rivier, since the southern part is an extensive vegetationless area of shifting dunes (Fig. 198). E. Kaiser (1926), who investigated the geology of the diamond desert south of Lüderitz Bay, has also described the origin of the dunes. The sand is marine in origin but, in part, also consists of the weathered products of the outcropping rocks. The sand is gradually shifted northwards because of the pre-

[1] A research institute was recently installed near Gobabeb, about 115 km south-east of Walvis Bay at the Kuiseb river.

vailing southerly winds. It can be returned to the sea only via the Kuiseb during high floods. However, since the water of the Kuiseb only reaches the sea very rarely (only 8 times in the last century), the dunes are able to cross the Kuiseb at Walvis Bay where it enters the sea. They

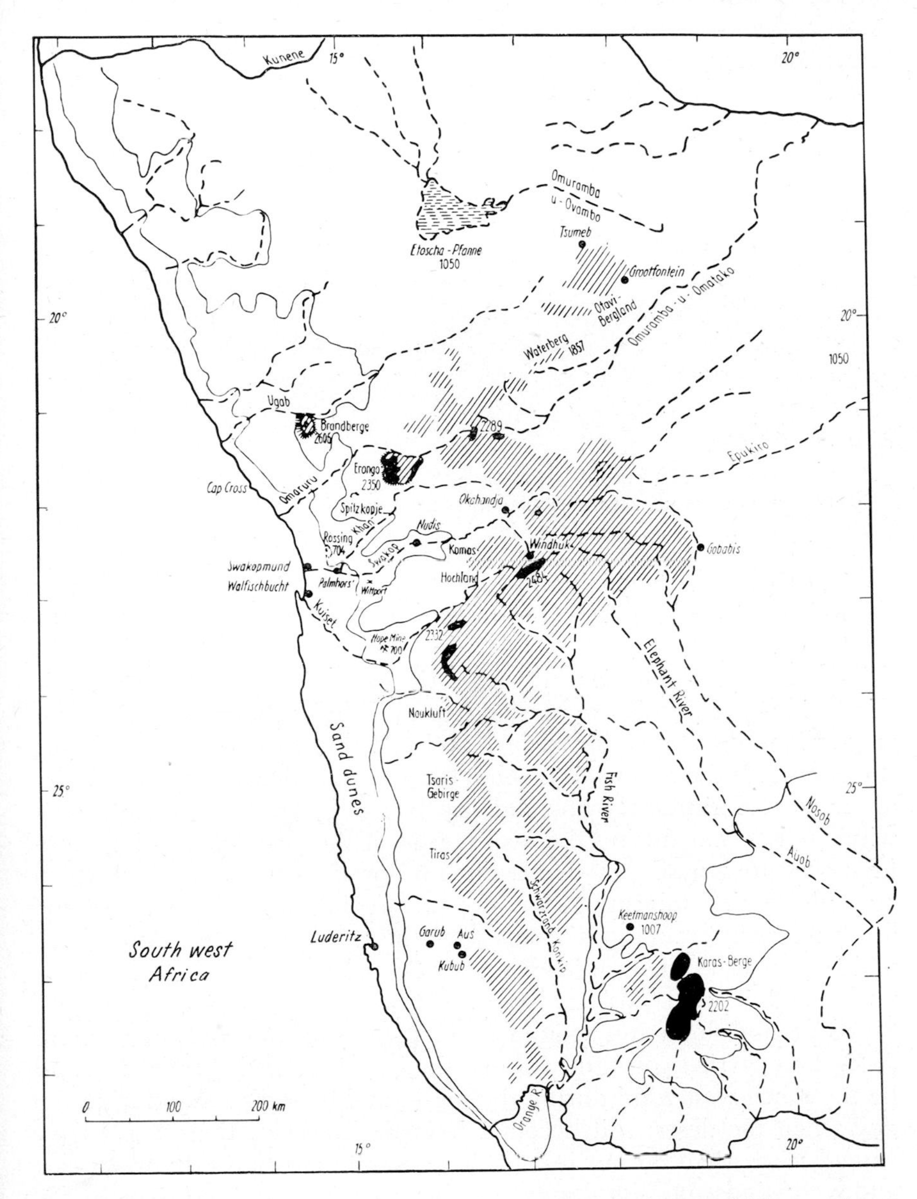

FIG. 198. Map of S.W. Africa showing the Namib desert, which occupies a strip about 100 km wide along the coast. The two contour lines (continuous) in the west correspond to 500 m and 1000 m above sea level, respectively. Hatched—raised transitional zone of the highland (1500-2000 m above sea level); black—areas above 2000 m; dotted lines—riviers; i.e. dry-valleys.

only come to a halt on the right bank of the Swakop rivier.[1] North of the Swakop a shingle and stone desert begins; here sand accumulation and dune formation occur only locally. This is the area that I investigated most closely.

The Namib rises rather uniformly, 10 m per kilometre, from the sea coast to the foot of the African highlands at about 1000 m in the east (Fig. 199). From there the scarp of the highlands rises (disregarding some irregularities) like a wall to over 2000 m at certain places. From

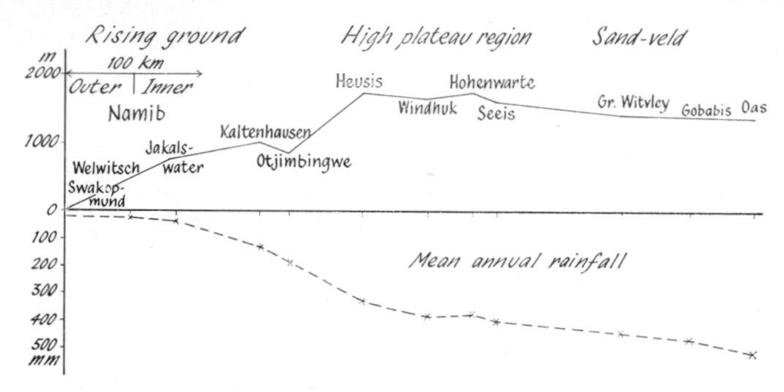

FIG. 199. Profile through S.W. Africa at about $22\frac{1}{2}$°S with rainfall graph. Rainfall decreases from east to west. The Namib is virtually rainless (after H. Walter, 1936).

here further east the highlands fall very gradually towards the large sand-basin of the Kalahari, the western part of which is known as the sand-veld.

The central Namib is well within the range of the south-east Trade winds. However, moisture carried from the Indian Ocean by these air masses is mostly precipitated in Natal and Transvaal before reaching the Namib. If the east wind reaches the Atlantic Ocean as a descending air mass it is extremely hot and dry but this is rarely the case. East winds only occur during the cool season of the year when a region of high pressure is maintained over the interior of the mainland and which coincides with a migrating, low-pressure region over the sea. Usually, a local pressure minimum forms over the strongly heated desert area which then sucks in oceanic air masses from the south-west. This south-west wind is cool and moist, for the sea-water temperature near Swakopmund fluctuates around 15°C as result of the cold sea current. A fog belt carried constantly above the cold current is brought inland by the south-west wind during the night, and only disperses during the day when the desert soil becomes heated. However, these fogs rarely extend more than 50 km inland. The continuous battle between east and west winds and, simultaneously, between the very hot (up to 40°C),

[1] The vegetationless dunes harbour a relatively rich insect world, e.g. the interesting ultra-psammophilous and wingless beetles (Tenebrionidea). They feed on the wind-translocated humus particles, which accumulate on wind-protected dune-slopes. Here the sand can contain up to 36·3% organic matter. In this connection see Koch (1961).

dry desert atmosphere and the oceanic, humid air with little temperature fluctuation, are peculiar features of the Namib desert and although the desert climate is occasionally ameliorated by its proximity to the cold sea current, this does not modify the extreme desert character of the land (Fig. 200). To understand this phenomenon we must consider fog formation in some detail. Fog is usually associated with dew. The

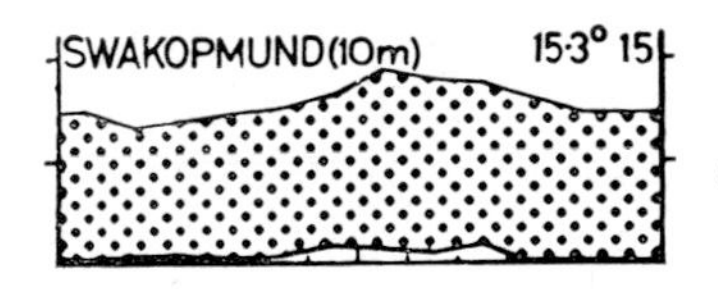

FIG. 200. Climatic-diagram of Swakopmund.

two phenomena are difficult to separate and are here treated together as 'fog precipitation'.

Gülland (1907) reported only 94 days with fog for Swakopmund in the year 1903, while he gave 215 days for 1905.

In 1927-28 143 fog-days occurred which were distributed among the individual months as follows:

VII	VIII	IX	X	XI	XII	I	II	III	IV	V	VI	Year
10	15	22	10	12	8	7	10	13	14	10	12	143

A longer series of observations is available for the Roessing mountains, 35 km from the coast near Swakopmund (1922):

1913						1914												1915	
VII	VIII	IX	X	XI	XII	I	II	III	IV	V	VI	VII	VIII	IX	X	XI	XII	I	II
8	15	14	8	2	—	—	28	23	7	10	12	9	24	30	30	25	7	25	12

This gives 127 fog-days from June 1913 to June 1914 and 204 days for the calendar year 1914.

The frequency of fogs does not decrease in the southern Namib. For example, Waibel (1922) shows for Anichab (30 km north of Lüderitz Bay) 295 dew-days for the period from July 1913 to June 1914, as follows:

VII	VIII	IX	X	XI	XII	I	II	III	IV	V	VI
30	27	22	27	17	27	27	22	28	24	27	17

I am indebted for further observations to G. Boss, who undertook numerous ecological investigations in Swakopmund during his many years as a teacher there and gave me his notes to evaluate (Walter, 1936). During 1934 and 1935, when he made measurements of evaporation, 134 out of 180 days showed fog. However, more important than fog-frequency are the quantities of water precipitated either by fog, or as dew. Here, I shall not give the individual measurements which were done with dew plates, with dry filter-paper packs and measurements of run-off from roofs. The correlation was good in all cases. I present only the general results.

Without undertaking exact measurements there is a tendency to over-estimate fog-precipitation because, at night, water drips constantly from the roofs in Swakopmund. Housewives collect this water, since the tap water is very hard. If one has to stay overnight in the desert, one's clothes are soaking wet in the morning.

However, all the measurements indicated that fog-precipitation only amounts to 0·7 mm at a maximum and, in most cases, it is less than 0·1 mm. This coincides with dew measurements from several different climatic regions.

Upon estimating the total fog-precipitation per annum, one arrives at about 40-50 mm, assuming about 200 fog-days occur. These 40 mm would be of extreme importance to the vegetation, if they were to fall as one or a few showers. However, as fog-precipitation, it is insignificant at least for the plains since the water hardly penetrates more than 2-3·5 cm into the sandy soil and only about 1-2 cm into a loamy soil. The precipitated water evaporates immediately the fog dissipates and the soil becomes heated by the sun. Flowering-plant vegetation does not occur under such conditions. The situation is different where fog, especially driving fog, is precipitated on cliffs. From these water runs off just as if from a roof and is able to penetrate rock crevices, where it is protected from evaporation. Plants can invade such places since a fog-precipitation of 0·1 mm results in a run-off from an area of 5 m^2 of 500 ml.

2. The ecological significance of fog

The question is raised time and again, whether there are plants that can absorb fog-precipitation directly with their leaves, so accumulating enough water to carry out their life-processes. A prerequisite for such an adaptation would be a mechanism to ensure rapid water absorption by the leaves and, simultaneously, to prevent water loss. This applies, for example, to the absorbing scales of the Bromeliaceae (Fig. 80, p. 131) or to the velamen of the roots of epiphytic orchids. However, no adaptation in this direction has been found in any of the plant species occurring in the Namib desert.

Obviously, some water absorption occurs when leaves are wetted, so long as they do not possess an especially thick cuticle or non-wettable waxy layers. For example, wilted plants regain their turgidity on prolonged exposure to dew. However, this does not really mean that they are plants adapted to dew or fog.

We can only regard as *fog or dew plants those that can survive through the wetting of the aerial parts by fog or dew without requiring water absorption through the roots.*

Marloth (1908) has provided data concerning water absorption by the above-ground plant parts in the Karroo. He distinguished water absorption through hairs, through an epidermis capable of imbibing water and through hygroscopic outgrowths and aerial roots in the leaf axils. However, proof of water absorption alone is not satisfactory. It must be shown that *such absorption is sufficient for the survival of the plant under natural conditions.* The peculiar aerial roots of *Cotyledon cristata* and the swollen epidermal cells of *Crassula decipiens* may suggest a similar function to that of the absorbing scales of the Bromeliaceae.

The Namib desert would, in particular, provide an excellent opportunity for the evolution of such a plant type, because measurable

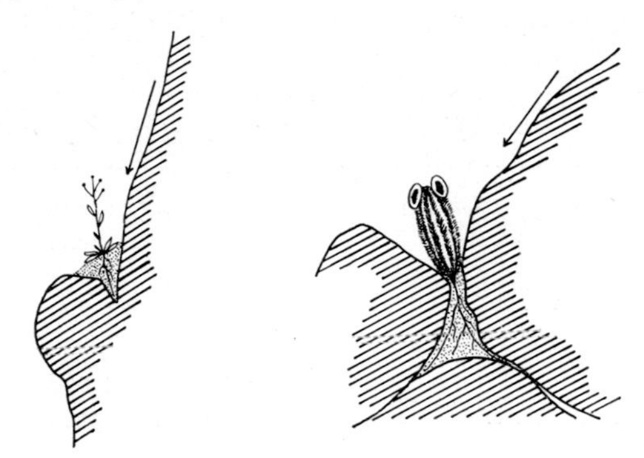

FIG. 201. Plant habitats in rocks determined by fog-condensation in the Namib fog-desert.

rainfall is exceptional in this region, which is characterised instead by great humidity due to fog. In spite of this, fog-plants are absent. The plains of the Namib, along the coast, are completely barren of vegetation except in rain-years and in habitats with ground water. Plant growth only occurs further inland with decreasing fog-frequency and increasing rainfall.

Different conditions occur at rock faces from which the fog-water runs off into crevices (Fig. 201). Here, plants can become established even though rainfall cannot be measured. Their main period of growth occurs during the cool season with frequent fogs and not during the summer, when there may be occasional rainfalls. Amongst them, water-storing succulents are particularly frequent but these plants absorb water in the normal way through the roots. They can grow just as well in other areas where there are no fogs but where they are wetted by occasional rains. Thus, one can here speak of a fog-vegetation, since fog is the factor permitting establishment of the plants at these habitats, but one cannot regard them as *specific fog-plants.* The plants merely require some wetness in the soil and it makes no difference whether this water is provided by rain, by run-off from fog or dew water, or by irrigation.

It is obvious also that when soil-water stress first develops, the hydrature of the plant is eased by the continued wetting of the leaf surfaces with fog or dew or by reduced insolation, or by complete prevention of transpiration. In this way the longevity of herbaceous plants can be extended, which suggests some advantage to the plant. It is, however, rather difficult to give an estimate of the magnitude of this advantage in any specific case.

In his investigations of this problem, Boss made some measurements with whole plants. Their roots were kept in rubber balloons and the plants were left in their natural habitat in the Roessing mountains. Water absorption through the roots was thus impossible. At nights the temperature dropped to 6·5°C and considerable fog and dew formation occurred. He experimented with two Acanthaceae, which were considered to be fog plants and with a thorny, succulent Asclepiad. Single, successive weighings gave the weights (in gms) for the individual plants (Table 53).

TABLE 53

	Petalidium variable	*Monechma deserticola*	*Hoodia currori*
17.VII 1900 hr	157·3	30·2	198
18.VII 0700 hr	166·0	35·5	200
0930 hr	135·5	—	—
1900 hr	123·7	20·0	192·3

It can be seen that the plants dried up rapidly during the day despite considerable water uptake during the night. In *Hoodia* water absorption was probably restricted only to its thorns and the outer epidermal wall.

The following measurements by Boss in September near Spitzkopje, on the inner Namib border, show how rapidly humid and arid conditions can alternate in the Namib:

An east wind prevailed in the evening; at 7 p.m. the relative humidity was 14% (abs. 2 mm) at 17°C. Then a south-west wind occurred, and the vapour pressure increased to 4·4 mm at 9 p.m., to 8·2 mm at 10 p.m.; the relative humidity increased to 94% at 9·5°C and the dew point was attained soon after. The overnight minimum was 2°C and dew formation was very strong. An east wind recurred next morning; the vapour pressure had already decreased again to 2·3 mm by 7 a.m. and the relative humidity to nearly 20% at 8°C.

Fog cannot have any lasting effect under such contrasting conditions. It is quite different in humid climates where the rainfall forms the basis for plant growth. Here, the non-measurable precipitation contributes significantly to the humidity of the climate.

It is often difficult to draw a line between fog and rain when one is dealing with wetting fogs or spray rain. This applies, for example, to fogs on Table Mountain, to the fog oasis, Erkawit, on the Red Sea, in the Sudan, or to the Garua in South Peru (see p. 379).

In all these examples, the humid air is forced to rise up steep slopes where it becomes over-saturated with water vapour so that measurable precipitation is recorded even by the usual rain gauges. However, the precipitation is greatly increased when wetting fogs are blown by the wind against upright obstacles resulting in the moisture condensing on the latter (Walter, 1960, pp. 155f.).

The preceding discussion does not relate to the lichens. They are fog plants in the true sense for they are not dependent upon moisture in the soil. In fact they cover every rock and stone which is relatively resistant to weathering, in the outer, fog-rich Namib while inland they practically disappear with increasing rainfall and low fog frequency.

Vogel (1955) has described particularly strange lichens from the southern Namib. Their thalli are only a few millimetres large. They extend into the soil with strong rhizoids on the underside. They can utilise the fog moisture. The algal layer is on the lower side of the thallus and protected from the effect of too much insolation by a thick, partially pigmented upper cortical layer. Vogel compares these lichens, with good reason, to the 'window plants' among the Mesembryanthaceae, for example *Lithops*, which also occur in the Namib. Perhaps even more remarkable are the 'window-algae', which use transparent quartz-stones as a 'window'. They form a green coat on the underside of such stones; that is, beneath the soil surface.[1] Here they can still obtain sufficient light and at the same time moisture from fogs. Condensed fog runs beneath the stones where the water is retained longer since it is more protected from evaporation (Fig. 202).

3. The vegetation of the Namib and its ecology

We may distinguish by habitats:[2]

(*a*) Plant communities of level plains

(*i*) in the outer, coastal Namib,
(*ii*) in the interior Namib.

[1] Similar algae were mentioned by Beadle (1948) in the 'salt bush region' of Australia. I noticed such algae in the central Namib and in the desert between Cairo and Suez. Such phenomena also occur apparently in the Negev desert (Evenari, personal communication).

From Nevada *Protosiphon cinnamomeus* and *Anacystes montana* are reported, beneath quartz-stones, also N-fixing *Nostoc* species. The N content of these algae is very high (7-8% of the dry weight). No algae were found within 1 km of the nuclear test site as they were probably killed by the heat (see Shields and Drouet, 1962).

According to Follman (1965) in the Atacama desert lichens still contained 44% water six hours after being wetted by dew with a soil temperature of 62°C.

[2] Photographs and a list of plants from the Namib near Gobabeb is given by W. Giess (1962). C. Koch (1963) took photographs of the Kuiseb in 1963 when it was carrying water for the 11th time since 1830.

(*b*) Plant communities of rocks

(*i*) on marble (limespar), quartzite and diabasic ridges,
(*ii*) on gneiss and granite ridges.

(*c*) Plant communities of erosion channels and dry valleys (here called riviers)

(*i*) riviers with slightly brackish water,
(*ii*) riviers with brackish water,
(*iii*) estuaries of the riviers.

(*d*) *Welwitschia* transition zone.

FIG. 202. Window-algae (after Vogel, modified). Algal growth in relation to light conditions beneath the quartz stones (algal layer dark). The algae only occupy a marginal zone on stones with tall vertical axes (left and right); where the thickness is less, they cover the whole underside (centre).

(*a*) *Plant communities of level plains*

The gravel desert occupies most of the outer Namib. Some 50% of the soil surface is covered by a stony pavement, which protects the fine micaceous sand beneath from being blown away. Small sand-tails of wind-blown sand are deposited on the lee side of obstacles. The soil is cemented into a rock-hard layer by deposition of lime and gypsum forming hardpan at a depth of 1-5 cm. Soil wetting normally occurs to this depth. The lime crust is only deeper in those places where slight micro-relief permits better wetting of the soil surface. All the stones are covered with colourful, foliose and crustose lichens, otherwise, the gravel desert is barren of vegetation.

This situation only changed after the heavy rains of 1934. Suddenly numerous plants appeared, their mean cover was 20% and, in little depressions, even 50-90%.

The typical plants were *Mesembryanthemum salicornioides*, *Hydrodea bossiana* (Fig. 203), *Drosanthemum paxianum* and, in addition, the dense cushions of *Aizoon dinteri* and *Zygophyllum simplex*.

Half of the plants were already dead by 1935. It is possible that fog-precipitation contributed to the prolonged persistence of the plants but, in the long run, this form of water supply was insufficient. These plants are all halophilous succulents with extremely low rates of transpiration. All the basal parts of *Hydrodea* were dried up and dead while the apical parts were still fully saturated. Thus, they continued to live on stored water reserves and further water absorption could not occur.

FIG. 203. *Hydrodea bossiana*, an annual, very succulent mesembryanthemum. Flowers yellow (Phot. K. Dinter).

The anatomical features of *Mesembryanthemum salicornioides* were investigated by Zemke (1939) at my institute. The plant consists of pencil-thick stem-segments 1 cm long. Elongate cylindrical, succulent, opposite leaves with an inflated, vesicular epidermis develop at the nodes. However, the leaves soon fall (Fig. 204). Photosynthesis is then carried on by the fleshy, often reddish coloured, young stem segments. Their surface is also covered densely with inflated vesicular cells (Fig. 205). The vesicles appear to be densely appressed but narrow cracks

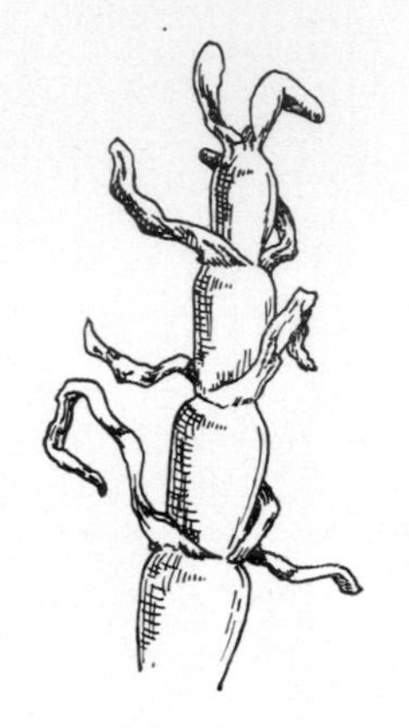

Fig. 204. *Mesembryanthemum salicornioides*. Part of a shoot with wilted leaves.

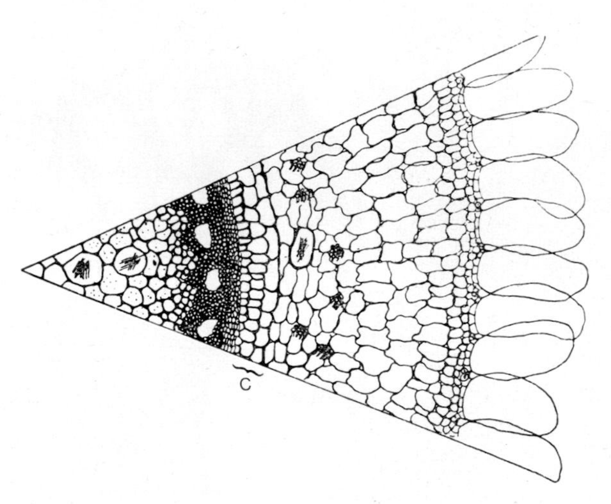

Fig. 205. Cross-section through an old shoot-section of *M. salicornioides*. At the outside, epidermal inflated vesicular cells. *C*, corky endodermis with cork tissue. The cortex dies back to the endodermis (after Zemke).

occur between them which lead to the stomata which number about 54/mm^2 (Figs 206 and 207).

The walls of the vesicular cells are thin and their cuticle is hardly visible. Despite this, cuticular transpiration is very low. Even larger vesiculate cells are found on thc leaves of *Cryophytum* (1·2-2 mm), the cuticle too is even thinner, often being hardly recognisable. The cell sap of the vesicles appears to contain more chlorides than the other cells to judge by their reaction with $AgNO_3$. The number of the stomata ranges from 45 to 50/mm².

The water-storage tissue of old stem segments of *Mesembryanthemum salicornioides* is empty and dies. The endodermis becomes corky and

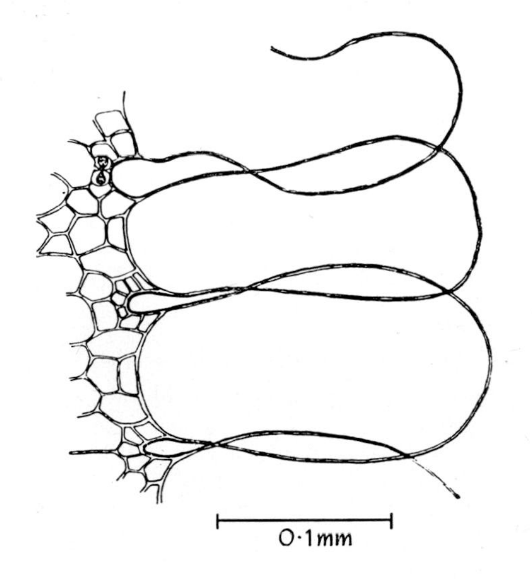

FIG. 206. Epidermal vesicular-cells of *M. salicornioides* with stomata seen in cross-section.

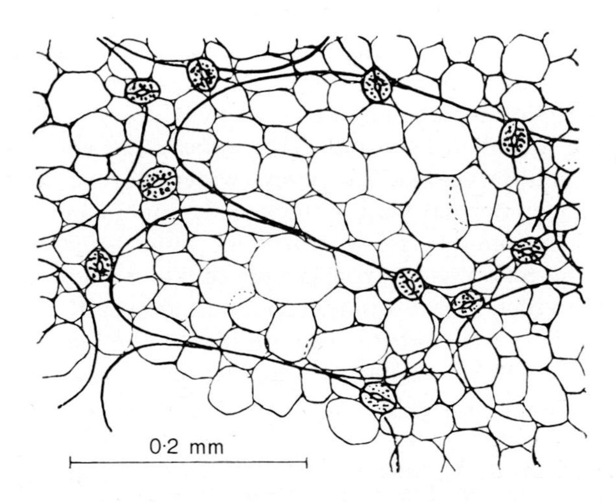

FIG. 207. The same seen from within (after Zemke).

a layer of cork forms within (Fig. 205). The cells of the central cylinder become strongly lignified so that the basal parts of the plant consist only of thin, woody stem segments.

The soil of the outer Namib contains readily soluble salts. In terms of dry weight the fine sand contained 0·274% NaCl and 3·34% of sulphate (mostly gypsum). The lime crust below contained only 0·1% NaCl. When plants germinated after the rain of 1934 the water content of the soil must have been very high.

In spite of the relatively low salinity, the plants store much chloride in their cell sap as do all obligatory halophytes. Osmotic values are shown in Table 54 (number of samples in parentheses).

As is always the case in halophytes, there is no clear relationship between osmotic value, salt content and growth. For example, in *Zygophyllum*, plants with both 16·3 and 33·5 atm showed vigorous growth. Initial signs of necrosis are found only with the highest values. The species just mentioned are ephemerals which only occur after rain. They have—particularly the Mesembryanthemums—their main distribution in the southern Namib, where a colourful cover of flowering plants can be seen after a good winter-rain.

In 1935 I also saw, in addition to these ephemerals, a few perennial species, which are normally restricted to habitats in damp hollows or to

TABLE 54

	Osmotic value (atm)	Proportion of chloride %
Hydrodea (2)	30·5–46·9	70–80
Mesembryanthemum (4)	28·6–31·1	30–40
Aizoon (8)	20·9–52·8	22–52
Zygophyllum simplex (5)	16·3–53·2	66–95

the estuaries of the riviers: *Zygophyllum stapfii* and *Arthraerua leubnitziae*. After the rain of 1934 they also colonised the level plains, where they were quite apparent in 1935 but were lost in subsequent years.

The mesophyll of the succulent leaves of *Zygophyllum stapfii* consists of thin-walled cells, which serve two purposes, water storage and photosynthesis. The outer epidermal wall is thickened, but the cuticle is very thin and the sunken stomata are few in number (about $10/mm^2$). The species belongs to the halophytes. In a sample of 50, the osmotic values ranged from 27·5 to 55·6 atm but plants with values over 40 atm were poorly developed. The contribution of chloride to the osmotic value is very high, about 70-80%. *Arthraerua* (Amarantaceae) has still higher requirements for water and achieves great abundance only on plains that are continually wet. It is a non-succulent species with completely reduced leaves, and strongly branching lateral shoots reminiscent of *Equisetum limosum* (Fig. 208). In a sample of 13 the osmotic values ranged between 24·2 and 33·2 atm. Boss found higher values, over 40 atm; the maximum value was 51·5 atm. However, these values were for the year 1932, thus they fell in the great period of drought. The chloride content is very low, only 9-48%, by comparison with the succulent halophytes. It was found that in contrast to the succulent halophytes, much sulphate is stored in addition to chlorides. It can amount to more than half the proportion of chloride, and it is also accumulated in the wood. *Arthraerua*, therefore, belongs to the class of non-succulent sulphate-halophytes. It is well known that sulphates have a strongly dehydrating action on colloids (Walter, 1960, p. 487).

Arthraerua acts as an efficient sand-trap and the plants always form a crown to dunes 1 m high, distributed across the plain. Wherever such a dune is dug up, one finds buried basal parts of the shoot-system.

FIG. 208. *Arthraerua leubnitziae* (or possibly a related form), Amaranthaceae (Phot. K. Dinter).

Adventitious roots are not developed from the stem in contrast to *Zygophyllum stapfi*, which is also a dune former. Both species have a tap root, which penetrates deep down to the wet soil horizons. Damp subsoil is always present in the estuaries of the large riviers.

According to Boss, the transpiration rate of both these species is about 10 times greater than that of the Mesembryanthemums, and their water deficit is small (around 10%). Thus, they maintain a water balance in damp habitats but have a poor chance in dry ones.

The next problem relates to the cause of the brackishness of the outer Namib. Salt-enriched marine sediments are not involved in the geomorphological make-up of the Namib. Larger salt deposits, as for example near Cape Cross, have originated in part from marine lagoons which have been cut off. However, fog must be thought to be responsible for the brackishness of the flat plains. Strong surf pounds the coast of South-west Africa resulting in the formation of a sea-water aerosol. This can be carried inland by the fog. E. Kaiser (1926) found a chloride content of 244 mg per litre in fog water. This corresponds to 0·4 g. NaCl. If we base our calculation on this figure and assume an

FIG. 209. Inner Namib. Desert plain with *Stipagrostis* (*Aristida*) *obtusa* on the way to the Naukluft; in the rear, mountains and escarpment (Phot. E. Walter).

annual fog-precipitation of 50 mm, i.e. 50 litre/m^2, we obtain an annual addition of 20 g NaCl per square metre (see also Boss, 1941). This is a considerable amount, particularly in view of the fact that redistribution of the salt only occurs to a very limited extent because of the low rainfall.

The salt's origin as blown drift enables one to understand why the outer Namib is brackish as far as the inland limit of coastal fog, while the inner Namib has no saline soils. The change-over occurs about 50km from the coast.

The inner Namib has a very different vegetation. Here, halophytes, such as *Zygophyllum stapfii* and *Arthraerua*, are absent; in their place occur grasses, several *Aristida* species. The density of grasses increases inland until they cover the entire high-plateau (Fig. 209). The sandy soil is alkaline (pH = 7·3-9), but it contains only traces of soluble salts. Dead tufts of grass were noticed on blown sand deposits in the outer Namib in 1935 where they were able to take hold after the rainfall of 1934, but they had already dried up within the first year. The osmotic values of the grasses are low (*Aristida hochstetteriana*, 16·6 atm); their shoots dry out as quickly as the soil. In perennial grasses the dry roots are separated from the surrounding soil by a sheath of sand grains (see p. 242). In associated herbaceous plants (Compositae, Acanthaceae) the osmotic values are also only high in the wilted condition. They belong to the class of malacophyllous xerophytes. Succulent-leaved species are only represented by *Sesuvium digynum* (Aizoaceae). However, its low osmotic values (14·6-18·8 atm, when dying 27·7 atm) and low proportion of chloride (2-15%) show that this species is not a true halophyte.

Trees and shrubs of the savannah and of the dry-forest are often found in favourable habitats, e.g. at the foot of granite-knolls, from which rain water drains, or in the drainage areas of the riviers. The low osmotic value of the tree *Terminalia porphyrocarpa* (23 atm), determined by Boss from its leaves, indicates a favourable water balance.

(*b*) *Plants of rocks*

Rocky habitats are relatively favourable ecological niches in deserts. Even the smallest amount of rain runs from their surface into cracks where it is protected from direct evaporation. This provides a real chance for the development of plants. Apart from ground water areas, these are the only habitats in the outer Namib to provide for the survival of perennials. Particularly good habitats are found on resistant quartzite and diabasic ridges and also on marble ridges (crystalline lime). As a consequence of the numerous recurring fogs, which only bring a small amount of water, the rocks are covered with lichens. The following dwarf-succulents predominate among the flowering plants:

1. Stem-succulents: *Trichocaulon dinteri* (Fig. 210) and *T. pedicellatum*, *Hoodia currori*, *Euphorbia brachiata*.
2. Leaf-succulents: *Lithops* spp., *Hereroa bossii* (Mesembryanthemaceae), *Anacampseros albissima*, *Aloë asperifolia*, *Cotyledon orbiculata*.
3. Stem-succulents with deciduous leaves: *Pelargonium roessingense*, *Sarcocaulon marlothii*, *Othonna protecta*, *Senecio* (*Kleinia*) *longiflorus*, *Adenia* (*Echinothamnus*) *pechuelii* (Fig. 211).

In 1935, in places with a little more soil accumulation there also occurred a few herbaceous species and, here and there, an isolated halophyte.

A similar, even richer succulent vegetation is found among the rocks near Lüderitz Bay, where the Mesembryanthemums are dominant. The anatomy of these strange succulents has often been investigated. The most recent study is by Zemke (1939), who has also cited the older literature.

These 'window-plants' were also studied by Schmucker (1931). Fig. 212 shows the structure of the epidermis covering the 'windows' in young and old leaves. Stomatal openings are rarely or never found. They are usually located in the side walls sunk below the surface where they are over the chlorophyll-containing cell layers. Their number varies from 3-15/mm². They are more numerous in *Hereroa* (23/mm²) whose leaf structure is very different (Fig. 213).

The transpiration rate of these stem-succulents is very low. This applies also to group 3 (p. 353) in the leafless condition. In *Sarcocaulon*

Fig. 210. *Trichocaulon dinteri* in rock-crevice. Witport mountains, Namib desert (Phot. E. Walter).

Fig. 211. *Adenia* (*Echinothamnus*) *pechuelii* (Passifl.) between rocks. Witport mountains, Namib desert (Phot. E. Walter).

the shoots are surrounded by a layer of cork, the cells of which are filled with resin. The resinous coat is retained after death and the plant burns like a smoky candle (hence the name 'bushman's candle'). Leaf-succulents also transpire little. I shall refer to experimental work on transpiration later in my discussion of the Karroo (Fauresmith).

Information on the structure of the epidermis of other succulents is given in Fig. 214 and in Table 55.

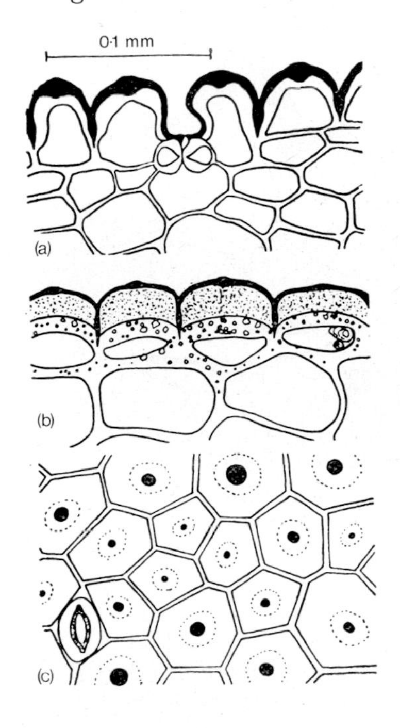

Fig. 212. *Lithops ruschiorum:* (*a*) young epidermis with thickened cellulose wall and cuticle, (*b*) old epidermis; the cellulose wall is impregnated with crystals; dotted—cutinised wall layer, (*c*) young epidermis and stomatas of the upper window surface from above. The spherical thickening of the cuticle can be recognised in the centre of each cell (after Zemke).

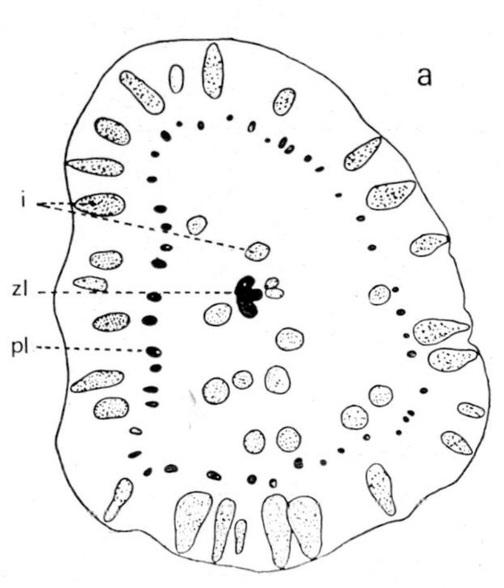

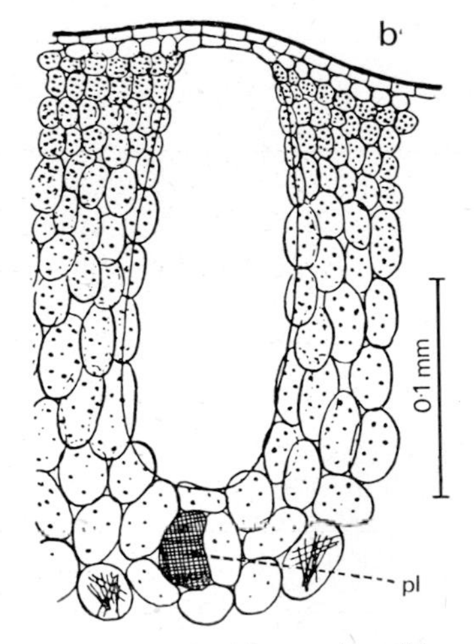

Fig. 213. *Hereroa bossii*: (*a*) cross-section of leaf with tannic acid cells (*i*); (*b*) Tannin-idioblast much enlarged. *pl*—peripheral and *zl*—central vascular bundles (after Zemke).

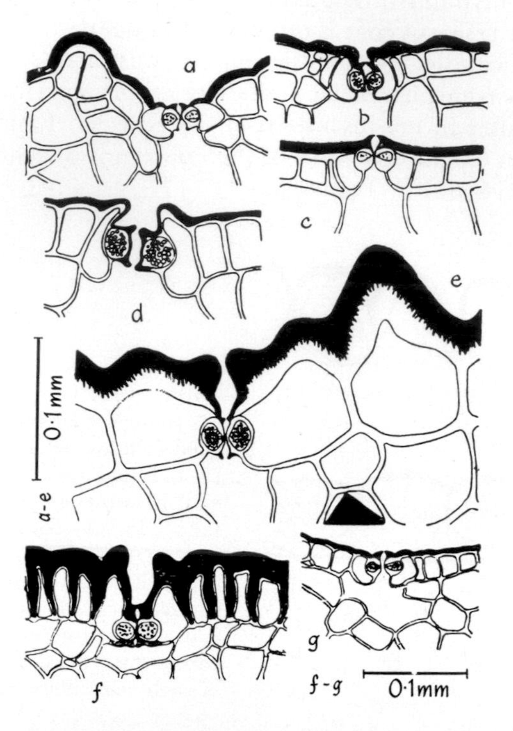

FIG. 214. Epidermis of different succulents (after Zemke): (*a*) *Euphorbia dinteri*; (*b*) *Euphorbia* sp. (Tsumeb); (*c*) *Hoodia currori* (Asclep.); (*d*) *Senecio longiflorus* (Compos.); (*e*) *Aloë asperifolia*; (*f*) *Sansevieria cylindrica*; (*g*) *Caralluma lateritia* (Asclep.). Cuticle shown black.

TABLE 55

	Thickness of the cuticular layer in μ	Thickness of the outer cellulose wall in μ	Stomatal openings No. per mm²	Stomatal openings Size (length × width) in μ
Trichocaulon dinteri	2·5–3·5	2·5	24	26 × 15
Hoodia currori	6	2·5	19	35 × 19
*Caralluma lateritia**	3·5	4–6	17	30 × 15
Tavaresia grandiflora	2	2	17	29 × 12
Euphorbia dinteri	6–8·5	3·5	40	28 × 15
Euphorbia lignosa	6–7	1	22	34 × 13
Euphorbia sp.*	3–6	2–3	23	32 × 14

* Originated from the damper north-east of the region near Tsumeb.

Among leaf succulents, the small *Aloë asperifolia* occurs in the Namib while the tall *A. rubrolutea* has its main centre of distribution in the interior of the country and *Sansevieria* in the north-east of South-west Africa. However, in the drier parts of this area they only grow in sheltered places in the rocks.

The number and size of their stomata can be seen from Table 56.

TABLE 56

	Stomatal openings		
	Number/mm²		Size (length × width) in μ
	above	below	
Aloë asperifolia	25	24	59 × 43
Aloë rubrolutea	41	51	54 × 26
Sansevieria cylindrica	39		53 × 26
*Senecio longiflorus**	25		—

* Regularly leafless in the Namib.

With strong insolation, the succulents may show extreme temperatures of 10-15°C above air temperature.

As one might expect, the osmotic values of these plants are very low (Table 57).

TABLE 57

Osmotic values of rock-habitat plants (Namib, in atm)

Plant species	after Boss	after Walter
Trichocaulon dinteri	5·3–5·7	—
Trichocaulon pedicellatum	6·6–8·6	—
Hoodia currori	5·8–6·4	9·4
Cotyledon orbiculata	2·4 (?)	7·0
Aloë asperifolia	6·6	6·5–8·0
Senecio longiflorus	8·5–10·8	7·2–9·2
Adenia pechuelii (stem)	5·2–8·0	—
Adenia pechuelii (leaves)	12·5	—
Pelargonium roessingense	8·8	8·4 (flowering)
Sarcocaulon marlothii	—	9·7–12·3
Euphorbia brachiata	—	11·8
Othonna protecta	21·1 and 21·4	—

Chloride was found in the extracted sap of all these plants but only in very low concentrations. An exception are the Mesembryanthemums (*Hereroa*, *Lithops*). Their osmotic values were 14·5-16·6 atm and the proportion of chloride was about 25%. Thus, they show pronounced halophytic tendencies.

The vegetation on granite, gneiss and mica-schist ridges is very different from that on the hard rock ridges. The softer rocks undergo much greater mechanical weathering resulting in the accumulation of large masses of detritus. Small succulents cannot maintain themselves on such substrates so that here one only finds the large, pillar-forming *Euphorbia dinteri*, which is branched at the base, and the candelabra-like *Aloë dichotoma*. The latter species is particularly characteristic of the desert-like areas further inland in the south of South-west Africa (Fig.

FIG. 215. *Aloë dichotoma* with bushy spurges (*Euphorbia*) near Keetmannshoop, S.W. Africa (Phot. E. Walter).

215). There, great areas are covered with bushes of *Euphorbia* 1 m tall, with thin, cylindrical shoots (Fig. 216). In addition, there are also halophytes. Otherwise, their ecological relations have not been further investigated to date.

A very peculiar arborescent life form occurs in rocky habitats near the inner border of the Namib. These are reminiscent of *Idria* and the Burseraceae of the Sonoran desert.[1] Burseraceae are also represented here, such as *Commiphora glaucescens*, *C. oblanceolata* or *Euphorbia guerichii*,

[1] Also compare these with similar types on the island of Socotra (Popov, 1957).

which looks deceptively like the former. Additional species are *Cissus cramerianus* and *Moringa ovaliifolia*. To the south, in the area of the Orange River, *Pachypodium namaquanum* occurs. The same types are found on dry lime and dolomite inland in the Otavi mountains (Figs 217 and 218).

In the marginal area of the desert one finds poikilohydrous ferns, such as *Actiniopteris*, *Notholaena* species (Fig. 219), and the phanerogamous *Myrothamnus flabellifolia* (Figs 220 and 221). This shrub, of about 50 cm, is distantly related to the Rosaceae. It may form a dense cover on

FIG. 216. Plain with bushy spurges (*Euphorbia gregaria*); in the rear the low Karas Mountains, S.W. Africa (Phot. E. Walter).

stony slopes and normally looks like a dry broom. However, only a little rain is required when the shrub becomes covered within half an hour with small leaflets, bright green on the dorsal side. Even after transporting it dry to Stuttgart the leaf cells could still be plasmolysed after a year. However, the twigs failed to form roots. The opposite leaves are by no means xeromorphic (Fig. 222). The epidermis is composed of large and small cells. These are arranged alternately on the upper and under sides. As a result the leaves become plicately folded (Fig. 223). On wetting, the leaves gain 62% both in weight and area. This anatomical structure does not explain the capacity of this flowering plant to endure complete desiccation (Walter and Kreeb, 1970). The number of stomata, which occur mostly in the furrows, is $64/mm^2$ on the ventral side and $78/mm^2$ on the dorsal side.

Ficus species are found in this area at sites with temporary springs, e.g. *F. guerichiana* which creeps over rock-walls. Between blocks, the

Fig. 217. *Cissus juttae* with a succulent stem growing among limestone rocks. Graslaagte farm, near Otjiwarongo, S.W. Africa (Phot. E. Walter).

Fig. 218. *Pachypodium lealii* (*giganteum*), Apocyn., in the background *Commiphora berberidifolia*, in the dolomite mountains on Heidelberg farm near Tsumeb, S.W. Africa (Phot. E. Walter).

rather sizeable tree, *Sterculia guerichii*, with large, delicate leaves grows frequently.

A very peculiar water plant can be found in small pans resulting from weathering in the granite mountains. After rain, they usually contain water for only a few days. The soil is covered with a layer of sand,

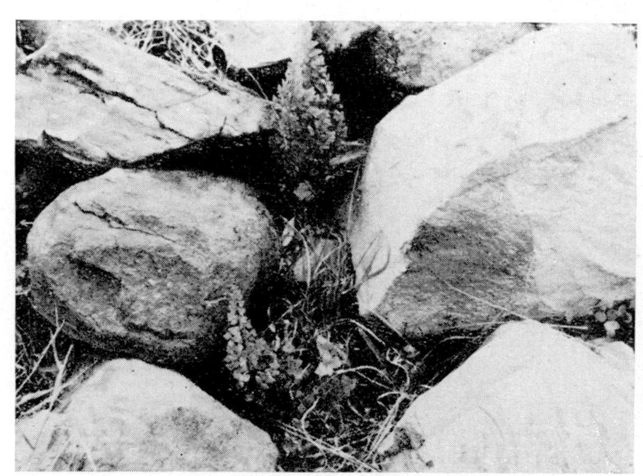

FIG. 219. *Pellaea hastata*, a poikilohydrous fern among rocks. Voigtland farm, S.W. Africa (Phot. E. Walter).

1-2 cm thick, which is densely covered by *Chamaegigas intrepidus*. Only the roots and the shrunken, tuber-like, submerged leaves of this species remain alive in the completely dry condition of the drought. Upon wetting, the leaves swell rapidly and attain 10 times their length after only a few hours. At the same time, they become green. Thereafter, floating leaves and flowers develop from buds within a few days. The seeds have ripened before the water has completely evaporated (Heil, 1925, Walter and Kreeb, 1970).

(*c*) *Vegetation of erosion channels and dry valleys*

As in all deserts, erosion channels and dry valleys (locally termed riviers) are the most favourable habitats. During intermittent flow of the riviers, the sand in the dry valleys becomes wetted in depth and then remains damp for several years from a little below the surface down.

The rivier-water may either be enriched with salts or have only a low salt content. This depends on the site of origin of the riviers and the geomorphological characteristics of their catchment areas. Smooth rock surfaces provide a rapid run-off, salts are carried along in the first wave of flood water, while subsequent run-off is rather low in salts. If, on the other hand, water seeps slowly from gravel or sandy strata to the surface, it evaporates considerably and the remaining water is increasingly concentrated. The schematic profile transect through the basal slope of a small valley illustrates these relationships (Fig. 224).

FIG. 220. *Myrothamnus flabellifolia* after a short rain with fresh green leaves. Naukluft, S.W. Africa (Phot. E. Walter).

FIG. 221. *Myrothamnus flabellifolia* in dry state on mica slate rocks. Escarpment in the Namib Desert near the highway to Walvis Bay, S.W. Africa (Phot. E. Walter).

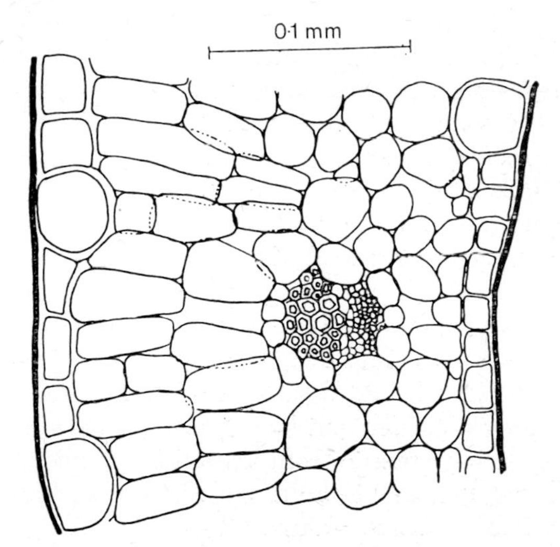

FIG. 222. Cross-section of leaf of *Myrothamnus flabellifolia* (after Zemke).

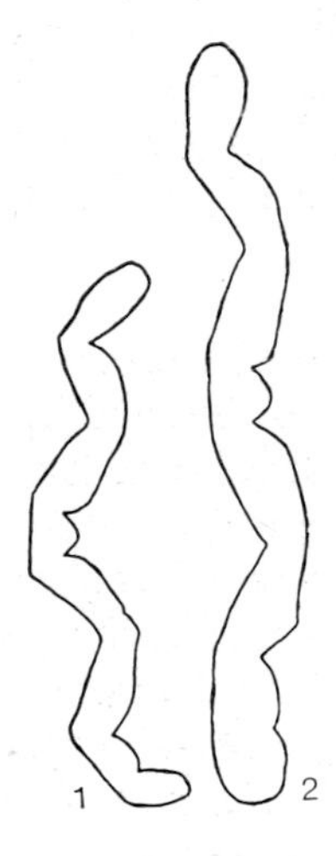

FIG. 223. Cross-sections of leaves of *Myrothamnus flabellifolia*; 1, dry; 2, fresh.

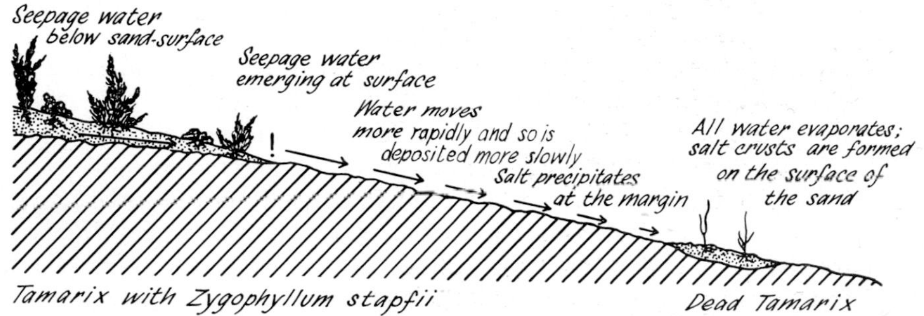

FIG. 224. Salt deposition in an erosion channel below a seepage outflow in the Namib (schematic, after H. Walter, 1936). Dotted—sand; hatched—rock.

Water seeps from the sand wherever the underlying bedrock outcrops. The sand itself is well wetted and covered by halophytes. The water then continues to seep along channels on the rock surface; evaporation occurs and the salt concentration increases so that some salt begins to precipitate at the margin. The rate of water movement decreases (indicated by the decreased lengths of the arrows) through continued evaporation. Finally salt is deposited in the centre of the channel and seepage ceases so that further down the channel remains dry. The zone of salt deposition gradually moves up-slope with decreasing amounts of seepage. The margin of the salt crust alters with the volume of water flow and evaporation, resulting in several banded salt zones. The pattern is best described as a miniature delta formed by salt. Evidence for an earlier and larger flow is provided by a few dead *Tamarix* plants which emerge from the midst of the salt-crusts. These plants must have become established when the then faster flowing water had not become so concentrated.

The same development, here seen on a small scale, is evident in the vicinity of many large salt pans in desert regions. In dry valleys too the ground-water flow always has a lower salt content in its upper reaches than in its lower reaches. Water is constantly evaporating as it flows wherever the soil surface is kept damp by capillary water. Free water need not be apparent at the surface for evaporation to take place.

A permanent ground water stream or lake is only found beneath the large riviers. In their upper part and in the erosion channels storage water only is retained by the soils. This, however, is sufficient to ensure the establishment of some vegetation. The greater the water stored, the taller the plants which endure the long dry spells. Thus, an equilibrium becomes established between the transpiring surface and the available water in these situations also. In small erosion channels with non-brackish water one finds *Citrullus ecirrhosus* and a number of annual herbs which are widely distributed across the country during the rainy season (*Tribulus* and *Cleome* species). Where the amount of storage water increases, certain shrubs join the association, such as *Bauhinia pechuelii*, *Parkinsonia africana*, *Commiphora dulcis*, and possibly certain *Acacia* species also. Only where ground water becomes readily available do *Rhus lancea* and *Salvadora persica* occur, and the trees *Euclea pseudebenus*, *Acacia giraffae* and *Acacia albida*. The latter form regularly spaced gallery forests with wide trunks along the margins of large riviers.

Introduced species easily become established in the riviers, e.g. *Nicotiana glauca* and *Argemone mexicana*, also ruderal species (*Ricinus*, *Datura metel*). A characteristic plant of sand dunes with subterranean ground water in the lower Kuiseb is the strange cucurbitaccaeous, *Acanthosicyos horrida*, known as the Naras gourd. It is leafless and covered with green thorns whose fruits are the preferred food of the natives and whose seeds are sold as a substitute for almonds (Fig. 225).

Many herbs which have their main centre of distribution in the east penetrate deep into the desert along the riviers.

FIG. 225. *Acanthosicyos horrida* (Naras) with fruits on sand dune, lower course of the Kuiseb Rivier, Rooibank, near Walvis Bay, S.W. Africa (Phot. E. Walter).

Table 58 shows that the density of the stomata in these plants is far greater than in succulent species.

TABLE 58

Number of stomata per square millimetre on the upper side (first number) and lower side (second number) of the leaves

Lyperia litoralis	56 56	*Helichrysum roseo-niveum*	94–126
Justicia arenicola	119–110	*Tripteris arborescens*	136–132
Petalidium variabile	193–183	*Gazania varians*	64–150

These rather high numbers indicate that these plants must have an active gas exchange with a favourable water supply.

Wherever the water becomes brackish the salt-excreting *Tamarix* shrubs are found, together with *Lycium tetrandrum*. On flood plains with ground water in the estuaries of the riviers typical halophytes occur. These are distributed in relation to increasing wetness and salinity as follows: *Zygophyllum stapfii*, *Cryophytum barkleanum*, *Arthraerua leubnitziae*, *Salsola* species, *Suaeda fruticosa*, *Arthrocnemum glaucum*. In depressions lacking outflows, resulting from blow-outs caused by the wind (Kaiser, 1926), salts accumulate to such an extent that they are barren of vegetation or even become covered by a solid deposit of salts: a clear demonstration that lateral salt transport always occurs in the desert from higher land into hollows after rain, even though it falls but rarely.

The formation of a surface salt crust gives no indication of the salt content of the ground water. Wherever ground water is less than 1 m below the surface, capillary water keeps the surface damp and continuous evaporation occurs, so that salt is left behind and deposited as a white crust. Sodium carbonate formation together with a black crust develops where the soil contains organic matter.

Around springs with slightly brackish water, salt swamps occur with *Phragmites*, *Diplachne paucinervis*, *Sporobolus pungens*, *Juncellus* (*Cyperus*) *laevigatus*, etc.

The plants of the erosion channels and dry valleys are not deeply rooted. They only have to reach the wet layers of sand. Even in the large riviers tree roots seldom extend deeper than 1-2 m, but instead they extend horizontally for many metres.

A napiform root of *Citrullus* was found to reach 15 cm where it branched and the finer roots extended downward at an angle. At 40 cm, the sandy soil contained 1·3% water, traces of chlorides and nitrates but no sulphate. The osmotic pressure of the soil solution was less than 1 atm. The cell sap concentration of the *Citrullus* leaves was 13·6 atm, whole shoots showed a value of 10·8 atm. The proportion of chloride was usually below 20%. Osmotic values were found to be between 12-18 atm in all other herbaceous rivier-plants in non-saline localities and only rarely did they exceed 20 atm. Shrubs with succulent stems showed values around 13 atm; all other shrubs showed higher values, between 20-22 atm. The only exception was the herbaceous *Lyperia litoralis* (Scrophulariaceae), which showed very high values of 35-50 atm. Chlorides alone were responsible for 23-26 atm. This plant again provides evidence for the observation made repeatedly that halophytes also store chlorides in non-saline habitats. *Zygophyllum stapfii* and *Arthraerua* are found in these non-saline rivier-habitats but they are more abundant and better developed in adjacent brackish rivier habitats.

All plants typical of non-saline riviers wilt very shortly after cutting. Thus, they have a high rate of transpiration. Yet, the low osmotic values indicate that their water balance remains stable without difficulty. Furthermore, the presence of woody plants showed that they had not become established as a consequence of the rainy season in 1934, since they had also survived the preceding drought in the years 1930-33. By contrast, the herbaceous plants disappear when the stored water in the erosion channels is used up and is not followed by rain for a long period.

The salt relations of the large riviers are rather complicated. The Swakop river rises in the savannah area to the east. When the ground water from wells at different locations along the course of the Swakop is sampled one finds that the originally salt-free water becomes increasingly brackish. Table 59 shows the chloride and sulphate content in grams per litre of well water.

TABLE 59

	Sampling location along Swakop rivier		
	Upper end (Okahandja)	Central part (Nudis)	Estuary (Swakopmund)
Chloride (NaCl)	0·008	0·23	2·28
Sulphate (Na_2SO_4)	0·012	0·05	0·39

The increased salt concentration has two causes:

1. Wherever ground water wets the surface soil of the rivier-bed, vigorous evaporation occurs. The significance of this evaporation is shown by the following observation. At the lower end of the rivier-bed, near Palmhorst, a sill forces the ground water to the surface so that every morning it runs as a rather broad stream. It falls off more and more during the course of the day and, in the evening, one can cross the rivier-bed without getting wet feet. Next morning, the broad stream is restored. Evaporation during the day, therefore, exceeds the water flow. This high evaporation naturally results in a corresponding increase in salt concentration. A similar type of salt accumulation also occurs on steep slopes and rivier terraces (see Fig. 172, p. 294).

2. The second reason for the increased salinity is that brackish water is fed in from tributaries. Whenever it rains in the Namib the flood sheet carries all the salt accumulated on the knolls into the riviers, which further transport it into the Swakop. Because of this, the first rains give rise to very salty seepage water but the salt content decreases with continued rain. During the rainy period of 1934 the salt content of the residual ground water of the Swakop, left from the preceding period of drought, fell from 2·27 g/l to 0·49 g/l at Swakopmund. However, since the specific gravity of water differs depending on whether it is salt-enriched or low in salts, complete mixing of the ground water does not occur. This explains why one can find streams of ground water with very different salt contents immediately adjacent to one another. The same explanation also applies to neighbouring wells, which often contain water with quite different salt contents.

In the estuary of the Swakop, tidal influences can be detected in the sedimentary plains which resulted from the flood water of 1934. There the ground water occurs at 62 cm and the soil is damp up to the surface. Little sand dunes form around individual plants. The sedimentary soil contains much organic matter, H_2S forms in lower horizons and the soil darkens due to ferrous sulphide. The plants root to a depth of 10-30 cm.

The vegetation is made up of halophytes or salt-tolerant species for which the osmotic value and content of chloride were determined (Table 60).

TABLE 60

	Osmotic value (atm)	Chloride proportion (%)
Shrubs		
Tamarix articulata	12·8	75 with excreted salt
Lycium tetrandrum	19·6	60
Suaeda fruticosa	26·5	50
Succulent halophytes		
Arthrocnemum glaucum	28·0	75
Grass-like plants		
Diplachne paucinervis	23·3	55 with excreted salt
Phragmites communis var.	15·1	35
Juncellus (*Cyperus*) *laevigatus**	13·8	60
Herbs		
Heliotropium curassavicum	12·1	60
Nesaea fleckii	16·7	40

* On the coast of Florida, Harris (1932) obtained 15·0 atm and 70% chloride.

Of these species, the non-succulent *Tamarix* again belongs to the sulphate-halophytes because the plant cell-sap contains much sulphate in addition to the chlorides.

Salt accumulates at the soil surface by evaporation. The osmotic value of the soil solution was 75 atm at 1 cm, but only 3·6 atm at 30 cm. The plants, therefore, have sufficient water, but germination can only occur in those seasons when the upper soil layer has a low salt content.

A similar flora is found at the coast on sand and decaying algae. However, here more nitrophilous plants occur in addition, such as *Atriplex muralis*, *Datura metel*, *Nicotiana glauca*.

On damp brackish soil the very succulent mesembryanthemum *Cryophytum barkleyanum* frequently occurs. It has large leaves covered with millimetre-sized, vesicular epidermal cells which look like crystals. Its osmotic value was 22·8 and 26·2 atm and the chloride content 40 and 65%, respectively. *Arthrocnemum* has leafless, green shoots reminiscent of *Salicornia*. Its cuticle is 5·4μ thick, the stomata are sunken and number 81/mm^2.

The *Salsola* species which resemble semi-shrubs are less succulent. Among them *S. zeyheri* is very pubescent and shows a xeromorphic-like habit. Species of *Salsola* form a cover on periodically flooded soils further inland although they are usually very dry and recall the Takyrs of the loam deserts in the southern U.S.S.R. (Ganssen, 1963). The osmotic values of these species were 48·3 and 61·9 atm, their percentage chloride 40%; they also store a good deal of sulphate. These values are not high for halophytes but this is understandable since, in 1935, their conditions were still very favourable.

Boss found much higher values in 1932.

TABLE 61

	Maximum osmotic values (atm)
Suaeda fruticosa	44·3
Salsola zeyheri	74·0
Arthrocnemum glaucum	33·0
Cryophytum barkleyanum	40·8
Heliotropium curassavicum	32·4

Still higher values are reported from Beni-Unif (Algeria, Sahara) by Killian and Faurel (1933, 1936) and from Montpellier (South France) by Adriani (1934).

TABLE 62

	Beni-Unif osmotic values (atm)	Montpellier osmotic values (atm)
Suaeda fruticosa	26·2–51·7	80 (max).
Arthrocnemum glaucum	41·2–49·8	40–100

This section has attempted to describe the anatomical and morphological features of several life-forms from the Namib. From it we can see how the variety of life-forms within a single macroclimatic area is often greater, because of the many different habitats, than in similar ecological niches in quite different macroclimatic regions.

(*d*) *The ecology of Welwitschia mirabilis*

This plant is the most famous of the Namib desert. In a taxonomic sense it is quite isolated among the gymnosperms (Markgraf, 1926). Morphologically it is unique with only two leaves that follow the cotyledons. These leaves have an unlimited capacity for growth even though plants may persist for more than 100 years (Fig. 226). The plant is endemic to the Namib, where it occurs only in a narrow strip some 50 km from the coast at the Hope-Mine between Kuiseb and Swakop. From here the population extends with interruptions to Mossâmedes in south Angola. The ecology of this plant has often been misinterpreted so that I shall discuss it more closely.

In the case of a male plant with an apical disk some 18-27 cm diameter, the turnip-shaped head of the root reached 20 cm above the soil and 30 cm into the soil. At this point it tapers abruptly and continues unchanged as a 1 cm-thick tap root, for at least 60 cm.

According to Boss the roots extend no deeper than 1-1·5 m. In addition to the tap root, which branches lower down, the plant also has lateral roots, which branch from the root head at 7-10 cm. These are spread profusely beneath the soil surface.

Thus, the plant has two distinct root regions. The lateral roots serve for water absorption after light rain which fails to penetrate

FIG. 226. Male plant of *Welwitschia mirabilis*, in flower. Diameter of the bi-partite apex 80 cm. Namib, 20 km inland of Swakopmund (Phot. Schultze-Jena).

deeply and the tap root absorbs soil water from lower layers. However the distribution of *Welwitschia* is in no way related to the presence of ground water. Moreover, absorption of dew or fog precipitation by its surface roots does not seem likely in *Welwitschia* since the plant occurs on the border between the outer and inner Namib and the fogs rarely extend so far inland. Light summer-rains can, however, be expected more frequently in this area.

The typical habitats for *Welwitschia* are the high plateaus of the outer and inner Namib. But it is not the plains as such that are its characteristic habitat, for these become covered with grass after a good rain, but the riviers which rise in this area as broad, flat channels hardly discernible in the plains. These channels receive extra water after heavy rain in the form of flood sheets from the higher areas so that the soil may become damp to a depth of 1·5 m. This sub-surface moisture is retained for years. It serves as a water reservoir for *Welwit-*

schia, which occurs as widely scattered individuals in these flat channels. The distance between the plants is usually over 20 m.

In 1935, many dead grass tussocks, scattered *Arthraerua* and *Zygophyllum stapfii* and a few non-halophytes occurred in addition to *Welwitschia*. In following the course of the channels one soon comes across the typical rivier vegetation with *Bauhinia*, *Parkinsonia*, *Salvadora* and, in brackish areas, *Tamarix*. These areas contain much more soil water than the others.

A dug-up *Welwitschia* showed the soil profile given in Table 63.

TABLE 63

Depth	Kind of soil	100 g dry soil contained (g)			
		Water	NaCl	Sulphate	$NaNO_3$
0–5 cm	coarse gravel	0·19	0·004	0	0·010
40 cm	coarse gravel	2·3	0	0	0·002
60 cm	hardpan	5·9	0	gypsum	0·005

The upper layers were dry as dust. By contrast, the coarse sand at 60 cm felt distinctly wet on handling. In other places also the root region was found to be wet at 55-60 cm depth. Here the soil contained 0·008 g NaCl.

Welwitschia also occurs in other habitats. For example, on slopes between coarse stone scree: in one case, 10 m above a bare erosion channel, in another, almost 100 m above the bottom of the slope. One example was found in a weathered pegmatite-crevasse. However, these last habitats are not typical; the plants are completely isolated and are not very vigorous. It is perhaps possible that they may receive run-off fog-water here.

A certain amount of water can be stored in the turnip-like part of *Welwitschia*, in spite of the absence of specialised water-storing tissue.

With lack of water the leaves become desiccated to their bases and only the meristematic layer remains active, regenerating new leaf tissue with the return of favourable water relations. I marked a leaf: after 3 weeks it showed a 10 cm increment after a rainfall of 16·5 mm. Boss estimated the annual increment as 30-40 cm. Individuals transplanted to Swakopmund showed a 15 cm increment in one year.

In 1934 the leaf growth of *Welwitschia* was unusually great but the inflorescences were attacked by mould (cf. *Aspergillus niger*). *Welwitschia* is dioecious, pollination being carried out by the Hemipteran *Odontopus sexpunctatus*. Reproduction of the species is poor and many seeds are empty. The sound ones germinate in 8-10 days. In spite of this Boss only noted 5 seedlings among a large number of adult plants. This may be related to the salinity and rapid drying-out of the upper soil layers. *Welwitschia* leaves are browsed by wildlife or grasshoppers in extreme cases.

Information on the anatomical structure of *Welwitschia* is given in Fig. 22 in H. Walter (1936).[1] The leaf has an extremely xeromorphic (sclerophyllous) structure. It is strengthened by many non-woody sclerenchyma fibres occurring in bundles and by large sclereids with small oxalate crystals in their walls. The cuticle is very thick and the numerous stomata are sunken. Their number on the upper side is 98/mm^2, on the lower side 144/mm^2.

This results in a high transpiration rate when the stomata are open and also a considerable reduction during conditions of reduced water supply. At times of water scarcity, its conservation is aided by the reduction of the leaf area through necrosis from the tips. Detached leaves show some signs of wilting. With an evaporation rate of 1·55 ml/h at noon, a leaf section (cut ends sealed) lost 16·5% of its fresh weight in 80 minutes. Cannon (1924) has shown by the cobalt paper method that the surface transpiration of *Welwitschia* is ten times that of succulents and corresponds to that of sclerophylls.

The structure of the leaf does not lend itself to absorption of dew and fog precipitation by the surface, as has been suggested repeatedly in the literature. The leaves are unwettable and when fog develops one can see numerous droplets on them as if on the surface of a sheet of metal in the morning. Only the stem disk, which is covered with cork, can absorb water like a sponge. Whether this has any ecological significance still needs to be investigated. *Welwitschia* has in common with sclerophyllous plants the relative constancy of its osmotic value (hydrostable type). Even drought and rain years only cause small variations. Boss determined the osmotic values (Table 64).

TABLE 64

	Old plant (atm)	Young plant (atm)
Drought-year 1932	39·6	35·4
Rain-year 1934	34·8	34·9

Table 65 shows data that I obtained on the composition of the expressed cell sap of *Welwitschia mirabilis*.

The analysis shows that the cell sap contains chlorides but its proportion, 20-25%, shows that *Welwitschia* is not a halophyte but that it can tolerate some salt. Sulphates are also present in some quantity amounting, in atm, to 1/5 of the chloride, i.e. about 4-5% of the osmotic value. The chloride content of halophytes growing adjacent to *Welwitschia* amounted, in the case of the sulphate-halophyte *Arthraerua*, to 35% and in the case of the chloride-halophyte *Zygophyllum* to

[1] See also Rodin (1958) who, with Zemke (1939), cites the earlier literature.

81-95%. This seems to clarify the essential ecological features of this strange plant.

TABLE 65

	Osmotic value atm	Chloride N/100	Chloride atm	Chloride %	Monosacch. mg/ml	Disacch. mg/ml	Sugar atm
17.2. *Welwitschia*-arca							
Leaf-strip 1. basal half	**32·3**	15·9	**6·6**	20	20·6	7·4	**3·1**
Leaf-strip 2. apical part	**33·5**	18·3	**7·7**	23	31·0	9·7	**4·6**
17.3. Road to Witt-port near the Zebra Mts.							
Whole leaf-strip	**34·0**	20·4	**8·5**	25	36·9	6·9	**5·2**
Same from another plant	**33·4**	17·5	**7·3**	22	37·0	2·3	**5·0**

Large individuals can exceed 100 years. A plant with a diameter of 1·5 m was found to be 500-600 years old as determined by the ^{14}C method (Herre, 1961). The stem of this plant was half split apart and only one leaf was still alive on one side.

References

ADRIANI, M. J. 1934. Recherches sur la synécologie de quelques associations halophiles méditerranéennes. *Sigma*, **32**.

BEADLE, N. C. W. 1948. *The vegetation and pastures of western New South Wales.* Sydney.

BOSS, G. 1941. Niederschlagsmenge und Salzgehalt des Nebelwassers an der Küste Deutsch-Südwestafrikas. *Bioklim.*, **1**.

CANNON, W. A. 1924. General and physiological features of the vegetation of the more arid portions of southern Africa. *Carnegie Inst. Wash. Publ.*, **354**.

FOLLMANN, G. 1965. Fensterflechten in der Atacamawüste. *Naturwissenchaften*, **52**, 434-35.

GANSSEN, R. 1963. *Südwest-Afrika, Böden und Bodenkultur.* Berlin.

GIESS, W. 1962. Some notes on the vegetation of the Namib Desert. *Sci. Pap. Namib. Des. Res. Stat.*, **3**.

GÜLLAND. 1907. Das Klima von Swakopmund. *Mitt. dtsch. Schutzgeb.*, **20**, 131.

HARRIS, J. A. 1932. *The physico-chemical properties of plant saps in relation to phytogeography.* Minneapolis.

HEIL, H. 1925. *Chamaegigas intrepidus* Dtr., eine neue Auferstehungspflanze. *Beih. Bot. Zbl. I*, **41**, 41-50.

HERRE, H. 1961. The age of *Welwitschia bainesii* (Hook f.) Carr: C^{14} research. *J. S. Afr. Bot.*, **27**, 139.

KAISER, E. 1926. *Die Diamantwüste Südwestafrikas.* Vol. 2. Berlin.

KILLIAN, CH., and FAUREL, L. 1933. Observations sur la pression osmotique des végétaux désertiques et subdésertiques de l'Algérie. *Bull. Soc. Bot. Fr.*, **80**, 775-8.

KILLIAN, CH., and FAUREL, L. 1936. La pression osmotique des végétaux du sud Algérien. *Ann. Physiol.*, **12**, *5*, 859-908.

KOCH, C. 1961. Some aspects of abundant life in the vegetationless sand of the Namib Desert dunes. *J. SW Afr. Sci. Soc.*, **15**, 76-92.

KOCH, C. 1963. *Der Kreis.* 2/3. Windhoek.

MARKGRAF, FR. 1926. *Welwitschiaceae.* In Engler-Prantl, *Natürliche Pflanzenfamilien.* 2nd ed. **13**, 419-29.

MARLOTH, R. 1908. Das Kapland. *Wiss. Ergeb. Dtsch. Tiefsee- Exped.*, **2**, Teil 3, 1-436.

POPOV, G. B. 1957. The vegetation of Socotra. *J. Linn. Soc. Lond. Bot.*, **55**, 706-20.

RODIN, R. J. 1958. Leaf anatomy of *Welwitschia,* I and II. *Amer. J. Bot.*, **45**, 90-5, 96-103.

SCHMUCKER, TH. 1931. Ökologie der Fensterblätter. *Planta,* **13**, 1-17.

SHIELDS, L. M., and DROUET, F. 1962. Distribution of terrestrial algae within the Nevada test site. *Amer. J. Bot.*, **49**, 547-55.

VOGEL, S. 1955. 'Niedere Fensterpflanzen' in der südafrikanischen Wüste. *Beitr. Biol. Pfl.*, **31**, 45-135.

WAIBEL, L. 1922. Winterregen in Deutsch- Südwestafrika. *Hamburg Univ. Abh. Auslansk.*, 9, *Reihe G. Naturw.*, **4**.

WALTER, H. 1936. Die ökologischen Verhältnisse in der Namib-Nebelwüste (Südwestafrika) unter Auswertung der Aufzeichnungen des Dr. G. Boss (Swakopmund). *Jb. wiss. Bot.*, **84**, 58-222.

WALTER, H. 1960. *Einführung in die Phytologie.* Vol. III. *Grundlagen der Pflanzenverbreitung.* Part 1. *Standortslehre.* 2nd ed. Stuttgart.

WALTER, H., and KREEB, K. 1970. Die Hydration und Hydratur des Protoplasmas. *Protoplasmatologia,* **2**, C/6, 95-108, Vienna.

ZEMKE, E. 1939. Anatomische Untersuchungen an Pflanzen der Namibwüste. *Flora N. F.*, **33**, 365-416.

X. *The Chilean-Peruvian Coastal Desert and its Fog-oases*

1. Fog as a source of water

The Peruvian sea current along the south Peruvian and north Chilean coast induces a desert climate just as the cold, Benguela current does along the coast of south-west Africa. In both situations cold water from greater depths moves to the surface. However, on the South American west coast the temperatures are a little higher, nearly 17°C in the winter and 24°C in the summer. Another important difference is that this coast is quite unaffected by the Eastern Trade winds, which cannot pass the Andean mountain chain. Moreover, a coastal mountain range, 500-800m high, lies just back from the coast (Fig. 235). By contrast, the Namib desert in south-west Africa only rises gradually inland with a gradient of about 1:100. There, the high mountains, such as the Brandberge, the Erongo mountains, the Khomas highland, the Naukluft and Tsaris mountains are so far inland that they are beyond the range of the fog-laden south-west winds. Only a few, smaller mountains, such as the Roessing mountains, are within 30 km of the coast. This different orographic situation in Peru is responsible for the much greater effect of the well-known coastal fogs, here known as 'Garua'. The steeply rising slopes of the coastal mountains are called 'Lomas'. On these, the fogs produce genuine fog-oases during the winter months and, at this time, the slopes are covered with 'Loma vegetation', standing out in the otherwise nearly rainless desert.

The coastal desert extends from about 6°S to about 27·5°S. From here southwards it grades into a semi-desert with winter-rains and into an area covered with sclerophyllous vegetation. By contrast, summer-rains become more prevalent to the north. This is accompanied by a change from desert to dry-scrub, then to rain-green and finally to semi-evergreen tropical forest. The vegetation types of the coastal desert have been described by several authors, particularly Reiche (1907) and Weberbauer (1911-45), and also by Berninger (1925), C. Troll (1930-56), H. and M. Koepcke (1953-54), Schmithüsen (1956) and Rauh (1951, 1958). A synoptic description is given by Koepcke (1961). However, more detailed ecological investigations became available only recently from Ellenberg (1959). He studied the Loma vegetation more closely and provided data on fog precipitation, which is not included with the usual rainfall records (Ellenberg, 1959).

Therefore, I shall discuss here this latter work more fully, especially since descriptions of the vegetation will be presented in a forthcoming monograph on South American vegetation[1].

[1] Now published. Hueck, K. 1966. *Die Wälder Südamerikas*, Stuttgart.

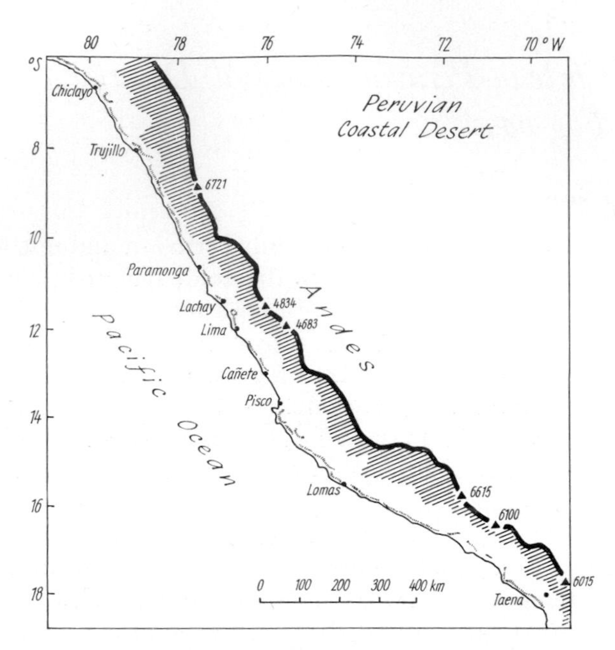

Fig. 227. Map of S.W. Peru showing locations of meteorological stations along the coast. Fog-precipitation records were taken at Lachay.

Table 66 and the climatic-diagrams (Figs 228-230) give information about the climate.

Table 66

Precipitation (in mm) at Peruvian coastal stations (after Ellenberg). See Fig. 227.

Stations immediately on the coast	Stations somewhat inland	Summer I	II	III	IV	Winter month V	VI	VII	VIII	IX	Summer X	XI	XII	Year (mm)
Chiclayo (31 m)		1	1	1	1	0	0	0	0	0	0	1	0	5
Trujillo (26 m)		0	0	1	0	0	0	0	0	0	0	0	0	3
Paramonga (15 m)		1	0	0	0	6	1	1	3	2	0	0	0	15
	Lima (137 m)	1	0	1	0	3	5	6	7	6	2	1	1	34
	Cañete (36 m)	0	1	1	0	9	3	4	3	3	1	1	12	40
Pisco (6 m)		0	0	0	0	0	0	0	1	1	0	0	0	2
Lomas (10 m)		0	0	0	0	1	0	0	0	0	2	0	0	3
	Tacna (475 m)	3	2	0	2	4	3	3	7	8	6	3	2	43

The data show that this area is virtually rainless, just like the Namib. So, too, vegetation is here found only in places with ground water. These are the estuaries of the rivers. The rivers originate in the Andes and carry much water. Halophyte associations are found in lagoons, while mangroves are entirely absent within the range of the cold Peruvian current.

The fog situation immediately on the coast is very much as in the Namib. Here, fog is relatively rare and, when it occurs, it only comes during the night. Correlated with this the coastal strip is barren of

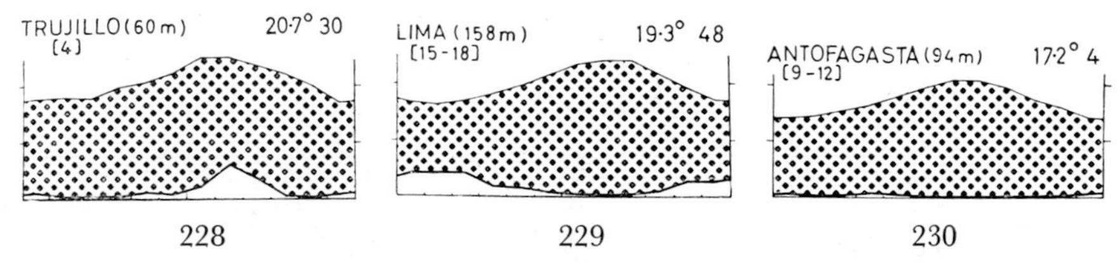

FIGS. 228-230. Climatic diagrams for stations in the coastal desert of S.W. Peru and N. Chile (Antofagasta). The rainfall data are different from that in Table 66. This is usually the case for arid areas, when the years in which observations are made are not the same.

vegetation. Only groups or colonies of large *Tillandsia* species grow in horizontal rows on the blown sand dunes. Sand builds up on these plants and the dunes become 1 m high and the Tillandsias often lie loosely on top. They have also been carefully described from Chile at 800-1200 m by Berninger (1925). Their roots are little developed and serve neither for anchorage nor for water absorption. Water is absorbed by the 'grey' leaves through absorbing scales, a characteristic feature of the Bromeliaceae. Absorption can occur only after the extremely rare rains, or after fog-condensation during the damp nights and mornings of the southern winter. Their transpiration losses are extremely low in spite of the great heating of the leaf surfaces, which reach 30°C even during dense fogs.

The water relations of these true fog plants have recently been investigated more closely by P. de T. Alvim and Andres Uzeda. Their as yet unpublished results were kindly made available to me.

The species investigated was *Tillandsia straminea*. The osmotic values of the leaves, determined by the cryoscopic method, were 7-8 atm. When leaves were suspended for 24 hours over solutions in a closed chamber in an atmosphere of 80-100% relative humidity, only a very small increase in weight was detected. This may have been caused in part by condensation of water. It follows that, essentially, the Tillandsias only absorb water with their leaves when they are wetted directly.

Water gained and lost can easily be determined in the natural habitat. Plants weighing about 3 kg were placed on a plastic sheet and left to grow *in situ*. Successive weighings were done at 9 a.m., when the dew had already disappeared, and at 5 p.m., after the plant had transpired the whole day. The investigation was done in July under the most favourable conditions. Fog and dew are particularly frequent at this cool season of the year. The mean temperature was 14°C and the mean vapour pressure 12 mmHg. The results are shown in Fig. 231. One can clearly recognise the increase in weight during the night and its loss during the day.

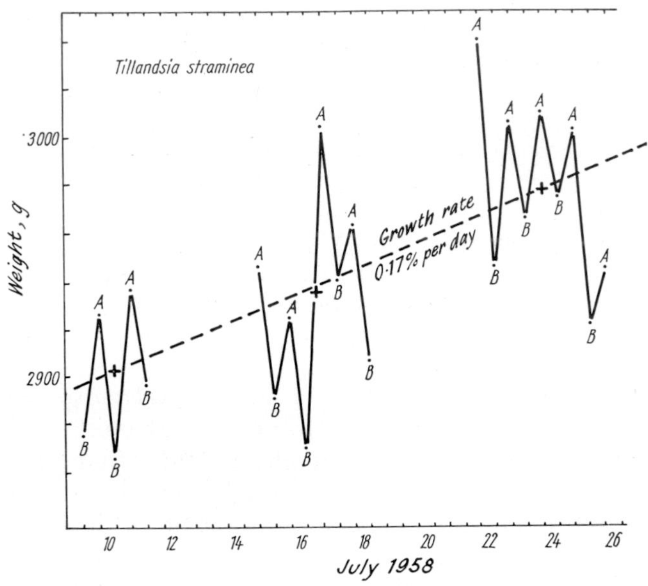

FIG. 231. Water balance of Tillandsias in the Peruvian coastal desert (after Alvim and Uzeda). *A*, weight at 9 a.m. after condensation; *B*, weight at 5 p.m. Further explanation in text.

Calculation of the mean fresh weight increase during July gave a growth rate of 0·17% per day. Thus, they grow at an extremely slow rate even in this favourable season. Growth probably ceases entirely in the unfavourable season. This investigation shows that the Tillandsias survive entirely on the water which they absorb from condensation of fog or dew with their well-known absorbing scales (Fig. 80). These forms of precipitation provide these plants with a means of growing in the practically rainless fog desert. They represent the only case of true 'fog-plants' so far known among angiosperms. Similar life forms have not yet been found in the Namib desert and South Africa, even though they may be found among the mist-absorbing plants mentioned by Marloth (1908). However, unequivocal physiological proof has not yet been given. In plants with functional roots the water can also be absorbed from the soil after a heavy wetting by dew or fog (see p. 343).

The ascending air cools off only a few kilometres inland on the west slope of the coastal mountains in Peru, the Loma. Here the frequency of fogs increases (Fig. 232). At about 600 m, where the cold, ascending air meets the layer of warmer air above, a dense fog cover is formed, the Garua. This can last for months, often over midday and so maintains the humidity of the air at a constantly high level. An additional factor,

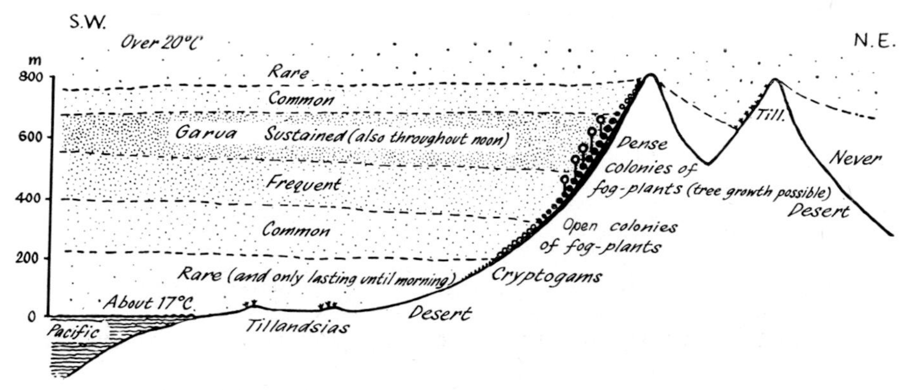

FIG. 232. Schematic profile through a Peruvian coastal Loma to show the frequency of fog (garua) and distribution of vegetation during the southern winter (after Ellenberg).

emphasised by Ellenberg, is the exceptionally favourable conditions for fog condensation prevailing on the South American coast. For this, the following features are important:

1. In the coastal area a light, but constantly blowing south to south-west wind prevails. Its speed rarely exceeds 4 m/sec but it is sufficient to bring in new fog-masses constantly.

2. The coastal mountain range rises steeply so that the air is forced upward, whereby it is cooled. In this process more water is constantly converted from the vapour form into fog droplets. Hence, condensation never ceases.

3. The temperature of the Peruvian current is significantly warmer than that of the Benguela current. As a result the water vapour content of the air above it is greater so that, upon equal cooling, more water occurs as droplets.

Condensation of the fog occurs on fixed structures past which the fog swirls. The larger the surface of the obstacle the greater the quantity of water precipitated. Because of this, a rain gauge will only catch a little water. Moreover, condensation is still relatively slight near the soil surface, it is greater at shoot height, so, of course, the amount increases with the size of the plant. Thus, trees collect the greatest amount of condensation resulting in constant dripping of water. They effectively 'comb' the water out of the passing fogs. Without air movement condensation would remain low. The amount of water as droplets in

1 m³ fog varies greatly (from 0·01-5 g) and averages 0·4-0·8 g. If about 0·5 g are 'combed out' by the branch-system of a tree this would result, with a wind speed of 1 m/sec (= 3·6 km/h), in 1·8 litre of water, i.e. nearly 2 mm precipitation per hour. Since the fog lasts throughout the day for many hours through about 120 winter days one can appreciate that considerable amounts of water may accumulate beneath trees.

Measurements taken between 1944-54 resulted, indeed, in very large annual accumulations beneath planted trees (highest and lowest annual amounts in parentheses) (Table 67).

TABLE 67

Average quantity of water (mm) collected annually in rain gauges

In the open	Under *Casuarina*	Under *Eucalyptus*
168 (121–219)	488 (262–819)	676 (252–1240)

The years with the two highest fog precipitations were in all three cases 1948 and 1949. By contrast, the years with the smallest amounts only rarely coincide.

The values found for soil water contents are also correlated with these precipitation measurements. At the start of the fog season in the winter the extent the soil water penetrates beneath the trees increases rapidly to a depth of 1·5 m (Fig. 233). The computed amount of water

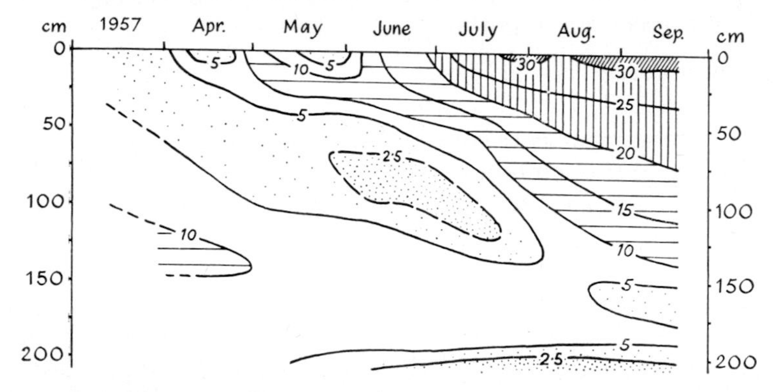

FIG. 233. Soil moisture (in %) beneath an open *Eucalyptus* stand in the Loma of Lachay 1957. In this year the fog season only began in June (after Ellenberg).

absorbed by the soil even exceeded that measured in the rain gauges. Rooting characteristics correspond to the distribution of the water (Fig. 234). A shallow penetration of water beneath herbaceous vegetation coincides with shallow roots extending only to 50 cm. By contrast, at 1 m the trees have assured soil moisture reserves throughout the whole year. On a foggy day, evaporation and, therefore, transpiration also is hardly detectable. On moderately cloudy days evaporation is quite marked (maximum 0·45 ml/h) and during the dry season it is very considerable (above 0·7 ml/h, air temperature nearly 28°C).

This shows that the water regime is very favourable to the development of lush vegetation during the winter months. These data only

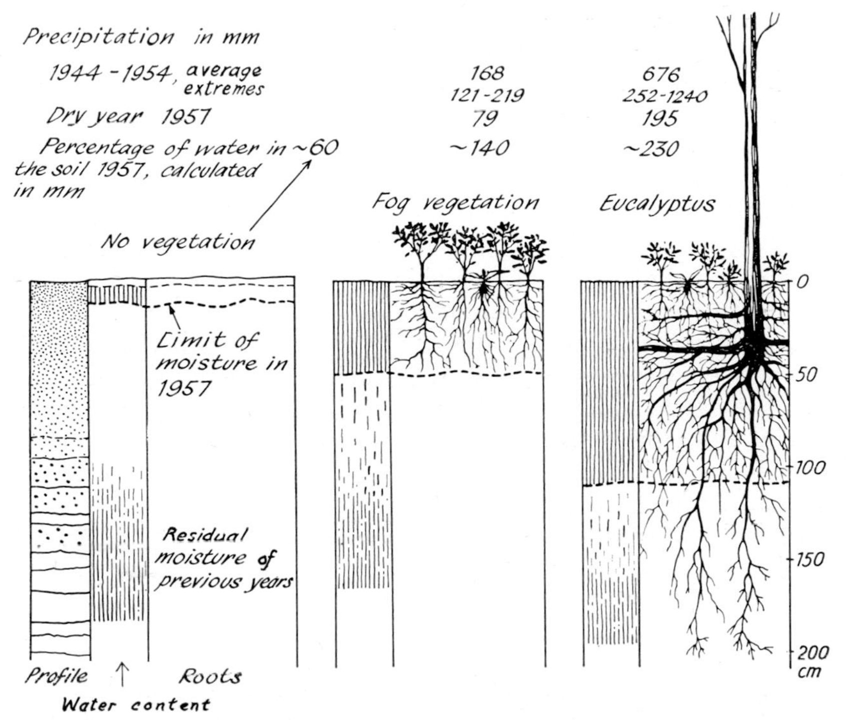

FIG. 234. Precipitation quantity and soil moisture in sandy soils beneath different vegetation in the Lomas of Lachay (after Ellenberg).

apply, however, to the highest altitudinal zones; water relations are less favourable at lower altitudes.

2. The Loma vegetation

When one leaves the coastal strip, with the Tillandsias, behind and crosses the desert for a few kilometres, the land begins to rise gradually. Here one first encounters cryptogamic communities. 'These consist in most cases of blackish, thick-as-a-thumb, jelly-like, blue-green algae (*Nostoc commune*, etc.). They cover at most one-third of the surface. In contrast to the Tillandsias, the algae cannot stand being covered by blown sand. Large colonies of grey fruticose lichens (*Cladonia* species) cover the stable sand surface in places, in the gently undulating landscape, e.g. south of Lima near Pachacamáe. They remind one of the grey-dunes in north-west Europe, but here, near Lima, they occur on a neutral to alkaline substrate. Wherever stones occur at some distance from the sea they are covered with colourful crustose lichens, e.g. at the Moro Solar south of Lima' (Ellenberg (1959), p. 51).

A little higher up, where the fogs have a greater wetting effect, the cryptogams become replaced by colonies of ephemerals. The endemic *Nolana* species are frequent here, amongst the annuals, while grasses do not occur as yet; they only become more numerous higher up. The herbs remain dominant, grow more closely together and form a dense carpet up to 60 cm tall. Noteworthy among them is *Loasa urens*, covered with stinging hairs and, in addition, *Nicotiana paniculata*, *Chenopodium* species, Malvaceae, Caryophyllaceae, *Galinsoga*, *Urtica* and European agricultural arable-weeds (*Sonchus oleraceus*, *Stachys arvensis*, *Erodium cicutarium*, *Stellaria media*, etc.) occur. Geophytes are also represented, such as Iridaceae, *Solanum*, *Oxalis*. Among these *Stenomesson coccineum* develops its leaves only in the foggy winter and opens its orange-red flowers only in summer.

Today woody plants are almost absent from these herbaceous communities. However, evergreen trees, such as *Casuarina* and *Eucalyptus*, have been successfully introduced. They survive the long period of drought and grow well. Endemic woody plants are only found growing singly in rocky habitats. These are *Acacia macracantha*, *Carica candicans* and *Caesalpinia tinctoria*. Their branches are covered in epiphytic mosses and lichens and even scattered, small, semi-succulent plants of *Peperomia crystallina*. The shrubs must be assumed to represent relicts from an earlier woody formation, as has been suggested by Ellenberg.

There is a similar, more succulent fog-vegetation type in north Chile.

The Loma area is densely populated in the winter months. At this time nomads migrate with their herds down from the winter-dry grazing areas in the Andes. For centuries wood has been cut for fuel and even the wood from roots has been used. This has resulted in the complete destruction of the forest. Regeneration by seedlings is prevented by browsing animals. Only the poisonous *Croton* shrubs and the thorny *Acacia macracantha* remain untouched by the animals. Destruction of the forests was probably aggravated as well by the need for charcoal by the coastal settlements.

Only a small relict forest near Atiquipa can provide a glimpse of the original vegetation. It consists predominantly of a small, evergreen *Eugenia* species with very tough leaves. Its crooked trunks attain heights of 5-8 m and form a dense crown cover. It is associated with *Capparis*. In this shady forest a few hygromorphic ferns occur.[1]

[1] Such fog-oases are also found on the west slope of the Coast Cordilliera in central Chile (precipitation as high as 250 mm) and near Erkawit on the Red Sea (precipitation 218 mm). These were recently discussed by: Kummerow, *et al.* (1961) and Kassas (1960), see also Follmann (1960). This area has 10-11 dry months but fog occurs throughout the whole year. Kummerow (1961) describes a similar association with *Anaptychia intricata* from north Chile (18-30°S) at 0-1600 m. In addition *Trentepohlia* and, in the forks of cacti, Tillandsias also occur.

The density of the cloud cover decreases above 700 m, its upper limit occurs at 800-1000 m. The Loma vegetation dies out at this altitude. In south Peru, Tillandsias occur in the upper fringe-zone, where the clouds pile up and also form colonies, just as on the coast, leading to the formation of small dunes (see Fig. 232). On the lee of dry northern and eastern slopes lichens and cacti occur. The fogs do not spread beyond the coastal mountain range so that the barren desert recurs again from there as far as the Andean chain. This desert occurs in its most extreme form in the Atacama desert in north Chile.

But the Atacama is not completely without vegetation. Autotrophic vegetation is represented by soil algae, which in the soil form richly branched, free, filamentous colonies: *Schizotrix atacamensis* lower down and *Calothrix desertica* higher up; 2 forms of *Cyanidium* live endolithically in diorite fissures on the Chilean coast (Schwabe, 1960).

Ellenberg's investigations show clearly that the fogs give rise to a lush vegetation only under conditions favouring heavy condensation. In these cases, the fogs simply replace rain and the plants show no particular adaptations except for the Tillandsias. Among annuals one finds hygromorphic and mesomorphic species. Among perennials, which have to survive a long dry period, one finds sclerophyllous forms. All these plants absorb their water in a completely normal way through their roots from the soil. Thus although one can speak here of 'fog oases' one should avoid the term 'fog plants'. Otherwise one would have to include *Sonchus arvensis*, *Stachys arvensis*, *Erodium cicutarium* and other European weeds in this group, which today form floristic members of the Loma vegetation.

A synoptic scheme of the vegetation zones from sea coast to the peaks of the Andes is shown in Fig. 235.

The high dry plateaus of the Altiplano east of the main ridge are known as Puna (see p. 189 ff.). They extend into Argentina and have been described from there by Cabrera (1958). In this area at about the Tropic of Capricorn very extreme conditions prevail between 3300-4300 m. The mean temperature amounts to 8-10°C, the minimum to −18°C, the maximum to 30°C. Frost occurs throughout the year, the snow limit lies at 5500 m. Precipitation occurs in the summer (December-March) and amounts to 103-324 mm per year. Precipitation decreases southwards and becomes very erratic. The driest, western part, the Puna de Atacama, is rich in salares, i.e. salt-pans. The eastern part is damper and cultivated in parts. Cultivated plants include *Vicia faba*, *Zea mays*, *Medicago sativa*, *Solanum andigenum* and *Chenopodium quinoa*. Introduced weeds are *Melilotus indicus*, *Rumex crispus*, *Convolvulus arvensis* and *Brassica campestris*.

Among the vascular plants of which the vegetation is composed, 23% of all species are composites, 13% grasses and 7% legumes. A typical plant cover of the dry Puna is a community of *Fabiana densa* (Solanaceae), *Psila boliviensis* (Compositae) and *Adesmia horridiuscula*

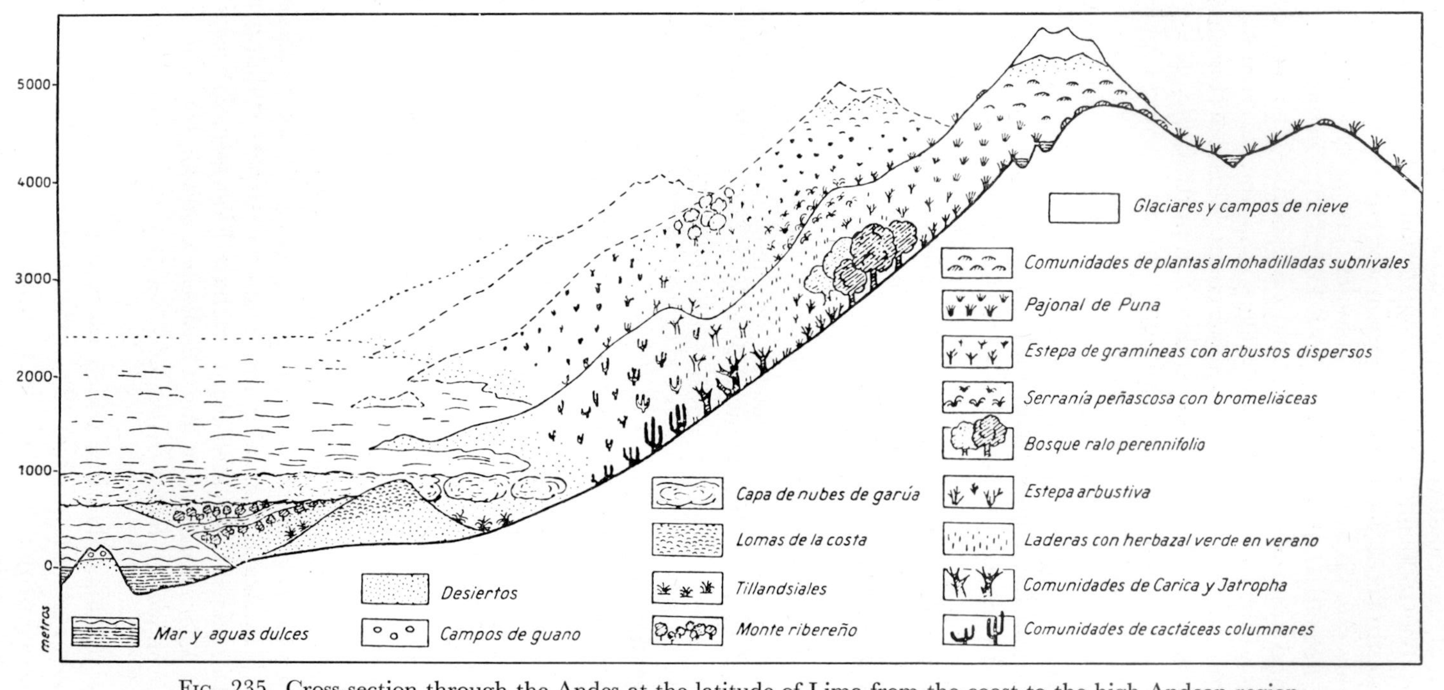

FIG. 235. Cross-section through the Andes at the latitude of Lima from the coast to the high Andean region after M. Koepcke in C. Troll (1959): 1. Sea- and fresh-water; 2. Deserts; 3. Guano fields on islands; 4. Fog of the Garua; 5. Lomas (fog-green slopes) of the coastal zone; 6. Stands of rootless Tillandsias in the desert; 7. River forest; 8. Glaciers and snow-fields; 9. Sub-alpine cushion-plant communities; 10. Puna grassland; 11. Grass-steppe with scattered shrubs; 12. Rocky slopes with bromeliads; 13. Open, evergreen forest; 14. Scrub-steppe; 15. Valley-slopes with summer-green herbaceous communities; 16. Stands of *Carica* and *Jatropha*; 17. Stands of pillar-cacti. Enumeration from left to right and vertically, from top to bottom.

(Leguminosae). It is a semi-desert with low shrubs and about a 10% cover. A little higher up *Prosopis ferox*, the pillar-cactus *Oreocereus* and the tree-forming *Polylepis tomentella* occur. Tussock grasses predominate in places, and also in the high Andean belt above 4300 m (*Festuca orthophylla*, *F. chrysophylla*, *Poa gymnantha*, etc.). Cushion plants are present as well, such as *Azorella yareta* (Umbelliferae), *Pycnophyllum* (Caryophyllaceae), *Opuntia nigrispina*, and, on salty soils, *Anthobryum triandrum* (Frankeniaceae) and *Salicornia pulvinata*. Halophyte communities are represented by stands of *Sporobolus rigens* with *Atriplex* species and the *Lycium humile-Distichlis humilis* community.

References

BERNINGER, O. 1925. Extreme Ausbildung einer Nebelvegetation in der nordchilenischen Wüste. *Zschr. Ges. Erdkunde*, **9-10**, 383-4.

CABRERA, A. L. 1958. La vegetation de la Puna Argentina, VI. *Revista Invest. Agricolas*, **11**.

ELLENBERG, H. 1959. Über den Wasserhaushalt tropischer Nebeloasen an der Küstenwüste Perus. *Ber. Geobot. Forsch. Inst. Rübel* 1958, 47-74.

FOLLMANN, G. 1960. Eine dornbwohnende Flechtengesellschaft der zentralchilenischen Sukkulentenformationen mit kennzeichnender *Chrysothrix noli-tangere* Mont. *Ber. Dtsch. Bot. Ges.*, **73**, 449-62.

FOLLMANN, G. 1961. Eine dornbewohnende Flechtengesellschaft der nordchilenischen Sukkulentenformationen mit kennzeichnender *Anaptychia intricata* (Desf.) Mass. *Ber. Dtsch. Bot. Ges.*, **74**, 495-510.

KASSAS, M. 1960. Certain aspects of landform effects on plant water resources. *Bull. Soc. Géogr. Egypte*, **33**, 45-52.

KOEPCKE, H. W. 1961. Synökologische Studien an der Westseite der peruanischen Anden. *Bonn. geogr. Abh.* H. **29**, 320.

KUMMEROW, J., MATTE, V., and SCHLEGEL, F. 1961. Zum Problem der Nebelwälder an der zentralchilenischen Küste. *Ber. Dtsch. Bot. Ges.*, **74**, 135-45.

MARLOTH, R. 1908. Das Kapland. *Wiss. Ergeb. Dtsch. Tiefsee-Exped.*, **2**, Teil 3, 1-436.

RAUH, W. 1958. Beitrag zur Kenntnis der peruanischen Kakteenvegetation. *Sber. Heidelb. Akad. Wiss.* (*math-naturw. Kl.*) *Jahrb.* 1951, 1 Abh., 1-542.

REICHE, 1907. Grundzüge der Pflanzenverbreitung in Chile. *Die Veg. der Erde* VIII, Leipzig.

SCHMITHÜSEN, J. 1956. Die räumliche Ordnung der chilenischen Vegetation. *Bonn. geogr. Abh.* H. **17**, 1-86.

SCHWABE, G. H. 1960. Zur autotrophen Vegetation in ariden Böden, IV. *Osterr. Bot. Z.*, **107**, 281-309.

TROLL, C. 1930-59;1932. Die tropischen Andenländer. *Hdb. Geogr. Wissen.* Bd. *Südamerika*, 309-462. Potsdam. 1959. Die tropischen Gebirge. Ihre dreidimensionale klimatische und pflanzengeographische Zonierung. *Bonn. geogr. Abh.* H. 25.

WEBERBAUER, A. 1911. Die Pflanzenwelt der peruanischen Anden. *Die Veg. der Erde* XII, Leipzig.

WEBERBAUER, A. 1945. *El mundo vegetal de los Andes Peruanos* (*Estudio fitogeografico*) Minist. de Agric., Lima.

XI. *The Karroo*

1. General features

By the Karroo[1] one understands the dry South African region, lying between the rising coastal plain with pronounced winter-rains and the region of summer-rain in the highlands to the north. It is not an extreme desert but rather a semi-desert comparable to the Sonoran desert around Tucson. The region is isolated from the coast by the chain of the

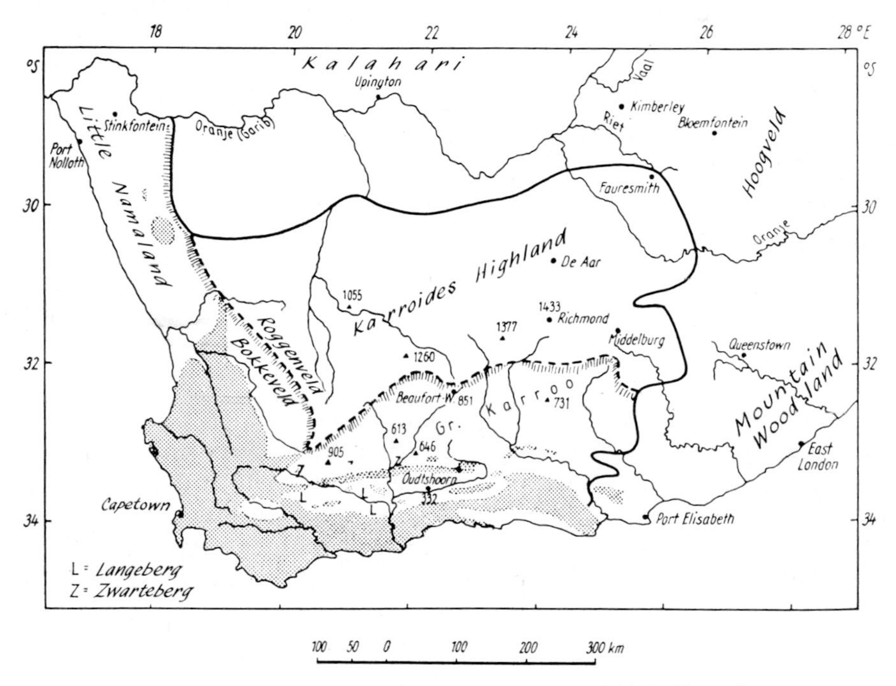

Fig. 236. Map of the Karroo region in South Africa. Dotted areas—Cape Province vegetation of the winter-rain region.

Swartberge and Langeberge, which are over 2000 m high, and it runs parallel to the south coast. The country lying between these mountains and the steep scarp of the highlands is the actual Great Karroo (Fig. 236). It is divided into several individual basin-like plains. The annual rainfall is usually a little below 200 mm and normally occurs in the autumn. Summer rainfall increases eastwards. The temperatures are very high. Just north lie the highlands of the Karroo known as the

[1] Karroo in the language of the Hottentots (Nama) means a dry region.

Upper Karroo. They extend across the Orange River into the southern Orange Free State and here, near the northern border, I had an opportunity to carry out ecological investigations for a long period (Walter, 1939; the experiments, described later, are taken from this work). This was at the Veld Reserve, Fauresmith, which was under the directorship of M. Henrici. The area consists of plateau-topped mountains interspersed with basins, which drain into the Orange River.

The Upper Karroo is cold in winter because of its altitude (1100-1300 m). Frosts frequently occur at night. The area is used almost entirely as pasture. Irrigated crops only occupy a small area. The maize-growing area occurs further north of Bloemfontein. The three climatic diagrams (Figs 237-39) give some idea of the climate of different parts of the Karroo.

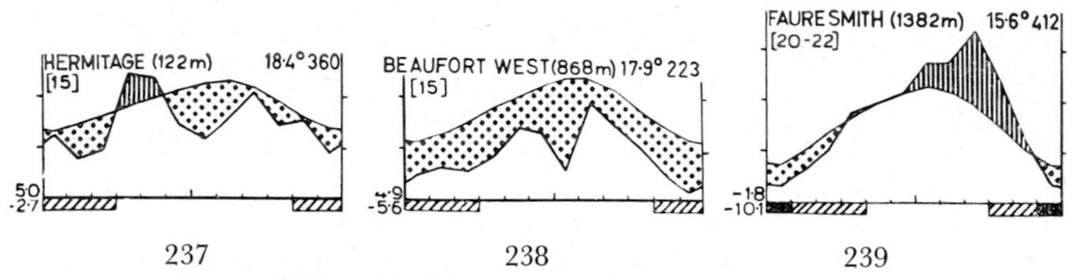

FIGS. 237-239. Climatic diagrams of stations in the Karroo (from *Klimadiagramm-Weltatlas*).

The arid areas that are transitional between winter- and summer-rain regions are usually typified by their rich succulent flora (cf. Sonoran desert). This also applies to the Karroo, which is well known for its many succulent forms. In addition, dwarf-shrubs are important and these are especially valued for sheep grazing: they are mainly composites. The Little Karroo south of the Swartberge (Fig. 240) begins to show an intrusion of plants from the winter-rain region of Cape Province. However, in the Great Karroo north of the Swartberge the succulents are locally dominant (Fig. 241). In addition, geophytes are also extremely numerous (Iridaceae, Amaryllidaceae, Liliaceae, and bulbous *Oxalis* species) so that after good spring rains they can transform the entire area into a sea of flowers for 1-2 months. Among the dwarf-shrubs which occur especially in the Upper Karroo are: *Pentzia*, *Pteronia*, *Eriocephalus*, *Chrysocoma*, *Euryops*, *Diplopappus*, *Tripteris*, etc. These form the composite—'Veld'. In addition there are *Lycium*, *Zygophyllum* and a series of *Mesembryanthemum* species with succulent leaves but woody twigs. Larger woody plants are not entirely absent, such as *Olea verrucosa*, *Rhus* species, *Euclea*, *Osyris*, *Ehretia* or even trees, such as *Rhus lancea*, a few *Acacia* species or *Salix capensis* and *Royena* (= *Diospyros*). However, these woody plants are severely restricted to habitats with locally better water supplies.

FIG. 240. Karroo with dwarf shrubs (*Pentzia incana*, *Lycium* spp.) and grasses (*Aristida congesta*, *Aristida curvula*), near Beaufort, West Cape Province, S.W. Africa (Phot. E. Walter).

FIG. 241. Karroo-vegetation on quartz reef; foreground: *Cotyledon wallichii*, *Cotyledon reticulata*; in the rear: Swart Mountains; in the gulley behind the bus only *Galenia africana*; near Laingsburg, Cape Province, South Africa (Phot. E. Walter).

Among the most typical succulents of the Great Karroo may be mentioned:

1. The stem-succulents, represented by *Euphorbia* species (*E. mauretanica*, etc.), the Asclepidiaceae *Stapelia*, *Caralluma*, *Hoodia*, *Trichocaulon*; in addition *Kleinia* (*Senecio*), *Sarcocaulon* and *Pelargonium*. The last two have mesomorphic leaves which they shed during drought. Another important plant is the butter tree *Cotyledon fascicularis*, which is somewhat transitional to the next group.

2. Leaf-succulents such as *Cotyledon* and *Crassula* species, *Portulacaria* and *Anacampseros* (Portulacaceae), Liliaceae-like *Aloë* species, *Haworthia*, *Gasteria*, *Apicra*; in addition, plants that look like stones such as *Titanopsis* *Lithops*, *Argyroderma*, *Didymaotus* and a few *Crassula* species.

3. Root-succulents with subterranean storage organs such as *Pachypodium bispinosum*, *Asparagus* species, *Fockea* species. The smaller Euphorbias, which are stem-succulents, commonly have, in addition, large subterranean water storage organs.

These succulents extend into Little Namaqualand and the southern part of south-west Africa for as far as the winter-rains show their effects (Fig. 242).

Fig. 242. Karroo, near Laingsburg with *Ruschia crassa*, *Pteronia* and much *Galenia africana*, Cape Province, South Africa (Phot. E. Walter).

Other members of the Karroo flora are *Testudinaria elephantipes* (Dioscoreaceae) with a semi-globose caudex covered by torn polygonal corky warts, and two *Encephalartos* species (Cycadaceae). In addition, there are a few poikilohydrous ferns and different, short-lived herbs. In the Karroo a few depressions with saline soils occur but larger salt lakes or salt deposits are absent since there are no young, marine sedimentary rocks. Brackish soils are indicated by the presence of halophytes such as *Tamarix* and *Salsola* species, *Atriplex halimus* and several mesembryanthemums.

A most detailed description of these areas has been published by Marloth (1908).

2. Vegetation of the Upper Karroo

The area near Fauresmith in the southern part of the Orange Free State represents a partial transition zone to the grassland of the High Veld with pronounced summer-rains.

The climatic-diagram (Fig. 239) shows that summer-rains predominate in Fauresmith, while the winter months (June-September) are dry. The mean annual rainfall is 412 mm. Frost can occur any time during June and July but usually only at night.

FIG. 243. Landscape near Fauresmith, Orange Free State, South Africa (Phot. E. Walter). In the valley, the grass cover has been destroyed through grazing; beginning of channel erosion (from Walter, 1960).

In such areas bordering different zonal vegetations, the edaphic factor determines the differentiation of the vegetation. In fact, one can find different zonal vegetation types side-by-side in the Upper Karroo. The area is also known as 'Broken Veld', because the 'Veld', i.e. the grassland, does not occur as a continuous area, but is interrupted by areas with a different vegetation or it is even split into a vegetation mosaic which is restricted to the high plateau-like dolerite ridges (Fig. 243). These inter-relations can best be understood by a schematic vegetation profile (Fig. 244).

The nearly level plains of the higher land consist of dolorite, occurring as blocks of rock outcrops with very little soil in between. They sustain typical Karroo vegetation consisting predominantly of succulents. The steep slopes descending from the plains are littered with rough rocks. In the past the commonest tree on these slopes was *Olea verrucosa*, which has dense, evergreen crowns. Its stands are much reduced around Fauresmith as a result of its utilisation by man. Shrubs such as *Rhus erosa*, *Euclea ovata* var. *undulata*, *Osyris compressa* (*abyssinica*), *Ehretia hottentotica* (Boraginaceae), *Royena microphylla*, *R. cuneifolia*, *Chilanthus arboreus* (Loganiaceae) now occur between the trees. The strange looking tree *Cussonia spicata*, with a tufted crown of huge, lobed leaves, appears rather foreign to the community. In addition, there are smaller shrubs, dwarf-shrubs and a few herbs. Poikilohydrous ferns are commonly found in shady places behind rocks. These include *Pellaea hastata*. *Ceterach cordata*, and *Cheilanthes hirta*. Associated with them are a number of *Riccia* species, growing in colonies as large as a hand. These liverworts become in-rolled when dry.

The scrub forest, which here represents depauperated forest, is best developed on southern exposures. These are the cooler aspects in the

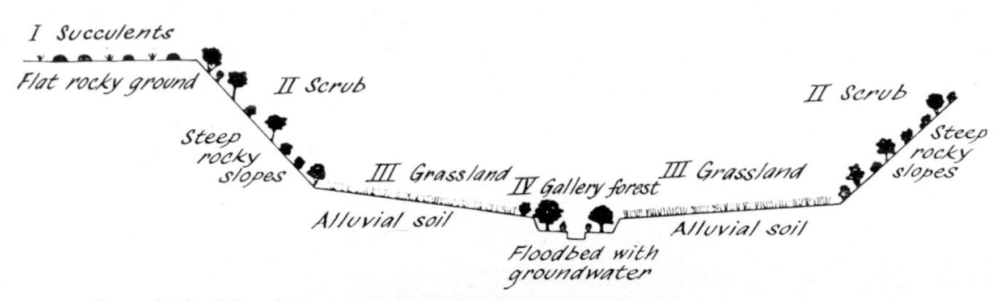

FIG. 244. The importance of soil and topography in the distribution of vegetation in the Orange Free State near Fauresmith: I: Mountain-ridges of dolorite rock with shallow soils occupied predominantly by succulents. II: Slopes of rocky outcrop, which are best suited to the development of woody plants with extensive root systems. III: Fine-grained sedimentary soils which hold much water during the rainy season. Here, grasses compete successfully. IV: Gallery forest formed from woody plants with deep-reaching tap roots that reach the water table (after Walter 1939).

southern hemisphere. As we know, in arid climates rocky habitats are the most favourable habitats with respect to water, especially for woody plants. This is also indicated by the presence of the grass *Themeda triandra* which occurs here in small, isolated patches, although it forms continuous grassland further to the north. Groups of tall *Aloë* species are often found on steep, hot, north slopes. This area represents the limit of their distribution which is shown by the signs of frost damage after cold nights.

The slopes become less steep near their bottoms. Here, the soil is at first sand, then loam. These are drier areas and are covered with short grasses, such as *Eragrostis lehmanniana* and in towards the centre of the valley with *Sporobolus ludwigii*. The cover is about 30%. Dwarf-shrubs and herbs only occur rarely.

Here, dry, river channels sharply dissect the alluvial flats and their banks are associated with the remains of former gallery forest. This probably contained *Acacia* species that have now been cut. Today there are only shrubs present such as *Rhus lancea*, *Zizyphus mucronata*, *Lycium hirsutum*, *Royena decidua*.[1] In damper places *Melianthus comosus* and *Salix capensis* occur.

The river bed itself, which is usually dry, always has ground water beneath it and is occupied exclusively by herbs, such as *Gomphocarpus fruticosus* and such ruderals as *Nicotiana glauca*, *Argemone mexicana*, *Tagetes minuta*. They are covered by sand at floodtime, but always become re-established. Brackish soils occur on flood plains behind natural levees where the water becomes trapped. Here *Salsola* species and *Sueda atramentifera* grow and, also mesembryanthemums, such as *Psilocaulon*, the strange *Nananthus vittatus* and *Lithops salicola* (Figs 245 and 246). In damp places the salt-excreting grass *Sporobolus tenellus* comes in. In fringes around salt-patches, where, the heavy clay soil has become too dry for grasses, are found the Karroo shrubs *Pentzia incana*, *Pteronia glauca*, *Eriocephalus glaber*, *E. spinescens*, *Tripteris pachypteris*, the peculiar *Thesium hystrix* and the thorny *Mesembryanthemum* cf. *spinosum*. I found that where there were salt soils the ground water was at 2-2·5 m during the dry season. The osmotic value of the ground water was 0·3 atm showing that it was not brackish. The soil developed polygonal cracks in places barren of vegetation. A pH of 7·4 was measured at the surface and at 10 cm, of 8·4. These, therefore, are solonetz soils with some sodium carbonate formation.

3. Ecological investigations

I carried out the investigations to be described during the dry season of 1937. In Fauresmith, 6 months may pass without rain. This prolonged drought was advantageous to my investigations because the critical period for the water relations of arid-zone plants is not during the favourable season, but during the dry season. The plants must reduce their water loss to the point where the water balance does not fall below the critical value.

(*a*) *The behaviour of succulents and halophytes*

This group is particularly typical of the Karroo. The stem-succulent Euphorbias behave like cacti and remain alive for many months without water absorption. The same is true for the *Aloë* species, whose beautiful, bright red flowers unfold during the dry season. In spite of their large surface area, the inflorescences show very small water losses. On one day the transpiration rate was 8·5-13·5 mg/g fresh weight. After a night frost, the rate doubled as a result of damage. The flowers' nectar, which is very abundant, had an osmotic value of 29·0 atm.

More detailed studies were made with smaller succulents. Water losses were so small that weighing had to be done at intervals of days. The transpiration rate as a rule, was reduced by one-half to one-third,

[1] *Royena* has recently been included in the genus *Diospyros*.

from the first to the second day. Then, however, the rate remained constant at about 1% of the original fresh weight per day. This is shown in Table 68.

FIG. 245. Colony of *Lithops salicola* with 'window-leaves'.

FIG. 246. *Nananthus vittatus*; only the leaf tips are visible.

FIGS. 245 and 246. Slightly brackish clay-pan. Rosemarie farm near Fauresmith, South Africa (Phot. E. Walter).

It was remarkable that *Mesembryanthemum* continued to flower unabated during the experiment with cut off shoots. Even new flower buds continued to open.

The decrease in transpiration from the first to the second day showed that the rooted plants probably continued to absorb a little water from the soil. However, on cutting they can reduce their transpiration markedly without any difficulty. When the plants are exposed to a very long dry period, water is translocated from the older leaves in favour of the younger ones; the dehydrated leaves shrivel up and are shed.

TABLE 68

Kleinia radicans	3·3%
Massonia (with bulbils)	1·7%
Crassula platyphylla	0·9%
Mesembryanthemum unidens	1·0%
Mesembryanthemum hamatum	1·0%
Bergeranthus glenensis	0·5%

Lithops salicola was investigated, ecologically, more closely. It grows wholly embedded in the ground (Figs 245 and 247).

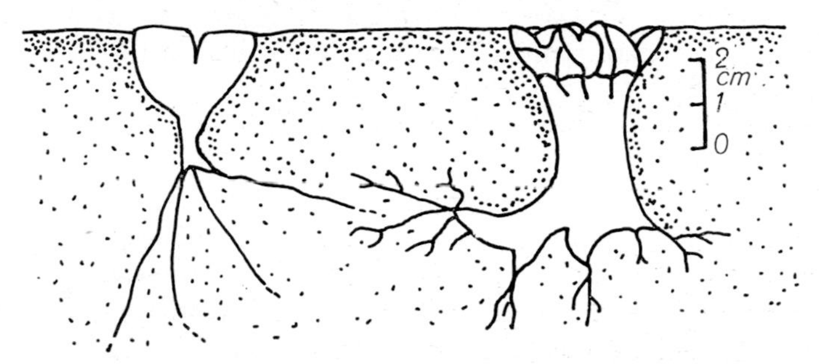

FIG. 247. *Lithops* (left) and *Nananthus* (right) with wholly embedded root systems (see also Figs 245 and 246).

The heavy clay soil of the *Lithops* habitat held only 4·5-5·4% water even in the region of the root. Thus, it was practically air-dry. The soil had an alkaline reaction of pH 8·6-8·8 and at 30 cm there was some accumulation of Na_2CO_3 and $MgCO_3$. NaCl was present, up to 1%, in the upper 15 cm, while sulphates and nitrates were present in much smaller quantities.

The succulents only have actively absorbing roots during the wet period, when the salt concentration is low. During a drought the plants live on their stored water reserves. In spite of this the osmotic value was only 19·2 atm; in plants kept dry in the laboratory for 10 days the values were 20·4 to 21·2 atm; in the case of a damaged specimen 24·1 atm. The chloride content, as in all mesembryanthemums, was high and exceeded 40% of the osmotic value.

To gain an insight into the water economy, the following experiments were made:

1. A whole *Lithops* plant was removed from the soil with its short root. It was wrapped in cotton wool on a watch glass in such a way

that only the 'window' leaves could transpire as is the case under natural conditions. The weight loss in 16 days had not reached 0·2 g (4·5% of the original weight).

2. Shrivelled plants with uninjured basal parts were immersed in water. They increased 15% in weight in 15 days but remained shrivelled. When, instead, a few small incisions were made in the epidermis, water absorption reached 30% after 24 hours and new, absorbing roots were formed by the plant after only 3 weeks.

3. When only the 'window' surfaces were brought into contact with the water, the daily water absorption was hardly detectable. However, absorption increased rapidly upon puncturing a few holes in the surface. Where two plants are connected to the same root, it is sufficient if only one of these absorbs water through puncture holes, for the other plant also becomes turgid in this case. This shows that direct absorption of dew or fog precipitation cannot be considered a factor in the plant's water relations.

After rain the plants must first form roots before they can absorb water.

Nananthus showed a very similar pattern. These plants also grow embedded in the soil and only the numerous leaf tips reach the surface (Figs 246 and 247). The plant had already formed new roots after being kept damp for 9 days in the laboratory. Then rapid water absorption began.

As an example of a root succulent form I investigated *Pachypodium succulentum*. The 20-30 cm tall, thorny shoots were leafless. They arise from a subterranean tuber from which the roots run fairly closely parallel to the surface. The tuber contained 266 g stored water (95% of the dry weight). The osmotic value of the cell sap was 4·2 atm. This indicates that the water stored in the tuber can provide the leafless shoot with all the water necessary to replace its small transpiration losses for a very long time. Marloth gives the weight of the tuber of an old specimen as 7 kg and that of the corresponding leafless shoot as 68 g. In *Asparagus* species, which remain green during the dry season, the daily transpiration of the twigs also only amounts to 7·5 mg/g fresh weight, while a good deal of water is stored in the subterranean tuber or in the fleshy roots.

This explains why the cell sap concentration does not increase much in all these plants during a long drought. Apart from the mesembryanthemums the osmotic values are all below 10 atm (Table 69).

Of the values given, only *Anacampseros* shows a somewhat higher cell sap concentration. However, this species has no succulent shoots, only its roots are fleshy. *Crassula*, *Kleinia* and a *Euphorbia* species also grew in habitats that may have been somewhat brackish; yet the osmotic values were hardly any higher (about 3-4 atm).

In contrast to these succulents the mesembryanthemums always have values above 10 atm, in brackish as well as non-saline habitats, e.g., Table 70.

The data show that the osmotic value in these species is notably influenced by chloride storage even on non-saline soils. Thus, mesembryanthemums can be classified as typical halophytes. The sulphate

TABLE 69

Osmotic values (atm) of succulents of the Karroo

Liliaceae		Euphorbiaceae	
Aloë grandidentata (leaves)	5·0	*Euphorbia* cf. *pulvinata*	8·5
Aloë grandidentata (flowers)	7·9	Compositae	
Aloë hereroensis	6·8		
Haworthia tessellata (fresh)	4·1	*Kleinia radicans*	8·3
Haworthia tessellata (wilted)	6·3	*Senecio longiflorus*	9·5
Crassulaceae		Portulacaceae	
Cotyledon decussata	7·6	*Anacampseros ustulata*	15·3
Crassula platyphylla	7·7		
Crassula harreyi	9·6		

TABLE 70

Samples from non- or slightly saline (the two last) habitats (— = not determined)

Plant species	Osmotic value atm	Chloride %
Mesembryanthemum saxicolum	16·4–17·4	—
Bergeranthus glenensis	16·5–17·6	—
Stomatium erminium	17·7–21·0	—
Mesembryanthemum unidens	18·2–20·6	38
Mesembryanthemum ferox	19·1	—
Mesembryanthemum hamatum	23·4	—
Mesembryanthemum cf. *spinosum*	27·6	—
Mallephora mollis	29·4	48
Lithops salicola	19·2–24·1	43
Nananthus vittatus	18·5–25·7	42

content is from 1-3·6% and only in *Mallephora* was 11%. On typical saline-soils the values are markedly higher and they are hardly lower in mesembryanthemums than in Chenopodiaceae which grow in the same habitat (Table 71).

However, the Chenopodiaceae differ by having very high osmotic values even on non-saline soils. This can be seen from Table 72 which may be compared with those of Table 70, which are from the same habitats.

Salsola and *Atriplex* are only slightly succulent and belong to the sulphate-halophytes. They also extend into drier habitats (xero-halophytes), while *Suaeda* is always restricted to wet, salty habitats.

TABLE 71

Samples from saline-soils

	Osmotic value (atm)
Psilocaulon (Mesembr.)	41·3–48·1
Salsola spp.	50·7–74·9
Suaeda atramentifera	54·8–66·7

TABLE 72

	Osmotic value (atm)	Chloride (%)	Sulphate (%)
Salsola spp.	51·7–65·3	23	17
Suaeda atrimentifera	35·9–47·2	—	—
*Atriplex nummularia**	51·8–57·1	29	16

* *Atriplex nummularia* is native to Australia.

(*b*) *Experiments with poikilohydrous ferns*

These plants survive the dry season in an air-dry condition. Therefore, their weight does not change when left to lie in dry conditions. At most, their weight may increase a few tenths of a gram on cool and damp days and decrease again on hot, dry days. This is characteristic of all hygroscopic bodies.

The water content of young, air-dry, dormant, living leaves amounts to 6·7 to 8·0% of the dry weight. *Cheilanthes hirta* was found in a semi-turgid condition at a somewhat damp place. In this specimen the water content of the leaves was 95% of the dry weight and the osmotic value of the cell sap was 20·5 atm.

By soaking dry plants in water, which are then briefly dried off on filter paper and allowed to transpire, one obtains typical drying curves. Under constant conditions in the laboratory, the point of change-over from evaporation of capillary water at the surface to real transpiration is not marked by a sudden change in the rate of water loss. Therefore, the transpiration resistance must be as low as in mosses. In the natural habitat the water losses are replaced by water absorption from the soil after rain. The desiccation curve only begins when no more water is available in the soil. On wetting, the desiccated roots cannot absorb water immediately; just as in cacti, they have first to form new absorbing rootlets (Oppenheimer and Halevy, 1962).

(c) *Water relations of Karroo dwarf-shrubs*

The transpiration of these dwarf-shrubs has been thoroughly investigated by Henrici (1940a, 1940b). In this connection one is again mainly interested in the situation during the dry season. Plants were investigated that had been cultivated for 5 to 10 years in experimental plots near the laboratory but they were grown under natural conditions, i.e. without irrigation.

The dense thickets of planted dwarf-shrubs were still completely quiescent in August-September. Their leaves had been shed except for small buds.

Osmotic values could be determined only for *Hermannia candidissima*, which gave 37·3 atm. By contrast with densely grouped plants, individuals with larger root areas, for example, those growing at the margin of roads, had already formed small, green leaves. They had lower osmotic values, approaching 22-30 atm.

Transpiration experiments also showed a very marked difference between dormant plants and those in flush. Successive hourly weighings of parts of shoots of the dormant dwarf-shrubs showed practically no fall in their rate of water loss. The values were low. By contrast, the shrubs in flush after cutting showed, initially, the highest values which successively decreased from hour to hour. This indicates that water loss was reduced during dormancy, but not during the flush. Their first growth was merely a function of their ability to absorb a certain amount of water from the soil. In the more sparsely covered places, the soil had a considerably higher water content but it was highest in the fallow-plots. A few scattered individuals of *Salvia cleistogama* that grew on these fallow-plots, lacking any root competition, were already full grown and in flower. These plants have a similar habit to *S. pratensis*, but belong to the mesomorphic, malacophyllous type. They had osmotic values from 15·2 to 18·7 atm, a water content of 410% of the dry weight and a very high rate of transpiration. One of the plants wilted within 20 minutes of being cut. Their roots must have had to absorb at least 60 g of water per hour from the soil, for rooted plants did not wilt in their natural habitat even at midday. The lateral roots reached over 40 cm horizontally and more than 25 cm downwards. The water content of the soil at 10 cm was 10·5-12·7%.

This shows that the plants have relatively sufficient water in the dry season when growing at low densities. Under these conditions the soil can appear completely dried up at the surface.

(d) *The response of sclerophyllous woody plants to lack of water*

The transpiration of sclerophyllous trees and shrubs, in spite of retaining most of their leaves, is so reduced during the dry season that one cannot employ the method of short-period weighing. Instead, one has to use large branches and work with exposure periods of an hour. If the experiments are carried on throughout the day with the same sections of branch one observes that the values do not decline continuously as in a dessication curve, but they alter in parallel with the rate of evaporation, i.e. they first increase to noon and then decrease again in the afternoon (Fig. 248). From this one concludes that the stomata of these

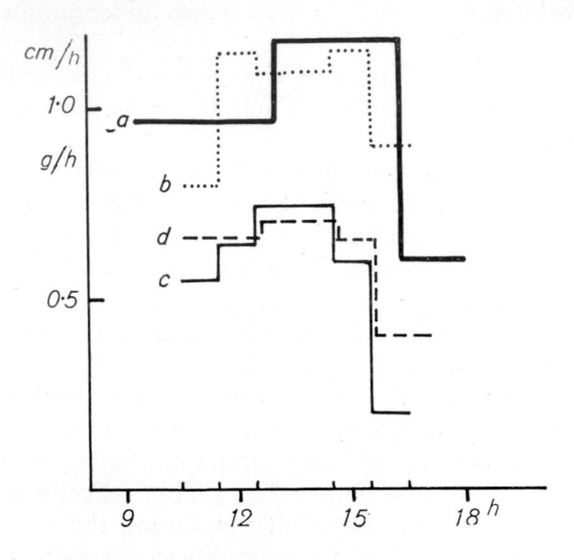

FIG. 248. Transpiration (in g/h) of sclerophyllous woody plants during the dry season (on Aug. 28, at the Veld Reserve, Fauresmith, South Africa): *a*, evaporation (in ml/h); *b*, *Rhus erosa*; *c*, *Rhus ciliata*; *d*, *Osyris compressa*. Weather: cloudless, calm air. Dry and wet bulb temperatures at 9 a.m., 15·4°—9·0°; at 1 p.m., 22·5°—12·4°; at 4 p.m., 22·0°—10·5°; at 6 p.m., 19·2°—19·0°; minimum temperature, preceeding night, −2°C; following night, + 1C°.

sclerophyllous plants are shut. A different response was only shown by individual, fresh-looking shrubs, or in those that showed some growth. In these, the transpiration values decreased slowly but continuously.

Information on their hydrature is given by the osmotic values (atm) in Table 73.

Since the values are always high in sclerophyllous plants, these values show no unusual increase. Thus, these plants maintain their water balance even under very unfavourable conditions.

Tarchonanthus has relatively soft, very pubescent leaves. By contrast, *Osyris* has somewhat succulent leaves but it is not a halophytic species (chloride content only 14%, sulphate content 15%). *Cussonia spicata* (Araliaceae) has a very different habit. Its stem is succulent and has a rosette of very large, incised leaves at the apex, each of which weighs more than 20 g. This *Cussonia* species is called the cabbage tree. The leaves, which together weighed 44 g, only lost 1 g water per day. *Cussonia* represents a special type of succulent tree.

TABLE 73

Olea verrucosa	37·4
Chilanthus arboreus	33·3
Euclea undulata	29·8
Osyris compressa	34·7–35·8
Tarchonanthus minor	29·0
Rhus erosa	27·5
Rhus ciliata	24·6
Rhus burchellii	36·8
Rhus burchellii (more fresh)	29·2
*Viscum capense**	30·7
Viscum rotundifolium†	57·1

* Growing on fresh *Rhus burchellii*.
† Growing on dry *Ehretia hottentotica*.

The main centre of distribution of this species is further east, in the hinterland of East London, which has a less pronounced dry season. *Cussonia* certainly survives the dry season easily on its stored water, particularly in the east. Yet some water uptake from the soil must also be necessary. This applies to all sclerophyllous plants. The water loss of an area covered with sclerophyllous plants can be estimated as 1/22 mm rain-equivalent per day, i.e. a water loss of 8·2 mm in 6 months. This is not much, yet it is 10 times the amount that I calculated for the *Catophractes-Rhigozum* community (see p. 244).

A daily water loss of 1/20 mm was calculated for the dry-grassland of the same area, i.e. 9 mm during the dry season. Determinations of the soil water content showed that there always appears to be a sufficient reserve of water available in the soil region exploited by roots.

Thus it can be seen that each different ecological life-form is well adapted to its survival during the dry season, i.e. each species is adjusted to its corresponding soil conditions.

References

HENRICI, M. 1940a. The transpiration of Karroo Bushes. *Sci. Bull. Union S. Afr., Dep. Agric. & For.*, **185**.

HENRICI, M. 1940b. The transpiration of large Karroo Bushes. *S. Afr. J. Sci.*, **37**, 156-63.

MARLOTH, R. 1908. Das Kapland. *Wiss. Ergeb. Dtsch. Tiefsee-Exped.*, **2**(3).

OPPENHEIMER, H. R., and HALEVY, A. H. 1962. Anabiosis von *Ceterach officinarum* Lam. et D. *Bull. Res. Coun. Israel, D*, **11**, 127-47.

WALTER, H. 1939. Grasland, Savanne und Busch der ariden Teile Afrikas in ihrer ökologischen Bedingtheit. *Jb. wiss. Bot.*, **87**, 750-860.

XII. *The Arid Regions of Central Australia*

1. A comparison of Australia with Southern Africa

In order to compare its latitudinal position with South Africa, Australia has been shifted 100 degrees to the west in Fig. 249. It can now be seen that Australia coincides with Southern Africa, including Madagascar, provided the latter were extended 14 degrees further south. The superimposed outlines show that west Australia coincides latitudinally with the Republic of South Africa, the southern half of south-west Africa,

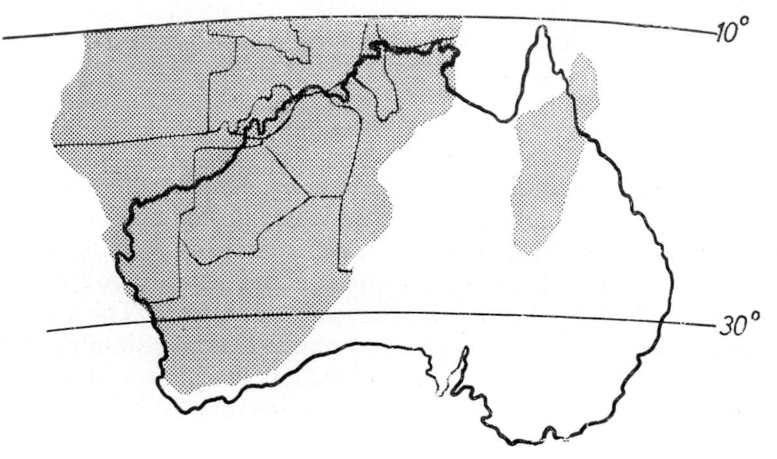

Fig. 249. A comparison of the latitudinal position of Australia with that of Africa (dotted).

Botsewana and Rhodesia. The Northern Territory of Australia can be compared with Mozambique, and east Queensland with Madagascar. The similarities in climates are obvious from the climatic diagrams (Figs 250-55).

The west coast of Australia is not quite as dry as the Namib desert on the west coast of South Africa. This difference is due to the absence of any cold current along the west coast of Australia corresponding to the cold Benguela stream bordering the Namib. Central Australia corresponds with the South African Kalahari, which is wrongly called a desert. It would better be called a 'thirsty-land' for the sandy deposits soak up all rain water. The Kalahari has more rainfall than the area to its west in south-west Africa. Consequently, the Kalahari vegetation is also more dense.

Climatic deserts are also absent in the interior of Australia. Its arid areas are, however, geographically more extensive than those in South

Africa. This is related to the longer east-west axis of the Australian continent. Geographical maps of Australia show several deserts in this region, for example the 'Simpson desert' north of Lake Eyre and west of this the 'Gibson desert'. This is really a problem of definition. Just

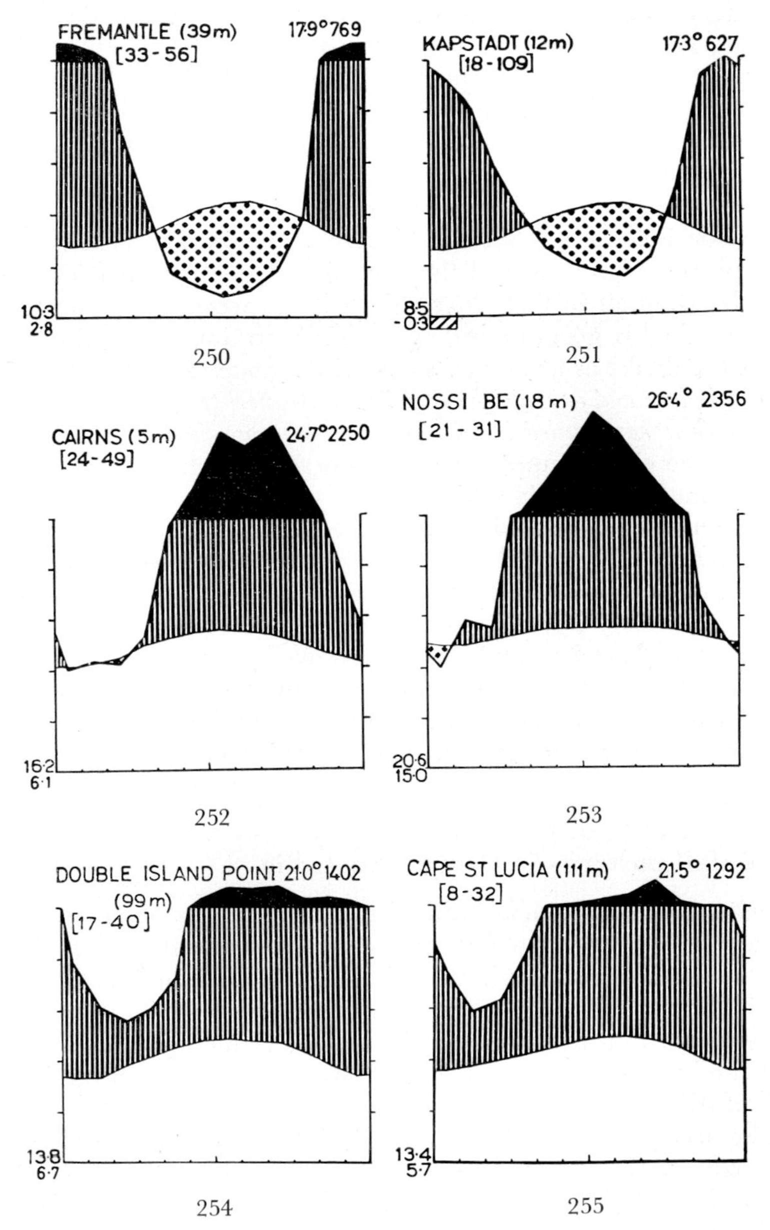

FIGS 250-255. Stations with corresponding climatic types. The stations in Figs 250, 252 and 254 are in Australia; those of Figs 251, 253 and 255 in Africa (from *Klimadiagramm-Weltatlas*).

as in the Kalahari, these so-called deserts are characterised by the absence of surface run-off. In part, they are wide sandy areas with parallel dune-ridges, in part, they are stony plains.

To the south is the Nullarbor Plain (*nulla arbor* = no tree) which extends more than 700 km from west to east and over 300 km from north to south. It is a highly effectively drained plateau ('Karstland') consisting of Tertiary lime outcrops (Adamson and Osborn, 1922). In the northern part Cretaceous stones also occur at the surface, in the south, Miocene deposits. These fall as coastal cliffs, 60 m high, to the sea. The layers of limestone are about 250 m thick and the small rainfall immediately percolates them. The limestone surface is, in part, covered by a red cover of soil, 30 cm deep. Doline formation has been observed and the erosion holes are connected to cave channels. The lack of surface water prevents habitation of this area, just as in the other 'deserts' of Australia (rainfall in these 'deserts' varies between 150 to nearly 500 mm). So, in his geographical work on Australia, G. Taylor (1951) defines the desert as *an area with low rainfall, with sparsely distributed and specialised plant and animal life, affording no possibility of fixed ranching even where ranchers have settled in adjacent areas for more than* 50 *years*. By this definition he clearly emphasises the time factor, which plays a role in the exploitation of newly settled areas.

By contrast to the arid inland the eastern slopes of the mountains have a very heavy rainfall both in South Africa and Madagascar, and the same applies also to Australia. Here, however, the mountains are not very high but east Queensland can be compared with the east coast of Madagascar (Figs 252 and 253, respectively). In the same way the eastern part of New South Wales can be compared with the coastal area of the Drakensberg mountains. By contrast, Victoria and, even more so, Tasmania are so far south that they have no analogous land areas in Africa.

2. Rainfall patterns in Central Australia

When analysing the annual rainfall in Australia one observes that the region receiving less than 250 mm extends to the west coast near Cape Carnavon and Onslow and, at Eucla, nearly to the south coast. To the east the border lies near Wilcannia and Broken Hill. The region with less than 200 mm of rain is much smaller. It includes the Nullarbor Plains and the area north of Port Augusta; the northern border runs south of Alice Springs.

The area with less than 150 mm rainfall is very small. It is only represented at the stations Charlotte Waters, Williams Creek, Cook and Mulka. The latter, lying 50 km east of the eastern shore of Lake Eyre, has the smallest measured mean value with 101 mm (Fig. 256).

However, we know that for the ecological evaluation of arid areas it is also important to know the seasonal distribution of rainfall in addition to the total amount. In this respect the climate of Australia is quite unique and totally unlike all other parts of the world.

The desert areas in the northern Sahara receive a little rain in winter, e.g. in Cairo (30°N); rainfall is practically absent from the

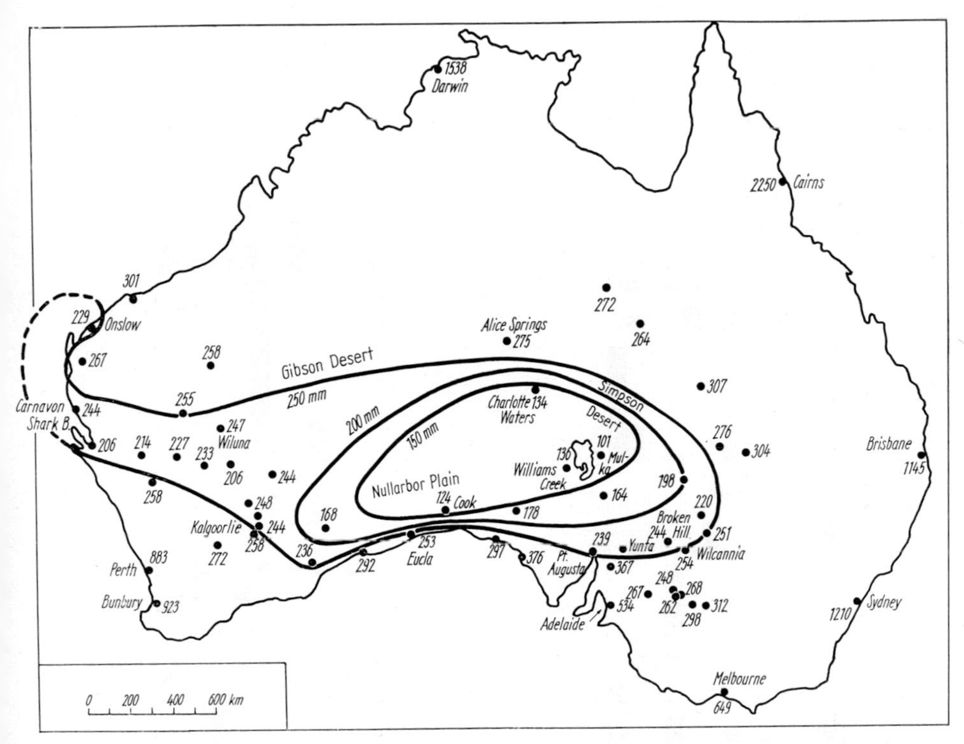

FIG. 256. Arid areas in Australia with annual rainfall below 250 mm, 200 mm and 150 mm respectively.

central part and, in the south (20°N), rain only occurs during the summer months. Australia is too small a continent to ensure a sharp distinction of the rainfall into annual seasons. The tropical cyclones commonly extend southwards beyond 30°S (Gentilli, 1955) and the Antarctic lows may just bring rains to the northern part of the arid zone. For this reason, we find predominantly winter-rain in the south and, in the northern part of central Australia, predominantly summer-rain. In between occur stations which show an almost level rainfall curve so that one gets the impression that there is some rainfall in every month of the year. However, this impression is false; it is due to the means being based over many recording years. In this region rainfall is very unpredictable. There are years with almost no rain and others with heavier rainy periods. However, when it does rain, even once, *it can occur at any time of the year*. No season is marked out in any way.

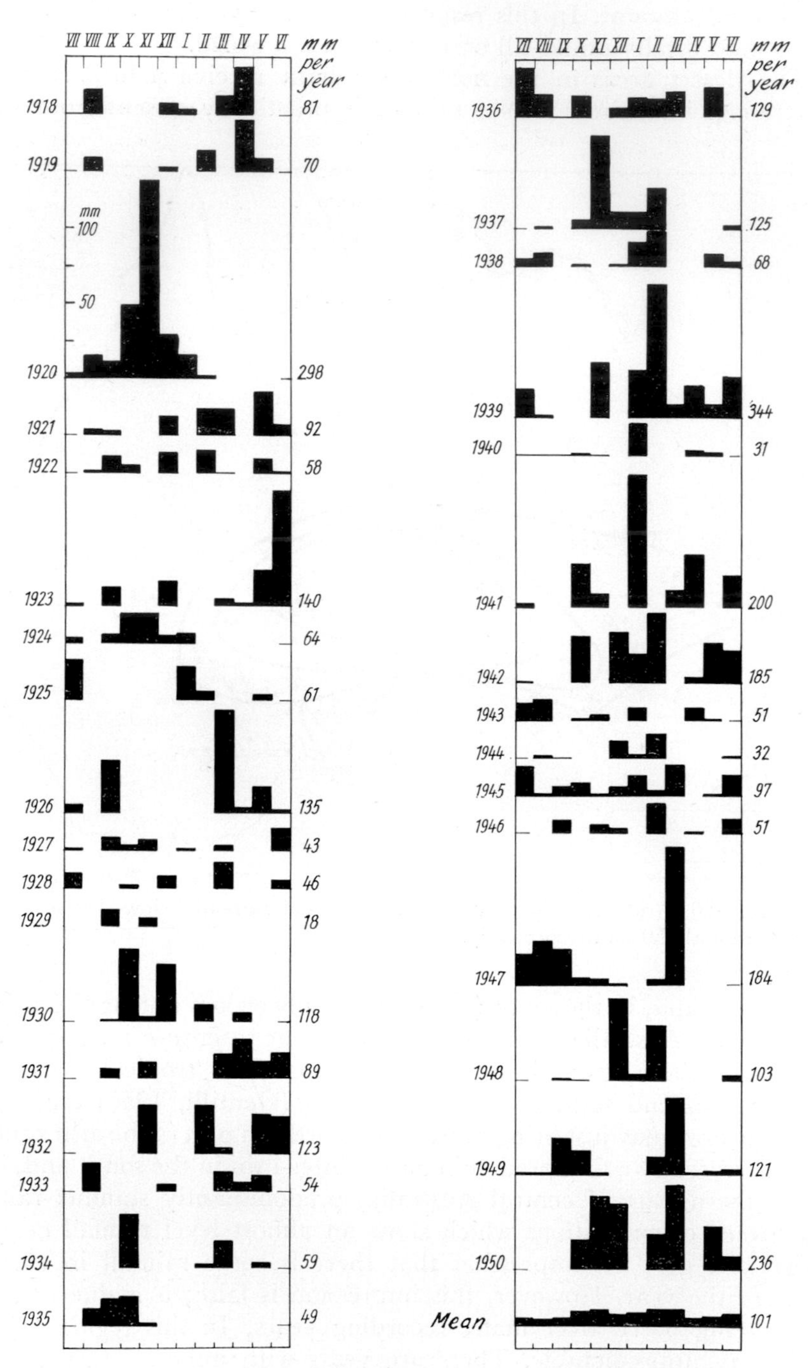

FIG. 257. Amount and distribution of rainfall for the years 1918-50 at Mulka (50 km east of Lake Eyre).

To provide an example I shall analyse the amount and distribution of rain for 1918-50 at the driest place (Mulka station; Fig. 257). The annual rainfall varies greatly from year to year. An annual fall of 18 mm in 1929 (only 18% of the mean) contrasts with 298 mm (nearly 300% of the mean) in 1920. Thus, the rainfall of the driest years compared with those of the wettest gives an approximate ratio of 1:16. If one asks whether the distribution about the mean of 101 mm is uniform one can see from Fig. 258 that this is not the case: 19 more dry years are balanced against only 14 wetter years. Dry years with a rainfall of between 40 and 60 mm occur most frequently.

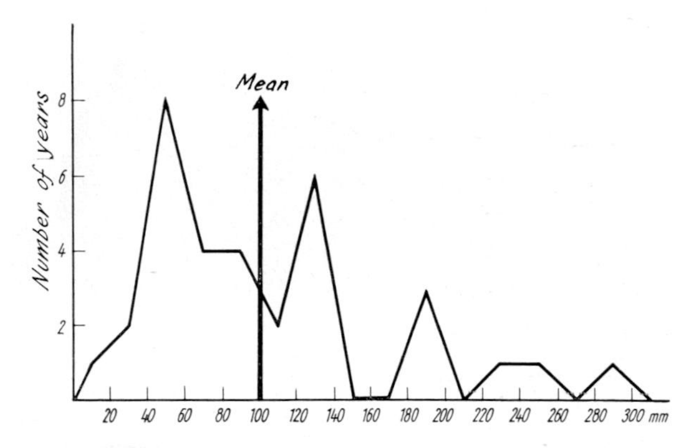

FIG. 258. Frequency curve of annual rainfall at Mulka (compare with Fig. 257 and 280).

Equally irregular is the distribution from month to month (Fig. 257). Heavy-rain months, with amounts in excess of 50 mm, occur in October, November, December, January, February, March and June while rainfall in July, August, September, April and May only rarely exceeds 25 mm.

Vegetation in this area, therefore, must be adapted to withstand periods of drought exceeding a year but it must also be able to utilise rainfall at any time of the year, whenever it happens to occur. In this area a special type of vegetation, the chenopodiaceous dwarf-shrub semi-desert, is found.

In the Namib, pure grassland occurs on sandy soils forming a transition zone from desert to savannah with a summer rainfall of 100-200 mm. Such grassland is absent from Australia, provided that we ignore the sclerophyllous *Triodia* plains (see p. 422). Pure grassland extending over wide areas is found in the summer-rain region but in a climate with an annual rainfall of 250-270 mm. These are the Mitchell grass plains (Fig. 259), which are dominated by *Astrebla* species. The Mitchell grass plains are restricted to Cretaceous marl or alluvial soils (Warrego and Balonne rivers in the interior of Queensland). These heavy soils are very wet during the summer-rain season but dry up hard during the dry season with cracks one metre deep. The soil profile was described by Everist as:

0-15 cm Grey or grey-brown clays, loose, friable;
15-75 cm Grey clays, firm, cracking apart into large blocks;
75-125 cm Yellow-brown clays with precipitation of gypsum. Weathered parent material lower down.

Thus, this grassland is edaphically determined and this is further indicated by its being surrounded by vegetations growing on other parent materials. These are *Eucalyptus* savannah or *Acacia* scrub.

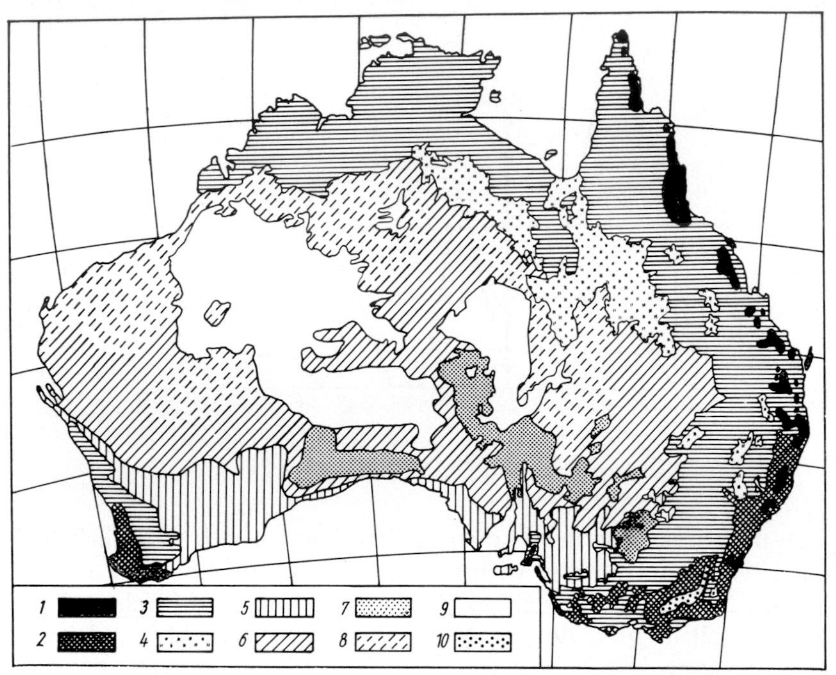

FIG. 259. Vegetation of Australia (schematic after several sample maps): 1. Tropical-subtropical rain forest with Indo-Malayan floral elements; 2. Wet *Eucalyptus* forest; 3. Dry *Eucalyptus* forest; 4. Mitchell (*Astrebla*) grassland; 5. Mallee (open *Eucalyptus* scrub); 6. Mulga (scrub formed by *Acacia aneura*, etc.); 7. Saltbush (*Atriplex-Kochia* association); 8. Plains with much porcupine grass (*Triodia*); 9. Unpopulated areas described as 'deserts'; 10. Alpine above the tree limit (only in the southeast, Australian alps).

The *Astrebla* species undergo a resting period in the winter months (April-September). When I travelled through this area in 1958, the rain was delayed and the grass plains looked completely dead until the beginning of December. After rain the new grass shoots emerge very rapidly, vegetative growth is completed in 2 weeks and the seeds have already ripened after 6 weeks. According to Everist, the associated plants are a few perennial herbs (*Boerhavia*, *Sida*, *Psoralea*, etc.), annual grasses (*Iseilema*, *Dactyloctenium*) and herbs (*Salsola kali*, *Amaranthus*, etc., and after sporadic winter-rains, *Lepidium*, *Helipterum* etc. also), and a few perennial grasses, such as *Digitaria* and *Dichanthium* (Davidson, 1954).

The bunches of *Astrebla* often only cover 4% of the ground with their bases. The rhizome is well protected during drought by clustered, dry, leafy shoots. The Mitchell plain grasses cannot withstand immersion for long and in such places one regularly finds *Marsilia hirsuta*.

3. The vegetation (Fig. 259)

Beadle (1948) has provided a thorough description of the vegetation and soils of the arid regions of the western part of New South Wales. Here, and in areas with the lowest rainfall, the saltbush vegetation occurs.

The most important species belong to the genera *Atriplex* and *Kochia*. Among these, *Atriplex vesicaria* (Saltbush) and *Kochia sedifolia* (Bluebush) are the most widely distributed species. Both dwarf-shrubs grow to about 30-60 cm. This vegetation type begins in the west near Kalgoorlie. Here, *Kochia* and *Atriplex* mostly occupy shallow hollows between very open stands of *Eucalyptus*, so forming a macro-vegetation mosaic. They also occur to some extent within the *Eucalyptus* stands.

The *Atriplex-Kochia* association is also found in places on the Nullarbor Plains where the limestone is covered by a thin layer of red loamy soil. *Kochia* appears to be more frequent here than *Atriplex*. In small hollows *Pittosporum phillyraeoides* occurs here and there as a small arborescent shrub. Otherwise, one only observes a few annuals after rain, such as *Salsola kali*.

The main distribution of the chenopodiaceous dwarf-shrub semi-desert is in the northern part of the State of South Australia and it extends from here eastwards to Broken Hill in New South Wales.

Kochia sedifolia and *Atriplex vesicaria* form both mixed and pure stands. *Kochia* is supposedly restricted to habitats with deep calcareous soils.

In order to understand the structure of the soils in this region, one must consider climatic conditions of the past. A. W. Jessup (personal communication) gives the following account:

'In the Tertiary the climate was much cooler in South Australia. For example, *Nothofagus* was widely distributed, as shown by fossil records. We also know from palaeomagnetic studies that Australia had a higher latitudinal position at that time. Then a warm period occurred associated with some fluctuations which continued into the Pleistocene' (see also Burbridge, 1960).

The post-glacial period was also characterised by several climatic cycles, so that the climate was at times arid and at others, wetter. According to Jessup the present climate in the northern part of South Australia is drier than in the earlier period.

As evidence for this one can cite:

1. The Nullarbor Plains are today barren of soil. Soil was present at an earlier date and subsequently it was blown eastwards with the increasing aridity.

2. Also on the 'Gibber plains' (stone desert) one finds in the heart of the stone pavement small stone-free hollows, covered with vegetation (*Atriplex ragodioides*, *A. nummularia*). The development of these plants may be explained as follows: sand collects around the shrubs under arid conditions, so that eventually they stand on small sand dunes. When a flood sheet sweeps across the plain after rain, carrying with it mud and stones, the latter become deposited in between the sand hills. The fine material is then blown away and the stones are left to form stone pavement. The scrub then dies off with increasing aridity and the sand is blown away, creating shallow hollows in place of the little dunes. With an increasingly damp climate water collects in the hollows which are invaded by new shrubs. Wherever the whole area is exposed to blown sand, new, little sand hills can form. Recurrence of flood sheets again leads to the formation of stone pavement between the sand hills. In fact, one can observe several layers of stone pavements alternating with sand in the soil profile. These are in turn interrupted by islands of vegetation below which stones are absent.

3. Salt pans are blown away during dry seasons. In this process dunes are formed around the salt pans in the direction of the strongest winds (not the prevailing winds). In Western Australia I noted these to be always on the west side despite the prevailing west winds; yet the east winds are the stronger ones. The dunes consist either of very loosely deposited gypsum, forming a surface crust after rain, or of sand deposits. This pattern indicates a change of climate. The easily soluble salts are carried a long way as dust and lead to a slight increase in salinity in surrounding areas through being incorporated into the surface soil in powder form. It is difficult to determine the origin of the salt in this area, because Cretaceous deposits occur around Lake Eyre (actually a pan) while, further to the south, Tertiary marine sediments are exposed. The salt content of the rain can be attributed to locally produced salt dust.

The chenopodiaceous semi-desert has, in part, become greatly degraded by grazing, since sheep prefer to feed on *Atriplex vesicaria* and *Kochia sedifolia*. According to reports, nine-tenths of the north-east area of South Australia has been cleared of saltbush through overgrazing.

The most important stand-forming species in the remaining arid areas are *Acacia aneura*, known as Mulga, and *Triodia* species, generally called Spinifex.

In Australia, *Acacia* species replace the dominant genus *Eucalyptus*, represented there by 600 species, wherever the climate becomes too dry for the Eucalypts. The Eucalypts are not tied to a certain type of climate. They grow in winter- and summer-rain areas, and in the east in climates that are wet all the year round. In this last climate, however, the Eucalypts are replaced on basaltic soils by tropical-subtropical rain forest (see p. 226). Individual species of *Eucalyptus* are very greatly specialised in regard to their habitat requirements but they can all grow on rather poor soils. They form giant trees in the wettest climate of their area of distribution, where they often exceed heights of 75 m (*Eucalyptus diversicolor* in West Australia, *E. regnans* in Victoria and in Tasmania). With decreasing rainfall they become progressively smaller until they are represented only by the mallee-form on the border of the winter-rain area (see Fig. 259). These are shrubs whose branches spring from a large, subterranean lignotuber. *Eucalyptus* penetrates the arid area about as far as one can recognise an annual wet season on the climatic diagrams. Within the arid area proper one finds the tree-like *Eucalyptus camaldulensis* (= *rostrata*), but there it always grows in dry valleys with ground water.

In addition to the predominant *Acacia aneura* (Mulga) many other acacia species are found in the arid region. They also penetrate together with *Casuarina* species from sand dunes into the saltbush area. All these are evergreen *Acacia* species with phyllodes. As a general rule, among the phyllode-bearing Acacias those with thin, cylindrical phyllodes are commonest in the driest parts, while those with flat phyllodes are most common in the less dry areas. There are two forms of *A. aneura* itself, with cylindrical or flat phyllodes. Both forms can occur together in the same habitat. However, on the wet border of its distribution, both in Western Australia and also in Queensland, one finds only *A. aneura* with broad phyllodes.

Wherever summer-rains occur in addition to winter-rains in the Mulga-region, *A. aneura* is replaced on sandy soil by some species of grass of the genus *Triodia* and, less frequently, by *Plectrachne*. These cushion-forming grasses are also known as Porcupine grasses, since their inrolled, mainly sclerenchymatous leaves project terminally as a sharp pointed thorn. They are generally called 'Spinifex' in Australia (not to be confused with the genus *Spinifex*, which occurs on tropical beaches).

Sclerophyllous grasslands occur as pure stands over wide areas in the west, inland from Sharks Bay and Cape Carnavon in the region of Murchison River. Here rain falls only in the summer (on average 250-300 mm). However, the summer-rains are very unpredictable. The grassland is made up mainly of *Triodia pungens* and the same species recurs also in the interior of Australia in climates with heavier summer-rains. In this area, however, it is restricted to dry, stony slopes.

So, we see that these sclerophyllous grasses behave quite differently from normal grasses (see pp. 242ff). Although they prefer summer-rain climates and sandy soils they are also found on stony habitats like the sclerophyllous woody plants. Because of the sclerophyllous structure of their leaves they are able to survive very long periods of drought and can invade areas with intermittent summer-rains. This applies also to the sclerophyllous grasses of south-west Africa, where they occur on dunes on the coast and in the southern Kalahari but, in addition, in the southern part of the country in areas receiving drainage water after rain (*Eragrostis spinosa*, *E. cyperoides*, *Aristida namaquensis*, *A. brevifolia*). During rainy periods, these have more favourable soil-water relations. *Stipa tenacissima*, in North Africa, can also be considered to belong to this group of sclerophyllous grasses.

Typical grasses with soft leaves that dry up during the dry season are also found in Australia but only where the summer-rains are more regular and also further to the north. These are mainly Andropogoneae and Paniceae and the introduced grass *Cenchrus ciliaris*.

Typical of the arid areas of Australia are hard pan soils. However, while hard pans in other arid regions of the world are formed through lime deposition, they are formed in lime-poor Western Australia from silica concretions of extreme hardness. Silica is extremely mobile in these arid soils and forms a hard layer at a depth of 15-30 cm. This layer is of a plate-like structure and can be of variable thickness. Soils with such hard pans provide habitats especially suited to *A. aneura* scrub. In addition, there are other shrubs and semi-shrubs, especially frequent are *Eremophila* spp. (Myoporaceae). Otherwise the soil is barren. It is only covered with ephemerals after a good rain. After winter-rains these are made up largely of composites, in particular the genera *Helipterum*, *Helichrysum*, *Waitzia*, *Schoenia*, etc.; after summer-rains grasses and *Trichinium* spp. (Amaranthaceae) occur. In wetter years the soil in shallow hollows is often covered with flowering ephemerals as during my 1958 expedition.

Salt pans are very widely distributed in Western Australia. There are at least 200 large ones and, in addition, numerous smaller ones. They overlie a granite shield and represent the remains of an ancient river system, from the wetter Pleistocene, which has become blocked in many places by drift-sand and dune formation. Lakes which form after rain can even become partially joined after very strong downpours (Fig. 260). The halophyte vegetation around such pans shows a characteristic zonation which becomes more complex when associated with gypsum dunes.

In conclusion I shall give a description of the Simpson desert based on Crocker (1946). The Simpson desert is the central sandy part of an unpopulated area north of Lake Eyre. This area was formerly known as 'Arunta desert'. It also includes areas covered with stones and gravel. The sandy desert is about 480 km long and 400 km wide. It consists of sandy ridges which are about 250 km long with a height of 7·5-30 m. These are orientated parallel to the south-east winds at distances apart of about 250-500 m. Only the uppermost ridge areas of the dunes are mobile, so that the east slopes are steeper than the west slopes. This is probably due to a westerly component of the wind. In the north the ridges grade down to a large sand flat. In between clay and salt pans occur.

This is considered to be the most extreme desert in Australia. Among the large sand deserts of the world it is ranked fifth in aridity. More extreme deserts are, in decreasing sequence: The central Sahara

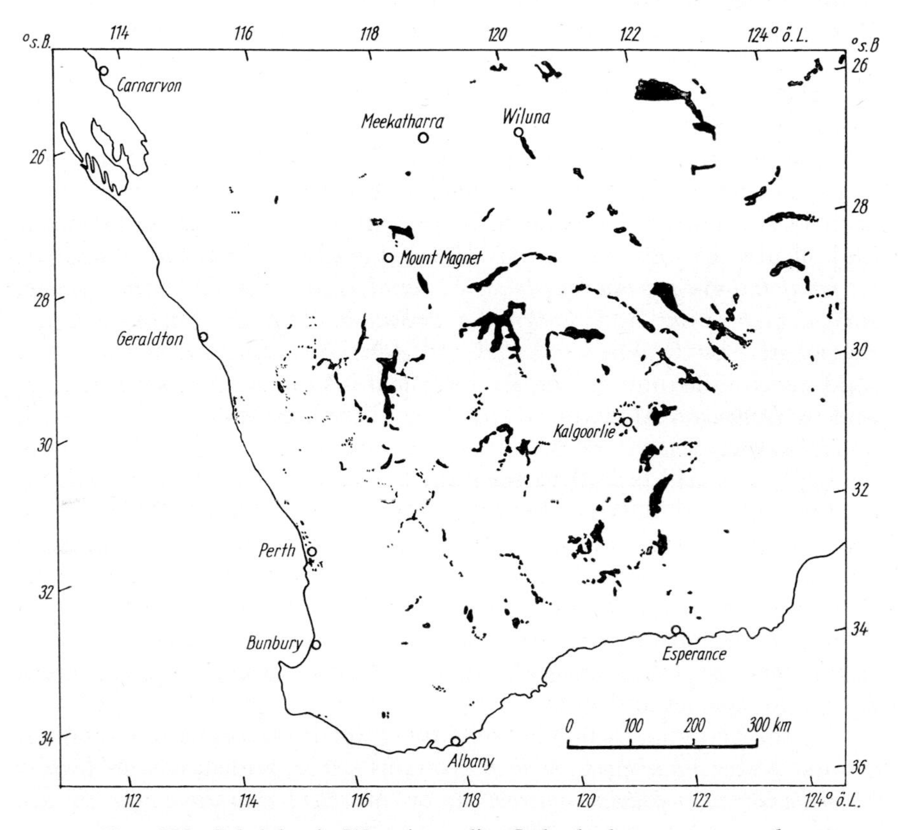

FIG. 260. Salt lakes in West Australia. Only the larger ones are shown. They are absent in the wet coastal zone and in the limestone area of the Nullarbor Plains. For the origin of the salt see p. 293ff.

including the Libyan desert; Takla Makan (south of Sin-kiang); Rub'Al Khali (Southern Arabia) and the north-western Sahara. Less extreme deserts are: Kara Kum (Turkestan); Chihuahua (Mexico) and Thar or Sind (India). However, no rainfall data are available from the Australian desert.

Rainfall is very irregular and has been estimated as between 100 and 250 mm per year with an average of about 150-200 mm. The sands have a high pH of 6·5-7·3 on the dunes and 6·4-8·8 in the dune hollows. The dunes are nearly lime-free and contain no soluble salts.

The sandy soil overlies a brown desert soil covered with stone pavement. These soils occasionally outcrop at the surface amongst the sand. Their pH is 7·6-8·7 and their salt content is 1·3-1·5%. Well-drained depressions are less salty. The grey and grey-brown alluvial soils contain 50% clay (pH 8·1).

The dune ridges are unstable and covered, usually loosely, with *Zygochloa paradoxa*. Associated perennials are *Ptilotus latifolius*, *Sida corrugata* var. *pedunculata*. On the dune slopes *Crotolaria cunninghamii* and *C. dissitiflora* grow. Annuals occurring after rains include: *Salsola kali* and several grasses (*Plagiosetum refractum*, *Eriachne aristidea*), *Trichodesma zeylanicum*, and, deeper rooted are *Blennodia pterosperma* and *Goodenia cycloptera*.

The most important community here is also formed by Porcupine grass, *Triodia basedowii*. It covers the lower dune slopes, the dune valleys and, in the north, the dune flats. In addition, scattered herbs occur and shrubs are also present: *Hakea leucoptera*, *Grevillea stenobotrya*, *G. juncifolia*, *Eremophila longifolia*, *E. latrobei*, also the Chenopodiaceae, *Rhagodia spinescens* and *Euchylaena tomentosa*. The Euphorbiaceae are represented by *Euphorbia wheeleri* and *Adriana hookeri*. *Triodia* acts as a sand trap on the dune slopes. In this role it is supported by some wattles and by *Dodonaea*. After the winter-rains there are also composites, such as *Helipterum*, *Schoenia rosea*. This vegetation, therefore, represents only a slightly less arid type than the sand vegetation of West Australia.

On the sand flats to the north one finds the *Triodia basedowii*—*Grevillea juncifolia*—*Eucalyptus pachyphylla* community. A community developing on flats with stone pavement includes *Astrebla pectinata*, *Atriplex vesicaria* and *Bassia* species. On stony soils the Chenopodiaceae dominate and on stony ridges, i.e. in particularly favourable habitats, *Acacia cambagei*, *A. tetragonophylla*, etc., *Cassia desolata*, etc., different *Eremophila* species and so on.

On the periodically flooded soils along the dry rivers with permanent ground water *Eucalyptus coolabah* grows as a tree, and stands of *Atriplex nummularia*, *Chenopodium auricomum*, or *Muehlenbeckia cunninghamii* are found.

The taxonomic diversity is still greater on level areas where overspill water from rivers penetrates after rain.

Lake Eyre is fringed with gypsum-containing dunes. These are characteristically occupied by the thorny zygophyllaceous, *Nitraria schoberi*, which forms large semi-globose bushes. On the dune ridges *Zygochloa paradoxa* occurs again. The only shrub to occur here is *Hakea leucoptera*; among annuals *Salsola kali* occurs in great numbers.

We see, therefore, that this most extreme 'desert area' still has a relatively dense plant cover totalling 350 species belonging to 50 families. Of these species, 76 are characteristic for the sand-dune areas, whose desert-like character is emphasised by the dunes. The sand with its sclerophyllous grasses that are so unsuitable for grazing is the main obstacle to settlement here.

Beadle and Tchan (1955) were able to isolate *Azotobacter, Clostridium* and Cyanophyceae (*Nostoc, Anabaena*) from clay soils in these arid areas and also from stony soils covered with some vegetation. The soils always contain nitrate. Those algae which grow beneath translucent quartz stones that outcrop at the surface are particularly well developed, just as described for the Namib. In places where sheet-erosion has occurred one can often find one-sixth of the soil surface covered with 0·5-3·5 cm, translucent, quartz pebbles, half of whose lower sides are covered with algae. Thus, the algal cover spreads over about one-twelfth of the total surface. These observers believed that these algae were able to fix nitrogen.

4. Ecological investigations

(*a*) *Saltbush semi-desert*

In 1925, Osborn (see also Osborn and Wood, 1923; Osborne *et al.*,1932, 1935; Wood, 1925, 1934, 1936) established a small research station in the *Atriplex-Kochia*—semi-desert in the north-east part of South Australia. He arranged for the fencing-in of a large tract of land in this area, which was badly degraded by over-grazing, and erected the research station within the protected property. Since that time he and Wood and several collaborators from the University of Adelaide have carried out ecological investigations at the station.

The experimental station lies within Koonamore station, a sheep farm of 300,000 ha size, 60 km north of the railway station at Yunta along the track to Broken Hill. The average rainfall amounts to a little over 200 mm.

The characteristic vegetation type is the *Atriplex-Kochia*—semi-desert on loamy soil with much lime in the **B** horizon. Scattered in between are sand dunes that contain $CaCO_3$ concretions, but no chlorides or sulphates. Sand flats are also present covering clay with iron-oolites to a depth of 150 cm. A gypsum horizon occurs at a depth of 90 cm. The sand dunes are covered by *Acacia aneura*, *A. burkittii*, *Eremophila sturtii* and *Stipa nitida* (a summer annual). On the sand flats *Casuarina lepidophloia* occurs with *Bassia uniflora* as undergrowth. On silty soils one finds tree-forming *Heterodendron* and *Myoporum* species, and also shrub-like *Eremophila* and *Cassia* species. The soil on which *Atriplex* and *Kochia* grow has a pH of 7·7 and a chloride content of 0·1%.

Atriplex vesicaria is a dwarf-shrub which can attain an age of 12 years. Like many Chenopodiaceae it has slightly succulent leaves which are shed during prolonged drought periods. The root system is very shallow as expected with the low rainfall and the roots grow at 10-20 cm depth above the lime hard pan and extend a long way laterally. Individual bushes, therefore, are well separated and only cover 10-25% of the surface. In very dry habitats they even cover far less surface area.

By contrast with *Atriplex*, *Kochia sedifolia* is said to reach a very old age and to have a deep root system that may reach vertically 3-4 m depth in crevices in the hardpan and to the same distance laterally. Therefore, these plants must grow in places where the rain water penetrates a long way. *Kochia* is said to be restricted to lime soils and to be able to survive longer droughts than *Atriplex vesicaria. Bassia patenticuspis* (Chenopodiaceae), which often occurs as a pioneer on over-grazed areas, has roots that also reach to a depth of 2 m.

TABLE 74

Salt content (as per cent of dry weight)

In the soil	0·05	0·07	0·12	0·68
In the plant	14·1	23·5	15·3	32·9

The question whether *Atriplex vesicaria* should be considered a halophyte has been discussed in the literature. Wood said not by pointing out that *Atriplex* does, indeed, occur on salt-soils but usually on those containing only about 0·1% chlorides. However, I believe that this question cannot be answered from observations of the soil situation but rather from the behaviour of the plants themselves. In this respect it has been shown that the leaves of *Atriplex* always have a high salt content, mainly of chlorides, even if the soil only contains traces (Beadle *et al.*, 1957).

According to Wood, *Atriplex vesicaria* contains 90% of water. The chloride content is always high. As shown in Table 74 no clear relationships exist between the salt content of the soil and that of the plant, yet an increase is indicated with a higher salt content of the soil.

TABLE 75

Soluble salts	Cl^-	CaO	K^+	Na^+	Sulphate
25%	7·78	2·20	6·20	6·00	traces

The salt content of *Atriplex* is higher in the dry period than in the rainy period.

An analysis of the leaves of *Atriplex nummularia* growing on salt-free soil in the Botanic Garden at Adelaide showed the composition (per cent/dry weight) given in Table 75.

Because of its salt content, *Atriplex vesicaria* showed an osmotic value of 40 atm; in *A. paludosa* it was 65 atm. Plasmolysis occurred in *A. vesicaria* at 50 atm, in *A. spongiosa* at 55 atm.

The fact that *Atriplex nummularia* stores salts was demonstrated in culture by Ashby and Beadle (1957). They compared the osmotic value of the expressed cell sap with that of the culture solution and found that the cell sap concentration constantly exceeded that of the salt solution by about 16 atm (Table 76).

The plants developed much better with the addition of salts, and addition of NaCl was more favourable than addition of Na_2SO_4.

All these observations and experimental results show that these *Atriplex* species are typical salt-storing halophytes. They are xerohalophytes which grow on soils having low salt contents on a dry weight basis. But, since these soils often have a low water content, the salt concentration in the soil is affectively very high.

This illustrates the pitfalls which arise in assessing the ecological status of plants after a short study of the habitat.

The leaves of *Atriplex* and *Chenopodium* have inflated, vesicle-like hairs. In Beadle's department it was shown that these hairs bring about loss of salts. The salt concentration in the cell sap of inflated hairs is

TABLE 76

NaCl in culture solution (mg/l)	Osmotic value of the culture solution (atm)	Osmotic value of the cell sap (atm)	Difference (atm)
0	0·3	16·0	15·6
50	1·7	17·6	15·9
200	7·9	24·0	16·1
400	14·7	31·6	16·9
600	23·3	41·5	18·2

very high. The hairs are short-lived and dry up rapidly but new hairs are formed constantly. In this way considerable quantities of salt can be removed from the leaves. Determinations of salt content and of the osmotic value using the cryoscopic method are subject to some error, since the salts from dead hairs remain on the leaf surface during dry weather.

Berger-Landefeldt also assumed that the inflated hairs functioned in salt secretion (cf. p. 456). The earlier contention of Wood that chlorides occurred mainly in the vascular bundles and in the chlorenchyma around the veins of the leaves and that the hairs, therefore, only contained small amounts would appear to be incorrect. The salts on the leaf surface would also explain the repeated observations of increased weight of the leaves in a damp atmosphere. This contrasts strongly with the response of *Eucalyptus* or *Acacia* leaves. The increase in weight is probably not caused by water absorption of the leaves but by a hygroscopic attraction for water by salts on the leaf surface (see pp. 157-8).

The transpiration of Chenopodiaceae under comparable conditions to that of Australian sclerophyllous plants (branch cuttings in water) is about 4-7 times smaller.

In the Chenopodiaceae the stomata are uniformly distributed on both sides of the leaf surfaces. The numbers of stomata per square millimetre (upper and lower sides) were determined (Table 77).

TABLE 77

Chenopodiaceae		Sclerophyllous plants (for comparison)	
Atriplex vesicaria	240	*Acacia aneura*	224
Atriplex stipitata	256	*Acacia* (8 species)	116–385
Kochia sedifolia	100	*Cassia sturtii*	220
Salsola kali	112	*Eremophila* (5 species)	164–249
Bassia (7 species)	46–104	*Eucalyptus* (5 species)	224–440

According to Wood the number of stomata in sclerophyllous plants are of the order of the figures shown in Table 78.

TABLE 78

Proteaceae	145±19·0
Leguminosae	277±21·7
Myrtaceae	307±32·0
Epacridaceae (on lower sides only)	302± 5·0

It was shown in his analysis that the pore area of the individual stomatal openings was inversely proportional to the stomatal number. Therefore, the total pore area would be more or less constant, irrespective of the stomatal number. However, this principle is probably not universally applicable.

Upon investigating the chloride content of other Chenopodiaceae, it was found that it was not as high as in *Atriplex*. Sulphates too remained below 0·5% of the dry weight. The osmotic values (atm) were recorded (Table 79).

TABLE 79

Kochia sedifolia	43	*Kochia pyramidata*	47
Kochia planifolia	35	*Kochia tomentosa*	38
Kochia triptera	47	*Bassia obliquicuspis*	47
For comparison:			
Acacia aneura	68	*Zygophyllum fruticulosum*	10
Acacia burkittii	55	*Zygophyllum prismatothecum*	7
Eremophila scoparia	50	*Lycium australe*	8
Myoporum platycarpum	43		

In order to check this earlier data I took several samples in the Koonamore Reserve and its nearby surroundings in 1958 on October 8 and 9 after a long, cool rainy period. The leaves were dried after determining their fresh weights and their dry weight and water content were determined. The dried material then served to determine the osmotic value and chloride proportion in Stuttgart-Hohenheim.

For this purpose water was added to the dried material until it reached the original fresh weight when it was kept at 90°C for 45 minutes in a closed container. After expressing the solution with a pressure of 100 atm, the osmotic value and chloride content of the press-sap could be determined. Control determinations with several species in Stuttgart showed that the error using this method is less than 1 atm. For example, in ivy a value of 16·5 atm was found as the mean of 5 samples determined by the normal method, while in 5 parallel samples treated by the method just described the mean value was 16·1 atm (see also Thren, 1934; Knodel, 1939).[1]

Since rain had occurred for a long period in the Koonamore Reserve before the samples were collected, the soil was thoroughly wetted. Therefore, the values can be considered optimal rather than extreme. Moreover, the rain had washed the salts, which would otherwise have been deposited on them, off the *Atriplex* leaves. Thus, this source of error need not be considered. The results are presented in Tables 80-82. In each case, Π stands for osmotic value in atm, Cl for chloride content (calculated as NaCl) in atm, and % for the proportion of chloride contributing to the osmotic value.

TABLE 80

Atriplex vesicaria, soil salinity estimated to increase from 1-6 relative units

Sample	Habitat description	Π	Cl	%
1	Sand dune with *Eucalyptus* (mallee), soil poor in salts	37·4	22·6	59
2	On sand, beneath *Casuarina*	53·5	32·0	60
3	*Atriplex* stand with *Myoporum*	55·4	42·4	76
4	Typical *Atriplex-Kochia* stand	65·4	45·2	69
5	Salt pan, adjacent to fresh-green *Arthrocnemum*	48·6	37·5	77
6	Sample of salt pan, Lake Way, West Australia (3.IX. 1958)	70·1	58·9	84

The values correspond to the expectations and discussion on p. 416. The low value of sample 5 can be attributed to a decreased salt concentration in the plant after rain. The corresponding values for *Arthrocnemum* are shown in Table 86, sample 2.

The constant, high chloride content shows that *Atriplex vesicaria* is a true halophyte. The leaves are markedly succulent; the water content amounts to 230-250% of the dry weight.

The osmotic values found for *Kochia* (Table 81) are only half as high

[1] As an average, 0·4 atm was given as the difference. This method is very useful when one does not have the use of a laboratory on an expedition. One only needs to weigh and dry the samples: with larger samples the accuracy of an ordinary pan balance as used in chemists' shops is quite sufficient.

as the single values shown in Table 79. However, my values were reproducible when taken as duplicate samples and correspond to the data of Harris from Utah. The chloride content is low and the proportion of chloride varies considerably. Therefore, we can consider *Kochia* only as a facultative halophyte. While the chloride content of

TABLE 81

Kochia species, Koonamore Reserve

Sample	*Kochia* species and habitat	Π	Cl	%
1	*K. sedifolia*, pure stand not far from Koonamore	25·0	5·4	21
2	*K. sedifolia*, same location, duplicate sample	24·6	4·6	19
3	*K. sedifolia*, adjacent to *Atriplex*, sample 4, Table 80	33·7	13·7	41
4	*K. sedifolia*, from Kalgoorlie, West Australia (28. IX. 1958)	26·3	3·3	13
5	*K. lanosa*, adjacent to *Atriplex*, sample 2, Table 80	22·3	10·2	46
6	*Kochia* sp., Lake Way (3. IX. 1958), salt pan	26·7	15·0	56
7	*K. georgeii*, Lake Way, rock-ridge	25·4	4·3	17

TABLE 82

Other species from Koonamore Reserve

Sample	Plant species and habitat	Π	Cl	%
1	*Casuarina lepidophloia*, within an *Atriplex* stand	25·8	8·5	33
2	*Casuarina lepidophloia*, sand dune, old plants	25·1	8·5	34
3	*Casuarina lepidophloia*, young root suckers	16·8	6·5	38
4	*Heterodendron oleaefolium*, in an *Atriplex* stand	27·5 (approx.)	3·0	11
5	*Heterodendron oleaefolium*, from top of a sand dune	24·2 (approx.)	1·1	5
6	*Myoporum platycarpum*, adjacent to *Atriplex*, sample 3, Table 81	33·5	6·7	20
7	*Acacia aneura*, sand dune, adjacent to *Atriplex*	23·0 (approx.)	5·0	22
8	*Acacia burkittii*, also on sand dune	24·1	6·4	26
9	*Eremophila sturtii*, beside *Acacia aneura*	26·1	3·6	14
10	*Eucalyptus oleosa* (mallee), sand dune, non-brackish	28·2	5·1	18

Atriplex (sample 4) amounts to more than 45 atm, that for *Kochia* in the same habitat only amounts to 13·7 atm, and that for *Atriplex* (sample 2) comes to 32 atm, that for *Kochia* (sample 5) to only 10·2 atm. The water content of *Kochia sedifolia* is similar to that of *Atriplex vesicaria*.

Table 82 contains values for other plant species typical of this area.

Probably none of these species is halophytic. However, the salt content of *Casuarina*, as determined in a slender shoot with reduced leaves, is remarkably high. *Casuarina* can at least be considered salt tolerant. This probably also applies to *C. equisetifolia*, which is commonly planted along beaches. The water content of my sample was 104% of the dry weight; in *Heterodendron* 80-90%, in *Myoporum* 154%, in *Acacia aneura* 87%, in *A. burkittii* 108%, in *Eremophila* 129% and in *Eucalyptus* 80%. Since *A. aneura* and *Heterodendron* also contain, in addition, mucilaginous material, they only gave a little press-sap. The values obtained are, therefore, not considered as very accurate.

In addition, I investigated two bright green, annual, decidedly succulent halophytes:

Zygophyllum iodocarpum, growing with flowers and fruits on the edge of an *Atriplex* stand, showed an osmotic value of 19·8 atm with a chloride content of 16·1 atm, corresponding to a proportion of 81%. In *Babbagia acroptera*, growing on a slightly brackish, drainage flat, the corresponding values were 20·8 atm, 8·7 atm (Cl) = 41%. The water content in both *Zygophyllum* and *Babbagia* amounted to more than 600% of the dry weight.

Similar relations to those in the *Atriplex-Kochia* semi-desert of Australia were found in the *Atriplex* and *Kochia* species of the arid region in south-western U.S.A. *Atriplex confertifolia* is commonly the only plant on the high plateaus of the Painted Desert, north of the Grand Canyon. The distribution of rainfall here can be very similar to that in the saltbush region of Australia.

Harris (1932) determined the osmotic values and, in part also, the chloride content of this species and of *Atriplex canescens*. The latter also occurs in the Sonoran desert in brackish habitats. I computed, in addition, the chloride content in per cent of the osmotic value (Table 83).

TABLE 83

Cell sap and chloride concentration of two *Atriplex* species (after Harris)

Atriplex canescens			*Atriplex confertifolia*		
Π	Cl	%	Π	Cl	%
21·7	8·4	39	24·8	4·5	18
24·5	9·6	39	28·3	12·1	43
27·7	6·5	23	38·6	36·2	94!
30·9	10	32	38·7	16·6	43
36·0	9·7	27	75·2	28·7	38
38·3	16	42	82·9	55	66
39·7	28·5	72	99·4	40·6	41
40·3	13·7	34	104·2	61·2	59
46·8	17·7	38	154·5	76	49
49·3	42·7	87!	156·2	93	59
107·7	51	47	202·5	80	39

These data show that these *Atriplex* species too are typical halophytes storing much NaCl in their cell sap. The proportion of sulphate is relatively small as estimated from only 3 determinations on *A. canescens*. They show that in cases of a low chloride content the sulphate content amounts to about half the former, while in cases of a high chloride content the sulphate content amounts to only one-tenth the chloride content. It is not surprising that there is no clear correlation between the level of the osmotic value and the proportion of chloride since the samples originated from salt-enriched and salt-poor soils which were both wet and dry. Water relations and salt relations are closely interrelated and determine the levels of the osmotic value in one case and of the proportion of chloride in the other.

In 75% of the samples the osmotic values were below 40 atm for *Atriplex canescens*. One sample with the extremely high value of more than 100 atm needs checking.

Atriplex confertifolia survives in much drier and more saline habitats. Here, the values were below 40 atm in less than 20% of the samples and, in nearly 50% of the samples, they exceeded 100 atm, in a few even 200 atm. However, these values are not completely unambiguous, since the salt of the dead, inflated hairs can cling to the leaf surfaces after a long dry season.

I established that *Kochia* species, in Australia, contain less salt in their press sap than the *Atriplex* species, this is true also for *Kochia vestita* in the U.S.A., according to Harris. These species have no salt-secreting hairs. The 29 samples from Utah gave osmotic values ranging from 18·9-54·6 atm and chloride proportions of 9-39%.

Thus, it seems that the vegetation in Utah and North Arizona is ecologically similar to that in the Australian saltbush. Climatically, the conditions should be analogous also but, in Australia, the winters are nearly frost-free while they can be very cold in the corresponding region in the U.S.A.

(*b*) *Spinifex grassland* (*Porcupine grasses*) *and Mulga* (*Acacia aneura scrub*) *in Central Australia*

The ecology of the interesting, sclerophyllous Porcupine grasses (*Triodia* and *Plectrachne*), which form the Spinifex grassland, has not been investigated much so far.

Among the 10 species of the most important genus *Triodia*, *T. basedowii* is distributed on the sand flats of the driest regions, while *T. pungens* occurs further north, from Western Australia to Queensland. The latter species also often covers the stony scree slopes of central Australia in pure stands.

In *T. basedowii*, the 50 cm-high cushions soon die in the centre and only the northward oriented runners continue to grow. Therefore, a cushion soon changes to a ring, then to a horse shoe open to the south, until, finally, the stand consists of partially disconnected arcs. Within the grassland the cover reaches 40%. By contrast, *T. pungens* maintains its cushion form for longer. The cushions are supposed to reach heights of up to 2 m with diameters of 4 m.

The internodes of the runners of these grasses alternate from long to short. Adventitious roots form at the internodes after rain, which root thoroughly in the upper metre of the sandy soil and extend down to 3 m depth (Burbridge, 1941, 1942, 1945, 1946).

The leaf blades are inrolled, as in *Ammophila*, and contain little chlorophyllous tissue but much sclerenchyma in several layers within the epidermis. The latter contains silica and resin-secreting cells, and the leaf surfaces are covered with resin. Such resin coats are especially characteristic in this area also for the genera *Eremophila* (Myoporaceae), *Dodonaea*, *Olearia* (Compositae), etc. The stomatal openings are always sunk on the outer as well as inner leaf surface.

The inner stomatal grooves are closed when the blades are tightly inrolled and the outer ones close through shrinkage of the parenchyma cells. Water losses are, thereby, reduced to a minimum.

These, therefore, are sclerophyllous grasses with a decidedly xeromorphic structure. Only such a structure enables the grass blades to survive extreme periods of drought. By contrast, a normal grass blade dries up quickly with shortage of water.

The young leaves of *Triodia* under favourable moisture conditions and high temperatures grow up to 3·5 mm per day. As a forage plant *Triodia basedowii* is useless, although the young leaves of *T. pungens* are eaten by cattle. As mentioned before a pure sclerophyllous grassland occurs north of the Murchison River (Western Australia), while in other areas there is only a macro-mosaic of Mulga (*Acacia aneura*) interspersed with *Triodia basedowii* plains on sandy soils. Borders between communities are not stable. If the annual rainfall is increased on average by 25-50 mm during a wet climatic period, one community may soon replace the other. Under such conditions, small soil differences or the routes of flood sheets after rain, can play a very important rôle. All these things influence the complicated, dynamic interrelationships of the different vegetation units. Anthropomorphic grazing practices also have enormous effects. Fire hardly plays a rôle in this open vegetation. Only the *Triodia* stands can burn, giving rise to dense, black smoke similar to that from burning resin.

Near Alice Springs a fire occurred after lightning, it extended over 50 km. After 18 months many young *Triodia* plants developed in the burnt area. In addition, much *Aristida arenaria* occurred and several herbs and numerous ephemerals. The absence of root competition from *Triodia* permitted other plants to become established. Otherwise *Triodia* takes all available water itself. The sclerophyllous leaves probably enable the grass to utilise the winter-rains as well. The leaves turn yellow after prolonged drought but they become green again after rain. A typical stand on red sandy soil showed the following composition:

Cover 50%, cushions 50 cm high, individual bushes up to 2m.

Triodia basedowii	3[1]
Plectrachne schinzii	1

in addition, a few scattered *Cassia* sp., *Eucalyptus gamophylla*, *Hakea* sp., *Grevillea juncifolia* and *Acacia aneura*.

The osmotic relations of these sclerophyllous grasses were investigated near Alice Springs by a research team under the direction of Dr R. O. Slatyer (1958). It seems that the grasses are very hydrostable and that they show high suction forces with small water deficits. It would be of particular interest to know whether negative turgor pressures develop in these plants, since these could then account for their drought resistance.

The most important species of the Australian arid areas is undoubtedly *Acacia aneura*, known generally as Mulga. This species dominates vast areas, which appear from a plane as an unending grey-brown sea, interrupted only by salt pans, sand dune areas and a few dry valleys.

A. aneura is very variable. Not only can the phyllodes be flat or cylindrical, but the growth form also differs, often markedly. Either a small main shoot is formed (up to 4-6 m tall), which gives off branches horizontally, or the stem divides at an early stage into a number of ascending branches, which in their turn develop secondary branches pointing upwards. Size and density of the stands frequently varies with the water relations of the habitat. In places the shrubs form an almost impenetrable thicket in small hollows. However, it is usually possible to progress slowly on foot through such thickets since the Mulga lacks thorns, in contrast to the African Acacias.

The phyllodes are also eaten by sheep. In extreme circumstances one can keep sheep alive without additional fodder by cutting the shrubs.

This *Acacia* has a very extensive root system, which penetrates cracks in the hard pan. In deep soils, the upper 2 m are thoroughly exploited by roots. Flat, radially extending roots serve for anchorage, others extend downwards at an angle and are strongly branched to provide the plant with water. The wood is very hard and the phyllodes appear to be varnished from their coating of resin. An ether extract can amount to 4-15% of the dry matter and the amount increases, in general, with the dryness of the habitat. The inflorescences consist of 2-3 cm long yellow spikes. Flowering is not related to any particular season of the year but always begins after a rain. Fruits and seeds only ripen after heavy rain. The Mulga cannot tolerate any salt but is otherwise very tolerant. In drier habitats, the area is often covered with rather open, 1·8-3 m tall, pure stands, which hardly have any under-

[1] On Braun-Blanquet's cover scale.

growth and only a few associated species (species of *Acacia*, *Eremophila* and *Cassia*, etc.). Numerous ephemerals only develop after heavy rain. The stands are then transformed suddenly into a sea of flowers.

The species composition varies with differing rainfall and time of year, so that one species may commonly predominate over a large tract of land.

The seeds of *A. aneura* only germinate when they can imbibe water for several weeks. Therefore, regeneration only occurs after especially good rain years. Seed viability is probably retained for quite a long time, as in most of the legumes with hard testas but no data are available. The trees mature in 30 years and are often parasitised by several *Loranthus* species.

The water relations of *A. aneura* were investigated by Slatyer (1960) near Alice Springs. In this area the mean rainfall is about 250 mm, but 75% falls during the 6 summer months. As in all arid areas, rainfall is very irregular. Dry periods in excess of 3 months are frequent. The measurements were carried out over 6 months and samples were taken at weekly intervals. He determined the relative water content of the phyllodes, the osmotic value and the suction tension. For the latter Slatyer used the method of relative vapour-pressure equilibrium that he had developed himself. The method was checked in Hohenheim (Kreeb, 1960; Kreeb and Önal, 1961) and was found to be quite reliable. The suction tension values are somewhat low, if respiration losses are ignored.

Slatyer only presented his results as curves to show the relationships between relative water content, suction tension and osmotic concentration. The maximum values of the latter appear to be calculated rather than determined. Suction tension values of 130 atm were endured by phyllodes without any damage. These values were greater than the osmotic values, which indicates a negative turgor. The strongly xeromorphic structure of the phyllodes may enable this to occur but a detailed discussion will have to await further research on this question.

I also took a few samples near Alice Springs after a long dry period, October 23 and 24, 1958. They were examined according to the method described on p. 418 ff, where I have already emphasised the difficulties with regard to *A. aneura*.

TABLE 84

Acacia aneura, 40 km north of Alice Springs

Sample	Appearance of the plant	Π	Cl	%
1	Old tree with needle-like phyllodes	25	4·5	18
2	Small bush with flat, grey phyllodes	27	4·8	18
3	Young bush with fresh-green phyllodes	26	3·5	13

This species appears to be extremely hydrostable (see also Table 82, sample 7), i.e. showing hardly any variation in its hydrature, so it must, apparently, be able to regulate its water balance very closely. The water content of the three samples as a percentage of the dry weight was 58, 65 and 80, respectively. However, these give no information about the water deficits. The differences are more likely to be related instead to structural differences in the mechanical tissues and cell walls.

It was not possible to take samples of *Triodia basedowii*, because of the construction of the leaves. However, samples were taken from 3 typical shrubs associated with the *Triodia* grassland. These were taken from a sand flat 100 km north of Alice Springs. The analysis of *Acacia kempeana* presented the same difficulties as in *A. aneura*.

TABLE 85

Shrubs in *Triodia* grassland

Sample	Species investigated	Π	Cl	%
1	*Eucalyptus gamophylla*, fertile juvenile form with succulent leaves	17·0	5·0	29
2	*Grevillea juncea*, with tough, cylindrical leaves	18·0	1·1	6
3	*Acacia kempeana*, with flat phyllodes	24·5	6·8	28

In spite of the long dry seasons the osmotic values of these species are low, as in *A. aneura*. A higher chloride content was not to be expected, since there are hardly any brackish soils in central Australia, despite the absence of outward drainage to the sea. The water content as a percentage of dry weight was 95 in *Eucalyptus*, 53 in *Grevillea* and 80 in *Acacia*.

Slatyer's research team observed an alternating pattern of strips of dense Mulga-scrub and grass, perpendicular to the slopes north of Alice Springs. It was found that the slope, on average 1:500, was not quite uniform, that of the mulga-strips was 1:600, while it is steeper in the intervening zones, 1:400. Consequentially, after rain, sheets of flood-water will be somewhat channelled on to the Mulga-strips. This is also indicated by the presence of washed-in litter on them. Thus, *A. aneura* receives a little more water than the intervening strips which are covered with *Eragrostis eriopoda* at their higher ends, but barren at the lower ends where most of the water runs off.

Stands of *A. aneura* lack undergrowth because of its effective root system. Only a few feeble *Eremophila* bushes or the poikilohydrous fern *Cheilanthes tenuifolius* may be present. The soil surface beneath the *Acacia* bushes is about 4 cm higher than that surrounding them. This is related to the secondary growth of roots and the accumulation of litter. Because of this, flood-waters do not reach the trunk; instead, 50% of the rain falling on the crown runs down the trunk. Therefore, the distribution of water is very unequal in the soil. Scrub-thicket formation through over-grazing (see pp. 254ff) cannot occur in this area, since cattle prefer to graze on the young, thornless *Acacia* bushes.

There is probably no other woody species occupying such a large area as does *A. aneura*. Commencing in the west it extends through the whole of central Australia and further to the east, deep into Queensland. In its distribution range in the east, near St George, I saw dense, nearly pure stands 10 m high. In undulating country, *A. aneura* always occupies the higher areas, so that it regularly alternates with the *Eucalyptus melanophloia*, or *E. populnea* savannahs of the lower areas.

In the same region there is another important stand-forming *Acacia*, *Acacia harpophylla*, known as Brigalow. It grows on heavy, somewhat solonised, pale grey, clay soils (containing sodium carbonate). It can grow 10-12 m tall.

The ecological conditions of this habitat are particularly interesting. As a result of alternate wetting and drying, associated with swelling and shrinkage, the soils show a polygonal pattern comparable to certain Arctic soils. In the Arctic, however, the pattern results from alternate freezing and thawing, and the soils are covered with stones. Stones are entirely absent on these Australian polygon soils which are known as Gilgay soils. The land is covered with pits, up to 1 m deep and having a diameter of 4-5 m. These are the gilgay or 'melon holes', which are separated by low ridges from one another. The *Acacia* shrubs only grow on these ridges. The pits are filled with water during the rainy season (600 mm annual rainfall) and are, therefore, barren of vegetation.

The formation of the gilgay is explained as follows:

During the dry season, large, metre-deep cracks develop in the soil which become filled with soil transported by wind or water. When the soil becomes wet again and swells there is no room for the increased volume. As a consequence the soil lower down is pushed upwards and pits are then formed in between.

Acacia harpophylla has roots which spread very widely at a depth of 5-10 cm. Sinkers extend deep down from the lateral roots, which also give rise to suckers. These contribute to the reproduction and denseness of the stands. Conversion of such areas into arable land is made difficult by the suckers, which need to be sprayed with chemicals from planes.[1]

(c) *Moist habitats in Central Australia*

As we know, in arid zones some salt accumulation occurs in all depressions which lack outflows (see pp. 293ff). In these salt pans the soil is always moist, even if a salt crust is formed at the surface. Moisture and salt content decrease from the centre of the depression to its margin. The vegetation shows a very pronounced corresponding zonation. Such salt lakes are found especially in the dry interior of the State of Western

[1] I had the opportunity of being introduced to this area under the excellent guidance of S. L. Everist (Brisbane).

Australia (Fig. 260) and in the northern part of the State of South Australia. Their size varies from very small pans to extensive ones such as Lake Eyre. The vegetation surrounding these pans is composed of hygrohalophytes, among which the Chenopodiaceae are the most important group in Australia. As an example, I shall describe a small salt lake near Wiluna, which was still partly filled with water on September 1, 1958. The following zones could be distinguished:

Zone 1: 30 m wide starting at the margin of the water; it was covered until recently with water and had, therefore, old, previously submerged, bushes of dead *Athrocnemum*; new seedlings were now appearing.

Zone 2: 80 m wide, *Athrocnemum-Salicornia* zone. These succulent saltbushes only covered 3% of the area in the lower 40 m, while in the higher 40 m the cover had already reached 10%. Here, other species had become associated, such as *Atriplex inflata*, *Kochia glomerata*, *Bassia* and *Frankenia*.

Zone 3: dwarf-shrub zone, about 1 km broad, loosely colonised with *Bassia* species, in addition scattered *Scaevola spinosa* (Goodeniaceae) and the composite *Cratystylis*; the annual grass *Aristida arenaria* occurred with high cover; it had already began to fruit and go over.

Zone 4: a non-brackish area, of shining green, mainly covered with annuals, in addition to *Kochia pyramidata* and *Eremophila* species.

More complicated situations occur where a salt pan is bordered by a steeply ascending, gypsum dune. Lake Way, south of Wiluna, is a mosaic of salt pans, gypsum ridges and sand dunes. Here, the bottom of a huge salt pan, last filled with water in the thirties, was now almost entirely covered with *Arthrocnemum* and *Salicornia*. On the margin I also found a very lush species of *Kochia*. Higher up, at the foot of a gypsum ridge there was first *Frankenia*, the strange Malvaceae *Plagianthus holmsii* and a large-leaved, succulent *Zygophyllum* species. One metre up, on the gypsum soil was a dense stand of *Melaleuca* with scattered *Casuarina* in between. Higher still, in addition to *Casuarina*, *Acacia longifolia* also grew and, on the dune itself, *Eucalyptus lesouefii* (mallee) with *Atriplex vesicaria*, *A. paludosa*, *Lycium* and a small-leaved *Zygophyllum*.

Table 86 shows the results of the analyses of samples taken at the pan at Lake Way on September 3, 1958.[1]

As can be seen from the data, non-halophytes with low salt contents are able to grow on the gypsum ridges a little above the salt pan. However, the true halophytes which grow on the sand dune show very high proportions of chloride (samples 9 and 10).

[1] I was able to visit this remote area by joining a land survey expedition (C.S.I.R.O.) under J. A. Mabutt. The botanist on this expedition was N. J. Speck. Permission to join this expedition is gratefully acknowledged. The many critical discussions and suggestions made on the trip were also much appreciated. Time did not allow more detailed ecological investigations but the few samples which could be collected were investigated in Stuttgart-Hohenheim by E. Bihler.

A few plants were sampled from non-brackish soils in the area around Wiluna on September 2, 1958. Beside *A. aneura* and *Triodia basedowii*, which are both very important here, there are several

TABLE 86

Lake Way, species in order from the lower to the higher habitats

Sample	Plant species and habitat	Π	Cl	%
1	*Athrocnemum* on salt pan	53·6	36·5	68
2	*Athrocnemum* for comparison, from Koonamore	48·7	36·5	75
3	*Kochia* species, from the border zone	26·7	15·0	56
4	*Zygophyllum* (large-leaved), upper salt-zone	16·4	13·1	80
5	*Plagianthus holmsii*, lower gypsum terrace	32·5	13·2	41
6	*Casuarina*, lowest tree	17·8	4·9	27
7	Myrtaceae, same location	27·7	5·4	19
8	*Eucalyptus lesouefii*, on sand dune	23·9	3·6	15
9	*Zygophyllum*, herbaceous, with small leaves	13·7	11·0	80
10	*Atriplex vesicaria* on sand dune	70·1	58·9	84

Eremophila species, which are especially characteristic of this area. They commonly grow in extremely dry habitats and their lacquered, shiny leaves feel completely dry. Yet, the water content of the leaves is 120% of the dry weight so it was of interest to learn something about their water relations.

TABLE 87

Samples from non-brackish soil, Wiluna region

Sample	Plant species	Π	Cl	%
1	*Eremophila freisii*, very dry habitat	15·4	1·3	10
2	*Eucalyptus oleosa*, in a dry-valley	22·8	1·4	6·1
3	*Bassia* sp. (Chenopodiaceae)	29·7	12·9	44

Table 87 shows that *Eremophila* has neither a high osmotic value, nor a large chloride content. Those *Eucalyptus* species restricted to dry valleys in arid zones also only show a slightly increased osmotic value compared with the *Eucalyptus* species occurring in the wet coastal areas (Π = 13 to 16 atm and Cl = 1·5 to 5 atm). Only *Bassia*, as a halophyte with succulent leaves, stores a relatively large amount of chloride on non-brackish soils. Its water content comes to nearly 470% of its dry weight.

In summary we may say that the non-halophytic species of the Australian arid zones are able to maintain a high hydrature (low osmotic values) even during prolonged periods of drought. One explanation is that their water supply is not as poor as is usually believed (see pp. 270ff). Another explanation is that their ability to reduce gas exchange and hence transpiration very drastically, as soon as there is a lack of water, ensures that they are able to maintain a water balance for long periods without incurring larger saturation deficits. The species of extreme-arid zones with droughts of many months' duration cannot afford to suffer large water deficits, as this would eventually lead to desiccation; rather they persist without increasing the dry weight. The question, whether respired carbon dioxide is re-assimilated in this situation, needs further study.

The region of Alice Springs, which is approximately in the centre of Australia on the Tropic of Capricorn, shows yet another unique aspect. Here, the usually rather monotonous and flat topography is interrupted by several mountain ridges (Macdonnell Ranges). This brings about an ever greater diversity in the differentiation of the vegetation. The Macdonnell Ranges represent an ancient Pre-cambrian formation, consisting of quartzites and sandstones, which rise from the level plains (650 m) to a height of 1200 m (highest peak, 1460 m).

As previously described (p. 422), a soil-conditioned pattern of Mulga (*A. aneura*) and Spinifex (*Triodia basedowii*) occurs on the level ground. However, the vegetation of the mountain ranges is entirely different. On the steep rocky slopes only a few very scattered small trees of *Callitris glauca* occur. In cracks between rocks one can see the cycad, *Macrozamia macdonnellii* and also *Capparis mitchellii*. It is only further down the slope, where it is not so steep, that *Triodia pungens* occurs with increasing cover. Still further down, *A. aneura* comes in.

The region includes a series of dry creeks. The Todd River belongs to the catchment area of the Finke River which empties in the direction of Lake Eyre. However, the water usually soaks into the soil or is lost by evaporation before reaching the lake.

On sand banks along the river bed *Nicotiana velutina* occurs. On the banks of the dry river bed trees of *Eucalyptus camaldulensis* (= *rostrata*) grow and at the river's edge, a few *Hakea lorea* and *H. intermedia*. In the river bed itself the grass *Themeda australis* grows, a plant characteristic of the moister areas in the east.

Atriplex nummularia is frequent on broad, clayey river terraces, while many *Eremophila* bushes grow along the bank. In the beds of smaller creeks *Eucalyptus camaldulensis* is replaced by the Ghost tree, *E. papuana*, with a shining-white bark that is very conspicuous at night. Associates are *E. dichromophloia*, *Acacia estrophiolata*, *Melaleuca* spp., etc. These trees and shrubs also grow scattered over the rocky plains. *Citrullus ecirrhosus* becomes established along very small channels, *Atalaya hemiglauca* spreads by root suckers on alluvial flats; vast plains may also be covered with *Acacia estrophiolata* and *Ventilago viminalis*, flood pans with *Themeda australis*, *Digitaria brownii* and the large *Aristida nitidula*.

The ground water stream of dry-river valleys occasionally becomes banked up before breaking through quartzite ridges and in such places permanent water holes are formed around which one can find a relict flora, such as *Ficus platypoda*, *Dodonaea viscosa*, the liane *Marsdenia australis*, on rocks *Sarcostemma australis* and, in the water, *Ricinus* and *Hibiscus*.

The Palm Valley near Hermannsburg is famous, for here, in the heart of central Australia, one finds the palm *Livistona murryae*. The normal range of *Livistona* is only a little south of Darwin.

According to information given to me by Chippendale this area also contains: *Typha angustifolia*, *Phragmites karka*, *Najas major*, *Xerotes dura*, *Cryptandra spathulata* (Lauraceae), *Rulingia magniflora* (Sterculiaceae), *Baeckea polystoma*, *Hibbertia glaberrima*, *Teucrium grandiusculum*; at other relict habitats there are also rare ferns such as *Adiantum hispidulum*, *Nephrolepis cordifolia*, *Cyclosorus gongylodes* and *Lindsaea ensifolia*.

In all deserts one finds an ecologically entirely different vegetation at water holes or in cracks in the rocks. In most cases it probably represents a relict flora, i.e. the remains of vegetation that was formerly widely distributed in the area when different climatic conditions obtained.

References

Adamson, R. S., and Osborn, T. G. B. 1922. On the ecology of the Ooldea district. *Trans. Roy. Soc. S. Austr.*, **46**, 539-64.

Ashby, W. C., and Beadle, N. C. W. 1957. Studies in halophytes III. Salinity Factors in the growth of Australia saltbushes. *Ecology*, **38**, 345-52.

Beadle, N. C. W. 1948. *The vegetation and pastures of western New South Wales.* Sydney.

Beadle, N. C. W., and Tchan, Y. T. 1955. Nitrogen economy in semi-arid plant communities. I and II. *Proc. Linn. Soc. N.S.W.*, **80**, 62-70.

Beadle, N. C. W., Whalley, R. D. B., and Gibson, J. B. 1957. Studies in halophytes II. Analytical data on the mineral constituents of three species of *Atriplex* and their accompanying soils in Australia. *Ecology*, **38**, 340-44.

Burbridge, N. T. 1941. Ecological notes on the vegetation of 80-mile Beach. *J. Roy. Soc. W. Austr.*, **28**, 157-64.

Burbridge, N. T. 1942. Ecological notes on the vegetation of the De Grey-Coongan Area. *Trans. Roy. Soc. S. Austr.*, **29**, 151-61.

Burbridge, N. T. 1945; 1946. Morphology of the Western Australian species of *Triodia* R. Br, I and II. *Trans. Roy. Soc. S. Austr.*, **69**, 303-08; **70**, 221-29.

Burbridge, N. T. 1960. The phytogeography of the Australian region. *Austr. J. Bot.*, **8**, 75-212.

Crocker, R. L. 1946. The Simpson Desert Expedition of 1939. The soils and vegetation of the Simpson Desert and its borders. *Trans. Roy. Soc. S. Austr.*, **70**, 2.

Davidson, D. 1954. The Mitchell Grass association of the Longreach district. *Univ. Queensland Papers*, **3**, 45-59.

Everist, S. L. Verbal communication.

Gentilli, J. 1955. Die Klimate Australiens. *Die Erde*, **6**, 206-38.

HARRIS, J. A. 1932. *The physico-chemical properties of plant saps in relation to phytogeography*. Minneapolis.

KNODEL, H. 1939. Über die Abhängigkeit des osmotischen Werks von der Saugkraft des Bodens. *Jb. Wiss. Bot.*, **87**, 557-64.

KREEB, K. 1960. Über die gravimetrische Methode zur Bestimmung der Saugspannung und das Problem des negativen Turgors I. *Planta*, **55**, 274-82.

KREEB, K., and ÖNAL, M. 1961. Über die gravimetrische Methode zur Bestimmung der Saugspannung und das Problem des negativen Turgors, II. Mitteilung. *Planta*, **56**, 409-15.

OSBORN, T. G. B. 1925. Introduction and general description of the Koonamore Reserve for the study of the saltbush flora. *Trans. Roy. Soc. S. Austr.*, **49**, 290-7.

OSBORN, T. G. B., and WOOD, J. G. 1923. On some halophytic and non-halophytic communities in arid South Australia. *Trans. Roy. Soc. S. Austr.*, **47**, 388-99.

OSBORN, T. G. B., WOOD, J. G., and PARTRIDGE, T. B. 1932. On the growth and reaction to grazing of the perennial saltbush *Atriplex vesicarium*. *Proc. Roy. Soc. N.S.W.*, **57**, 377-402.

OSBORN, T. G. B., WOOD, J. G., and PARTRIDGE, T. B. 1935. On the climate and vegetation of the Koonamore vegetation reserve to 1931. *Proc. Linn. Soc. N.S.W.*, **60**, 392-427.

SLATYER, R. O. 1958. *Annual Progress Report for* 1957-58. Climatology Unit, C.S.I.R.O.

SLATYER, R. O. 1960. Aspects of the tissue water relationships of an important arid zone species (*Acacia aneura* F. Muell) in comparison with two mesophytes. *Bull. Res. Counc. Israel, D*, **8**, 159-68.

TAYLOR, G. 1951. *Australia*. 6th ed. London.

THREN, R. 1934. Jahreszeitliche Schwankungen des osmotischen Werks verschiedener ökologischer Typen der Umgebung von Heidelberg. *Zschr. Bot.*, **26**, 449-526.

WOOD, J. G. 1925. The selective absorption of chlorine ions and the absorption of water by the leaves in the genus *Atriplex*. *Austr. J. Exp. Biol. Medic. Sci.*, **2**, 45-56.

WOOD, J. G. 1934. The physiology of xerophytism in Australian plants. *J. Ecol.*, **22**, 69-87.

WOOD, J. G. 1936. Regeneration of the vegetation on the Koonamore vegetation reserve 1926-36. *Trans. Roy. Soc. S. Austr.*, **60**, 96-111.

XIII. *The Sahara*

1. Introduction

THE largest desert in the world, the Sahara in North Africa, stretches from the Atlantic Ocean to the valley of the Nile (Fig. 261). To the east, the desert belt reaches beyond the Nile to the Red Sea and from there through Arabia to the Great Indian (Thar) desert in north-west India (Schiffers, 1950).[1] To the north, the desert belt abuts at its western end on the Mediterranean sclerophyll forest region, while in the east it

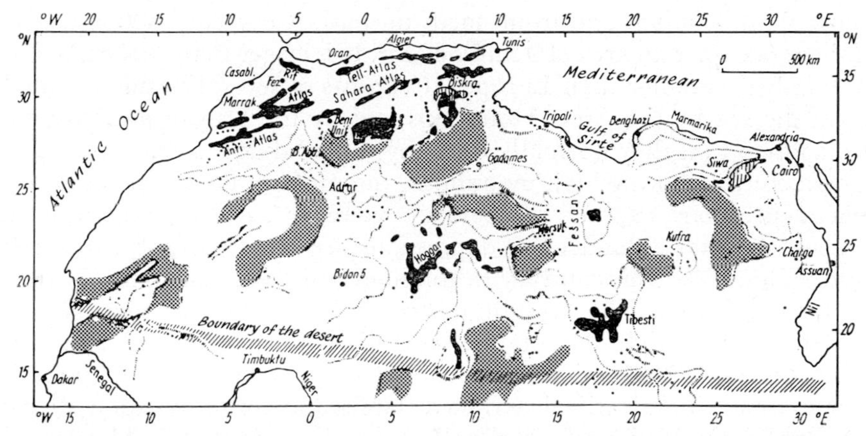

FIG. 261. Sketch map of the Sahara showing the features mentioned in the text. Black—mountains; stippled fields—large sand areas; vertically hatched fields—depressions below sea level; lines shown by fine vertical dashes—steep slopes of high plateaux; small dots—water-places; B. Abb.—Beni Abbes (modified from Schiffers).

borders the Mediterranean directly, the desert belt following round, eastwards, and so deep into Syria and Mesopotamia. The arid zone then stretches beyond the high mountain ranges of Iran to the north and to the Gobi desert in the east. However, these Iranian-Turanian and Mongolian desert regions have quite a different character, since their winters can become quite cold.[2]

The northern border of the sub-tropical desert regions can be drawn wherever date-palm cultivation ceases to be economically feasible. Date-growing for export is only possible in areas with sufficiently high temperatures, complete absence of cold and very dry atmospheres.

[1] The word 'sahara' in Arabic stands for 'yellow' or 'red', also for 'desert plain'.

[2] An outline of the floristic elements of the entire desert complex is given by Lavrenko (1962).

The boundary of date-cultivation coincides in the west with the southern slopes of the Anti Atlas and the Sahara Atlas. Further to the east the date groves often reach the coast.

It is much more difficult to define the limit of the desert region to the south. The southern Sahara grades very gradually into the Sudano-Sindic belt, which stretches across South Iran to the Indian peninsular (Zohary, 1963). The transition zone (with a rainfall of 100-200 mm) between the southern Sahara and the savannah in the Sudan is known as Sahelian zone.[1] This zone is somewhat affected by summer rains and the vegetation consists primarily of an open grass community with few woody plants. Its western limit in Africa with the Sahara can be regarded as approximately the 19th parallel and its eastern border as the 18th parallel. The Sahelian zone lies between 20 and 16°N.

The axis of the Sahara from the Atlantic Ocean to the Red Sea is more than 6000 km and from north to south it exceeds 1800 km. The Sahara occupies an area of 9 million km^2. It is larger than Australia and only a little smaller than Europe. It consists of several basins, isolated from the sea by the Atlas Mountains and by the Libyan and Nubian uplands so that they normally lack any outward drainage.

In contrast to the high mountains, the uniform landscape of basin-like depressions exposes geologically younger surfaces, which form extensive plains. The deepest depressions are often covered with sand and dunes. However, contrary to common belief, sand only covers 20% of the total area. The crystalline bed rock only breaks through the surface in the high mountains. The highest, such as the Ahaggar Plateau, reach 3000 m, or even exceed this height, as do the Tibesti mountains. The stratified terraced landscape forms a stone desert (Hamada), divided up by dry-valleys (wadis, oueds) (cf. Meckelein, 1959). Shallow depressions are filled with Pleistocene deposits. They are either gravel deserts (serir, reg) or sand deserts (erg, areg). Along the northern margin of the Sahara the run-off is channelled through the valleys into inland basins, where it collects in the deepest depressions. Here, the finest sediments are deposited. Salt accumulation occurs simultaneously as a result of evaporation and large salt and clay pans (sebcha, schott) are formed, which are only temporarily covered by salt lake. Scattered throughout the vast desert sea are outlets for ground-water, which form the oases. These are all used in cultivation (Fig. 262).

According to the rainfall distribution, three zones can be distinguished in the Sahara: a northern, central and southern zone. In the northern zone, rain only falls during the winter months and does not exceed 200 mm per year. Rainfall continues to decline more and more, southwards. In the central region, rain occurs only episodically, i.e. not every year. This area contains but few plant species. The following

[1] 'Sahel' means 'shore' in the language of the inhabitants, i.e., this zone forms the shore of the great desert-sea to the north.

species are cited by Killian (1945) as the only ones from the central Saharian Fezzan:

Zygophyllum simplex, *Z. album*, *Fagonia bruguieri*, *Traganum nudatum*, *Cornulaca monacantha*, *Alhagi camelorum* and *Aristida pungens*. Perennials can only grow at a few places where water collects after an episodic rain and where the soil remains damp for several years without rain.

FIG. 262. Small oasis with a spring, date palms, *Tamarix gallica* and *Lycium arabicum* between Gabès and Kebili, Tunisia (Phot. E. Walter).

The high mountains of the central Sahara receive hardly any rain. Vast areas, e.g. of the southern Libyan Desert, are practically rainless. I did not see a single plant in this area for a distance of over 200 km.

Further south again rainfall increases, but this is summer-rain. The rainfall pattern is illustrated in the following diagrams (Figs 263-65).

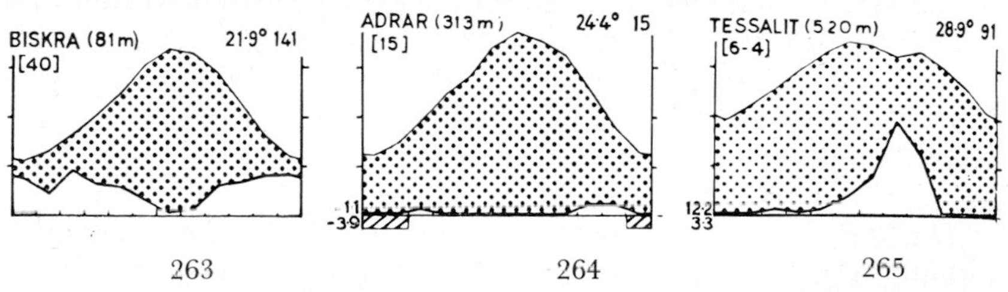

FIGS. 263-265. Climatic diagrams from the north, central and south Sahara.

This distribution of rainfall results in an entirely different floristic composition for the vegetation in the north to that in the south

(Ozenda, 1958; Quezel, 1965). The border dividing the two large floristic realms, the Holarctic and Palaeotropic, runs through the central Sahara. More than half of the species belong to the Saharo-Arabian element, which show clear relationships with the Holarctic floristic realm. In addition, Mediterranean elements occur from the north, deep into the central Sahara, e.g. the genera *Astragalus*, *Reseda*, *Plantago*, *Salsola*, while Sudanian elements of the south do not reach very far north. The latter include species of the genera *Indigofera*, *Tephrosia*, *Hibiscus*, *Cleome*, *Acacia*, *Eragrostis*, *Cyperus*.

Among the Saharo-Arabian species one can distinguish species with a very wide distribution and others, whose presence is restricted to Africa. The latter are the Saharian elements in the narrower sense. Among these one finds some species represented more in the western area and others more in the east. There are also species with very limited ranges. These include, for example, *Fredolia* (*Anabasis*) *aretioides*, the well-known pillow-cushion plant of the Sahara south of Oran. Endemism is fairly typical of many Saharian biota.

However, these are usually vicarious taxa of Mediterranean species or micro-endemics. Yet, even ignoring these, Ozenda was able to enumerate 162 endemic species for the entire Sahara, a figure representing about 25% of the total Saharian flora. This shows that the desert brings about the isolation of taxonomic groups.

Among the different families the Chenopodiaceae (17 genera), Cruciferae (31 genera), and Zygophyllaceae (7 genera) are particularly well represented.

Not all deserts are floristically poor. For example, the floras of the Karroo and the Sonoran desert are both very rich, especially in succulents. Yet, the flora of the Sahara is not in the same category; on a basis of equal area it is ten times as poor in comparison with the flora of south Europe. According to Ozenda, 450 species of higher, and 75 of lower plants have been recorded for the central Sahara. For the Tibesti mountains the corresponding figures are 350 and 45. The paucity of mosses and lichens is remarkable. Poikilohydrous ferns are entirely lacking in the Sahara and this indicates the extreme drought conditions.

Only the northern border area and the fog-rich Atlantic coast are rich in lichens; here also occur typical succulents (*Euphorbia*, *Senecio*). Two *Caralluma* species occur in the Hoggar mountains. In an area of 10,000 km^2 in the tropics one can find about 3000-4000 species, in Europe about 1000-2000 but, in the Sahara, hardly 150.

The floristic relationships of the Sahara, particularly the penetration of Mediterranean elements far south into the central zone, are explicable only if one considers its floristic history. During the Pleistocene the Sahara had a climate much more like that of the Mediterranean with a considerably higher rainfall. This is shown clearly by deeply eroded valleys and river terraces which could have originated only during a pluvial period. It is also shown by the skeletons of giraffes, elephants and rhinoceroses and traces of habitation by man (e.g. rockwall paintings) in regions that are today lacking fauna and human habitation. It is possible also to find pollen of *Pinus halepensis*, *Quercus ilex*, *Callitris*, *Cupressus* and *Juniperus* in neolithic deposits from the Sahara. All of these are Mediterranean elements, which today are completely absent from the desert. Some can only be found as relicts in the central Saharian mountains. All this suggests a marked contraction northwards of the southern limit of the Mediterranean region since the Pleistocene.

By contrast, the northern limit of the tropical elements appears to have changed much less or, perhaps, not at all. This has resulted in the largest desert region of the world.

2. Soil and vegetation relations

Since there is no humus accumulation in the desert, the soils represent strictly the raw products from weathering of geological bedrocks. Physical weathering predominates but some chemical weathering occurs under the influence of readily soluble salts. The lack of humus results in a practically lifeless condition in the soils. However, Killian and Fehér (1935) claim to have observed considerable microbial activity in all desert soils and even continuous soil respiration. These findings are difficult to understand and require substantiation. Soil respiration is limited by the organic matter produced from vegetation (Walter and Haber, 1957). If there is hardly any production of organic matter over vast areas of the desert, soil respiration and microbial activity cannot be of any significance. Exceptions can occur only where organic matter has accumulated locally, for example, in dense cushion plants or from accumulated litter beneath a solitary shrub or bush. Here active microbial life could develop under temporarily favourable water relations. Nitrification could also take place which would explain the relatively high content of nitrates found in many desert soils (Vargues, 1952).

The properties of desert soils are determined largely by the parent material. The most influential factor in a desert is the wind, and water plays a minor role. As a result of wind action we can distinguish between wind-eroded sites and sites determined by wind-blown deposits.

(*a*) *The stone desert or hamada*

All fine weathering products are removed by the wind from the tops of plateaux. In this process sand-blasting causes wind abrasion of all projecting rocks. Stone fragments about the size of a hand accumulate on the surface to form a 'stone pavement'. In this way stone desert or hamada originates (Fig. 266).[1]

All surface strata of Cretaceous and Tertiary origin form stone

[1] Hamada = dying off.

deserts. These occupy relatively the largest areas in the Sahara. Sea salt is only found in rocks from marine deposits. Beneath the stone pavement is a brownish-red dust soil of irregular depth which has originated through weathering of limestone and marl. The stones prevent the dust soil being blown away at its surface. Bivalent and monovalent cations become concentrated in this soil in the form of strings of gypsum and salt sheets because the low rainfall is insufficient to remove these salts from the soil. However, small quantities of water due to condensation during cold nights are sufficient to bring about, with the

FIG. 266. Wadi Hoff near Helwân, Egypt. Hamada or stone desert with steep slopes (Phot. E. Walter).

salts, some weathering of the stones and leads to an accumulation of iron and manganese compounds on the evaporating surfaces of the stone fragments. Because of this they become coated with a dark, brown-black 'desert varnish', which gives the dark coloration to the hamada.

The hamada plateaux are incised by deep erosion channels in a canyon-like manner, especially near terraced beds, forming steps. The steep slopes of the canyons are covered with rock debris.

The actual hamada plateaux are usually completely barren of vegetation because of the low water reserves in the soil and the, usually, rather high salt content.

Plants grow only in rock crevices and cracks in regions with a light rainy season.

(*b*) *The pebble or gravel deserts* (*serir or reg*)

Where the parent material consists of heterogeneous deposits, e.g. conglomerate, the more easily weathered, cementing material is removed by the wind and pebbles accumulate at the surface. These are called autochtonic gravel deserts.[1] However, gravel flats in the desert can also be allochtonous when they result from large alluvial deposits

[1] Seghir = small.

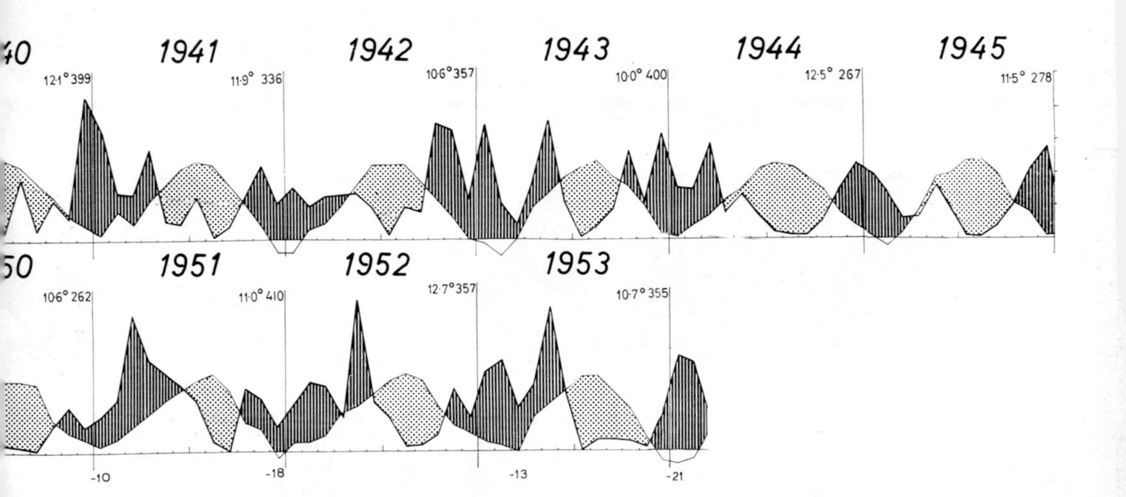
40
1941
1942
1943
1944
1945
12·1° 399
11·9° 336
10·6° 357
10·0° 400
12·5° 267
11·5° 278
50
1951
1952
1953
10·6° 262
11·0° 410
12·7° 357
10·7° 355
-10
-18
-13
-21

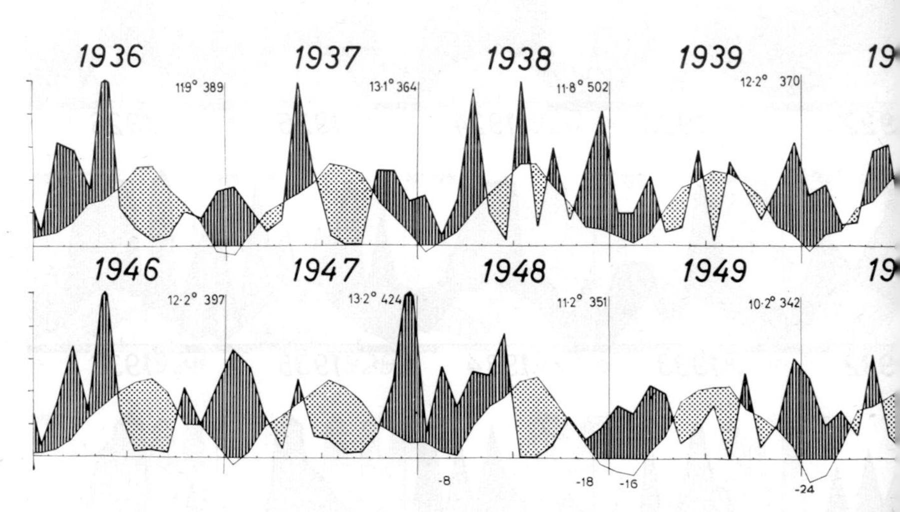

FIG. 18. Climatogram for Ankara, 1936-53. Scale and symbols as Fig. 17.

from which all the fine material has been removed by the wind. Such plains were formed by the Nile in north Egypt during the Pluvial period.

By contrast, the gravel deserts between Cairo and Suez are formed from river deposits of the Oligocene and the non-marine Miocene and Plio-pleistocene, respectively.

Gypsum accumulates as a fine powdery mass beneath the superficial layer of gravel. At a somewhat greater depth (about 50 cm), crystalline gypsum can be found strongly cementing the sandy gravelly strata. The origin of these hard pan horizons may be explained in two possible ways:

1. Rain water transports $CaSO_4$, in dissolved form, downwards where it becomes crystallised at the lower limit of water penetration. With evaporation, some of the dissolved gypsum rises once more with the water and is precipitated as a powder at the surface when the water evaporates rapidly.
2. Another possibility is that the crystallised gypsum layer originated under earlier, wetter conditions, while the powdery horizon may be of recent origin.

Gravel desert is generally salt-poor, yet the chloride content in the gypsum-containing layers may reach 1% or more.

The surface gravel is commonly so compact that roots can hardly penetrate it. Thus, the gravel desert of the Sahara is without vegetation, particularly in the drier parts. The surface is either flat or of shallow hills separated by sand-filled depressions of irregular width. These may enlarge as very broad sandy plains. The vegetational cover of the slopes is extremely sparse and restricted primarily to channels with northern exposures. A cover of shallow sand favours the establishment of perennial species.

The vegetation soon becomes somewhat richer in sandy erosion-valleys and on sand-covered plains. In the central Sahara the gravel desert is the most inhospitable habitat for organisms. For the somewhat damper south Algerian Sahara, Ozenda reports very diffuse stands of *Haloxylon* (Chenopodiaceae) with only a few other associated plants.

Wherever a layer of sand is present, *Aristida* species, geophytes and some annuals can be found. Where the clay content of the soil increases the thorny *Cornulaca* (Chenopodiaceae), *Randonia* (Resedaceae) or *Zygophyllum album* and *Hyoscyamus muticus* may be found.

(*c*) *The sand desert* (*erg or areg*)

Sand plays the main role amongst the accumulated materials in depressions. It is blown from higher regions and deposited everywhere in the lee of small obstacles or in front of steep terraces. These sand deposits are usually only shallow; erg, however, relates to the extensive dune areas which form vast seas of sand. The two ergs in the Algerian

Sahara cover areas which are equal in size to France. Where the wind blows from opposite directions, but predominantly from one side, crescentic dune or barchans are formed, with gentle slopes on their windward, and steep ones on their leeward sides. Where one of the two wind directions is only slightly more frequent, the dunes are nearly stable and their ridges only are constantly shifted to respond to the current wind.

Sand grains are not necessarily derived from quartz, in places they may be composed of up to 70% lime. The surfaces of the sand grains can also be coated with iron oxide. Such sands are bright red in colour. However, where the climate is somewhat moister, for example near the coast, the coatings and also the sand itself are of yellow-brown colour.

A cover of sand favours water storage in arid areas. In the deep sand layers of the large erg near Beni Abbès (300 km south-south-east of Beni Unif), the water content is very high in spite of very low rainfall. The distribution of episodic rains exceeding 5 mm during 1946-50 is shown in the following table (after F. Pierre, 1955):

TABLE 88

Episodic rains in excess of 5 mm in the central Sahara

	Month												Total for Year
	1	2	3	4	5	6	7	8	9	10	11	12	
1946	18	—	—	—	—	—	—	—	—	—	—	—	21
1947	—	—	—	—	—	—	—	—	—	—	—	—	7
1948	—	—	5	9	—	7	—	—	—	25	—	7	55
1949	—	—	20	30	—	—	—	—	—	—	—	22	75
1950	—	—	—	—	56	—	—	—	—	—	22	—	80

In addition to the low rainfall another factor is the high temperatures showing a July maximum of 43·3°C, a July minimum of 27·3°C, a January maximum of 18·8°C and a January minimum of 3·1°C. Evaporation, which increases with increasing distance from the Mediterranean coast is correspondingly high.

TABLE 89

Annual potential evaporation in the north and central Sahara

Place	Distance from the coast (km)	Potential evaporation (mm)
Oran	on the coast	1383
Ain Sefra	*c.* 300	2916
Colomb Béchar	*c.* 425	4052
Beni Abbès	*c.* 575	4809
Adrar	*c.* 850	5033

Of course the shifting dunes, whose sand is constantly moving and, therefore, exposed to desiccation, are barren of vegetation. However, on the more stable sand, plants can become established if their roots extend into the damp layers. The most characteristic species is *Aristida pungens* with tough, thorn-tipped leaves. In addition, there are *Tamarix articulata*, *Ephedra alata*, *Retama raetam*, *Genista saharae* and *Caligonum comosum* as well as *C. azel*, which is endemic to the north-west Sahara. Herbaceous species are reported as *Cyperus conglomeratus*, *Moltkia ciliata* and the endemic grass *Danthonia fragilis*. Since the perennials form an obstacle to the wind, sand is deposited either behind them or between their branches. Therefore, they frequently stand on small dunes (Fig. 267). Because of their wide-spread root system they can absorb sufficient water from the damp sand layers throughout the whole year. Thus, they show no marked periodicity in their development.

FIG. 267. Dunes with *Ziziphus lotus* and *Lygeum spartum* between Sfax and Gabès, Tunisia (Phot. E. Walter).

(*d*) *The dry-valleys* (*wadis or oueds*)

In addition to the wind, water plays some role in the desert, even though it is of lesser importance. The stone and gravel deserts dominate the landscape, yet erosion channels are also present. They are cut out of these plains and join to form larger watercourses and valleys. After rain, water runs from the plateaux along these routes, in which gravel and sand is deposited. Moreover, the water dissolves and removes salts and wets the alluvial soil relatively deeply. These habitats are, therefore, the most favourable for plant growth. The larger the watersheds of the dry valleys, the larger the water masses that move through them. Even ground-water streams can be found occasionally that are active throughout the year. In such localities small woods are found developing on the margins of valleys and colluvial slopes. They even occur in the central Sahara at the foot of mountains. The valley centres and the run-off slopes are not usually vegetated since, in these places, the soil is washed away during flash floods. Thus, perennials do not gain a foothold here (Fig. 268).

FIG. 268. Depression in the desert between Cairo and Suez, Egypt, with *Haloxylon salicornicum* (Phot. E. Walter).

The characteristic plants of the wadis are *Acacia* species and the grass *Panicum turgidum*. However, occasional associates may be *Zizyphus*, *Maerua* and *Balanites* or *Pennisetum dichotomum*, *Zilla* (Fig. 269), *Ephedra*, the leafless umbellifer *Pituranthos*, etc. Wherever the soil becomes somewhat more saline, this vegetation is replaced by *Tamarix articulata* (*T. aphylla*) with *Zygophyllum*, or by *Salsola* and other Chenopodiaceae. In the south tropical elements can be found such as *Calotropis*, *Aerva* and *Ipomoea*.

(*e*) *Pans, dayas, sebchas or schotts*

Water moving through the wadis carries fine clay particles in suspension and deposits them wherever the water percolates down or where it collects in depressions lacking outflows.

Where these depressions have subterranean drainage, as for example in the doline-like formations of 'karst' landscapes on limestone, clay pans are formed with well-wetted, non-brackish soil. These habitats are preferred by plants and here one finds woody forms like *Pistacia* or *Zizyphus*, smaller perennials and many ephemerals.

Annuals (therophytes) are particularly abundant after rain in non-brackish depressions. Lemée (1953) distinguishes three communities near Beni Unif:

1. The *Althaea ludwigii-Trigonella anguina* community in loamy depressions.
2. The *Lotononis dichotoma* community in sandy channels.
3. The *Asphodelus pendulinus* community on sandy wind-blown deposits.

Therophytes can attain a cover of 5-20% and they only root to a depth of 20 cm. Twenty days after 8 mm of rain, the upper 10 cm had no more water available, thus, their roots must penetrate very quickly to a greater depth.

FIG. 269. *Zilla spinosa* (Cruc.), flowering shrub, Wadi Hoff, near Helwân, north of Cairo, Egypt (Phot. E. Walter).

The situation is different where water evaporates from depressions without drainage. Here, very considerable salt accumulation takes place. Salt pans are formed which are often covered with a white crust and are known as sebchas or schotts.

Salt-soils play only a minor role in the Sahara proper. The rainfall is so small that salts are neither removed nor accumulated in depressions. Salt pans are more common in the transition zone towards the Mediterranean coast and along the coast itself. Here, a whole sequence of salt-containing rock strata has been found in all kinds of geological formations. At the same time, the ground-water table is very high. The chlorides of sodium and magnesium predominate amongst the salts but sulphates are also present. Formation of sodium carbonate can occur through ion exchange in the presence of clay and lime. This increases the alkalinity very markedly.

However, in the desert, crusts are formed through evaporation in every locality where ground-water wets the surface by capillary action. These can consist of calcium, gypsum or chlorides and sulphates depending on the kinds of salts dissolved in the ground-water. Such white salt sheets can often be seen in the depressions between dunes.

The vegetation of these damp and brackish soils varies in relation to their water and salt content. On the somewhat drier soils one finds *Salsola* species, *Traganum nudatum* and *Zygophyllum album*. On somewhat damper soils one finds *Tamarix* species, on salt-poor soils *Atriplex halimus*, on damp, salt-enriched soils *Suaeda* species, *Frankenia*, etc. The greatest salt-resistance appears to be shown by *Halocnemum strobilaceum* and *Salicornia* species.

A rather peculiar situation exists in the Eygptian desert in the Wadi Natrun as described by Stocker (1927-28).

This is a lowland depression 23 m below sea level, 5-20 km wide and 160 km long, lying west-north-west of Cairo. Within the area are 16 large and small salt-lakes fed by a subterranean water supply through layers of gravel from the Nile. The Nile water dissolves salt in its passage through the salt-bearing desert strata. The lakes reach their highest water mark in January, about 2-3 months behind the period of high-water of the Nile. Yet the water level variation is only 30 cm. Wherever the ground-water appears at the surface, swamp vegetation is commonly developed with *Typha latifolia*, *Phragmites communis*, *Eragrostis bipinnata* and *Juncus acutus*. The surface between the plants may be covered with a crust of white salt. However, the plants root in the somewhat deeper, wet soil layers which have only low salt concentrations. According to Stocker, the ground-water 30 cm below *Juncus* had a salt content of 0·9%.

However, in the undrained lakes salt accumulation occurs through evaporation, resulting in its precipitation during the dry season. Because of chemical transformations besides NaCl there is also Na_2SO_4 and soda (Na_2CO_3).

The clay soils covered with a salt-crust which dry up completely at periodic intervals are barren of vegetation. They constitute the actual sebchas. Where the salt concentration is not too great, extreme halophytes can survive. However, where salt-free sand is deposited locally on salty clays, the conditions for germination are immediately improved and small dunes can form around the invading plants. Here we find *Zygophyllum album*, *Z. simplex*, *Alhagi maurorum* and *Tamarix* species.

Since the investigations of Stocker, the area has changed much. I visited it in 1960. The ground-water stream from the Nile is now diverted before discharging into the lakes and it is used for the irrigation of crops. The water contains 0·03% of dissolved salts. Wadi Natrun has been converted into an oasis with a dense population, who, in addition to agriculture, also have a few industries. So there is virtually nothing left of the original vegetation.

(*f*) *Oases*

In places where water of low salt content issues as springs in the desert, even hydrophilous plants can become established. When permanent water basins are formed the same vegetation is found everywhere.

In addition to several algae it includes *Potamogeton*, *Ceratophyllum*, *Utricularia* and *Lemna*. At the edge a semi-aquatic vegetation of *Typha*, *Phragmites*, *Scirpus* and *Erianthus* develops, which grades into stands of *Tamarix gallica* and *T. nilotica*.

The original vegetation of the oases consists of Doum palms (*Hyphaene thebaica*), *Acacia albida* and other wattles, *Maerua* and *Capparis* species, *Calotropis procera*, *Citrullus colocynthis*, etc.

Today, the original vegetation has been entirely replaced by the Date palm, *Phoenix dactylifera* (Fig. 262) together with other cultivated plants such as barley (*Hordeum*) and common bean (*Vicia faba*). *Ficus sycomorus* and *Albizzia lebbek* have also probably been planted by man. Many obnoxious weeds and ruderal plants have been introduced. The danger of salinisation of soils at oases is great, even where the water has a low salt content, because of the high evaporation rates. Salinisation can only be prevented by drainage and continuous watering of the soil. Thus, coarse textured, permeable soils are more favourable than clay soils, which are also usually more salty. In this connection, the purity of the water is of great importance.

The question arises as to what plants can be cultivated at oases with extreme climates provided there is a favourable soil and good quality water available. To answer this question I shall base my observations on the driest oasis in the southern Libyan desert, El Charga (rainfall in 20 years 0 mm!) and Baris, as well as Aswân on the Nile (with averages of 3 mm per year). Crop plants which can be cultivated without protection from the sun are wheat, barley, *Vicia faba*, Egyptian clover (*Trifolium alexandrinum*, in winter), *Medicago sativa*, cotton and *Ricinus*.

In the winter months with a mean temperature of 12-15°C nearly all forms of vegetable can be cultivated but of more interest are the perennial crop plants which have to withstand a mean summer temperature of 30°C.

The most famous tree is the Date palm. Trees planted for shelterbelts are *Casuarina*, *Prosopis*, *Dahlbergia* and *Eucalyptus camaldulensis* (= *rostrata*). The latter tree species also survives in Australia in the driest climates in habitats with ground-water. In the protection of shelterbelts olives, *Citrus* species, vines, apricots, almond, guavas and figs are cultivated.

Conditions are even more favourable where crops are cultivated beneath a cover of crowns of Date palms. Shading and wind-breaks produce a microclimate which not only allows *Citrus* species to grow better but, moreover, provides an opportunity of cultivating pomegranates, papayas and even bananas. I also saw banana fields without shade trees in Jericho (Jordan valley) but here they were planted in deep dykes so that only the tips of the leaves extended above the surface. The creation of a mild microclimate in a dense stand is the most important requirement for sensitive crops. This is easier in the centre of an oasis than at its fringe. Trials in the botanical garden on Kitchener

Island (a Nile island near Aswân) have shown that nearly all tropical plants can be cultivated in the most extreme desert climate if these prerequisites are met, including abundant watering (in the winter every two weeks, in the summer every week). The barren desert extends here right up to the river bed. On this island Kitchener arranged for the planting of trees, mainly Indian, which today form a high, shady forest. This experiment is so significant that a list of all the tropical trees which occur on this island is given. Most of them produce flowers and fruits. One can see from this list that practically all plants cultivated in India would also grow at the oases. Amongst tropical crop plants even coconut and oil palms, the breadfruit tree, mango, clove-tree, manioc, banana, coffee,[1] etc. are included:

Plants growing on Kitchener Island near Aswân (Egypt).
Crop plants in bold print

GYMNOSPERMAE: *Cephalotaxus drupacea*, *Cycas*.

MONOCOTYLEDONES:

Musaceae: *Strelitzia angusta*.
Palmae: *Areca lutescens*, *Arenga saccharifera*, *Borassus flabelliformis*, *Calamus indicus*, *Caryota urens*, ***Cocos nucifera***, *C. romanzoffiana*, *Cryptostegia madagascariensis*, ***Elaeis guinensis***, *Kentia balmoriana*, *Latania borbonica*, *Livistona chinensis*, *Raphis flabelliformis*, *Roystonia* (*Oreodoxa*) *regia*, *Sabal palmetto*, *Thrinax barbadensis*, *T. elegans*.
Zingiberaceae: *Alpinia nutans*.

DICOTYLEDONES:

Acanthaceae: *Sanchezia nobilis*.
Anacardiaceae: *Anacardium occidentale*, ***Mangifera indica***, *Spondias dulcis*, *S. mangifera*.
Apocynaceae: *Alstonia scholaris*, *Beaumontia grandiflora*, *Carissa grandiflora* *Tabernaemontana coronaria*, *Thevetia nereifolia*, *Trachelospermum jasminoides*.
Aquifoliaceae: ***Ilex paraguajensis***.
Araliaceae: *Aralia papyrifera*, *Brassaia actinophylla*.
Bignoniaceae: *Bignonia magnifica*, *B. purpurea*, *Kigelia pinnata*, *Markhamia* (*Dolichandrone*) *platycalyx*, *Spathodea nilotica*, *Tabebuia chrysotricha*, *T. grandiflora*, *T. guajacana*, *T. pentaphylla*.
Bombacaceae: *Bombax malabaricum*, *Eriodendron* sp., *Pachira excelsa*.
Boraginaceae: *Cordia alba*, *C. heterophylla*, *C. myxa*.
Caesalpiniaceae: *Bauhinia heterophylla*, *Cassia artemisioides*, *C. fistula*, *C. nodosa*, *C. siamea*, *Saraca indica*, ***Tamarindus indica***.
Capparidaceae: *Capparis zeylanica*.
Caricaceae: ***Carica papaya***.
Combretaceae: *Terminalia arjuna*, *T. catappa*.
Convolvulaceae: ***Ipomoea batatas***.
Ebenaceae: *Diospyros ebenus*.
Euphorbiaceae: *Acalypha marginata*, *Aleurites moluccensis*, *Antidesma bunius*, *Hura crepitans*, ***Manihot utilissima***.
Flacourtiaceae: *Flacourtia cataphracta*, *Fl. indica*, *Xylosma longifolia*.

[1] According to Alvim (personal communication) even cocoa can be cultivated with irrigation in the Peruvian coastal desert. Crop production is very high there because the cocoa plant is not affected by fungal diseases in this dry atmosphere.

Guttiferae: ***Garcinia dulcis.***
Lauraceae: ***Cinnamomum camphora.***
Loganiaceae: *Buddleia asiatica.*
Malvaceae: *Chorisia speciosa* (Bottle tree).
Meliaceae: *Amoora rohituca, Cedrela toona, Khaya senegalensis, Melia azedarach, Swietenia mahagoni, S. microphylla.*
Mimosaceae: *Acacia catechu, A. sundra, Adenanthera pavonina.*
Moraceae: ***Arthrocarpus incisa, A. integrifolia***, *Ficus nitida, F. macrophylla, F. misorlusis, F. pandurata, F. platypoda, F. repens, F. sycomorus, F. vogelii, Maclura aurantiaca.*
Myrtaceae: *Callistemon lanceolatum, Eugenia gambolana, E. javanica*, ***E. rosea***, *E. uniflora, Feijoa sellowiana, Melaleuca armillaria, M. styphelioides.*
Nyctaginaceae: *Bougainvillea precatorius.*
Oleaceae: *Jasminum sambac.*
Papilionaceae: *Butea frondosa, Dalbergia sericea, Indigofera pulchella, Machaerium tipu, Sophora secundiflora.*
Pittosporaceae: *Hymenosporum flavum.*
Proteaceae: *Stenocarpus stenaptera, Grevillea robusta.*
Punicaceae: ***Punica granatum.***
Rosaceae: ***Eryobotrya japonica.***
Rubiaceae: *Catesbaea spinosa*, ***Coffea arabica***, *Ixora parviflora, Randia dumetorum.*
Rutaceae: *Murraya exotica.*
Santalaceae: *Santalum album.*
Sapindaceae: *Dodonaea viscosa, Nephelium longana, N. tomentosa.*
Sapotaceae: *Bumelia* spec., *Chrysophyllum cainito, Mimusops caffra.*
Sterculiaceae: *Sterculia foetida, S. lurida.*
Verbenaceae: *Citharexylum quadrangulare, Clerodendron siphonanthus, Tectona grandis.*

3. On the biology of Saharan plants

The Sahara is the most extreme desert and it is exceedingly poor floristically. For this reason, the number of ecological life forms is thus much more restricted than in other desert regions. People familiar only with the Sahara cannot usually imagine the great diversity of life-forms adapted to arid climates elsewhere. Thus one should not generalise from experience only of the Sahara.

Other extreme deserts and somewhat less extreme arid regions are of much greater interest ecologically.

The large group of perennial succulents, which is so diversified in the American and South African deserts, is entirely lacking in the Sahara. The same is true also for the poikilohydrous ferns and mosses, and even lichens and soil algae are rare. The succulent species of the Sahara are halophytes (Chenopodiaceae, Zygophyllaceae, etc.).

The annuals are particularly well represented in the Sahara, i.e. ephemeral species that utilise the irregular rainy periods. Their growing season is extremely short. Their seeds germinate very rapidly and some species can produce mature fruits in 8-15 days. They do not show any special physiological peculiarities and they can be readily cultivated in central Europe in the summer. After heavy rain, the soil becomes covered with a green carpet of annuals, known as 'acheb'. It only lasts for a short time, yet it is of great importance to range management.

Some of the annuals, such as the halophilous *Mesembryanthemum* species, are succulent. They can extend their growth period because of their stored water and can remain alive up to six months even if the soil dries up completely.

Among the ephemerals there are also geophytes. These appear equally suddenly after a rain and remain alive in the soil for a long time after only a short period of growth. They are more frequent on the margins of the desert. The group includes among the Liliaceae some onions and *Pancratium trianthum* (Amaryllidaceae) as well as some rhizomatous plants, such as *Ferula*.

It is characteristic of the perennials—the true xerophytes—that they are concentrated in the few habitats with relatively favourable water relations (contagious distribution), where the water penetrates deeper into the soil. Their aerial parts are only weakly developed by comparison with their below-ground organs. Their roots penetrate as far as the depth of the wet soil and their lateral spread is often quite large. The plants have a far-reaching tap root wherever ground moisture or ground-water is present at some depth below the surface. Typical root systems are shown in Figs. 270-72 (p. 453). The reduction of the leaf area is especially pronounced. The leaves are either shed with shortage of water, or are never formed at all, as in the rod and thorn shrubs (*Zilla*, *Launaea* = *Zollikoferia*, *Retama*, *Pituranthos*, etc.). In times of emergency, transpiration can be reduced to a minimum. In this way, the species maintain their hydrature through long periods of drought despite small water reserves in the soil. In habitats with constantly damp subsoils or ground-water, the plants have no difficulties with their water supply. Here one even finds trees. Ozenda (1954) cites the following trees or tree-like plants:

Ephedra alata, *Cupressus dupreziana*, *Phoenix dactylifera*, *Hyphaene thebaica*, *Populus euphratica*, *Ficus salicifolia*, *Maerua crassifolia*, *Acacia raddiana*, *A. seyal*, *A. arabica*, *A. albida*, *Cassia lanceolata*, *C. obovata*, *Balanites aegyptiaca*, *Pistacia atlantica*, *Tamarix* (6-10 species), *Periploca laevigata*, *Calotropis procera*, *Salvadora persica*, *Olea laperrini*. In addition, a number of shrubs occur which can become quite large. In such habitats one also finds hygromorphic species, such as *Citrullus colocynthis*. These are of tropical origin and give the impression of being alien elements in this community. They are not killed by the strong insolation because of their high transpiration rates (Lange, 1959). A special group are the xero- and hygrohalophytes. Their number is probably greater than is usually assumed because the salt content of the cell sap has not been investigated in Saharan plants. This group probably includes most of the Chenopodiaceae, Aizoaceae, Zygophyllaceae, Plumbaginaceae, etc. Even where some of these occur on salt-poor soils they may accumulate salts in their cell sap.

A unique plant in the north-west Sahara is *Fredolia* (*Anabasis*) *aretioides*, which forms 50 to 60 cm tall, semi-globose, tough pillow-like cushions. This is a life-form found more frequently as an Antarctic floristic element in Patagonia, in the alpine region of South America and New Zealand.

Wind also plays an important role in the life cycle of desert plants. Very many species are anemogamous, such as the grasses, and Chenopodiaceae. The pollination relations in others have not yet been elucidated. The wind is particularly important for the distribution of the seeds or fruits. Beside anemochorous species, only epizoochorous species are strongly represented. By contrast, endozoochorous species are completely absent since the bird fauna is very poor. The strange imbibitional movements of the Rose of Jericho (*Anastatica hierochuntica*) are well known. Its branches curve around the inflorescences when it is dying from drought. It can be torn away by the wind when it rolls over the ground. Its water-imbibing branches only recurve and expand after rain, when the seeds are released.

4. Ecological investigations in the Algerian Sahara and in Libya

In the north-western Sahara experimental ecological research began with H. Fitting (1911), who published his investigations at Biskra in 1911. While his findings did not subsequently receive complete support in every respect, it stimulated much later work on those lines.

Fitting was the first to emphasise the importance of the osmotic value in desert plants. His results relate to several plants near Biskra, a region with a mean rainfall of 299 mm, where he determined the osmotic value plasmolytically with KNO_3 and NaCl solutions. He obtained very high values and interpreted them as a favourable adaptation to allow plants to absorb water from the near-dry soils.

Later however, Killian and Faurel (1933, 1936) showed, when they investigated the same plants at Beni Unif during an unfavourable season at a drier location with only 110 mm rainfall, that the plasmolytic method gave values which were too high.

Table 90 shows the plasmolytically-determined values of Fitting compared with the cryoscopically determined values of Killian and Faurel.

It can be seen from this table that salt-plants, such as *Atriplex*, *Statice*, *Limoniastrum* and *Frankenia pulverulenta* show similarities but for other plants, the plasmolytically-determined values are much too high.

Today we know that the soil water relations are not so unfavourable as was once thought at sites occupied by desert plants. However, a prerequisite for their persistence is a root system that has penetrated to the water-holding soil layers (see pp. 270 ff.).

Moreover, we should not interpret their high osmotic values as a favourable adaptation to low water supply but rather as a reflection of a disturbed water balance. This applies even although the plant may be able to extract a little more water from the soil because of its increased suction tension. The marked drop in their hydrature is not a feature favourable to them. No one would consider the decrease in body weight of a starving person as a favourable adaptation, even if the rate of his metabolism was reduced in consequence, so permitting a better utilisation to be made of an insufficient quantity of food. Any soil,

TABLE 90

Osmotic values (atm)	Biskra (after Fitting) Plasmolytic values (atm, approx.)	Beni Unif (after Killian & Faurel) Cryoscopic values (atm)
Citrullus colocynthis	11	5·5–7·7
Erodium pulverulentum	15–21	9·9
Erodium pulverulentum	27–53 desert	
Asteriscus pygmaeus	15–18 cropland	12·5
Atriplex halimus	36–53	35·5–70 (October)
Atriplex halimus	18–24 dunes	
Atriplex halimus	42–53 gravel-desert	
Fagonia glutinosa	70–100+	9·7–17·3
Statice delicatula	30–53 salt-swamp	20·7–48·6 (October)
Peganum harmala	42–70	14·9–34·4 (October)
Nerium oleander	21–24	12·4–22·3 (January)
Capparis spinosa	53–70	18·4–21·3 (November)
Capparis spinosa	27–42 cropland	
Phoenix dactylifera	42–53 salt-swamp	16·8–19·5 cropland
Phoenix dactylifera	36–42 Qued Biskra	28·9–32·6 dunes
Limoniastrum guyonianum	53–70 salt-swamp 70 desert	49·6–53·4 salt-swamp
Traganum nudatum	100+	20·2–30·7 dunes
Traganum nudatum		22·9–29·4 salt-swamp
Suaeda vermiculata	100+ salt-swamp	24·9–27·4 dunes
Suaeda vermiculata	desert	22·9–29·4 salt-swamp
Frankenia pulverulenta	21–30	40·6
Frankenia pulverulenta	27–53 salt-swamp	
Frankenia thymifolia	100+ desert	37·1

and particularly the sandy soil of the desert, contains virtually no water at the permanent wilting percentage of water potential, 15 atm, so that even a marked rise in the osmotic value of the plant could not produce an appreciable, additional water uptake. In later studies it has been found repeatedly that the osmotic value of desert plants—disregarding halophytes—is only a little higher with a favourable water supply than in plants of wet regions. Extreme desert plants have such a high osmotic stability—i.e. they are so hydrostable—that they only show a slight increase in their osmotic values during long-lasting droughts.

As we have seen, desert plants can reduce their water loss so markedly that they survive on extremely small soil water reserves. An additional factor is their very well developed root system and this has been more closely investigated by Kausch (see p. 452 ff).

We are not yet clear about the water potentials which have to be overcome by the roots for water uptake in desert soils. They cannot be very high because of the coarse textures of the soils. Determinations by Harder (1930) in Beni Unif appear to contradict this statement since they gave rather high values. However, the six samples investigated were not taken from the root zone. Moreover, water is very irregularly distributed in the soil, especially in the desert, so that Harder's values cannot provide any information in the local suction tension adjacent to living roots. Furthermore, the determinations were made by the capillary method which gives values which are too high in dry soils.

A most important question is whether the desert plants can absorb CO_2 during the long period of drought, or whether they survive in a dormant condition.

In this connection, the experiments of Harder, Filzer and Lorenz (1932) are very illuminating. These authors determined the CO_2 absorption of whole plants *in situ*. The plants were enclosed in cellophane chambers and air was sucked through. The excessive heating of the enclosed plants and the poor ventilation probably created conditions for photosynthesis far inferior to natural ones. In spite of this, it was found that all plants could absorb CO_2 even after almost a year of drought. Thus, the plants did not undergo a resting period.

Among the plants investigated in August, *Haloxylon articulatum* showed a sufficient excess of assimilated products to be able to produce flowers and fruits. The amount of assimilated material was much less in *Limoniastrum feei*. Both species showed a marked lack of response after rain in September. By contrast, the pillow-like cushion plant, *Fredolia aretioides*, showed a greatly reduced gas exchange during the drought; however, it became greatly enhanced after rain.

The differences before and after the rain were even more pronounced in *Zollikoferia arborescens*, which showed a great deal of yellowing and loss of parts during the drought. However, after a rainfall of 20 mm, very active CO_2 absorption developed.

The water balance of the plants before and after rain were, unfortunately, not investigated. One could assume that the hydrature showed no important differences in the first two cases, but that it improved considerably in the latter two. *Haloxylon* and *Limoniastrum* are probably salt-accumulating halophytes.

Water relations always control the dry-matter production of desert plants. Because of this, individual plants of the same species in favourable and unfavourable habitats, respectively, show quite remarkable differences in appearance and development.

In the same region, Stocker (1954) studied a community in 1953, in which he investigated, simultaneously, micro-climate, water relations and photosynthesis as well as respiration. The results have so far only been published in preliminary form.

Nine daily curves are shown for April to May for *Zilla macroptera* (wet soil, but also from a dry habitat), *Zizyphus lotus*, *Fredolia* (*Anabasis*) *aretioides*, *Retama raetam*, *Citrullus colocynthis*, *Capparis spinosa*, *Phragmites communis* and *Frankenia thymoides*. All species show an active transpiration, which is not reduced even at midday. However, the curves of CO_2 absorption, determined by instantaneous measurements, appear rather strange. With the exception of *Zilla* (wet soil) and *Phragmites*, which showed a positive net absorption over the whole day, the curves for all the others fell steeply in the morning and were below the zero point around midday. One can hardly resist the suspicion that the values do not portray the actual situation. April and May are the most favourable months for the desert plants. They grow intensively at this time and this is correlated with active transpiration which indicates favourable water relations and open stomata. If plants cannot show a high production of dry matter in this season, when can they produce the necessary organic substances?

Berger-Landefeldt (1959) who obtained similar results to Stocker when carrying out CO_2 absorption experiments with halophytes in the Great Syrte, said in this connection: 'One can hardly suppress the thought that every failure observed is merely the result of unnatural experimental conditions'. He means by this the drop in CO_2 absorption around midday, which is commonly associated with CO_2 being given off.

Recently, Bosian (1959) pointed out the great danger and the errors which result from overheating even in the mild climate of south Germany. However, in answer to this Stocker and Viehweg (1960) emphasised the short duration of their experiments, which lasted only a minute at a time. They also stated that 'the experimental error is much greater than the instrumental error'. Since no replicated determinations were made it is not possible to evaluate the experimental errors. It is possible that in the desert overheating occurs even within one minute at noon.

Kausch (1959), a collaborator of Stocker, worked more particularly with the roots of the plants near Beni Unif. This region is characterised by the contagious distribution of vegetation. Thus, plants are only found in favourable habitats, for example, in erosion channels and in dry river beds or in shallow hollows and deeper depressions. The bed-rock of the soil consists predominantly of sandstone which is partially broken-up or covered with stony, sandy or clay materials. Lime-crusts are a characteristic feature in the soil profile. As emphasised by Kausch, one gets the impression that 'there are particular localities in the terrain where water can penetrate to the bedrock whence it then spreads laterally at a greater depth'. For example, such places are stone cracks after rain and plants are found to grow in such localities. Fig. 270

indicates these relations schematically; Figs 271 and 272 show the actual root developments in two examples.

One can see clearly that part of the root system is concentrated at a shallow depth (10-40 cm), to provide an opportunity to absorb water from the upper soil layers after relatively light rains. The other part of the root system penetrates the soil or rock crevices relatively deeply.

FIG. 270. Schematic root profile (black sandstone) from the South Algerian Sahara near Beni Unif (after Kausch).

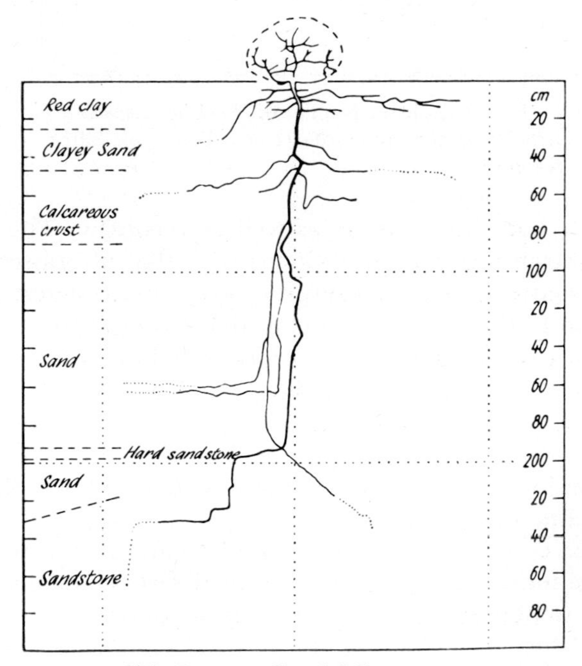

FIG. 271. Root profile of *Zilla macroptera* growing in a loam pan near Beni Unif (after Kausch).

The roots reach into local, permanently-damp soil pockets which are always topped up again in good rain years. This further supports the principle that root depth increases with soil wetness in 'contracted vegetation' (i.e. in island-like plant communities of a discontinuous vegetation cover).[1] However, these deeper, damper soil horizons have nothing to do with the ground-water horizon.

Lemée (1954) studied the water relations of perennial grasses near Beni Unif. These grasses are, in general, well supplied with water. Their root system shows the characteristic 'sand-sleeves' (p. 242), and are

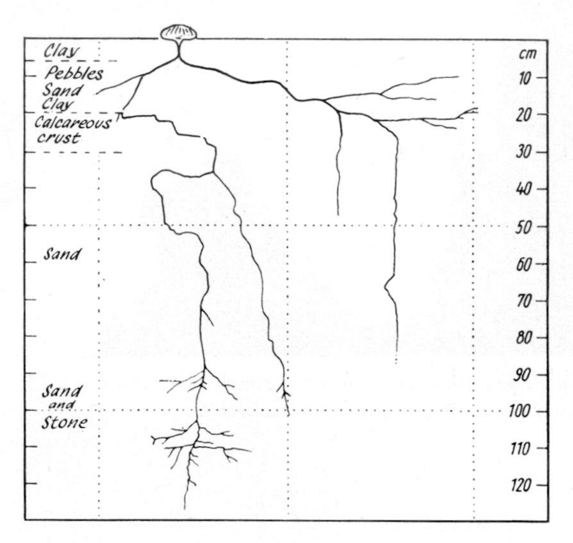

FIG. 272. Root-profile of *Fredolia* (*Anabasis*) *aretioides* in the gravel-desert near Beni Unif (after Kausch).

spread out far both horizontally as well as vertically. The amount of water lost by transpiration corresponds with that of grasses in central Europe in absolute terms; i.e. transpiration loss in the desert is relatively less in relation to the increased evaporative power of the environment. Transpiration can be reduced markedly by folding of the leaf blades which occurs at a specific water deficit (see Table 91).

In response to a favourable water supply the suction tension values are always low (6-15 atm). This is true also for the osmotic values with the exception of the slightly halophilous *Lygeum* and the strongly halophilous *Aeluropus* (Table 92).

The latter two species are also found in areas transitory to halophytic communities—at the margins of the desert. In areas with rainfall high enough to permit run-off of large masses of water into depressions

[1] A summary of maximum rooting depths in desert regions, but without more precise habitat descriptions, is given by Zohary (1961).

followed by evaporation, salt pans are formed with a characteristic halophyte-vegetation.

Ozenda (1954) described such pans on the high plateau of south Algeria, where he also investigated the mechanical composition and salt content of the soils. Killian (1941, 1951, 1953) referred to the indicator value of several halophilous species and described the biology of the halophyte *Frankenia pulverulenta.*

TABLE 91

Transpiration reduction due to folding of the grass leaves

	Reduction of transpiration (in %) on folding	Water deficit at which folding starts (in %)
Lygeum spartum	70	8–20
Aristida pungens	70	6–24
Stipa parviflora	75	18–40
Andropogon laniger		
upper leaves	30	40
lower leaves	60	90
Andropogon annulatus	83	50 upper, 90
Cynodon dactylon	50	30 upper, 80
Pennisetum dichotomum	leaves dry up	18–38
Aeluropus litoralis	leaves dry up	80

TABLE 92

Osmotic values of desert grasses (atm)

Aristida pungens	9·6–21·3	*Lygeum spartum*	23·6–26·6
Aristida obtusa	18·2	*Aeluropus litoralis*	25·2–55·3
Andropogon laniger	6·4–13·0		

A short summary of the plant communities on damp salt-soils (Solonchak) based on their own studies is also given by Dubuis and Simonneau (1957).

Large salt pans are found along the Mediterranean coast in Libya. A very intensive ecological investigation by Berger-Landefeldt (1957, 1959) was concerned with the largest of these salt pans, Tauorga, which covers an area of 5000 km² on the west side of the Great Syrte. The mean annual rainfall at three stations in the south-west, south-east and north of the salt pan (sebcha) were 60, 127 and 132 mm, respectively. The amount of rainfall is important in determining the degree of salinisation. The rain water transports salts from the highlands inland into the pan where the water evaporates. The soils are wet throughout the year but the salt concentration of the soil solution is subject to great fluctuations. During the rainy period, in winter and spring, the salt concentration is low as a result of the inflowing water masses. Plants can germinate at this time. The salt concentration then increases steadily during the summer with continued evaporation. A salt crust is formed at the surface, but the osmotic value of the soil solution in the rooting zone never exceeds 34 atm. The osmotic values of the plants were correspondingly high, with 35·4 atm in *Nitraria retusa*, 69·9-72·4 in *Salicornia fruticosa* (chloride proportion equivalent to 50·9 to 54·5 atm). The values in *Atriplex mollis* were exceptionally high with 123-156 atm. However, Berger-Landefeldt points out that these values may be doubtful, since salts excreted from

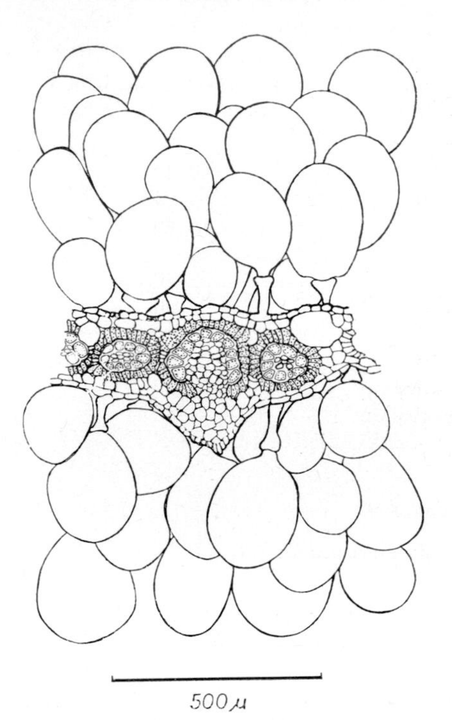

Fig. 273. Leaf cross-section of *Atriplex mollis* with inflated, vesicular hairs (after Berger-Landefeldt).

the leaves into the inflated hairs are included in the press-sap (see p. 417). In the press sap of *Atriplex mollis*, the proportion of chloride alone reached 103-132 atm. An idea of the contribution of the inflated hairs to the leaf tissue can be obtained from Fig. 273. Fig. 274 shows leaf tissue of *Nitraria retusa* with enclosed tannin cells (dotted) and large idioblasts (clear), which rapidly lose their turgidity on wilting but regain it immediately on addition of water. In the palisade parenchyma of the internodes of *Arthrocnemum glaucum* are thick sclereids, which serve as strengthening tissue (Fig. 275). The water losses through transpiration are always low during the day in *Nitraria*, *Atriplex*, *Salicornia* and *Arthrocnemum*. Yet the water contained in the shoots is replaced several times during the day. The plants named earlier are the most important components of the perennial vegetation. In addition, other Chenopodiaceae and Plumbaginaceae also occur and, in the spring, a group of annuals also.

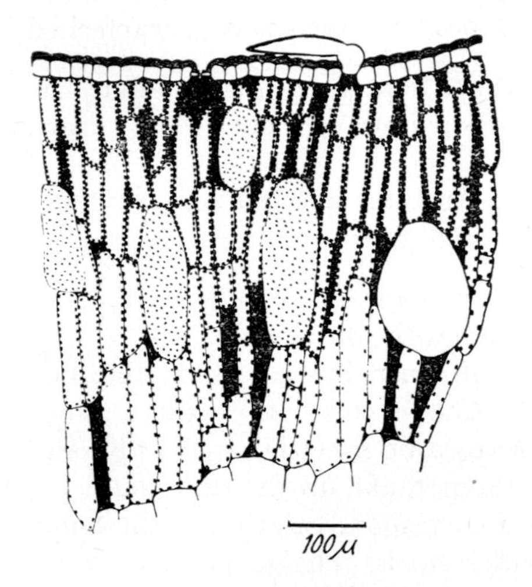

FIG. 274. Section through a leaf of *Nitraria retusa* (after Berger-Landefeldt). The dotted cells contain tannin. The large cell is an idioblast.

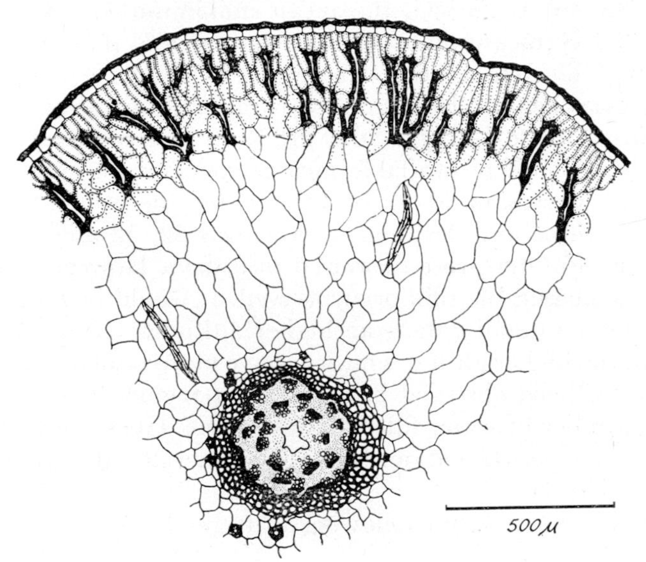

FIG. 275. Section through a stem internode of *Arthrocnemum glaucum* (after Berger-Landefeldt).

5. Relationships of the Vegetation in the Tibesti mountains

The Tibesti mountains occupy a very isolated position in the centre of the eastern Sahara somewhat south of the Tropic of Cancer. They rise from a 200 m high plateau to an altitude of 3400 m. A conspectus of the flora and vegetation is given by R. Maire and Th. Monod (1950).

The total of plants collected in the Tibesti mountains is 369, of which 40 are cryptogams. From the Hoggar mountains (Ahaggar) further to the west, 350 species are known so far. The two floras are fairly similar, yet that from the Tibesti mountains is characterised by:

1. A small number of northern (Mediterranean) elements,
2. An increase in southern (Sahelian) elements (45% instead of 8%), and
3. Numerous eastern (Saharo-Arabian and Sudano-Sindian) elements.

This can be explained by their geographical position. The proportion of these elements differs in each altitudinal zone.

The Tibesti mountains are built on a base of crystalline bedrock (granite and crystalline shale). Above this base rises a sandstone plateau. The highest parts are of volcanic material, such as basalt, tuffs and lava, and sharply pointed craters make up the summit.

The climate of the mountains corresponds to that of the Sahara in general. There are no long-term records. The mean temperature of Bardai (1000 m) is 25·1°C, the mean maximum 37·8°C, the mean minimum 2·8°C (measurements refer to a total of 31 months distributed over 4 years). The temperature can drop to below −10°C at an altitude of slightly over 2000 m. However, following night frosts of −12°C, the temperature can rise during the day to 14°-19°C. Records of rainfall are less complete. The few available measurements show that an annual rainfall peak occurs in May. In May 1934, the amount of rainfall measured at three low-elevation stations was 32, 52 and 455 mm. In the last case, 450 mm were measured in 5 days. However, years with virtually no measurable rain occur as well. It would be most desirable to obtain some rainfall data from higher altitudes. To judge from the vegetation, the higher altitudes must receive more rainfall. Strong cloud formation has been observed in the summit region. There are a considerable number of water-filled basins in the Tibesti mountains and some even have permanent outflows. Yet, generally, the mountains are of a desert character.

Monod distinguishes the following altitudinal belts which also apply to the Hoggar mountains:

Saharo-Mediterranean belt < upper 2400–3400 m
lower 1700–2400 m
Saharo-Tropical belt: below 1700 m

The Saharo-Tropical belt is characterised by a very dry *Acacia-Panicum* savannah, which surrounds the mountains. Higher up in the mountains, relict elements become associated which have originated further south. In particular, a 'Sahelian island' occurs on the south-west slope, because there one finds *Acacia laeta*, *A. stenocarpa*, *Boscia salicifolia*, *Carissa edulis*, *Rhus villosa*, *Cordia gharaf*, *Ehretia obtusifolia* and even the palm *Hyphaene thebaica*. In the dry-valleys and on sand flats in the area at the foot of the slope elements, essentially alien to the Sahara, are found such as *Cleome*, *Cassia*, *Crotalaria*, *Buchea*, *Geigeria*, *Cenchrus biflorus*, etc.; Mediterranean elements, such as are found in the northern Sahara as well, e.g. *Diplotaxis acris*, *Rumex*, *Erodium malacoides*, *Astragalus*, *Globularia alypum*, *Lavandula*, *Salvia*, *Ephedra altissima*, only come in at greater heights.

Plants characteristic of the lower Mediterranean belt are:

Helianthemum lippii, *H. ellipticum*, *Artemisia judaica*, *Teucrium polium*, *Myrtus nivellei*, *Argyrolobium abyssinicum*, *Echinops bovei*, *Cuscuta planiflora*, *Andropogon hirtus*.

Characteristic of the upper Mediterranean belt are:

Ballota hirsuta, *Sisymbrium reboudianum*, *Nepeta tibestica*, *Silene* spp., *Oxalis corniculata*, *Malva parviflora*, *M. rotundifolia*, *Cheilanthes pteridioides*, *Senecio coronopifolius*, *Bromus rubens*, *Stipa tibestica*.

These species indicate a somewhat wetter climate.

Some of the species are probably relicts of former epochs with wetter climates, for example *Ficus salicifolia* and *Adiantum capillus-veneris*, which grow by springs. Even a marine jelly-fish was found in a deep water hole! Maire and Monod cite a complete list of all species found and give information on their altitudinal and distributional features.

References

Berger-Landefeldt, U. 1957; 1959. Beiträge zur Ökologie der Pflanzen nordafrikanischer Salzpfannen. *Vegetatio*, **7**, 169-20; **9**, 1-47.

Bosian, G. 1959. Zum problem des Küvettenklimas. *Ber. Dtsch. Bot. Ges.*, **72**, 391-97.

Dubuis, A., and Simonneau, P. 1957. Les unités phytosociologiques des terrains salins de l'Ouest Algérien. *Trav. Sect. Pédol. Agrol. Bull.*, **3**, 1-29.

Fitting, H. 1911. Die Wasserversorgung und die osmotischen Druckverhältnisse der Wüstenpflanzen. *Zschr. Bot.*, **3**, 209-75.

Harder, R. 1930. Über den Wasser- und Salzgehalt und die Saugkräfte einiger Wüstenböden Beni Unifs (Algerien). *Jb. Wiss. Bot.*, **72**, 665-99.

Harder, R., Filzer, P., and Lorenz, A. 1932. Über Versuche zur Bestimmung der Kohlensäureassimilation immergrüner Wüstenpflanzen während der Trockenzeit in Beni Unif (algerische Sahara). *Jb. Wiss. Bot.*, **75**, 44-194.

Kausch, W. 1959. *Der Einfluss von edaphischen und klimatischen Faktoren auf die Ausbildung des Wurzelwerkes der Pflanzen unter besonderer Berücksichtigung einiger algerischer Wüstenpflanzen.* Dissertation. Darmstadt.

Killian, C. 1941. Sols et plantes indicatrices dans les parties non irriguées des oasis de Figuig et de Beni-Ounif. *Bull. Soc. Hist. Nat. Afr. N.*, **32**.

Killian, C. 1945. Observations sur la biologie des quelques plantes fezzonaisses. *Publ. Inst. Rech. Sahara*, **4**, 1.

KILLIAN, C. 1951. Observations sur la biologie d'un halophyte saharien, *Frankenia pulverulenta*. *Trav. Inst. Rech. Sahar.*, **7**.

KILLIAN, C. 1953. La végétation autour de Chott Hodua, indicatrice des possibilités culturales, et son milieu édaphique. *Ann. Inst. Agric. Algérie*, **7**, 5.

KILLIAN, Ch., and FAUREL, L. 1933. Observations sur la pression osmotique des végétaux désertiques et subdésertiques de l'Algérie. *Bull. Soc. Bot. Fr.*, **80**, 775-8.

KILLIAN, Ch., and FAUREL, L. 1936. La pression osmotique des végétaux du Sud-Algérien. *Ann. Physiol.*, **12**, 5, 859-908.

KILLIAN, C., and FEHÉR, D. 1935. Recherches sur les phénomènes microbiologiques des sols sahariens. *Ann. Inst. Pasteur*, **55**, 573-622.

LANGE, O. L. 1959. Untersuchungen über Wärmehaushalt und Hitzeresistenz mauritanischer Wüsten- und Savannenpflanzen. *Flora*, **147**, 595-651.

LAVRENKO, E. M. 1962. The characteristic features of the plant geography of the Euro-Asiatic and north African deserts (in Russian). *Akad. Wiss. Leningr.*, **15**.

LEMÉE, G. 1953. Les associations à therophytes des dépressions sableuses et limoneuses non salées et des rocailles aux environs de Beni Ounif. *Vegetatio*, **4**, 137-54.

LEMÉE, G. 1954. L'économie de l'eau chez quelques graminées vivaces du Sahara septentrional. *Vegetatio*, **5/6**, 534-41.

MAIRE, R., and MONOD, T. 1950. Études sur la flore et la végétation du Tibesti. *Mem. Inst. Fr. Afr. Noire*, **8**, 1-140.

MECKELEIN, W. 1959. *Forschungen in der zentralen Sahara*. Braunschweig.

OZENDA, P. 1954. Observation sur la végétation d'une region semiaride: Les Haut-Plateaux du Sud Algérois. *Bull. Soc. Hist. Nat. Afr., N.* **45**, 134-69.

OZENDA, P. 1958. *Flore du Sahara septentrional et central*. Centre Nat. Rech. Scient, Paris.

PIERRE, F. 1955. Le peuplement entomologique et ses rapports avec la végétation dans les sables vifs de la zone aride. *Proc. Montpellier Symp. Unesco Arid Zone Res.*, **5**, 107-16.

QUEZEL, P. 1965. La végétation du Sahara. *Geobotanica Selecta*, **2**.

SCHIFFERS, H. 1950. *Die Sahara*. Stuttgart.

STOCKER, O. 1927-28. Das Wadi Natrun. *Veg. Bilder*, **18**, Plates 1-6.

STOCKER, O. 1954. Der Wasser- und Assimilationshaushalt südalgerischer Wüstenpflanzen. *Ber. Dtsch. Bot. Ges.*, **67**, 289-99.

STOCKER, O., and VIEHWEG, G. H. 1960. Die Darmstädter Apparatur zur Momentanmessung der Photosynthese unter ökologischen Bedingungen. *Ber. Dtsch. Bot. Ges.*, **73**, 198-207.

VARGUES, H. 1952. Étude microbiologique de quelques sols sahariens en relation avec la présence d'*Anabasis aretioides* Coss. et Moq. (Trav. Lab. Beni-Ounif, **30**.) *Proc. Internat. Symp. Desert Res.*, Jerusalem.

WALTER, H., and HABER, W. 1957. Über die Intensität der Bodenatmung mit Bemerkungen zu den Lundegardschen Werten. *Ber. Dtsch. Bot. Ges.*, **70**, 275-82.

ZOHARY, M. 1961. On hydro-ecological relations in the Near East desert vegetation. *Proc. Madrid Symp. Unesco Arid Zone Res.*, **16**, 199-212.

ZOHARY, M. 1963. On the geobotanical structure of Iran. *Bull. Res. Counc. Israel, D*, **11** (suppl.) 1-113.

XIV. *The Egyptian-Arabian Desert including Sinai and Negev*

1. Habitats and plant communities

This desert region (Fig. 276) is the most eastern part of the Sahara, thus the conditions resemble those of the Sahara proper. The eastern region is readily accessible from the Nile. This has contributed to its early exploration by Volkens (1887), who studied some ecological problems and the anatomical features of certain desert plants. Volkens believed that the leaf hairs of plants function in providing water from dew. Later, during a longer expedition, Stocker (1928) made intensive ecological investigations. These included numerous measurements of the

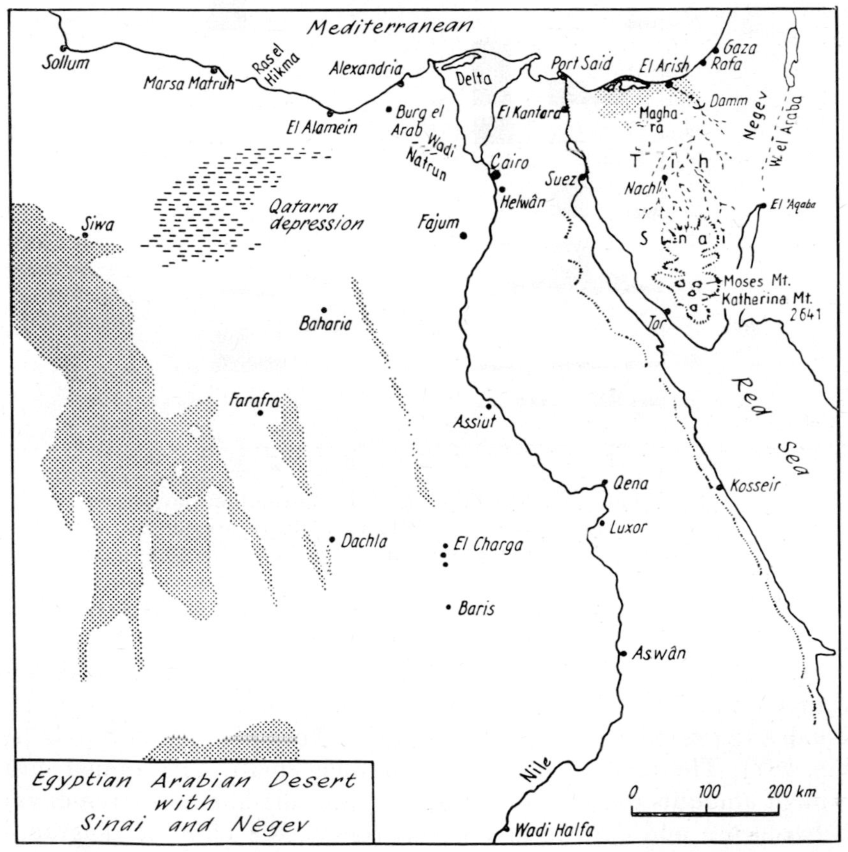

FIG. 276. Map of the eastern Sahara and the neighbouring desert regions. Dotted—sand-covered areas; dashed—depressions below sea level.

transpiration and water economy of the plants and also the water and salt relations of the soils.

However, long-term investigations have only begun within the last two decades after a station was established in association with the University of Cairo on the Cairo-Suez road near kilometre 34.

Several ecological problems, relating mainly to water relations, were investigated by A. M. Migahid and A. A. Abd el Rahman. These

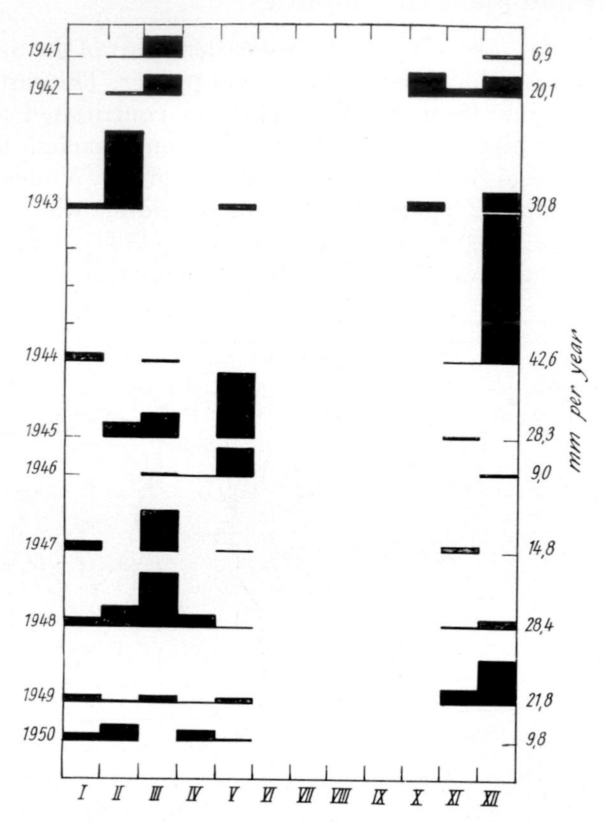

FIG. 277. Monthly rainfall distribution in Cairo during the years 1941 to 1950. Figures at right refer to annual rainfall.

provided much comprehensive information. M. Kassas studied the general vegetation relations. I shall refer later to these works.

Cairo lies on the 30th parallel and only receives occasional, light winter rains. The rainfall occurs between October and May yet it is usually restricted to the period between November and March (see Fig. 277). The amount of rain is extremely small. The annual mean rainfall amounts to 21·3 mm (1941-50), and the annual extremes vary between 6·9 and 42·6 mm. For this reason, the climatic diagram of Helwân near Cairo (Fig. 278) gives no indication of a wet period. The number of showers are rarely more than 5-6 per year and the rain-free period usually lasts 9-10 months. As in all arid regions, the annual rainfall is here also typically irregular. In analysing the mean annual rainfalls from 1906 to 1955, i.e. for a period of 50 years, one finds a very great range with a mean value of 27 mm $\pm 15{\cdot}2$. The yearly variation is shown as a curve in Fig. 280. This curve is very similar to that of Mulka in Australia (Fig. 258).

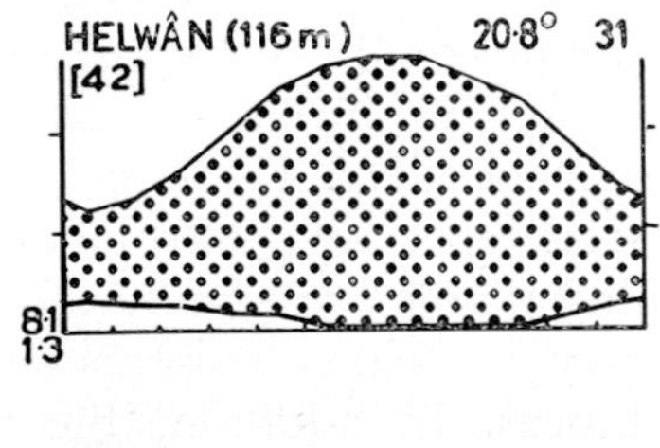

Fig. 278. Climatic diagram of Helwân in the desert near Cairo.

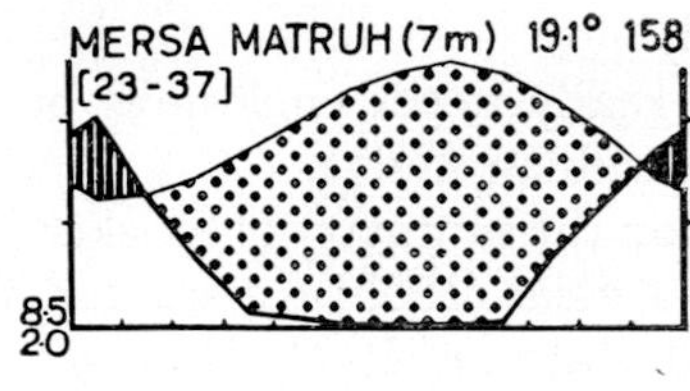

Fig. 279. Climatic diagram of Mersa Matruh (Marsa Matruk) on the Egyptian Mediterranean coast.

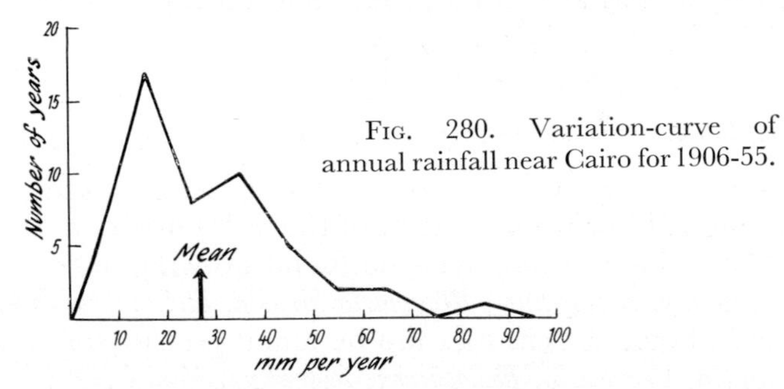

Fig. 280. Variation-curve of annual rainfall near Cairo for 1906-55.

Dry years with rainfall below normal are more frequent than heavy rain years. However, the latter may provide relatively large amounts. The mean temperature maxima of the warmest months is 35°C, the absolute maxima exceed 40°C. In the winter months, however, the temperatures can drop to 3°C. Dewfall has been observed on such nights. The relative humidity can decrease to 4% on hot days and the evaporation rate is very high in the summer months. In the hottest months it amounts to about 22 ml/day (standard Piche) and in the coldest months to about 8 ml/day. Evaporation is much increased by the often continued and strong winds of 5-10 m/sec. Kassas and Imam (1957) did micro-climatological temperature measurements on a south and north slope. As expected, the differences were greatest at low sun position in January. This is important in so far as the main growing season is in the winter. Only in this season can the soil become covered with a reasonably dense carpet of ephemerals after a good rain.

The soil conditions are just as unfavourable as the climate. The hamada south-east of Cairo near Helwân is derived from yellow-grey limestone and marl from the Eocene. These materials contain much salt. Towards the east, along the road to Suez, lies the pebble desert with a gypsum-rich soil. Conditions for the development of vegetation are only favourable in broad valleys covered with deposits of sand.

In describing the vegetation, one must take into consideration the changes in plant cover due to man's influence. The desert areas serve as range for sheep, goats, donkeys and camels. Since the plant cover is very sparse, grazing has a very pronounced effect. Therefore, the best forage plants—which include the grasses especially—have often completely disappeared. The few remaining plants are usually those that are not taken during grazing, i.e. they are either poisonous or unpalatable, or have rough hairs or thorns. In addition, the woody plants are used as fuel by the nomads.

To study the original vegetation really needs enclosures but these are entirely lacking in the Egyptian-Arabian desert.

The following description is based on the summary of Stocker (1926), the work of Davis (1953) and that of Kassas (1953; Kassas and Imam, 1952, 1953, 1954, 1957; Kassas and Zahran, 1962; Kassas and El-Abyad, 1962; Kassas and Girgis, 1964) and on my own field notes of 1960.[1]

(*a*) *Vegetation of the stone-desert*

The hamada is generally barren of vegetation because of the high salt concentration and low water content of the soil. Only in areas of rocky outcrops can a few chasmophytes be found growing in rock crevices, such as *Erodium glaucophyllum*, *Reaumuria hirtella*, *Helianthemum kahiricum*, *Fagonia mollis*. Some lichens can also be found on the stones (*Lecanora*, *Lecidea*, *Physcia*, *Verrucaria*, *Endocarpon*, *Ramalina*, *Usnea*).

During the winter months, following rain showers, run-off moves from the level stone pavement along vaguely indicated rain channels. These are covered by loosely held products of weathering and they anastomose to form erosion channels, which are usually no deeper than 2 m. There is some leaching of salts from these channels, so permitting the growth of scattered plants. Since the soil is only wet for short periods in these sites, the plants are either ephemerals, such as *Anastatica hierochuntica*, *Diplotaxis acris*, *D. harra*, *Centaurea aegyptiaca*, *Zygophyllum simplex*, *Trigonella stellata* and *Plantago ovata*, or winter-green perennials, such as *Asteriscus graveolens*, *Iphiona mucronata*, *Farsetia aegyptiaca*, *Erodium arborescens*, *Echinops spinosus*, *Calligonum comosum*, *Reaumuria hirtella* and *Ephedra* (Figs. 281 and 282).

[1] Everything that is known, in addition, about individual plant species can be found in the greatwork of Täckholm, V. and Drar, M. (1949-54), published in 4 volumes. A handy dentification manual, *Students Flora of Egypt*, has been published by the same authors.

FIG. 281. *Diplotaxis harra* (two flowers) and many *Plantago* plants in the sand bed of Wadi Hoff near Helwân, Egypt (Phot. E. Walter).

FIG. 282. *Centaurea aegyptiaca* in the Wadi Hoff, near Helwân, South of Cairo, Egypt (Phot. E. Walter).

Gorges with block-littered slopes and rocky walls are formed where the water falls from an erosion channel down a terrace-step. *Rumex vesicaria* (with delicate leaves) can be found on the damp soil between the blocks of rock, while *Capparis spinosa* may be found growing in the rock crevices.

Where a 50 m waterfall had produced a permanently wet, water hole 4 m deep, two relict plants of a damp climate were found. These were *Ficus pseudosycomorus* and *Adiantum capillus-veneris*. The latter is found in all subtropical regions in such permanently wet localities.

(*b*) *Vegetation of the wadis*

Plant growth is somewhat denser in the larger channels that join the broad sand-filled dry valleys or wadis (Fig. 268, p. 442) as steeply sloping pebble roads (Fig. 266, p. 432).

Kassas distinguishes the following communities in the several wadis around Helwân, for which he presents complete plant lists:

1. Stachietum aegyptiacae on crystalline limestone bedrock in valleys: here chasmophytes can maintain themselves in rock crevices. The dominant species is *Stachys aegyptiaca*, which becomes very old (roots with diameters of 5 cm). Also characteristic are *Limonium* (*Statice*) *pruinosa*, *Fagonia mollis*, *F. kahirina* and *Erodium glaucophyllum*. Many others are associated. *Asteriscus graveolens* and *Reaumuria hirtella* are occasionally dominant.

2. Zygophylleto-Anabasietum on stony-pebble soils only 50 cm deep which contains less than 30% of fine soil: the two dominant species, *Zygophyllum coccineum* and *Anabasis setifera*, have succulent leaves and stems. Yet they are halophytes rather than succulents, i.e. they have no large water storing tissues and probably accumulate chlorides. Many ephemerals occur as associates.

3. Zilletum spinosae on deep soils, which still contain 3% water at 60-70 cm. The salt content is around 0·2-0·8%; the organic matter content is usually below 1%. According to its annual rings, *Zygophyllum coccineum* does not live beyond 5 years. Because of this it may become replaced in such habitats by the thorny shrub *Zilla spinosa* (Fig. 269), which can reach heights of 1·5 m and diameters of 2 m. The soil profiles show irregularly alternating horizons of coarse and fine-textured materials. This is related to the changing velocities of flood waters after rain. There are all transitions between the *Zygophylletum* and the *Zilletum*.

4. Grassland on silty sand-soils exceeding 100 cm: the two most important tussock grasses are *Panicum turgidum* and *Pennisetum dichotomum*. *Panicum* is more heavily grazed. The cover can reach 80% in places. The grass tussocks cause sedimentation of finer soil particles during floods. The number of associates is smaller here.

5. Gallery-scrub on valley-terraces that are not covered by floods: the most important representatives are *Nitraria retusa* (*tridentata*), *Lycium arabicum, Atriplex halimus* and *Tamarix* species. The shrubs can become 1-2 m tall. *Tamarix articulata* is a tree of 1 to 10 m. However, large stands are rare, since they are cut for fuel. *Atriplex* is also strongly grazed. The formation of dense gallery forests is only possible where permanent ground-water horizons are present to at least 15 m depth. Near the coast permanent ground-water horizons may be as close as 2 m below the surface.

(*c*) *Vegetation of the pebble desert*

A vast pebble desert stretches along the Cairo-Suez road between the Mokattan hills (300 m above sea level) in the west and the Gebel Ataqa (700 m) in the east. The landscape is undulating and its surface is covered with very dark pebbles due to their coating of desert varnish. Small, sand-filled erosion channels, light in colour, stretch down the gentle slopes. They join up in the valleys to form, in part, extensive wadis or basins. The sands are low in salt content and the density of their plant cover increases with the depth of the deposit. Several plant communities can be distinguished: (*i*) That in which *Haloxylon salicornicum* is dominant with a cover usually less than 5%. (*ii*) Stands predominantly with *Panicum turgidum*, also in part with *Lasiurus hirsutus* in small channels. (*iii*) Communities with much *Zilla spinosa*.

The relatively damper wadis, with *Panicum turgidum*, may also be occupied by somewhat larger shrubs of *Retama raetam* and even by small trees of *Acacia raddiana* or *A. tortilis*. However, the latter have, today, nearly all been eliminated by felling.

Kassas also gives plant lists from sample plots which include the calculated constancy and frequency in per cent for individual species. However, here I shall only list (Table 93) plants that are found near Cairo University's Botany Department research station, i.e. the plants of the ecologically best known area.

The annuals grow during the short favourable season and thus correspond in their growth period to the winter ephemerals of Arizona. The perennials last throughout the long period of drought. Since there are no real succulents among them (*Haloxylon* is only somewhat succulent), they must take up water throughout the whole year. This problem will be discussed in the next section.

The list does not show *Zygophyllum coccineum*, which is especially frequent in the wadis near Helwân. However, it is also found on the Cairo-Suez road near kilometre 101. On the other hand, *Haloxylon salicornicum* is absent there and also from the wadis. It appears that this floristic difference is caused by the lime content of the soils. *Zygophyllum coccineum* is an indicator of lime soils which are avoided by *Haloxylon*.

The vegetation in the wadi near kilometre 101 had a cover of up to 20%. In addition to a few individual trees of *Acacia tortilis* and shrubs of *Lycium arabicum*, the following perennials were present as well: *Zygophyllum coccineum* and *Launea* (*Zollikoferia*) *spinosa* very abundant, in addition *Retama raetam*, *Zilla spinosa*, *Panicum turgidum*, *Echinops spinosus* and *Farsetia aegyptiaca*.

TABLE 93

Perennial species (in order of decreasing frequency):		
Haloxylon salicornicum	*Panicum turgidum*	*Artemisia monosperma*
Citrullus colocynthis	*Farsetia aegyptiaca*	*Daemia cordata*
Hyoscyamus muticus	*Heliotropium luteum*	*Cleome arabica*
Zilla spinosa	*Odontospermum graveolens*	*Pancratium sickenbergeri*
Pituranthos tortuosus	*Convolvulus lanatus*	*Halogeton alopecuroides*
Retama raetam	*Echinops spinosus*	*Astragalus forskalei*
Crotalaria aegyptiaca	*Fagonia arabica*	*Linaria aegyptiaca*
Pennisetum dichotomum	*Artemisia judaica*	*Gypsophila capillaris*
Biennial or annual species:		
Euphorbia cornuta	*Centaurea aegyptiaca*	
Annual species (in order of decreasing frequency):		
Mesembryanthemum forskalei	*Ifloga fontanesii*	*Zygophyllum simplex*
Trigonella stellata	*Plantago ovata*	*Malva parviflora*
Schismus calycinus	*Anthemis melampodina*	*Erodium pulverulentum*
	Matthiola livida	*Tribulus alatus*

If the long lists of plant species in the different communities is examined, one cannot but be amazed by the taxonomic richness of the vegetation in a climate with a mean annual rainfall of only about 25 mm. However, one should not forget the low density of vegetation in the area as a whole. All the plants together cover far less than 1% of the total surface of the desert. Because of this, the water supply to individual plants may still be regarded as quite favourable (see pp. 270ff.). This view was supported by detailed investigations.

2. Ecological researches near Cairo

Researches on desert plant ecology by the Botany Department of Cairo University were done in part in the pebble desert on the Cairo-Suez road (see p. 462) and in part in the Wadi Hoff near Helwân. This latter cuts deeply into the hamada for a distance of 15 km and joins the Nile valley about 20 km south of Cairo (Fig. 266).

(*a*) *Investigations near the Cairo-Suez road*

The area studied (Migahid and Abd el Rahman, 1953a, b, c) is not quite flat. Vegetation is mainly restricted to wide valleys and depressions, where perennials reach a 10% cover in places. Here, *Haloxylon* usually plays the main role. The soil is of permeable, coarse sand. Capillary soil water can easily be absorbed by the plants and less than 1% is non-available water.

Water content of the soil. It was found that the upper 25 cm was wetted almost regularly by the winter rains, but that the surface dried up completely in the summer. However, the soil horizon from 50 cm to more than 1 m contained some available water throughout the whole year. This horizon receives its water as gravitational water in some years after especially heavy rainfalls. In such rains, run-off water also becomes important, moving from the higher terrain into the depressions. The soil between 25-50 cm depth becomes temporarily wetted after strong rain but it dries up completely in years with little rain. Thus, a dry zone may commonly occur in the soil between the damp upper zone established in winter and the continuously moist lower zone.

The data of Migahid and Abd el Rahman can be cited as an example of soil water contents at different depths during 1949-50. One can see from Table 94 that the light winter rains of 1948-49 only wetted the upper soil horizons to less than 50 cm, in contrast to the more penetrating heavier winter rains of 1949-50. In February 1949, a dry zone was established below 25 cm. The authors interpret certain small irregularities as due to internal condensation but these could represent unavoidable random fluctuations that occur in soil investigations. Clear proof of the significance of internal condensation is difficult to obtain. It might be possible using plaster of paris soil-moisture blocks.

Drying of the soil by direct evaporation is only effective to a depth of 30 cm in coarse sand. Thus, there is a constantly damp zone from 50 cm down. The sparse plant cover only uses the water in this zone very slowly and the water reserves are always replenished in the more rainy years, such as 1950.

With diurnal cooling of the soil, some condensation from dew occurs at the surface. However, this phenomenon is not significant as a means of supplying water to the plants, since dew-water evaporates very rapidly during the day and, furthermore, the increase in surface soil-moisture does not even restore the percentage at permanent wilting (for more precise information see p. 473ff).

Root relations and plant development. When the root systems of annuals were investigated it was found that they only reach a maximum depth of 21 cm but laterally up to 35 cm. The annuals, therefore, are entirely dependent for their water supply upon the upper soil layers which are only wet during the rainy period. Their seeds germinate rapidly after

Table 94

Water content of the soil (in per cent dry weight) in the desert east of Cairo. Soil sandy, non-available water 0·8% of the dry weight. Last rain in November 1948 = 0·7 mm. Upper part of the table—rainfall 1949 and 1950

Year	1949											1950					
Rain-day	7.II	10.II		13.V						23.XI	19.XII	26.I		17.III			
Rainfall (mm)	2·9	0·4	—	0·2	—	—	—	—	—	6·1	6·9	9·6	—	9·2	—	—	—
Sample-day	17.II	17.III	7.IV	5.V	23.VI	12.VIII	—	19.IX	—	25.XI	22.XII	—	19.II	23.III	10.IV	6.V	8.VI
Sampling depth (cm) 5	2·7	1·5	0·6	0·4	0·3	0·2	—	0·4	—	5·5	2·7	—	1·7	2·8	0·9	0·5	0·3
10	2·0	2·0	1·1	0·5	0·3	0·5	—	0·7	—	2·9	4·3	—	1·7	3·3	2·7	0·7	0·4
25	1·0	1·2	1·6	2·1	1·0	0·8	—	0·8	—	1·0	1·2	—	8·4	2·5	3·5	3·9	1·3
50	3·2	3·3	3·3	3·3	3·2	3·3	—	2·6	—	2·9	2·7	—	2·9	6·0	6·4	6·4	5·2
75	2·9	2·7	2·7	2·7	2·8	2·7	—	3·5	—	2·7	3·4	—	2·4	2·7	3·2	—	4·0

the first rains. When rainfall is scarce during the rainy period only a proportion of the plants develop fully and dwarf plants are formed. For example, in 1949, only *Trigonella*, *Malva* and *Erodium*, were found from those listed in Table 94. In 1950, however, all 11 species developed vigorously. The annuals die as soon as the upper layers dry out since they never reach the continuously wet horizon with their roots. Only *Zygophyllum simplex* can accomplish this occasionally; then this species survives the dry period.

All the perennials root in the deeper, wetter layers. They usually have a tap root, but also develop lateral roots further down. An average rooting depth is 130 cm. In wadis with even more accumulated water, the roots can penetrate still deeper (Fig. 283). The roots extend laterally, for example, in *Haloxylon* up to 220 cm, and even to 440 cm in *Hyoscyamus*. As usual, the grasses root very compactly in a small volume of soil, while other species have extensive root systems.

The seeds of perennials showed delayed germination after wetting since inhibitors must first be leached out of the seed or fruit coat. At the same time, the soil must remain wet as far as the lower layers. If this is not so the perennials die, together with the annuals, as soon as the drought sets in.

The aerial parts of the perennials are very weakly developed relative to their root systems. Thus the water supply to the leaves is relatively good in view of the rather low water content of the soil. For example, the transpiration rate of *Citrullus* was very high. In calculating the transpiration rate per gram of water content, a daily maximum of 8·5 mg/min was found to occur in November. Yet, the maximum was achieved by 1000 hr and thereafter, declined greatly. The transpiration rate of the succulent *Haloxylon salicornicum* was much lower. Its maximum was only 5 mg/min and was reached between 0070 and 1000 hr; the minimum values were at 1 mg/min (based per g water content).

The smallest transpiration rate was found in *Mesembryanthemum*. Even in most extreme conditions only 4% of the water content was lost per hour, and, after reduction, only 1/5 of this amount. This explains how this annual survives for quite a long time after the soil round its roots has completely dried out.

Thus, we see that the water relations of the perennials are more or less balanced throughout the whole year, despite the extreme conditions in the desert. This is also shown by their relatively low osmotic values. Only in *Haloxylon*, which is probably a salt-accumulating halophyte, do the values vary from a minimum of 30·4 atm in January to a maximum of 43·2 atm in August. The annual variation approximately parallels the evaporation curve. Extreme values for the other species are shown in Table 95 (p. 473).

The data show that the assertion, repeatedly made, that desert plants have very high osmotic values, is not even applicable to this extreme desert. In spite of this, the assertion still survives in modern texts.

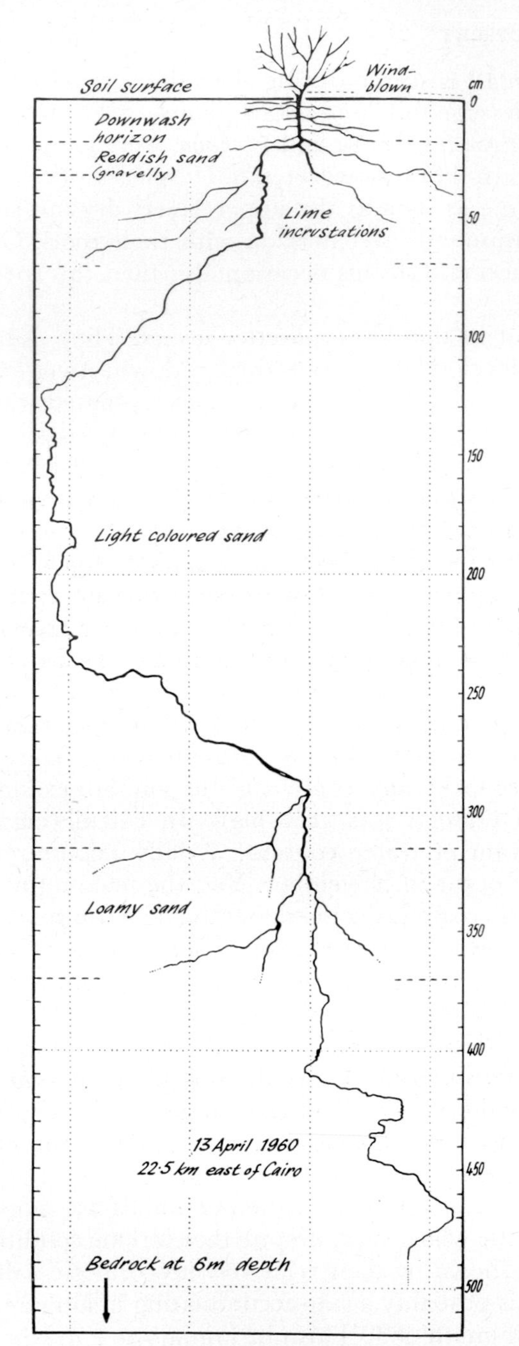

FIG. 283. Root system of *Pituranthos tortuosus* (after Kausch).

Investigations have also shown that wherever growing plants occur, water was available even under most severe drought conditions. Non-succulent species depend on some water uptake even when they have markedly reduced their transpiration and gone into a resting stage. This occurs as soon as the reserves of soil water become very restricted and when, perhaps, only a few roots still have access to the available

TABLE 95

Lowest and highest osmotic value (atm)

Hyoscyamus muticus	18·7–26·9
Pituranthos tortuosus	15·4–22·9
Zilla spinosa	16·7–18·7
Citrullus colocynthis	11·1–13·6
Echinops spinosus	15·9–18·3
Heliotropium luteum	16·8–17·0
Retama raetam	24·1–25·5
Artemisia sp.	15·2

water. This stage marks a certain point in drying out, which is indicated by a constantly increasing cell sap concentration. A few parts, less well supplied with water, begin to die off at this stage. The plant eventually dies unless this process is interrupted. However, desert perennials have a water balance that is adapted to long-lasting droughts and they can usually live for several decades, perhaps even over 100 years. Perennials, therefore, can only maintain themselves in the desert where the water supply to the root zone is never completely exhausted. Where rainfall is only episodic or irregular, as for example in the central Sahara, only ephemerals can find suitable living conditions. Their seeds can remain dormant without loss of germination capacity for more than a decade until the arrival of rain.

The most extreme deserts are practically barren of vegetation. Yet, after rain even these may become covered with ephemerals. The ephemerals are not, however, uniformly distributed over the whole area but occur predominantly in run-off channels and shallow hollows. These localities are wetted more deeply than the rest and seeds are also carried to them in run-off.

The significance of run-off in such deserts became clear to me during a trip through the Syrian desert from Baghdad to Jerusalem in March 1955. The entire desert was practically barren of vegetation but green strips of vegetation lined the paved road on either side. It had rained a little earlier and the water had drained off the road completely. As a result, the soil along the edge of the road received about 5-10 times as much water as the area round about and this provided sufficient for plant growth.

The effect of dewfall in the desert near Cairo is described as follows (Abd el Rahman and el Hadidy, 1958): no variations in soil water content can be detected at a depth of 5 cm during the night. Measurable variations can only be detected at the soil surface. The maximum usually occurs between 0600 and 0700 hr, the minimum in the early afternoon. The maximum variation was less than a few tenths of one per cent and the water content remained below the wilting point. Only once, on January 27, 1956, was a water content of 0·99% measured in the morning, but this, however, decreased rapidly to below 0·1% during the day. This shows that dewfall, or the hygroscopically absorbed soil water, is of no significance for plant growth in this desert.

Water loss from the plant communities. Determinations of water loss from the whole plant cover per 100 m² soil surface are very interesting. They show how far rain showers are utilised by the plants, i.e. how much dry matter they produce in relation to the amount of water lost through transpiration.

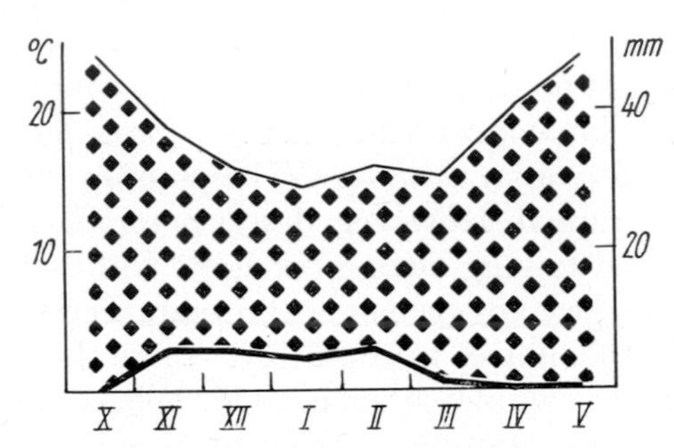

FIG. 284. Climatic diagram for Cairo for the winter of 1955/56.

This problem was studied by Abd el Rahman and el Hadidy (1958) in 1955-56 in the Suez road area. A total of 23·4 mm rain fell during the winter months (Fig. 284), which only effectively wetted the upper 25 cm of the soil so that the moisture could only be used by ephemerals. The perennials survived on the water supply of the lower horizons which had been topped up in earlier, wetter years. Dewfall took place, but it merely dampened the upper 2 cm, increasing the water content by only 0·16 to 0·91%. We can, therefore, ignore this factor.

TABLE 96

	kg
Panicum turgidum	1·960
Haloxylon salicornicum	1·245
Mesembryanthemum forskalei	9·718
Erodium pulverulentum	0·070
Schismus calycinus	0·046

The fresh weight of plants on the 100 m² area is shown in Table 96. Other species were present only as very few individuals.

Thus, 75% of the fresh weight was contributed by *Mesembryanthemum*. Daily transpiration curves were determined for individual species and, from these, the water loss per day was calculated for the total fresh weight of plants.

The hourly water loss from the entire vegetation on November 11-12 per 100 m² soil surface was maximally 1·02 kg (at 1400 hr) and minimally 0·5 kg (at 2000 hr). The rate of loss was approximately parallel to the evaporation. A maximum loss of 1·38 kg (at 1300 hr) and a minimum of 0·31 kg (at midnight) was measured on April 25-27.

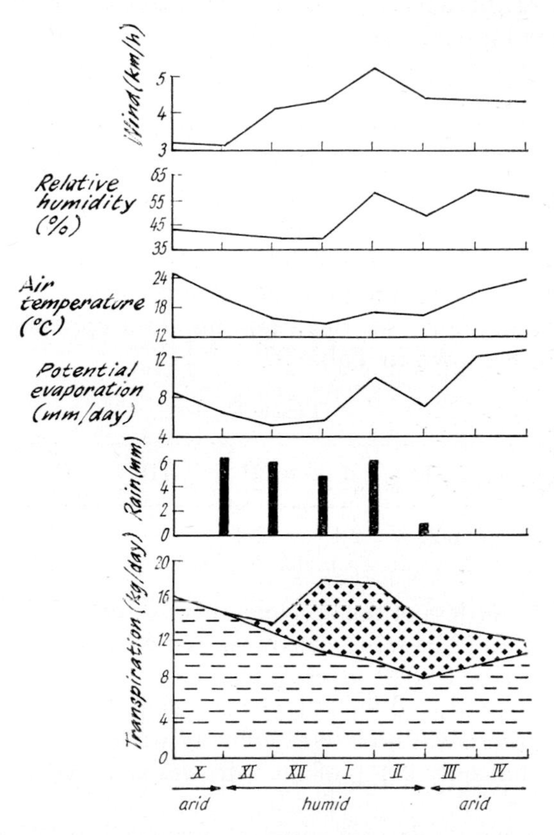

FIG. 285. Water loss through transpiration of the plant cover from a 100 m² area in the desert near the Cairo-Suez road in the winter 1955/56 (after Abd el Rahman and el Hadidy). Dotted—water loss of ephemerals, dashed—water loss of perennials. Further explanation in the text.

Daily water loss before the rain in October, i.e. without ephemerals, was 16·0 kg per day per 100 m² (Fig. 285). The amount decreased in November and December and was correlated with a decrease in the evaporative power of the environment. The ephemerals had not yet developed fully. A value of 18·05 kg was attained in January-February when the cover of vegetation was at its most dense. After that it decreased to 11·75 kg with the dying of the ephemerals. The ephemerals have the highest rate of transpiration.

For example, comparative daily maxima in *Schismus* were 599 mg/g/h, in *Erodium* 325, in *Panicum* and *Haloxylon* only a little above 100, while in *Mesembryanthemum* transpiration was lowest with only 28 mg/g/h. In the latter species, daily transpiration maxima decrease with progressive development from 41 mg/g/h in the seedling stage to 19 mg/g/h near the end of the developmental period. Simultaneously, the water content of the plants also decreases steadily.

In spite of the high fresh weight of *Mesembryanthemum* contributing some 75% to the total, the water used by this species only amounts to 37% of the total plant cover. It corresponds roughly to that of *Panicum* and is twice as high as that of *Haloxylon* in the same area. The ephemerals germinated after the first rain on November 9-10. Their contribution to the total water loss amounted to 5·4% in December, 40·5% in January, 44·1% in February, 39·2% in March, and in May, to 9·7% (see Fig. 285). Ignoring the perennials, since they extract their water from the lowest soil horizons, the water loss of the ephemerals, expressed in millimetres, is shown in Table 97.

TABLE 97

Month	XII	I	II	III	IV	V
Water loss	0·2	2·1	2·3	1·6	0·8	0·3

The total water loss of the ephemerals was about 7·3 mm; i.e. some 32% of the rainfall (23·4 mm) was transpired during their development.

The rains yielded 2340 litres of water per 100 m² on flat ground. The ephemerals produced 9·834 kg fresh weight and 518 g dry weight. Assuming that there was no lateral influx of water into the flat area one can conclude that only 0·4% of the rain was retained by the tissues of the plants.

Since the ephemerals transpired altogether 730 kg of water from 100 m² with a resultant production of 518 g dry matter, the productivity of the transpiration was 518 : 730 = 0·71. Thus, 0·7 g dry matter was produced per kilogram of water. If the amount of water in litres needed to produce one kilogram of dry matter is calculated, one obtains the value 1409. This is very high compared with values found in central Europe, which are between 400-600. However, high values must be expected in arid areas.

In south-west Africa I estimated that about 1000 kg/ha of dry matter are produced per 100 mm of rain. This, with a rainfall of 23·4 mm, would imply a production of about 234 kg/ha or 2·34 kg/100 m². Yet, in the Egyptian desert, only 0·52 kg was found, i.e. only one quarter of the theoretical value. This is quite understandable, since a much greater proportion of the rain water is evaporated directly when the rainfall decreases. Moreover, in south-west Africa, the vegetation investigated was pure grassland, while in the Egyptian desert the vegetation is dominated by the succulent *Mesembryanthemum*.

The development of the ephemerals and their slow death is shown in Fig. 286.

Competition. To follow the development of the vegetation during a year, Abd el Rahman (1953) made exact maps of the cover for all plants on two 100 m² areas three times in 1952, i.e. in January, February and July. Eight perennial, 2 biennial and 7 annual species occurred on these two plots. The cover of the perennial species was a little over 10%, with *Panicum turgidum* dominant. The annuals were very abundant in February, while they had disappeared completely by July.

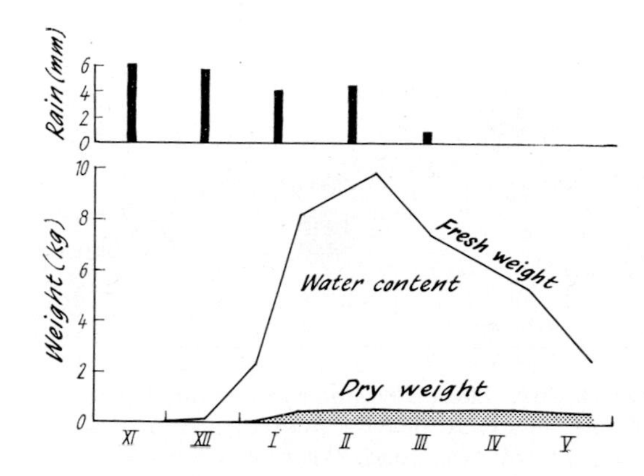

FIG. 286. Changes in the fresh and dry weight of ephemerals on an area of 100 m² during the winter months 1955-56 (after Abd el Rahman and el Hadidy).

It was shown that delicate annuals could be killed by sandstorms· Abd el Rahman and Hadidy found that more than 50% of all the plants were destroyed by uprooting, desiccation and sandblasting after a sandstorm, with a wind velocity of 15 m/sec, on January 27-30; the degree of cover was reduced by 31% (Abd el Rahman and el Hadidy, 1959; Abd el Rahman and Batanouny, 1965). By contrast, perennial species are wind-resistant. One plant of *Salvia aegyptiaca* and two of *Haloxylon* of the seedlings of perennial species, were left on the first sampling area; on the second, only one plant of *Centaurea aegyptiaca*. In spite of the large open area, multiplication is so ineffective that the roots of the perennial species do not touch each other. Thus, there is no competition. In contrast to earlier findings, this habitat is not, therefore, saturated. However, the authors mention that the experimental areas were subject to grazing and that many plants had died from grazing and trampling. Grazing has a particularly severe effect at the extreme limits of plant growth. However, it is not improbable that multiplication is so reduced, under very extreme environm ental cond tions, that it is insufficient to saturate the sand with plants (see Fig. 287

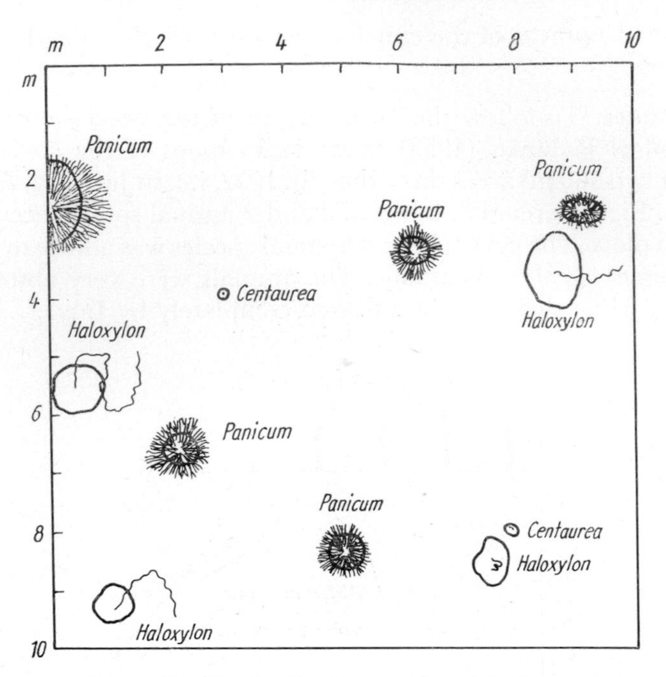

FIG. 287. The distribution of perennial plants on a 100 m², relatively heavily stocked, area in the desert near the Cairo-Suez road. The root systems of the plants are supposedly not touching each other but it is very difficult to isolate the most delicate root-ends (after Abd el Rahman and el Hadidy).

(*b*) *Investigations in the Wadi Hoff near Helwân*

The mean annual rainfall in this area is 31 mm. But, even with a shower of 10 mm, water runs from the vegetation-free limestone plateau first into shallow rain channels, then along anastomosing, deeper channels into the broad wadis. The beds of these wadis are only filled with water for a brief period in very rainy years. Some of the run-off penetrates the sand completely until it has disappeared. In this way some water accumulates permitting development of vegetation. The density of the plants increases from the shallow rain channels of the plateau to the wadis. Here, even the growth of trees (*Acacia, Tamarix*) becomes possible. The plant cover would be even thicker were it not subjected to grazing by the herds of camel of the nomads, or decimated for use as fuel. The water economy of these plants was investigated over a complete year by Batanouny (1963).

A cross-section through the wadi (Fig. 288) enables 6 habitats to be recognised: (*i*) The plateau with shallow rain channels. (*ii*) The cliff with a steep slope, at the foot of which plants only receive direct sunlight in the morning. (*iii*) The bottom part of the wadi. (*iv*) The first, low terrace. (*v*) The second, higher terrace. (*vi*) The scree-slope of the

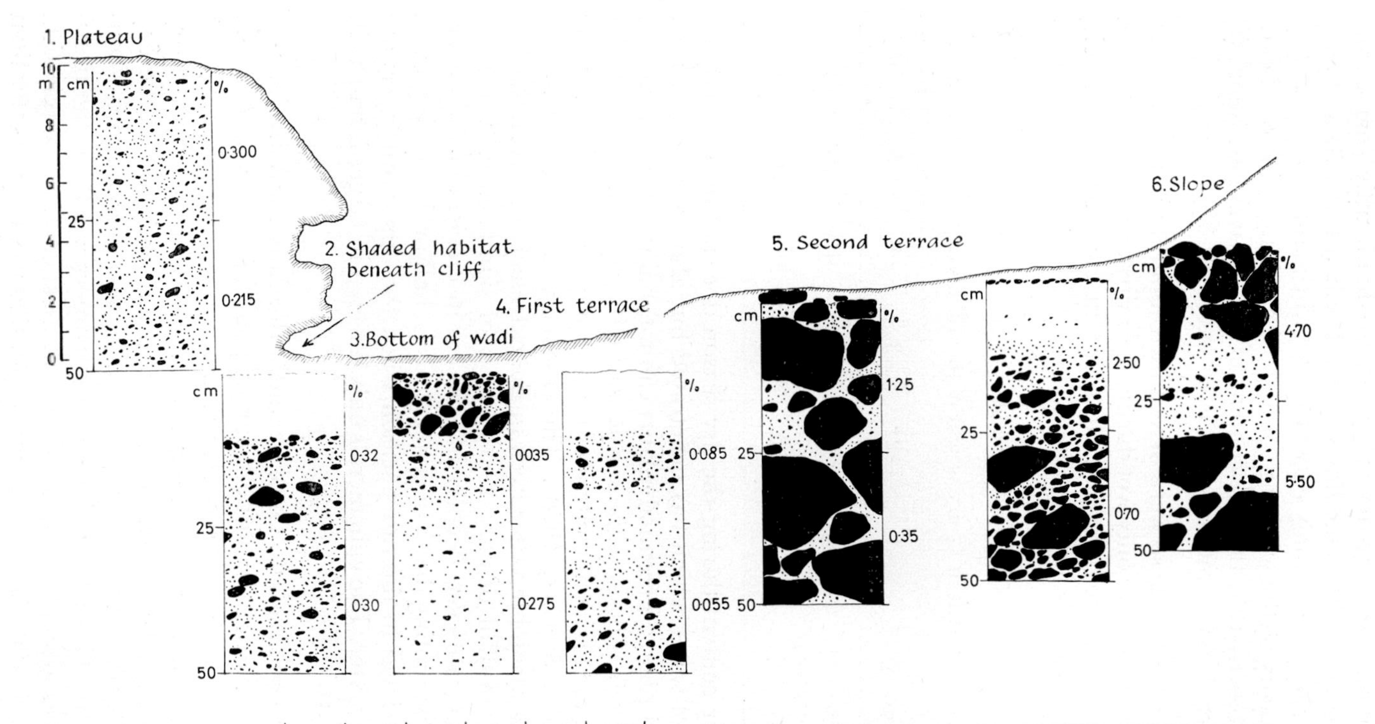

FIG. 288. Cross-section through the Wadi Hoff (after Batanouny, modified). 1-6: The habitats investigated with their soil profiles (depth-scale on the left side of each profile and total salt content % of soil dry weight on the right side at 0-25 cm depth and 25-50 cm depth).

Note the reduction in horizontal scale by one-fifth on the right half of the profile. The vertical scale (extreme left) refers to the whole profile.

hamada. The soils are raw desert soils of predominantly coarse sand between the stones, with some fine sand and only a little silt and clay. The silt and clay content only exceeded 10% in habitats (*ii*) and (*iv*). The field capacity of the soils was not high, the wilting point was below 1%. At habitats (*i*)-(*iv*), the salt content of the soil was low, but at (*v*) it was between 1-2·5% and at (*vi*) it even exceeded 5%.

Soil-water contents. These were determined monthly in the upper 50 cm. The upper layers are only wet during the rainy season in the winter. In the summer they are bone dry. However, in the profile as a whole, some available water is always present.

TABLE 98

Lowest water content in % dry weight of the soil during the dry season

Habitat	1	2	3	4	5	6
Min. water content (%)	1	2	2·3	1·4	0·8	2·5
Approx. cover degree of the plant (%)	12	50	0	20	1	0

The water content during the dry season does not depend only on the amount of water which has penetrated but also on the water used by the rooted plants. Perennial plants cannot maintain themselves on the wadi bottom (*iii*) since the soil is washed away in each flood. On the slope (*vi*), the high salt content of the soil prevents the growth of any plants (see also Stocker, 1928). The discovery that the soil always contains some available water even in the extremely dry year studied, which had a rainfall of only 19·5 mm, is of significance. The roots probably penetrate even deeper than 50 cm in crevices in the underlying rock. The rooting depth of ephemerals is shallow, but their growth period is restricted to the rainy season when the upper soil layers are charged with water. The ephemerals only develop in larger numbers in good rain years.

Water loss from the plants. Once a month, the daily transpiration was determined for the most important, annual plants, *Pennisetum dichotomum*, *Zilla spinosa* and *Zygophyllum coccineum*. Once more, these data support the view that plants absorb water from the soil throughout the dry season and always keep their stomata open for some part of the day which permits photosynthesis. But the water loss decreases with continued drought. This can be detected by temporary falls in the transpiration curves around noon, or by the occurrence of a transpiration peak as early as 1000 hr. Summer transpiration rates are greater than winter rates, yet, the summer rates do not increase as much as do evaporation rates. On irrigating the plants in summer, the transpiration rates of *Pennisetum* and *Zilla* about double. This shows that the transpiration rate was reduced as a result of impeded water absorption. *Zygophyllum* has succulent leaves, because of this, its transpiration values are much lower, since they are based on fresh weight. *Zygophyllum* shows a very pronounced regulation of its water balance. Its summer transpiration rates are not much higher than its winter rates, in spite of the much greater evaporation in the summer. Similar rates are also shown by the annual *Zygophyllum simplex*. This plant can store sufficient water in its leaves to enable it to survive the drought season in some years. The highest transpiration rates are found in the thin-leaved ephemeral species (max. 14-25 mg/g/min). But, malacophyllous plants also show similar high rates, e.g. *Stachys aegyptiaca*, *Artemisia fragrantissima* and *Diplotaxis harra*. A reduction in water loss during the dry season is achieved not only by closing of the stomata but also by shedding leaves or by developing small summer leaves or thorns, respectively. In this way the transpiring surface is reduced, for example in *Zilla*, *Pennisetum*, *Lavandula coronopifolia*, *Iphiona*, *Diplotaxis harra*, *Pituranthos tortuosus*, *Stachys aegyptiaca*.

Batanouny has also calculated the water loss from the total plant cover per 100 m² soil surface at 5 different habitats in March and June on a basis of the current fresh weight and the corresponding transpiration rates (Table 99).

TABLE 99

Plant mass (fresh weight) and water loss per 100 m² soil surface and day

Habitat	Month	Fresh weight of perennials (kg)	Water loss (kg)
1. Plateau	March	15·2	44·8
	June	15·3	51·0
2. Beneath the cliff	March	47·4	45·9
	June	42·8	48·7
4. First terrace	March	46·8	114·2
	June	44·3	120·0
5. Second terrace	March	6·1	8·3
	June	5·5	8·0

From these figures the approximate water loss per year in millimetres can be calculated for habitats (*i*) and (*ii*) as about 180 mm, for habitat (*iv*) about 400 mm and for habitat (*v*) about 30 mm. In addition, there is, of course, the water loss due to transpiration by ephemerals during the winter months, which varies from year to year with the rainfall, and the water lost by evaporation of surface soil moisture. It may be noted that the calculation concerning the water supply of the 'contagious' vegetation shown on p. 274 can be applied here to the vegetation in habitat (*iv*). The low water loss relative to the plant mass at habitat (*ii*) is due to the dominance of *Zygophyllum coccineum*, which has a low transpiration rate, and by the shaded position of the site which also decreases the water lost from the plants. We can, therefore, conclude from these data that the water supply of desert plants in this extremely arid region is by no means poor.

Osmotic values. The satisfactory water relations are also shown by the osmotic values of the cell sap. With the exception of the halophytes these are relatively low and they show no great differences between summer and winter (Table 100).

TABLE 100

Lowest (1st column) and highest (3rd column) osmotic values (atm) of species in the Wadi Hoff: columns 2 and 4, months in parentheses. (After Batanouny)

Species	1	2	3	4
1. Perennial species				
Farsetia aegyptiaca	18·2	(I)	23·0	(X)
Achillea fragrantissima	17·0	(I)	22·7	(X)
Lycium arabicum	19·1	(III)	21·3	(XI)
Asteriscus graveolans	16·3	(II)	19·0	(IX)
Stachys aegyptiaca	13·8	(IV)	19·3	(XI)
Zilla spinosa:				
Habitat 1 (plateau)	14·2	(II)	16·6	(IX)
Habitat 2 (below cliff)	12·1	(II)	14·4	(IX)
Habitat 4 (1st terrace)	14·2	(II)	14·9	(XI)
Iphiona mucronata:				
Habitat 1 (plateau)	13·1	(III)	23·3	(XII)
Habitat 2 (below cliff)	11·6	(II)	16·0	(XII)
Habitat 4 (1st terrace)	13·6	(IV)	17·8	(XI)
Pennisetum dichotomum	11·2	(III)	14·7	(XII)
Ochradenus baccatus	19·7			
Diplotaxis harra	22·0			
Retama raetam	14·9		18·2	
Pituranthos tortuosus	13·5		17·0	
Heliotropium arabianense	14·3		15·1	
Pergularia tomentosa	16·3			
Fagonia kahirina	12·2			
Lavandula coronopifolia	11·2			
Scrophularia deserti	12·1			
Centaurea aegyptiaca (biennial species)	13·1		14·4	
2. Ephemerals				
Diplotaxis acris 13·6; *Trigonella stellata* 13·2; *Plantago ovata* 11·6.				
3. Halophytes				
Nitraria retusa	25·3	(I)	33·0	(VII)
Zygophyllum simplex (annual)	28·9			
Zygophyllum coccineum				
Habitat 1 (plateau)	29·1	(IV)	35·6	(IX)
Habitat 2 (below cliff)	20·6	(II)	29·3	(X)
Habitat 4 (1st terrace)	23·6	(IV)	32·1	(X)

The proportion of chloride contributing to the osmotic value was not investigated. However, it is always very high in *Zygophyllum* species (see pp. 350 and 429) and in the case of *Nitraria*, it is known that the plant is always restricted to saline habitats (see pp. 414 and 456).

Diurnal variations in osmotic values, so far as they have been determined, are very small, particularly in *Pennisetum*. They amounted to about 3 atm in *Zilla* and *Farsetia*. Thus, the water balance is rather stable. Even after watering, the osmotic value in *Pennisetum* only decreased by 2·3 atm and in *Zilla* by only 1·7 atm. This shows that the water deficits are not large even in the summer. All these data indicate clearly that desert plants have no need to develop cytoplasmic drought-resistance. Their adaptations are of a more morphological-anatomical nature, as illustrated, for example, by a strongly developed root system and a reduction of the transpiring surface.

3. The vegetation along the Mediterranean coast of Egypt

When driving from Cairo to Alexandria through the desert, one notices the increased density of vegetation. This change is related to increased rainfall.

At the outset the desert is almost barren of vegetation. Along the edges of the road certain plants are more frequent (run-off effect). These include *Aristida plumosa*, *A. pungens*, *Panicum turgidum*, *Zygophyllum coccineum* and, here and there, a few others. In a sandy hollow *Aristida pungens* forms dunes 60 cm high. Then, half-way along the road, firstly *Mesembryanthemum forskalei* can be seen to form small, green patches in the depressions during spring time (March 4, 1960). These patches extend into the plains further north and coincident with this increase in population size *Artemisia monosperma* appears as an associate.

A vegetation analysis at kilometre 112 (counted from Alexandria, about 84 km from the coast) gave: ground cover 5%, with *Artemisia monosperma* and *Pituranthos tortuosus* co-dominant, and, partly dominant, *Panicum turgidum*. In addition, the following ephemerals: *Mesembryanthemum forskalei*, *Trigonella*, *Ifloga*, *Erodium*.

Soon after *Thymelaea hirsuta* can be seen and, at kilometre 88 (about 60 km from the coast), an *Artemisia monosperma–Thymelaea hirsuta* semi-desert begins, which extends with a cover of about 10% over the whole area.

Still further north new species appear, such as *Plantago albicans*, *Gymnocarpos decander*, *Zygophyllum album* and *Asphodelus*, and, among the ephemerals, *Matthiola*, *Salvia lanigera*, *Malva*. Near kilometre 53 barley fields are seen. The nomads first sow this cereal crop in depressions, but, a little further north, they scatter it over the whole surface. A crop can be harvested in good years; in poor years the barley fields are used for grazing.

At kilometre 29 one arrives at Maryuat Lake, shortly beyond which the dunes rise at the coast. Here, the uncultivated fields are already covered with a carpet of flowering ephemerals, dominated by *Chrysanthemum coronarium*.

The coastal strip west of Maryuat Lake shows the following topographic sequence:

At the back of the beach there is a wall of dunes derived from coarse $CaCO_3$ sand with a pseudo-oolitic structure. The dune zone is from 0·5-2 km wide with occasional stretches of stabilised sand. Behind the dunes lies a depression, whose deepest parts contain moist salt soils covered with halophytes. Further inland is another limestone ridge which is clearly composed of pseudo-oolites. In some places one can recognise yet a third ridge of rock and behind this rises the Libyan plateau. The coastal strip is derived from Pleistocene deposits whose rocky ridges hardly rise 35 m above sea level. It is bordered on its south side by the Miocene limestone strata of the desert plateau.

The plant communities were described by Tadros (1956, 1958; Tadros and Atta, 1958). Vast plains are covered with *Thymelaea hirsuta* growing in association with *Zygophyllum album* and *Asphodelus microcarpus*. Centuries-long grazing has degraded the vegetation and only species avoided by the grazing animals are now dominant. A strange phenomenon was observed in this region after the Second World War. The plains became green with a species identified as *Kochia indica*.

The first report of the spread of this annual species was given by Drar (1952). When studying the germination capacity of the seed, Shishiny and Thoday (1953a, 1953b) found that it is rarely retained for more than a year. Draz (1954; Thoday *et al.*, 1956) investigated the value of *Kochia indica* as fodder; it is very satisfactory. Since this species spread in the battlefields after the war, the local inhabitants believed it to have been introduced by the German troops. In the first few years after the war, *Kochia* covered vast plains and was the preferred grazing plant. Investigations showed, however, that seeds of this species, which is endemic to north-west India between Delhi and the Indus river, were sent in 1945 by the Waite Agricultural Research Institute in Adelaide (Australia) to Col. Hatton Bey in Egypt, the ex-chief advisor of the Egyptian frontier forces. Since *Kochia indica* is an abundant seed producer and a 'prairie-runner', the species spread very rapidly in the following years. This expansion was aided by the high rainfall of these years (in part, over 200 mm) and by protection from grazing in the areas originally occupied by troops. However, in the course of time, the species disappeared as a result of heavy grazing and several dry years. This is understandable, since *Kochia indica* is an annual which only germinates in late spring and which has its main development in the summer when there is no other green forage plant available. Even one extreme drought year is sufficient to prevent regeneration, since the germination capacity of the seeds is maintained only for a year.

Under strong grazing pressure the species will also have little chance of fruiting.

More intensive ecological investigations were made in the coastal region near Ras El Hikma. This is a cape that juts out 2 km into the sea, with a steep rocky coast on its west side and a flat, sandy beach on its east. It is 225 km west of Alexandria and east of the nearest meteorological station Mersa Matruh, whose climatic diagram is shown in Fig. 279. On comparing this with that of Helwân near Cairo (Fig. 278), one can see that the climate at the coast is not quite so extreme. The climatic diagram of Mersa Matruh shows two moderately wet months, December and January. The mean annual rainfall of 158 mm (extremes 49 and 275 mm) is six times higher than that near Cairo. The temperatures are not so high in the summer but the relative humidity is much greater with a maximum of 82% in July. At the cape itself, Abd el Gawad Ayyad (1957) carried out measurements from July 1955 to June 1956. Rainfall was recorded during this period from November to March (Table 101).

TABLE 101

Rainfall (mm) near Ras El Hikma 1955-56

XI	XII	I	II	III	Total
87·2	47·2	11·6	1·8	11·8	159·6

The absolute maximum temperature was 41·0°C, the absolute minimum 5·0°C. The diurnal temperature variations are not very high at the coast; 6°C in the winter and less in the summer. This leads to very great humidity, particularly in the summer (July, 82% relative humidity, by contrast with February, 61%). Only under exceptional conditions, during the chamsin (wind from the desert), does the humidity drop to 10%.

The potential evaporation is about half that of the desert. The potential evaporation values per year are given:

Alexandria	Helwân	Wadi Halfa
1825 mm	3830 mm	5660 mm

The frequent dewfall, 102 nights of the year is very interesting. It is especially pronounced in the rain-free season, particularly in May and June.

Since there are only a few dew observations for arid areas on record, the measurements are reproduced here in unabridged form (Table 102). The total dewfall in 1955-56 was 11·0 mm, yet no month had more than 3 mm. The highest individual dewfall on any day was 0·35 mm; only seven times did individual dewfalls exceed 0·2 mm, and mostly they were below 0·1 mm.

TABLE 102

Dewfall on individual days in microns (0·001 mm) at Ras El Hikma[1]

1955 Day	1	2	3	4	5	6	7	8	9	10	11	12	13	14	15	16	17	18	19	20	21	22	23	24	25	26	27	28	29	30	31	Total in mm	Days with dewfall
July	10			45	10	150	200						20	110	110																	0·655	8
August														150		10															45	0·205	3
September													100	150	200	200	200	200						75	20	75						1·220	9
October	150	150	110							150	200			200			20		45	75	75	110	75	110	20	75	110		20	45		1·740	18
November	completely absent!																															0	0
December																					45	20										0·065	2
January		10																		270	20											0·300	3
February																								20								0·020	1
March	completely absent!																															0	0
April					75		45	270	200		75		75	110	75	150	270					110		270		110						1·835	13
May				45		110	45	45	75	20	110	45		75	75		150	110	45	20	200	270	270	270	45	350	45	110		150		2·680	23
June	110			75	75	150	45	45	75		270	150	110	110	110	45		110	150	75		200	45	75		75		110	75			2·285	22
1956																											Total in one year					11·005	102

[1] Measurements by the Duvdevani method: Horizontal dewplate, quantity determined on drop-number and size.

Dew measurements were also taken 60 km west of Alexandria near Burg el Arab and 5 km from the coast by I. Arvidsson and B. Hellström (1955). However, their measurements refer to the months with the least dew, December to February. They used various funnels of wood, aluminium or plastic as dew collectors. The greatest dew run-off from 1 m^2 was 400 ml. Thus, the dewfall did not exceed 0·4 mm. The highest values were measured with a wooden funnel. They were still somewhat below those obtained with the Duvdevani method. Dewfall measurements were made also on leaves of *Asphodelus microcarpus*, *Olea europaea* and *Hordeum vulgare*. They amounted to one-fifth to one-quarter of the values obtained with the wooden funnel. In another publication, Arvidsson (1958) compared the dewfall on plants in Burg el Arab with those at Öland (Sweden) and found no significant difference. The maximum dewfall on a plant was 0·1 mm, on a single leaf 0·2 mm. Especially high dewfall occurred on *Chenopodium album* and *C. murale*, also on *Beta*. The number of dew nights at Burg el Arab is given as 130, with two minima, one in April to May and another in August to October. At Ras el Hikma, dewfall was minimal practically through the whole winter from November to March.

In this region, some available water is always present at some depth in the soil. Predominantly sandy soils have a maximum field capacity of 13% and a wilting point of about 1%. Here, the upper soil horizons are only wet during the rainy period but below 50 cm the soil always contains water available for plant growth. The water content of the soil never decreased below:

3·4% at 50 cm depth
4·1% at 75 cm depth
4·0% at 100 cm depth

All horizons were well wetted from November 7 to April 10. Dewfall rarely penetrates to a depth of 5 cm and, even then, the wilting percentage is hardly ever exceeded, even after heavy dewfall. Usually only the upper 2 cm are damp, thereafter the condensed water is rapidly evaporated from sunrise so that no utilisable water is left after 1000 hr. This is shown by the soil water contents sampled after dewfall on 3 days in April, May and June (Table 103).

The significance of dewfall for plant growth is at least questionable. It is most unlikely that plants would have absorbing roots with root hairs at 2 cm capable of absorbing water in the early morning, while, for the rest of the day, remaining unaffected in the soil.[1] Dew may have some importance for the ephemerals. They begin to dry out from April on, when the available water in the upper 25 cm of the soil becomes limiting. Wetting of the leaves by dewfall in the morning and dampness of the upper 2 cm of the soil can delay desiccation in April and May;

[1] According to Masson (1948) *Polycarpaea nivea* and *Sporobolus spicatus* can absorb dew in the fog-rich coastal desert near Dakar. They have little roots that develop within 1 cm of the surface. However, dew alone is not a sufficient water supply for these plants (see pp. 342 ff.).

TABLE 103

Evaporation of dewfall. Soil water content as per cent of dry weight at 0 cm, 2 cm and 5 cm depth

Time	16–17.IV.1956			20–21.V.1956			27–28.VI.1956		
	0 cm	2 cm	5 cm	0 cm	2 cm	5 cm	0 cm	2 cm	5 cm
1400 hr	0·42	0·75	0·96	0·38	0·45	0·55	0·26	0·34	0·42
0500 hr	**3·84**	**2·02**	**1·38**	**4·08**	**1·27**	1.14	—	—	—
0600 hr	3·68	1·86	1·35	3·89	**1·27**	**1·23**	**2·91**	**1·26**	**0·88**
0700 hr	1·86	1·85	1·32	2·30	1·03	0·93	2·50	0·87	0·76
0800 hr	1·49	1·83	1·29	1·40	1·05	0·73	1·34	0·75	0·75
1000 hr	0·82	1·51	1·27	0·73	1·03	0·76	0·43	0·69	0·67
1200 hr	0·53	1·16	1·25	0·37	0·77	0·75	0·32	0·65	0·65
1400 hr	0·55	0·87	1·00	0·29	0·76	0·74	0·29	0·42	0·48

yet dew cannot prevent death. In the deeper horizons of the almost invariably sandy soils, there is so little water condensation in the dry season that this form of wetting fails to recharge the soil even up to the wilting point. Thus, water condensation is of no significance to the plants. Corresponding with the higher winter rains, the vegetation is much more lush at Ras El Hikma than around Cairo. In particular, the number of perennial species is much greater. The composition of the various communities and their cover depends upon the soils (Migahid *et al.*, 1955; Abd el Gawad Ayyad, 1957).

We can distinguish the following communities (frequency scale 1-5):

1. *On rock-soils*: cover 5-10%; pH 8·1; lime content 25%; humus and salt only present as traces

5 *Gymnocarpos decander*
5 *Thymelaea hirsuta*
5 *Globularia arabica*
4 *Pituranthos tortuosus*
4 *Echinops spinosus*
4 *Dactylis hispanica*
3 *Helichrysum conglobatum*
2 *Helianthemum kahiricum* and others

The cover is less than 1% on oolitic rock cliffs that fall to the sea where plants occur only in crevices and holes. The following were found:

Limonium (*Statice*) *pruinosa*, *Frankenia revoluta*, *Inula crithmoides*, *Atriplex portulacoides*. On wind-blown sand deposits at the base of these rocks occurred *Agropyrum junceum*, *Euphorbia paralias*, *Nitraria retusa*.

2. *In alluvial depressions* on deep loamy soils: pH = 7·4-8·2; calcium content 15-20%; humus 0·6%; salt as traces; the vegetation is more dense since these habitats receive seepage water from the slopes

5 *Artemisia herba-alba*
4 *Asphodelus microcarpus*
4 *Noaea mucronata*
3 *Plantago albicans*
3 *Allium roseum*
2 *Lycium europaeum*
2 *Salvia lanigera*
2 *Stipa lagascae*
2 *Anabasis articulata*
2 *Matthiola humilis* and other ephemeral species

However, in hollows with gypsum-soils (pH = 8·0-8·5) containing 0·34% soluble salts the sclerophyllous grass *Lygeum spartum* is found.

3. *On sand dunes*: pH = 8·0-8·7; calcium content 30-35%; soluble salts 0·2%. Where the sand is still somewhat unstable one finds the typical dune grasses *Ammophila arenaria* and *Agropyrum junceum.* On older dunes the number of species is larger.

5 *Ammophila arenaria*	2 *Launea* sp.
5 *Echinops spinosus*	2 *Cyperus capitatus*
3 *Echium sericeum*	1 *Centaurea pumila*
3 *Agropyrum junceon*	1 *Crucianella maritima*
2 *Ononis vaginalis*	1 *Juncus acutus*
2 *Silene succulenta*	1 *Pituranthos tortuosus*

Where the sand cover is less deep species of other habitats become associated.

4. *On salt-soils* with ground water at 80-200 cm depth: pH = 8·3-8·7; calcium content 20-30%; humus 0·25%; soluble salts 2-5% including 0·5-3% chlorides, occasionally white salt-crusts

5 *Arthrocnemum glaucum*	3 *Frankenia revoluta*
4 *Aeluropus lagopoides*	3 *Suaeda pruinosa*
3 *Zygophyllum album*	1 *Atriplex* sp., etc.

Only one ephemeral (*Sphenopus divaricatus*) is associated, in the spring. *Halocnemum strobilaceum* also occurs here and, on larger salt-soil flats, *Suaeda fruticosa* and *Salsola tetrandra.* In small brackish valleys with up to 1% soluble salts one finds, in addition to *Suaeda fruticosa*, *Mesembryanthemum crystallinum* and *M. nodiflorum.*[1] Characteristic plants on damp sand at the coast are *Aeluropus lagopoides*, *Frankenia revoluta*, *Nitraria retusa.* These grow mixed with sand-dune plants.

5. A stand of *Phragmites communis* occurs in a *reed swamp.*

Part of this area was fenced in and protected from grazing. The change in vegetation was studied in 23 permanent 1m² quadrats. After 4 years the following changes were observed (Abd el Rahman and Hammouda, 1960):

i. On soils derived from rocky outcrops plant cover increased from 2% to 13·5%. Plants were more strongly developed especially prostrate species.

ii. In sandy hollows cover increased from 15·1% to 20%. An increase occurred, in particular, in the number of individuals of the favourite forage plant *Ononis vaginalis*, which increased from 113 to 788. The cover of this species increased from 0·2% to 10·6%.

iii. On dune ridges *Echinops spinosus* decreased, while *Helianthemum ellipticum* and *Crucianella maritima* increased. New, additional species

[1] More detailed information on the different salt communities and the composition of the soils in the area west of Alexandria is given by Tadros and Atta (1958).

recorded were *Echium sericeum*, *Zygophyllum album* and *Ononis vaginalis*. These probably appeared in response to the elimination of animal movement on the dune vegetation.

4. Outside the enclosure the plants preferred as forage such as *Echium sericeum*, *Ononis vaginalis*, *Lotus creticus* and *Echiochilon fruticosum*, developed less vigorously, while all species that were not selected as forage showed a much better development.

The vegetation shows clearly that the area at Ras El Hikma is in no way part of the desert. The *Artemisia herba-alba* stands, on deep loamy soils, belong to the zonal vegetation of the *Artemisia* semi-desert, which extends over much larger areas in north-west Africa and in Palestine and Syria (see pp. 510, 512 and 514).

A few determinations of osmotic values and chloride contents were done by Kreeb (1963) in connection with his investigations on suction tension and negative turgor pressure. He made these determinations at

TABLE 104

Osmotic value (Π) and proportion due to chloride (Cl) in atm

	Π	Cl
1. Species on weakly salty soils		
Thymelaea hirsuta, rock-ridges	34	14
Pituranthos tortuosus, sand-dune	25	20
Ammophila arenaria, sand-dune	40	21
Acacia saligna (planted), sand-dune	19	12
2. Species planted in gardens		
Olea europaea (non-irrigated)	59	9
Olea europaea (non-irrigated)	52	5
Olea europaea (irrigated)	33	3·5
Olea europaea (irrigated)	22	7
Amygdalus vulgaris, garden	39	14·6
Zizyphus jujuba, garden	30	18
Ceratonia siliqua, garden	31	9
Phoenix dactylifera, cemetery	26	13
Pinus cf. *halepensis*, near cemetery	65	33
Schinus terebinthifolius, cemetery	30	20
Acacia saligna, near mosque	15	7
3. Halophytes on salty soil		
Suaeda fruticosa, near a pump (soil only a little salty)	40	33·5
Salicornia fruticosa, salt-swamp (weakly saline)	79	67
Salicornia fruticosa, salt-swamp (weakly saline)	85	66
Salicornia herbacea (soil very saline)	113	102
Suaeda sp. (soil very saline)	130	100
Halocnemum strobilaceum	116	102

the end of the dry season (27.VIII-11.IX) near Burg el Arab, 7 km from the coast. The osmotic values were made by the method described on p. 419 and calculated at a temperature of 31°C. The chloride content of the press-sap is given in atm (based on NaCl, and likewise calculated for 31°C; Table 104).

In testing salt-excreting species one can never be quite sure that salts already excreted may contribute to the sample. Thus, the following values are not wholly free from error: *Limoniastrum monopetalum* gave $\Pi = 70$ and $Cl = 37$; *Tamarix articulata*, the values ranged between $\Pi = 49$ and $Cl = 26$ in a salt swamp, and $\Pi = 120$ and $Cl = 97$ in a cemetery.

The high proportion of chloride, even in non-halophilous species, is remarkable in all these values. The only exception is *Olea*.

4. Rainfall in Egypt in the past

From historical records it is known that the Mediterranean coast in Libya and Egypt once had a relatively dense population. This is demonstrated by the ruins of many former cities. Because of this it has been asked whether the climate was not far more favourable.

This problem was very carefully investigated in all its aspects by Butzer (1959). He considered both geological and geomorphological as well as prehistorical and archaeological findings.

If the changes in climate over the last 300,000 years are assessed there is no doubt that we are forced to assume a more favourable climate for the Palaeolithic period. In this period cool wet and warm dry climates alternated frequently. With a rainfall of 200 mm and somewhat cooler temperatures than today, one can imagine vegetation corresponding somewhat to that found today in Cyrenaica replacing the present-day desert. At that time the desert was still populated. Tools and agricultural implements are found all over the desert plateaux and even amongst the dunes of the Libyan sand desert. However, the rainfall had already declined in the Palaeolithic period. The populace migrated from the desert to the Nile valley which, at that time, was not so deep. Simultaneously, the course of the Nile was modified by the changes in climate, and mud of Abyssinian origin was deposited by the summer floods. According to Butzer the arid climatic-climax began about 18,000 years ago. Abundant wildlife could still maintain itself in the gallery forests or in river and swamp forests. Moreover, man could enjoy fish and clams, the latter living in large numbers in the mud of the flooded areas.

Further changes in the Nile valley were brought about by the rise in water level of the Mediterranean. As a result of the melting of the continental glaciers the sea level rose more than 50 m and reached its highest level, 4 m above its present height, about 4000-3000 B.CS Consequentially, the Nile changed from eroding to depositing. The average depth of Nile deposits is 9·7 m near Cairo and 6·7 m even between Qena and Aswân. About 60% of the mud cover was probably deposited before the development of the Egyptian nation (*c.* 2850 B.C.). Subsequently, the sea level fell again and was +2 m in 2500-1500 B.C., almost −2·5 m in 400 B.C., then rose to −2·0 m a century after Christ and to its present-day level only in the second century.

The climate was probably more favourable once again during the Neolithic period, compared with the present-day climate. The desert cannot have been covered with such sparse vegetation and it would be found in more favourable habitats, such as in the wadis, where there may even have been an open stand of trees. Butzer speaks of a Subpluvial period that occurred from 5000 to 2350 B.C.

Only during the sixth dynasty was the degree of aridity of today's climate reached. From the period of decline of the Old Empire to about 500 B.C. the climate was extremely arid. Amongst the wall paintings of the Old Empire portrayals of elephants, rhinoceroses and giraffes were already absent; lions and long-haired sheep less frequent. In paintings of the Middle and New Empire a steppe vegetation is no longer shown in hunting scenes. In the New Empire the wildlife of the swamps alone becomes important, such as the hippopotamus, the crocodile, the wild ox and species of birds and fish.

The period from 500 B.C. to A.D. 300 is characterised by recurrent high floods on the Nile, and dunes in western Middle-Egypt, on its margins, became covered with mud during Ptolemaean and Roman times.

This led to a relatively favourable period for agriculture in the Nile valley with an equable climate. Other variations occurred later (Table 105).

TABLE 105

Period (A.D.)	Climatic Indices
300–800	Number of floods less frequent, extensive wind blown deposits
800–1200	Further large floods
1200–1450	Dry, old dune formations
1450–1700	Large floods, but locally dry
since 1700	Recent dunes in Middle Egypt

Thus, there are no indications of a significant change in climate since the decline in the Subpluvial during the Neolithic period, which lasted until the time of construction of the pyramids. The total rainfall was no greater during the Old Empire than it is today. One should never underestimate the destructive potentialities of man during the last two thousand years, especially the losses due to grazing and felling of timber. Regeneration of the plant cover was greatly inhibited by the long-lasting, summer droughts. This should be considered in relation to the discussion on agricultural systems based on run-off (p. 275).

5. Ecological situations on the Sinai peninsula

The Sinai peninsula reaches, at its northern side, from the Suez Canal to Rafah (Rafa) or Gaza (Ghasa), respectively. In the south it is bordered by the Gulf of Suez and the Gulf of 'Aqaba. Three physiographic provinces can be distinguished: (*i*) the narrow coastal strip which, at its southern landward-side, grades into (*ii*), the table-land of the Isthmus or Tih desert, and (*iii*) the rough mountain area, which occupies the southern third of the peninsula (Fig. 276, p. 461).

The coastal strip of north Sinai has little in common with the coast of west Egypt. The north Sinaian coast is an extensive area of wind-blown sand with high dunes (up to 40-50 m tall), which are, however, relatively stable. The macadamed road from El Qantara (at the Suez) to El Arish and the railroad tracks may become covered with sand locally but its removal causes no particular difficulties.

Rainfall in this area is the lowest on the entire Mediterranean coast and amounts to less than 100 mm between Port Said and El Arish. Rainfall suddenly increases to the north-east (El Arish 100 mm, Rafah *c.* 200 mm, Gaza *c.* 400 mm). Local shifting sands maintain the dunes almost barren of vegetation. Thus, about 40% of the water that penetrates the sand after rain is either stored or transferred to ground-water. So, a stream of ground-water flows beneath the sand towards the sea, and it has been calculated that it delivers 50-164 m^3 fresh water per day per kilometre of coastline to the sea (Paver and Jordan, 1956). The fresh-water stream usually moves along the coast as a layer above the salt water. The latter has a higher specific gravity and moves inland from the sea into the substrate. Ground-water lies only a few metres below the surface of the dune valleys. Thus, the conditions are here suitable for the cultivation of date palms. To establish date palms, lateral shoots are removed from the base of old trees and buried in the ground. They are fixed in such a way that their lower ends are in contact with capillary water and their bound leaf tips stick out of the ground. Shoot growth begins after adventitious roots have formed. Salt accumulation occurs by evaporation from places where the water table is so high that the surface is kept damp by capillary water. Halophyte vegetation then grows in the dune valleys; at the deepest locations, *Arthrocnemum*, *Halocnemum*, *Salicornia*, *Juncus acutus*, *Suaeda*; somewhat higher *Nitraria*, *Tamarix*, *Limonium*, and finally *Zygophyllum* and *Thymelaea*, where the influence of the salt has become insignificant. Salt lakes may even be formed in dune valleys.

Date palms frequently grow so close to the surf-line along the beach that the waves can reach their bases during storms. One gets the impression that the trees are growing in sea water but this is not the case. Fresh water is moving through the sand into the sea and the trees root in these seepage layers; were it not so, one could use salt water to irrigate date palms. These ground-water relationships are shown schematically in Fig. 289.

Some water is stored in the dune sand after the rainy season. The amount is not very great but it is used by *Ricinus*, which is grown as an annual crop. In the spring, when the sand is wet after the winter-rains, seeds are planted in the sand on the seaward side of the dunes. After germination, the roots continue to penetrate deeper into the wet sand. In this way they maintain a constant supply of water until the seed is ripe even if there is no further rain after planting. A prerequisite for this practice is that the sand must be relatively stable so that roots are not exposed by wind erosion. According to my observations the plants crop very well. As a protection against blown sand *Acacia saligna* and *Eucalyptus* are planted along the railroad tracks. These plants can also grow without irrigation.

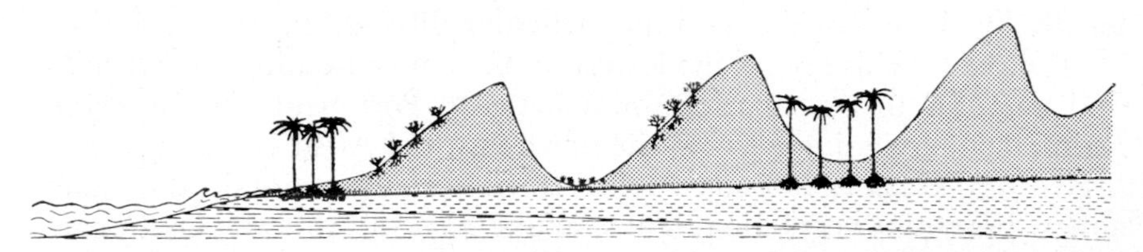

FIG. 289. Groundwater relations and the distribution of date palms and *Ricinus* plantations on the Mediterranean coast in the dune area near El Arish (schematic). Dotted—sand; limit of capillary water (short vertical hatching) above fresh groundwater (short dashes) and below; saline groundwater (long dashes) penetrating the ground from the sea beneath the fresh ground-water. Halophyte-vegetation in the first dune valley.

The largest settlement on the peninsula, El Arish, is located at the estuary of the Wadi El Arish. The catchment area of this wadi is 22,500 km². It occupies nearly 50% of the entire surface area of Sinai. However, south of the watershed of this river system, the rainfall is probably less than 25 mm. The wadi only contains rain water on rare occasions. In 1948 the amount of water flowing over the Ruafa dam (50 km from the coast) was measured as 21 million cubic metres; in 1951 only 3 million flowed, and nothing has been recorded since. Yet, along its lower course, the wadi is fed everywhere by ground-water seepage, in part from the dune area. This provides some well-water for irrigation purposes in El Arish so that some fruit trees can be cultivated in gardens, such as apple, pear, guava (*Psidium guajava*), fig, *Olea* and *Citrus*. In the bed of the wadi cereals are grown on a small scale during the winter months. Groups of *Tamarix* trees occur scattered along the margin of the valley.

Inland, the sand dunes grade into the gently rising hamada plains of the Isthmus desert. Here I found scattered vegetation, with a cover of up to 10%, including *Zygophyllum dumosum*, *Thymelaea hirsuta* and *Artemisia monosperma*, *Haloxylon salicornicum* and, in places with some sand accumulation, *Panicum turgidum*. The last species was represented by many dead clusters, since the preceding winter had had almost no rainfall.

A few flat-topped mountains rise abruptly above this plain extending to 1087 m. Kassas reports that the summit of the Gebel Ras-el-Ahmar is 90% covered with lichens. This indicates frequent dew or fog. The mountain rises 100 m above the desert plain, to an altitude of 256 m and is 20 km inland from the coast. The same flowering plants occur here as described earlier. *Anabasis articulata* is dominant in the highest part and at its base on the sandy plains there are, in addition, *Panicum turgidum* and *Aristida scoparia*.

The Gebel El Maghara is still higher (highest point 750 m). It is formed from Jurassic rocks and lies 100 km south-west of El Arish. The El Hamma station to the south-east foot of the mountain has a rainfall of 90 mm, in January and February. Boulos (1960; Range, 1921) presents a list of species that have been found on the 45 km long and 20 km broad mountain ridge. The most remarkable among these are *Caralluma sinaica*, *Sedum* cf. *viguieri*, *Juniperus phoenicea*, *Ephedra alata*, many *Astragalus* species, *Colutea*, *Iris sisyrinchium*, many Liliaceae. A fern, *Notholaena vellea*, also grows here and, in addition, many Mediterranean species such as *Lavandula pubescens*, *Teucrium polium*, *Thymus bovei*, *Globularia arabica*. This shows that rocky habitats provide more favourable conditions for growth. Probably, rainfall is also somewhat higher on the mountain summits.

The entire middle part of the Sinai peninsula is made up of the dry Tih table-land, with an average altitude of 700 m above sea level. El Nakhl (400 m above sea level), located on this plateau, has a rainfall of 26 mm. The vegetation shows no special features compared with that of the Egyptian-Arabian desert in the west and the Negev desert in the east.

Zohary (1944) gave a brief outline of the vegetation on the basis of two transects. The actual hamada plains are nearly free of vegetation. Where sand covers hollows, erosion channels, or wadis, one finds different communities, whose composition is strongly dependent on the depth of the sand, the topography, the fissuring of the rocks, etc. The most important plant species are *Anabasis articulata*, *Zygophyllum dumosum*, *Zilla spinosa*, *Reaumuria hirtella*, *Noaea mucronata*, *Panicum turgidum*, *Haloxylon salicornicum* (locally very dominant).

Characteristic of stable or only slighty mobile sand are: *Aristida scoparia* and *A. plumosa*, *Artemisia monosperma*, *Retama raetam*, *Panicum turgidum*. In those wadis that empty into the Gulf of Suez there is, in addition to *Acacia* species, *Ephedra alata*, commonly as a dominant, but also *Tamarix mannifera*, whose secretions form the desert-manna.[1] *Nitraria retusa* is characteristic of places with high levels of brackish ground-water.

[1] The assertion that the manna of the Sinai desert was the wind-blown lichen *Lecanora esculenta*, is not correct according to V. Täckholm.

This plant can form large hills through accumulation of sand. *Tamarix articulata*, *T. tetragyna* and *Zygophyllum species* can also occur there. On more strongly saline, wet soils *Suaeda vermiculata* grows.

The southern part of the Sinai peninsula is a wild mountain area, heavily faulted (mostly of granite) and rising over 2000 m with four peaks. Among these is Moses' mountain (2285 m), and the highest is Katharina Peak (2641 m). On my flight over the desert between Suez and el 'Aqaba one could see these summits stretching far above the dust layers.

Recently, the vegetation of this area was described by Migahid *et al.* (1959) as well as by Zohary (1944).

The mountainous part forms a triangle pointing south. In the north it is separated from the El Tih table-land by a level sand plain containing Nubian sand-stone. The El Tur station on the south part of the Gulf of Suez has reported an annual rainfall of 13 mm and a potential evaporation of 3500 mm. Snow can remain for several months on the mountains in winter. The precipitation is not known for this area. On the basis of the vegetation Zohary estimates the precipitation as 300 mm, while meteorologists estimate it as hardly more than 50 mm. Nevertheless, occasional rain showers are supposed to change the dry gulches into wild streams for brief periods. The rainfall is sufficient to supply enough water to maintain continuously-flowing springs at the foot of the mountains. Near one of these springs lies the well-known Katharina monastery. The plant cover becomes closer wherever water is present. In damp gulches scattered shrubs and trees occur, such as *Ephedra alata*, *Crataegus sinaica*, *Moringa peregrina*, *Ficus pseudosycomorus*, *F. carica* var. *rupestris*, *Cupressus sempervirens*.

On swampy soils and at water holes *Scirpus holoschoenus*, *Juncus arabicus*, *Veronica anagallis-aquatica*, *Equisetum ramosissimum*, *Mentha microphylla*, *Origanum syriacum* and even *Adiantum capillus-veneris* all grow. In irrigated gardens olives, pomegranates, almond trees, prunes, apples, pears, peaches and vines are grown, in addition to the date palms.

In the mountains one can recognise certain altitudinal belts:

Above 1000 m *Artemisia judaica* and *Zilla spinosa* still persist but *Artemisia herba-alba* is already coming in and with it a series of Irano-Turanian elements able to tolerate the low winter temperatures. A complete floristic change occurs above 1600 m, which includes the appearance of Mediterranean species. *Phlomis aurea* dominates at first with *Pyrethrum santolinoides*, but disappears completely above 2000 m and, instead, *Artemisia herba-alba* becomes dominant. In addition *Poa sinaica*, *Nepeta septemcrenata*, etc. occur.

The isolated position of the Sinai mountains has led to the development of a sizable number of endemics: *Galium sinaicum*, *Anarrhinum pubescens*, *Phlomis aurea*, *Thymus decussatus*, *Nepeta septemcrenata*, *Otostegia sinaitica*, *O. kaiseri*, *Convolvulus spicatus*, *Centaurium malzacianum*, *Primula boveana*, *Rosa arabica*, *Cotoneaster orbicularis*, *Arabidopsis kneuckeri*, *Dianthus sinaicus*, *Silene leucophylla*, *S. schimperiana*. They all show marked Mediterranean relationships and probably represent relicts.

6. Ecological investigations in the Negev desert

The Negev desert stretches from a line drawn from Gaza on the Mediterranean to the Dead Sea and south, to the Gulf of 'Aqaba. Here, rain falls in October to April. The average amount is 150 mm in the north, decreasing to 20 mm in the south (Zohary, 1953). In the north

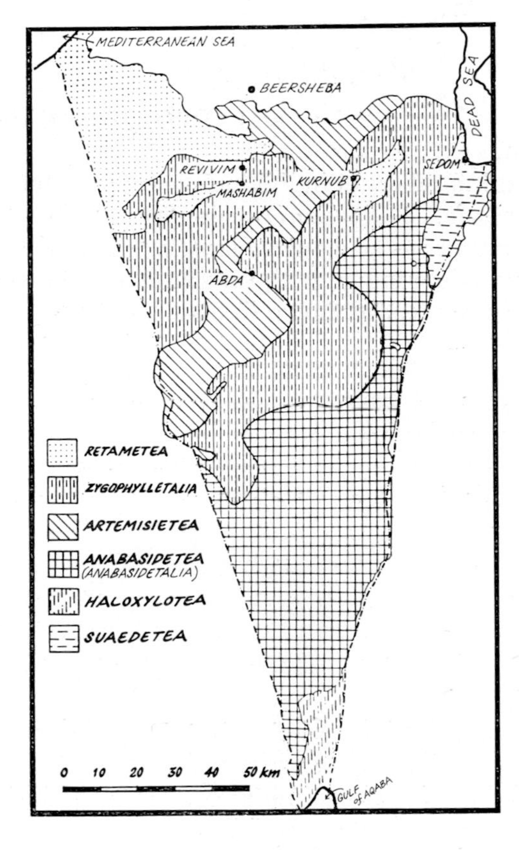

FIG. 290. Vegetation map of the Negev desert (after Zohary).

the mean temperature of the coldest month (January) is 11·5°C, that of the warmest month (August) is 27·3°C. The average relative humidity varies between 37% (July) and 55% (January). Dewfall was recorded on 163 days of the year and is mainly concentrated in the summer months. Fog is very rare and thus, ecologically insignificant. Therefore, in the north the climate shows some Mediterranean relationships; for the rest of the area the climate resembles that of the northern Sinai peninsula (Fig. 290). The high plateaux form an almost barren hamada, stocked only with a few scattered plants of *Reaumuria palaestina*. *Zygophyllum dumosum* 2-3 m high, dominates the stony slopes. On the banks of the wadis *Haloxylon articulatum* grows, while the wadi-beds are occupied by *Retama raetam* (Fig. 291).

The ecological relations of a hamada community, 30 km south of Beersheba, dominated by *Zygophyllum dumosum* were studied during the

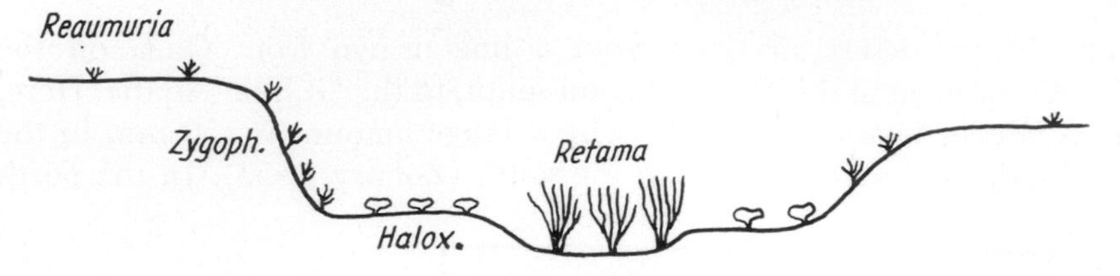

FIG. 291. Vegetation profile through a wadi (after Zohary and Orshan).

dry summer season in August 1951 (Zohary and Orshan, 1954). Here the mean rainfall is 80 mm.

The following species and numbers of individuals grew on a 100 m^2 area: 39 *Zygophyllum dumosum*, 8 *Reaumuria palaestina*, 2 *Atractylis serratuloides* and 258 of the summer annual *Salsola inermis*. The water loss from the total plant cover was determined with the exception of that of *Reaumuria* and *Atractylis*. The results are shown in Table 106.

TABLE 106

Transpiration measurements in the Zygophylletum dumosi in August 1951

Determinations during summer drought	Mean transpiration (mg/g/h)	Leaf weight of all plants (g)	Water loss per hour (g)
Zygophyllum dumosum	72	875·5	60·4
Salsola inermis	192	256·4	49·3
Total		1131·9	109·7

If one considers a daily transpiration period of 10 hours, one obtains a water loss of 1·1 kg per day or 33 litres of water per month.

The summer annual *Salsola inermis* germinates after rain during November to February but it only develops from March to September. During the growth period the long winter leaves are replaced by short summer leaves. The summer annuals only grow in disturbed areas where water remains in the soil. Otherwise they cannot cope, after germination, with the competition for water from the winter annuals or with that of the perennials during the summer. Areas that are entirely occupied by annuals show an increasing density from year to year until they are so weakened by intra-specific competition that they eventually disappear (see p. 29) (Negbi and Evenari, 1961). In this winter-rain region, summer annuals are represented only by ruderal plants.

These measurements were repeated in the spring. At this time a much larger number of ephemerals were present in addition to the 39 *Zygophyllum* plants. Thus, the water losses were much greater (Table 107).

TABLE 107

Transpiration measurements in the Zygophylletum dumosi in March 31, 1952

Determinations in spring	Mean transpiration (mg/g/h)	Leaf weight of all plants (g)	Water loss per hour (g)
Zygophyllum dumosum	198	6448·7	1276·8
All ephemerals	297	6610·0	1963·2
Total		13,058·7	3240·0

Zygophyllum sheds its two pinnate leaflets in the summer so that only the succulent petiole remains. Thus, the area of transpiring leaf is seven times greater in the spring than in the summer months. Moreover, since the transpiration intensity in the spring is 2·75 times greater per gram fresh weight, *Zygophyllum* alone contributes 20 times the water loss in March as compared with August. This is further aggravated by a greater water loss from the annuals so that the transpiration of the total plant cover is 30 times greater in March than in August.

Water losses in the spring amount to 32·4 kg per day or, roughly, 1000 litres per month.

Soil moisture determinations showed that, in the spring, the upper layers are well wetted while in the summer available water only occurs below 60 cm, provided that one corrects for the salt content in the calculation of availability. Calculations of changes in the water content in the soil showed an actual evapo-transpiration in the winter months of *c.* 6400 litres but, in the summer months, of only *c.* 300 litres. This compares well with the transpiration values already cited. Thus, the total transpiration per year is equivalent to a rainfall of 60-70 mm, of which less than 2 mm are transpired during the period of summer drought.

Since the soils contain readily soluble salts, the species mentioned above must be assumed to be xerohalophytes. This assumption is further supported by their high osmotic values (Table 108).

The salt content of the cell sap was not determined.

The development of vegetation in the very unfavourable hamada habitat is only possible by greatly reduced water losses during the summer months. Such a reduction can be accomplished only by:

1. A great reduction in the cover-density of living plant material through the disappearance of the ephemerals.

2. A great reduction of the transpiring leaf area in *Zygophyllum* which sheds its pinnate leaflets.

3. A decrease in the transpiration intensity of one-third, irrespective of the much greater evaporative power.

TABLE 108

	Osmotic values (atm)	
	23.VIII.1951	31.III.1952
Zygophyllum dumosum	72·5	40·0
Salsola inermis	116·8	59·5
Gymnarrhena micrantha	absent	39·5

The wadis that dissect the hamada frequently still carry water in the northern part of the Negev desert since the rainfall is higher there. A rainfall of 10 mm falling on a catchment area of 50 km² can produce a volume of 30,000 m³/h that flows through a wadi for several hours. The economic value of the wadis is very small if they are kept in their original condition. However, their value can be increased by impounding the run-off and so forcing the water to penetrate the soil, where it can be stored. Work of this kind has already started on an experimental scale in the Negev (Evenari and Koller, 1956). The valleys must be terraced immediately below the watershed. As a result, water penetrates the soil more deeply and promotes a denser plant cover. Large valleys must be subdivided into several flood areas by dykes with overflow-controls. Accumulated ground-water is used to irrigate cultures on terraces above areas affected by floods. However, it is essential for this type of cultivation that floods occur at least once a year, otherwise the outlay is not feasible economically.

The largest wadi in the eastern part of the Negev desert is the Wadi Araba 5-20 km wide, which represents the southern continuation of the Jordan Graben. This wadi starts at the south end of the Dead Sea 396 m below sea level, from where it rises gradually to 240 m above sea level. The watershed is 70 km from the gulf of 'Aqaba. The ecological investigations done in this area relate to the west bank of the Jordan Graben (Evenari, 1938a, 1938b; Evenari and Richter, 1938) and to the north shore of the Dead Sea (Zohary, 1947; Shmueli, 1948; Zohary and Orshansky, 1949). In this, the deepest hollow on the earth's surface, a desert climate prevails with a rainfall of 65-140 mm and mean monthly temperatures of 9·8°C (January) to 38·1°C (June). The annual mean temperature in Jericho is 23·1°C. Apart from a few oases, all the soils are exceedingly saline. They are: (*i*) autogenic or automorphic salt-soils that have originated from salt-containing parent rock material. They show a maximum salt concentration at a characteristic depth; (*ii*) hydrogenic or hydromorphic salt-soils, in which salt accumulates at the surface by evaporation of brackish ground-water. The osmotic values were investigated: they are only low in deep-rooting plants that reach into non-saline ground-water, such as *Prosopis farcata* and *Alhagi maurorum.* By contrast, the halophytes show values, which are never below 25 atm and which may exceed 100 atm. It was shown that the increase in osmotic values of the plants was correlated with an increasing salt concentration in the soil solution.

Annual fluctuations in osmotic values are slight in halophytes growing on constantly wet soils, but they were considerable in *Salsola tetrandra* (Fig. 292).

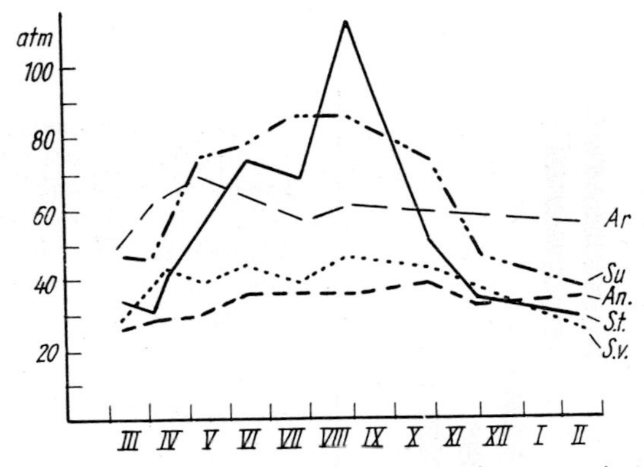

FIG. 292. Annual fluctuations in osmotic values of some characteristic halophytes from the Dead Sea area: *Ar, Arthrocnemum glaucum*; *An., Anabasis articulata*; *Su, Suaeda palaestina*; *S.t., Salsola tetrandra*; *S.v., Salsola villosa* (after Zohary and Orshansky).

The conditions further south are of particular interest, in the Wadi Araba proper, which is much drier. Here, rainfall decreases from 60 mm in the north to 20 mm in the south. However, rainfall varies greatly from year to year. Temperatures vary between 3°C and 45°C and the relative humidity from 20 and 60%. The soils only have a low salt content in the wadi itself (Karschon, 1953).

Zohary and Orshan (Zohary, 1945, 1952; Zohary and Orshan, 1957) investigated two communities: (*i*) The *Acacia-Anabasis* community in wadis with shallow soils and in erosion channels which are formed in the hamada of the southern part of the Wadi Araba, and (*ii*) the *Haloxylon persicum* community on sandy plains and dunes.

1. At an *Acacia* site the available water was found at 1 m, with a maximum water content of 6·1% at 2 m. By 2·5 m the water content had declined again to 2·4%. *Acacia* roots were observed to a depth of 1·5 m. Transpiration rates were determined eight times during the year for the following species: *Acacia spirocarpa*, *A. raddiana*, *Anabasis articulata*, *Ochradenus baccatus*, *Lycium arabicum*, *Haloxylon salicornicum*. The two

Acacia species showed particularly high rates of transpiration which were not reduced, even during the summer. Mean daily rates varied through the year from 500 to 1000 mg/g/h. In August, *A. spirocarpa* even exceeded 2500 mg/g/h around noon. Thus, with a continuous water supply the osmotic values remained uniformly near 15 atm throughout the whole year. The salt content of the soil is low at such sites.

The other four small shrubs constantly showed a much lower rate of transpiration (see Fig. 293). The water balance of *Ochradenus* (Resed-

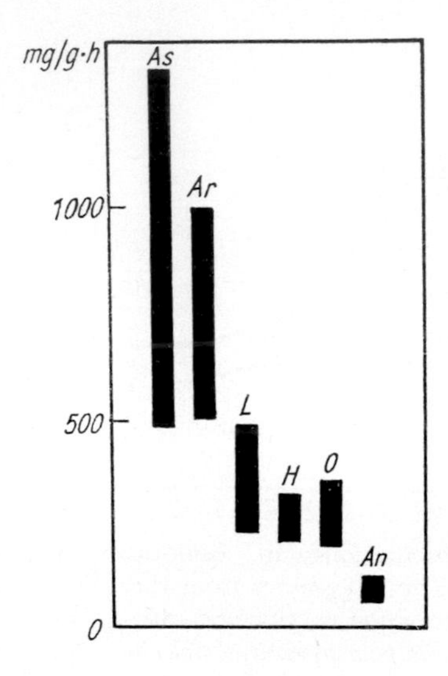

FIG. 293. Range of mean hourly transpiration rates of species of the *Acacia—Anabasis* community in the southern Wadi Araba. *As*, *Acacia spirocarpa*; *Ar*, *Acacia raddiana*; *L*, *Lycium arabicum*; *H*, *Haloxylon salicornicum*; *O*, *Ochradenus baccatus*; *An*, *Anabasis articulata* (after Zohary and Orshan).

aceae), which sheds its leaves early in the summer, remained uniform (always near 20 atm), while very steep increases (up to 50 atm) were observed in the other species (Fig. 294). *Lycium* also loses its leaves during the summer and this may be related to reduced root-activity. *Lycium*, *Anabasis* and *Haloxylon* must be assumed to belong to the salt-accumulating halophytes. However, the proportion of chloride in the cell sap was not determined.

Somewhat different relationships are found in the *Haloxylon persicum* community.

Haloxylon persicum (Chenopodiaceae) is a tree 3-5 m tall with slender, branched, succulent twigs. It occurs together with the leafless bush *Calligonum comosum* (Polygonaceae). Both species belong to the Irano-Turanian flora, and their centre of distribution is in the sand deserts of middle Asia. Vassiljev (1931) obtained the highest transpiration values for these plants in Turkmenia. However, two years later, Kokina (1932) found that Vassiljev's values were 2-3 times too high. The other two species investigated from these habitats in the Wadi Araba, *Retama raetam* and *Zilla spinosa*, belong to the Saharo-Arabian flora.

As is known, in the desert, sandy soils possess the most favourable water relations. However, in this region soil moisture determinations did not show any available water above 1·2 m. Only in March, was a water content of 2·4% found at 30 cm. The four species investigated are all deeply rooted and they must satisfy their water requirements from the deeper soil layers since the transpiration on a basis of the fresh

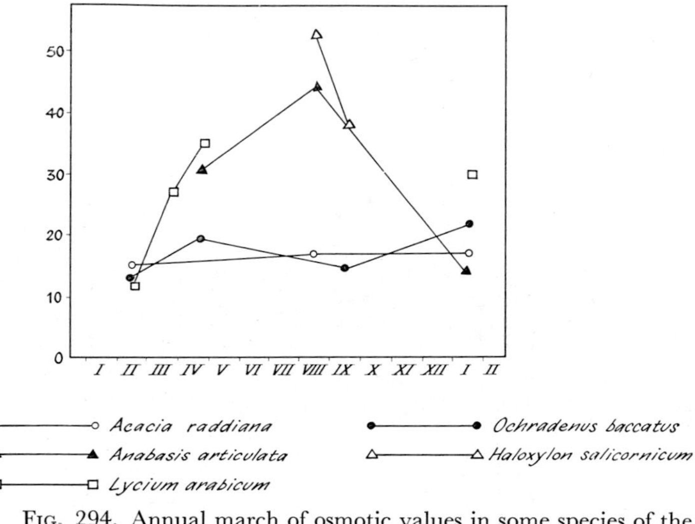

FIG. 294. Annual march of osmotic values in some species of the *Acacia—Anabasis* community in the southern Wadi Araba (after Zohary and Orshan).

weight is very high in all of them. In *Haloxylon persicum* transpiration on a fresh weight basis reaches 700 mg/g/h in June and it is only a little less in the other three species. However, these shed their leaves in the summer and continue to carry out photosynthesis with their green stems. Thus, transpiration is decreased (Table 109).

TABLE 109

Transpiration intensity in mg/h of plants from the *Haloxylon persicum* community; F calculated per g fresh weight and W—per g water content

Date	*Haloxylon persicum*		*Calligonum comosum*		*Retama raetam*		*Zilla spinosa*			
							Stems		Leaves	
	F	W	F	W	F	W	F	W	F	W
24.III.	396	—	456	792	402	—	—	—	954	1794
27.IV.	654	978	522	738	510	852	606	774	870	996
20.VI.	696	—	462	—	642	—	432	—	—	—
15.VIII.	558	882	270	570	372	996	324	618	—	—
16.IX.	444	750	300	462	234	1038	294	546	—	—
28.X.	318	516	192	282	210	420	384	552	—	—
23.I.	84	120	138	222	186	408	192	180	450	564

The osmotic values of *Zilla* and *Calligonum* are only a little higher than 10 atm. In *Retama* the osmotic value rises to 30 atm in August. In *Haloxylon* the values are around 30 atm in the spring and then show a steady rise to a maximum value of 56 atm, an odd feature which occurs in January (Fig. 295). *Haloxylon* is probably a salt accumulating halophyte. The high value is possibly produced by salt accumulation in the

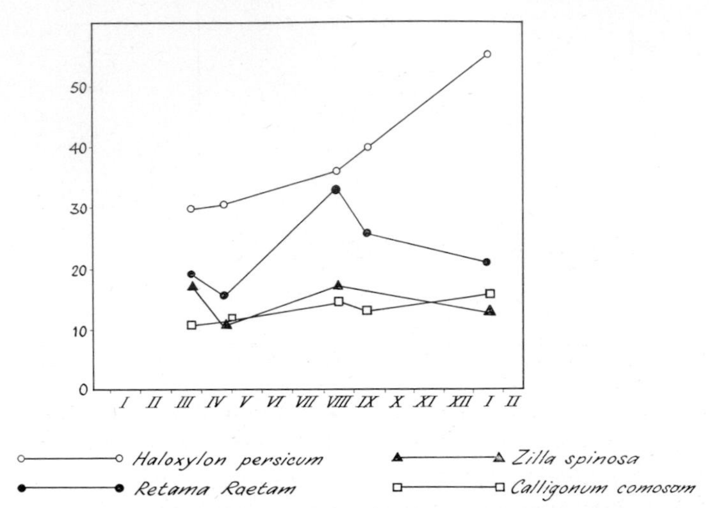

FIG. 295. Annual march of osmotic values of species of the *Haloxylon persicum* community (after Zohary and Orshan).

older tissues as a result of the high transpiration rate. It would be of great use to have chloride determinations of desert plants.

One cannot make good use of Raunkiaer's life forms to provide biological spectra. The positions of perennating buds are not the important criteria in deserts but rather, the adaptations to survival during drought. Zohary (1954), therefore, distinguished the following types with respect to their water relations (Fig. 296):

1. (*a*) Short-lived ephemerals, which complete their development in the three winter months and are already dead by March (*Filago* type, many species).
   (*b*) Annuals, crypto- and hemicroptophytes, which last till June (*Launaea* type, many species).
   (*c*) Summer annuals, which germinate only in the spring (March), lose their large leaves in the summer but retain their small, green scale-leaves and remain alive until winter (*Salsola* species).
2. Winter-green shrubs, which shed all their leaves in the summer, such as *Lycium arabicum*, *Anagyris foetida*.
3. Dwarf shrubs which reduce their transpiring surfaces at the onset

of the dry season: this group includes *Reaumuria palaestina*, *Salsola villosa*, *Suaeda palaestina*, *S. asphaltica*, *Artemisia monosperma*, *Zygophyllum dumosum*.

4. Evergreen switch plants which have lost their small leaves by the winter months, yet retain their green, photosynthetically active

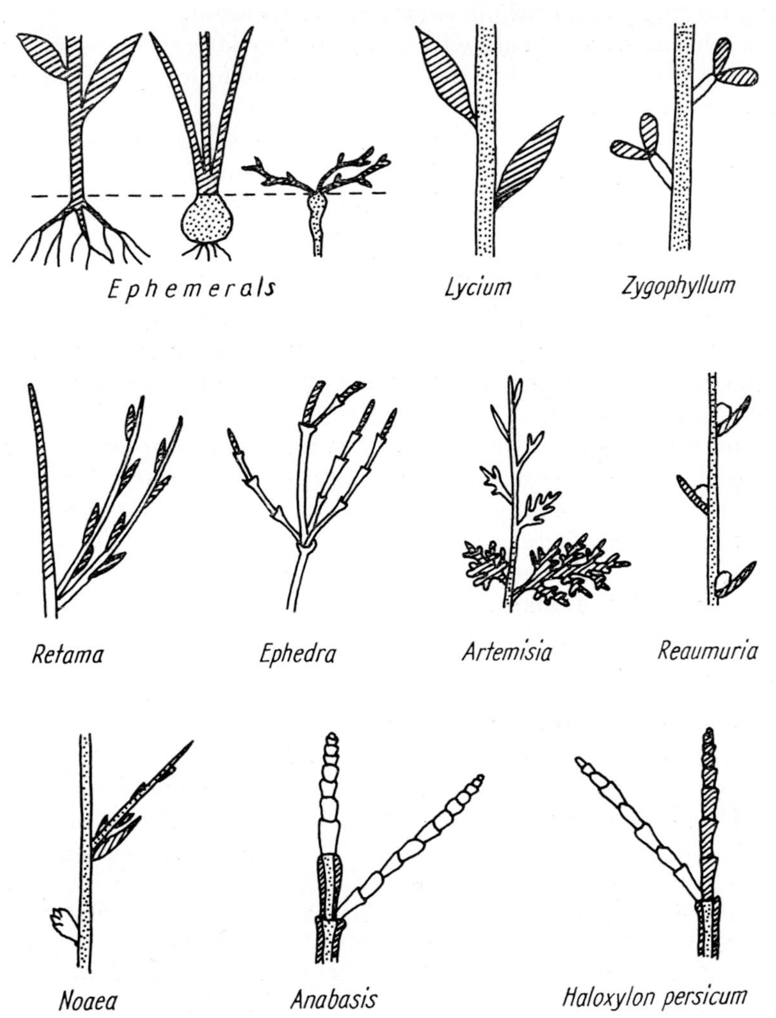

FIG. 296. Morphological plant types of the arid areas of the Near East as defined by the criteria affecting the water economy (after Zohary). Hatched—photosynthetically active organs shed during drought; dotted—perennating, photosynthetically non-effective parts.

shoots throughout the year. Only a small proportion of these green shoots may die off during the summer. This group includes such plants as *Retama raetam*, *Calligonum comosum*.

5. Leafless xeromorphic species which shed some of their green, fragile twigs, as for example *Ephedra*.

6. Malacophyllous xerophytes which, during the summer, lose their

lower, grey, hygromorphic leaves, retaining only small xeromorphic leaves, e.g. *Artemisia herba-alba*.

7. Dwarf shrubs which form bud-like, short shoots in the leaf axils of the young shoots and which, after shedding their leaves, remain active throughout the whole summer, as *Reaumuria*.

8. Thorny dwarf shrubs which form thorny lateral shoots with small leaflets in early summer but the latter die off completely during the dry season, e.g. *Noaea*.

9. Leafless, branched halophytes, in which the green cortical tissue of the previous year's shoots dies off in the summer, detaching itself in the form of rings, as in *Anabasis articulatum*, *Haloxylon articulatum*.

10. As in the preceding group except that part of the old shoots are also shed in the summer, e.g. in *Haloxylon persicum*.

Finally, the trees should be mentioned which shed their old leaves only when the new leaves are fully developed in the summer. To this group belong the *Acacia* and *Tamarix* species that are restricted to ground-water habitats.

Thus, we see that all types resistant to summer-drought are characterised by a great reduction in their transpiring surfaces (cf. p. 291). The reduction is probably due to a lethal disturbance of their water balance in the peripheral tissues.

In ephemerals, the entire shoot system dies off and only the seeds or subterranean parts remain.

A simpler classification was suggested by Orshan in 1953. He distinguished the following types:

I. Drought-escaping plants:	1. Therophytes;
	2. Entire shoot system deciduous.
II. Drought-resistant plants:	1. With deciduous branches;
	2. With deciduous leaves.

In increasingly dry climates the proportion in the first subgroup increases as the plants adopt a more effective form of protection. This becomes clear if one compares the percentage proportions in the geobotanical provinces by increasing aridity (Table 110).

In addition to reduction of surfaces, water loss is also greatly reduced through a decrease of the transpiration rate per gram fresh weight of the transpiring parts in the summer. The reduction is often so considerable that the summer transpiration rate is less than that in spring, despite the much greater potential evaporation of the environment during the summer (Fig. 297). This is particularly pronounced in *Artemisia* species, in *Noaea* and in *Salsola villosa*. Exceptions are *Acacia* species, whose water supply is maintained through the summer and also includes *Haloxylon persicum* and *Atriplex halimus*, which are associated with round-water habitats.

TABLE 110

Biological spectrum of prevailing species in the geobotanical provinces of Palestine employing Orshan's types

		I		II	
	Number of species	1. % Therophytes	2. % Shoot-deciduous	1. % Branch-deciduous	2. % Leaf-deciduous
Mediterranean province	90	30	37	17	16
Irano-Turanian province	70	47	30	19	4
Saharo-Arabian province	71	60	13	27	0

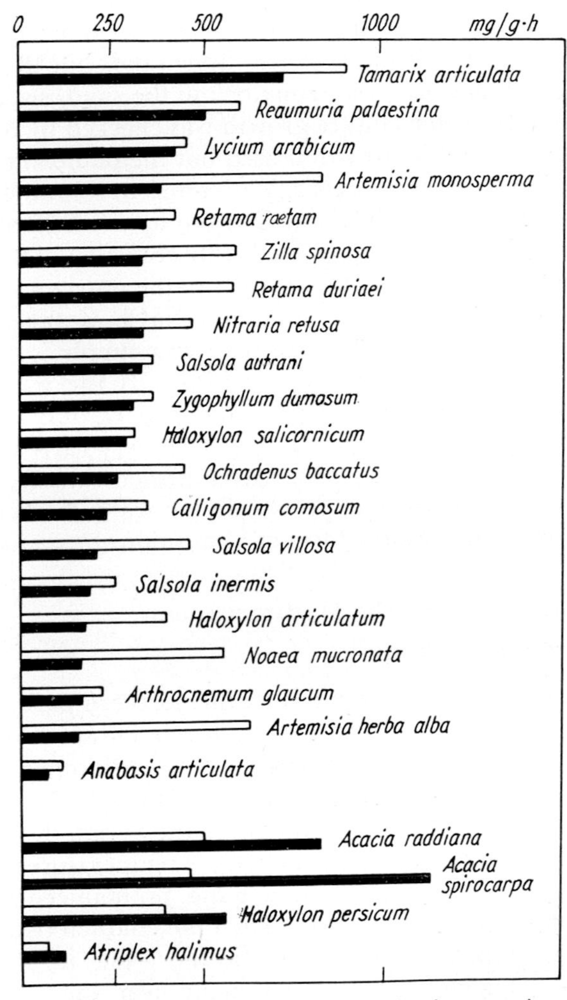

FIG. 297. Comparison of transpiration rates in spring and summer (black) of species from desert areas in Palestine (after Zohary).

However, neither the reduced rate of transpiration nor the reduction in surface area can prevent a worsening of the water balance. Osmotic values always show an increase in the summer. As we know, this increase is very small in some plants, particularly in *Acacia tortilis* and *Artemisia monosperma*, while in others it may be very considerable, as in *Zygophyllum dumosum* or *Salsola tetrandra*. The latter are halophytes and, in these, an increase of the osmotic values does not always imply a deterioration of the conditions for growth. However, extremely high values, in excess of 50 atm, usually begin to be reflected by reduced vital activities.

The ecological physiology of *Tamarix aphylla* (= *articulata*) has been investigated more fully (Waisel, 1960, 1961). The species grows in Palestine in non-salty areas with a ground-water supply. They can, however, survive saline conditions. Young trees become white with lime when growing on non-salt-impregnated dune sands but the roots of old trees, which can reach the salty zone below the sand, usually contain crystals of NaCl. In water-culture growth was checked by 0·1 and 0·2 M NaCl (0·6-1·2%), in 0·3 M growth ceased but the plants remained alive while they died in 0·4 or 0·5 M NaCl. With nutrient solutions of different compositions the amount of Ca and K taken up is constant in relation to the amount of Na and Cl up to a salt concentration of 0·2 M NaCl in the nutrient solution; with 0·3-0·4 M NaCl in the nutrient solution both growth and transpiration were less but more NaCl was taken up as well. Transpiration is relatively high when the water supply is good and the daily-curve shows a distinct peak at midday. If the water supply is reduced the curve shows a double peak and, finally, only a peak before midday, i.e. the transpiration rate is restricted very early in the day.

Although the salt deposits on the leaves attract water from damp air and causes them to deliquesce, practically no water finds its way into the branches as an alternative form of absorption.

7. Transitional areas to the Mediterranean sclerophyllous zone

The Sinai peninsula and, even more so, the northern part of the Negev desert represent typical transition areas. Irano-Turanian elements and a few Mediterranean elements occur here in addition to the Saharo-Arabian elements. The Irano-Turanian elements begin to dominate rather abruptly further to the north-east, in the Syrian desert. They are characteristic of deserts in which the winter temperatures can be very low. A cold annual season becomes quite pronounced further into middle Asia and at higher altitudes. Then the characteristic tree of the Sahara, the date palm, disappears. In the west, in Palestine, the Mediterranean element becomes dominant near the Mediterranean and this is correlated with an increase in winter rainfall. The whole character of the landscape changes as soon as the annual rainfall exceeds 400-500 mm. Sclerophyllous woody plants replace the xerophilous dwarf-shrubs and bushes and, finally, evergreen woodlands occur.

Of course, the Saharo-Arabian, the Irano-Turanian and the Mediterranean regions are not sharply distinguished. The map (Fig. 298) shows the limits of these vegetation regions in a very general way (Zohary, 1952). In reality there are several transition zones between the

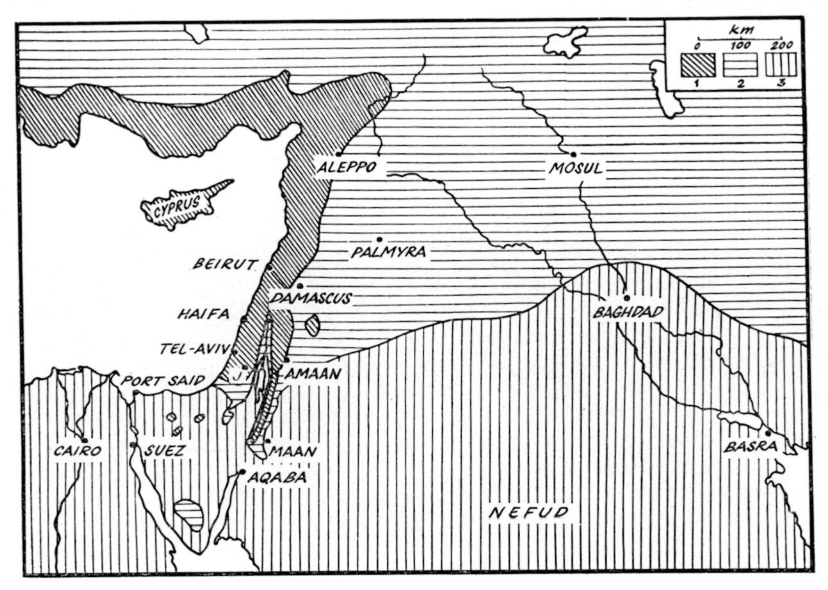

Fig. 298. Plant geographical subdivisions of the Near East: 1. Mediterranean province; 2. Irano-Turanian province; 2. Saharo-Arabian province (after Zohary).

Mediterranean sclerophyllous zone and the Saharo-Arabian and Irano-Turanian desert zones, respectively. These are the semi-desert zone with *Artemisia* species, the steppe zone with grasses and the zone with woody plants showing transitional characteristics. This is shown by the classification of Boyko (1954) and his vegetation map of Palestine, which shows how many different vegetation types are represented in Palestine (Figs 299 and 300).

In addition to these floristic elements, the Sudano-Sindian[1] element also reaches the Jordan Graben far to the north. The Sudano-Sindian element includes about 39 species, in particular different *Acacia* species, *Eragrostis bipinnata* (which grows 3 m tall and covers several square kilometres of the south-east shore area of the Dead Sea), and several other species. The best developed community is the *Zizyphus spina-*

[1] Zohary (1963) notes that the Sind desert contains Sudanian floristic elements of the Palaeotropics. So it is more correct to distinguish a Saharo-Arabian and a Sudano-Sindian floristic province. Typical palaeotropical (Sudanian) genera in south-east Pakistan (Sind or Thar Deserts) with 350 mm annual rainfall, are: *Euphorbia*, *Acacia*, *Commiphora*, *Salvadora*, *Grewia*, *Crotalaria*, *Aerva*, *Callotropis*, *Eleusine*, *Panicum*, *Cymbopogon* (cf. Wright, 1964).

christi—Balanites aegyptiaca community, which also contains *Callotropis procera*, *Solanum incanum*, and *Acacia tortilis*. In the southern part of the Dead Sea area *Salvadora persica*, *Moringa aptera*, *Cordia gharaf* and *Maerua crassifolia* also occur and *Acacia albida* is found scattered in pure stands in relict habitats.

It is interesting that even a relict tropical element occurs, namely *Cyperus papyrus* in the Huleh swamps of the upper Jordan (Oppenheimer, 1938; Zohary and Orshansky, 1947). These swamps at an altitude of 100 m are characterised by their high water temperatures: 25-35°C in September and 12-17°C in December. The pH of the water is slightly alkaline (6·95-7·65).

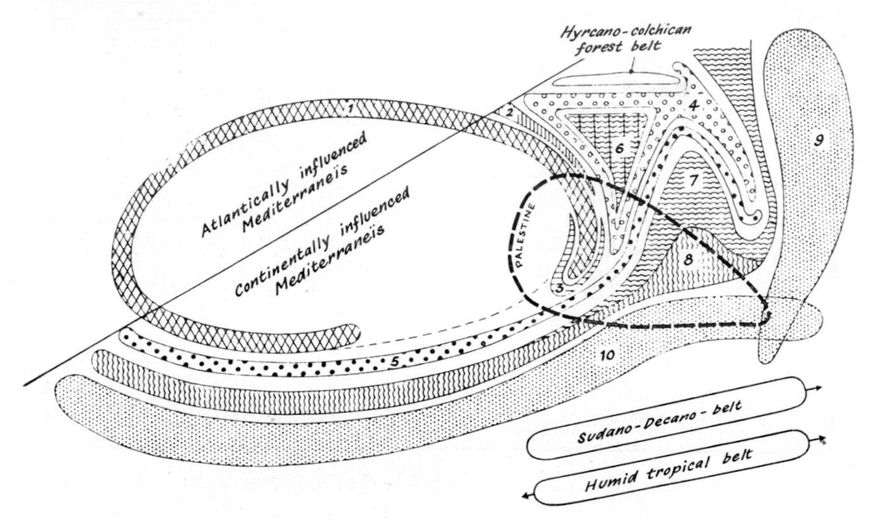

FIG. 299. Schematic classification of the eastern Mediterranean area into vegetation zones (after Boyko). Legend on Fig. 300. Instead of Sudano-Decano belt, the most recent term is Sudano-Sindian belt.

The number of geo-botanical studies made in Palestine is very large. A review of the literature was given by Oppenheimer (1952) and the vegetation and soil relations have been mapped by Zohary (1947a, 1947b; Feinbrun and Zohary, 1955; Gradmann, 1934).

The semi-desert zone, which begins to develop with an annual rainfall of over 150 mm, is represented on loess soils by stands of *Artemisia herba-alba*, while on sandy soils there are mainly bushes of *Zizyphus lotus*. These retain sand and, therefore, they usually rise from small sand dunes. The two species occur throughout the entire area from Morocco to Iran. Thus, they belong to the Mauritanian-Iranian or Mauritanian-Turanian elements, respectively.

By contrast, the grass-steppes, which occur next to the semi-deserts with increased rainfall, are occupied with Halfa-grass, *Stipa tenacissima*, in north-west Africa and, in Asia Minor, with tall *Stipa* species (*S. lagascae*, *S. szovitziana*) and *Aristida* species. Thus, they are clearly

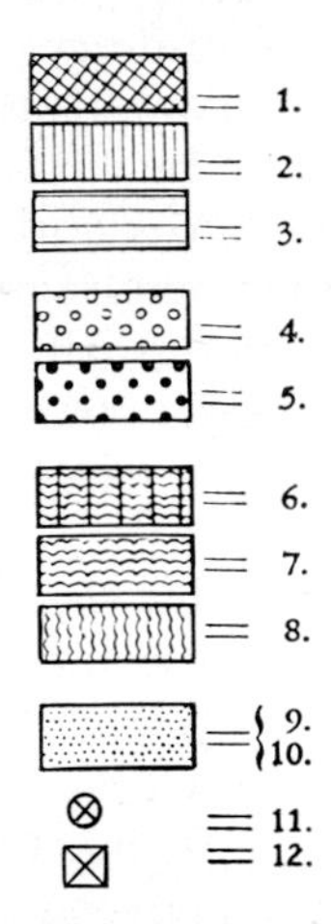

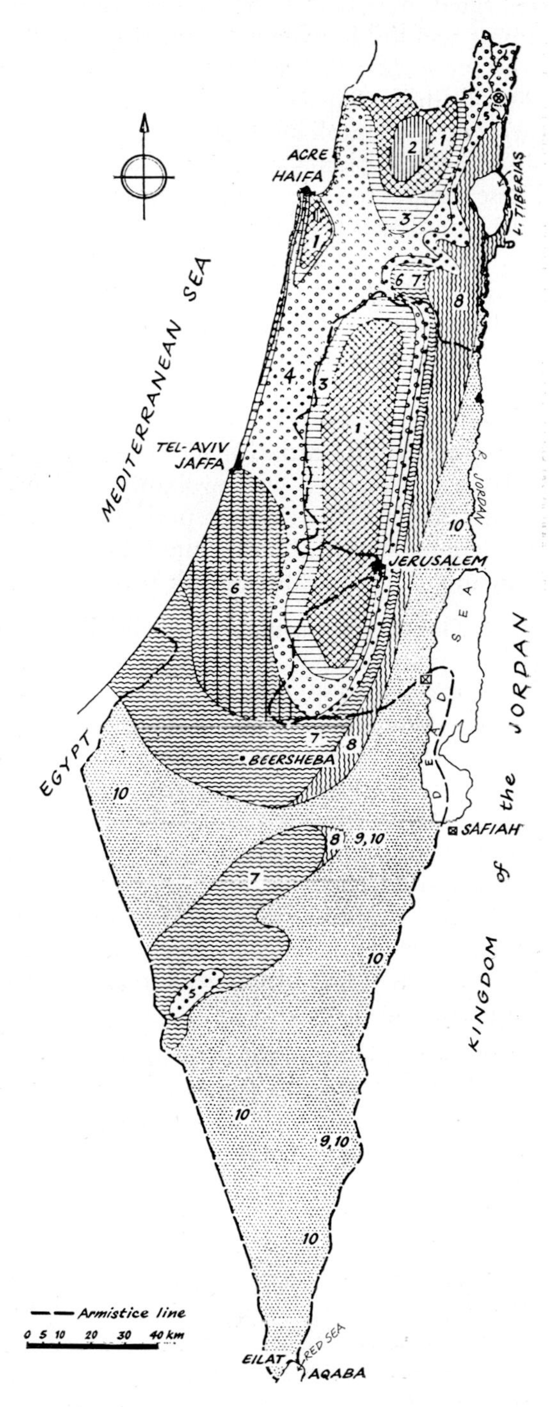

FIG. 300. Vegetation map of Palestine (after Boyko).

Mediterranean woodland zone: 1, typical; 2, semi-wet; 3, sub-arid.

Arid-scrub transition zone: 4, Anatolian-Iranian; 5, Maurctanian-Iranian.

Steppe zone: 6, Anatolian; 7, Irano-Turanian; 8. Mauritano-Iranian (thermophilous).

Desert zone: 9, Central Asiatic or Irano-Turanian; 10, Saharo-Arabian.

Sudano-Sindian zone: 11, *Cyperus papyrus* stands (Huleh swamps); 12, Outliers of the Dead Sea.

separated floristically. However, the steppes of the Near East are almost wholly destroyed by grazing. In their place, semi-desert with *Artemisia herba-alba* has become established and spread far beyond its natural limits. Those steppe types that are considered to be outliers of the central Anatolian steppe vegetation are richer in herbs and geophytes. These plants are able to withstand lower winter temperatures. These steppes too are almost completely destroyed by over-grazing and cultivation, and they persist only in the form of small relict communities. The steppes have been replaced by an *Artemisia fragrans* semi-desert (Walter, 1956).

There are no ecological investigations available from this transition zone. Studies on *Artemisia herba-alba* and the different grass species would be of especial interest.

Between the open steppe areas and the closed sclerophyllous woody vegetation communities of scattered trees or shrubs are found. Originally, a grass cover was probably present below and between these woody plants. These areas, probably of a savannah-like nature, are nowhere present in their original form. They only exist as fragmentary remnants. One can distinguish several climatically conditioned types:

1. The thermophilous type with *Pistacia mutica*, which represents a Mauritanian-Iranian element together with the north-west African *P. atlantica*. Today one only finds widely scattered, solitary trees, instead, *Artemisia herba-alba* has spread over all.

2. Closer to the sea, the semi-evergreen trees related to *Quercus aegilops* (in Palestine *Q. ithaburensis*, in Anatolia *Q. macrolepis*) are found. This tree must have a rainfall in excess of 400 mm; with less rainfall, one finds a scrub of summer-green species of the genera *Amygdalus*, *Rhamnus* and *Crataegus*.

3. In continental areas with cold winters there are, in addition to the summer-green oaks (*Q. infectoria*), tree-junipers (*Juniperus excelsa*, *J. foetidissima*). The distribution of these zones in Palestine can be seen from Fig. 300.

Oppenheimer (1949) investigated the water relations of *Quercus ithaburensis* on a shallow limestone habitat. This species is representative of the woody plants of the transition zone. It shows an extremely high transpiration rate in June and also in September after 5 months drought (Table 111).

TABLE 111

28 June Time	Transpiration (mg/g/h)	21 September Time	Transpiration (mg/g/h)
1527 hr	925	0830 hr	1000
1635 hr	1095	1100 hr	500
sundown	490	afternoon	650

This suggests a regulated water balance. The osmotic value was only 20·3 atm on October 8, which was well before the onset of the rainy season. Thus, the osmotic value had not changed greatly.

Such effective water utilisation is made possible by an especially good conducting system with very large vessels. These are several times wider than those of the evergreen *Q. coccifera* var. *calliprinos*. The provision of sufficient water from the soil is rendered possible by the wide spacing of the trees.

An especially interesting group are the lithophytes. These were studied by Zohary and Orshansky (1951). While lithophytes are well distributed in Mediterranean climatic areas with 400-700 mm of winter-rain, they occupy such extreme habitats that I must discuss them here.

Lithophytes differ from chasmophytes by rooting not in crevices but by rooting right into the rock, by dissolving the rocky materials (cf. Oppenheimer, 1956; Litav, 1965). Thus, these plants are found on dolomite, on crystalline and lithographic limestone and on chalk and surface lime deposits (incrustations). The seedling roots can penetrate rock only when the seeds germinate in small hollows or holes on the rock face, i.e. at those spots where the surface has already been somewhat softened by water. However, subsequently the roots grow right into the rock and may reach, in irregular twists, with frequent changes in direction, a length of 50 cm. Branching only occurs in softer parts of the rock. In addition to this tap root, other, even more slender little roots are formed on the rock surface. These spread to small hollows filled with fine soil.

The behaviour of these plants during a very hot and rainless summer was surprising. *Varthemia iphinoides* (Compositae) and *Podonosma syriacum* (Boraginaceae) were investigated. Both are small dwarf shrubs. *Varthemia* has large leaves in winter from November to July and small summer leaves from June to December. It flowers from September to October. In *Podonosma* the winter leaves remain from December to the end of August, the summer leaves from June to February. The flowering period lasts from February to the end of August. The transpiration rates of both species show no decline either in the summer or in the winter and are, with rates of 1000-1400 mg/g/h, much higher than those of species growing nearby, namely *Poterium spinosum* (644 mg/g/h) or *Cistus salviifolius* (336 mg/g/h). The total water loss of a plant of *Varthemia* reached 60-120 g per day or 10-20 kg during the summer months. In spite of these high losses the water balance of these species is remarkably uniform. This also applies to the associated lithophyte *Stachys palaestina*. In the summer the osmotic values in *Varthemia* were 23-29 atm, in *Podonosma* 16-19 atm and in *Stachys* 15 atm. The water requirements of these plants are met because the material of the rock absorbs water very easily. On putting 1 ml of water on the surface of the rock it is absorbed into the capillaries in less than one minute. The maximum water-holding capacity of the rocks amounts to 8-22%. Even in mid-summer the water content of stones with plant growth was determined as 2·5%. The roots can absorb this water which reaches the root tips through capillaries. In this way a plant can use the water reserve stored in less than 2 m^3 of rock. This corresponds, approximately, to the density of plant growth on such rock surfaces.

A somewhat arid semi-desert and steppe area, forming a transition zone between the desert zone and the originally-wooded Mediterranean zone in the eastern Mediterranean area of Palestine and Syria, is also found in north-west Africa. However, here the inter-relationships are much more complicated because of the high Atlas mountains (4000 m above sea level), which affect various climatic factors. To make a comparison we may briefly examine conditions in north-west Africa. The increased precipitation towards the Atlantic ocean on the one hand and increased altitude on the other, the greater dryness in the lee of the mountain ranges and the decrease in temperature with increasing height from the montane to the alpine belt, produces a highly variable vegetation mosaic which has been of concern to botanists for some time. Since the vegetation often forms woodlands in this area, I shall discuss their ecological relations in connection with the discussion of the Mediterranean plant cover. Here, it need only be pointed out that *Artemisia herba-alba* semi-desert and a grass steppe, with *Stipa tenacissima*, the Halfa-grass and *Lygeum spartum*, the Esparto-grass, predominate over wide areas, especially on the high plateau between the Maritime Atlas and the Sahara Atlas.

Precipitation in this area is around 250 mm but the rainfall occurs rather irregularly. The mean temperature of the warmest month is still 28°C but the winters are cold. Snow occurs and the number of frost days is recorded as 48. The vegetation and soil relations are more fully discussed by Ozenda (1954) and by Dubuis and Simonneau (1954).

Artemisia herba-alba generally occupies hollows with silty clay soils, but only those that are non-saline. In addition to the dominant, *Artemisia*, numerous annuals and a few geophytes also occur.

According to Dubuis and Simonneau the cover is 35-70% and the height of the vegetation over 25 cm. Two soil samples were taken to 70 cm (Table 112, figures as a percentage).

TABLE 112

Sample	pH	Coarse sand	Fine sand	Silt	Clay	Lime	NaCl
I	8·8	0	29·8	53	15·7	13	0·037
II	8·7	4·8	44·3	26·9	22·7	12·8	0·033

The soils can be even more sandy but the composition of the community is not then changed, only the degree of cover decreases (30-40%) and the height, to 10-15 cm.

A soil profile gave the results in Table 113 (% composition).

TABLE 113

Depth	pH	Gravel	Coarse sand	Fine sand	Silt	Clay	Lime	NaCl
0–15 cm	8·4	8·4	0	76·7	13·8	8·3	30·1	0·01
15–40 cm	8·7	0	0	42·3	52·8	4·1	75·8	0·07
40–100 cm	8·8	0	0	66·8	13·7	18·8	73·5	0·11

Wherever the salt content increases, *Artemisia* becomes associated with the halophytes *Atriplex mauretanica* and *Suaeda fruticosa*. With even greater salinity pure halophytic communities occur which show a very pronounced zonation. Such strongly saline habitats are very common on the high plateaux.

By contrast with *Artemisia*, *Stipa tenacissima* prefers physiographically higher positions and sloping sites with more or less stony soils forming the upper layers at least. The vast *Stipa tenacissima* plains are regularly used for cropping as the Halfa-grass provides an important export. These plains stretch eastwards to Homs in Libya.

Lygeum spartum prefers more sandy soils with good aeration but it tolerates a greater salt content in the soil. It occurs frequently at the edge of halophytic communities. Ozenda points out that, while these typical communities can be found, transition zones with mixed communities are often more frequent.

Unfortunately, the ecology of these three interesting and very important species has not been investigated so far. The Artemisias belong to the malacophyllous xerophytes. The two grass species will probably not be found to behave like typical grasses but will rather belong to the special type of sclerophyllous grasses, referred to earlier (see p. 412). A thorough investigation of these grasses during the rainy season should result in particularly valuable information.

Stipa tenacissima occupies areas already degraded to a certain extent which, under natural conditions, would be occupied by evergreen oaks and other woody plants. However, these habitats really belong to the Mediterranean region.

The Mediterranean zone proper fringes the Mediterranean sea. It only has one gap, the dry Egyptian coast. One can distinguish more oceanic and more continental subzones (Fig. 299). The latter includes the coastal area of Algeria to Cyrenaica, the eastern part of Greece and the coastal area of Asia Minor. In the last area one can distinguish with increasing wetness of the climate a further semi-wet subzone with *Laurus nobilis*, which then forms a transition to the summer-green, sub-Mediterranean hardwood forests. In the more continental mountains *Cupressus horizontalis* (on Cyprus from 700 m) and *Cedrus libani* (in the

Taurus mountains from 1200 m; on Crete from *c*. 900 m) occurs in the next highest belts. Above these there is the alpine thorn-cushion belt.

The characteristic tree of the eastern Mediterranean zone is *Quercus coccifera* (in Palestine var. *calliprinos*) together with other sclerophyllous woody plants (*Phillyrea*, *Arbutus andrachne*, *Myrtus*, *Pistacia lentiscus*, *Smilax*). In warmer habitats *Ceratonia siliqua* predominates. The more important conifers are *Pinus halepensis* and, further north and at higher altitudes, the closely related *P. brutia*. On sandy soils at sea level one may also find *Pinus pinea* which, however, is rarely found in natural stands.

The Mediterranean zone itself occupies an intermediate position between the Subtropical arid zone and the Temperate wet zone. However, since the Mediterranean zone represents an originally forested zone, its ecological relations would require to be discussed separately, together with the temperate zone.

References

ABD EL GAWAD AYYAD, M. 1957. *An ecological study of Ras El Hikma district*. Ph.D. Thesis. Cairo University.

ABD EL RAHMAN, A. A. 1953. Studies in the water economy of Egyptian desert plants. IV. Establishment and competition. *Bull. Inst. Désert Egypte*, **3**, 84-93.

ABD EL RAHMAN, A. A., and BATANOUNY, K. H. 1965. The water output of the desert vegetation in the different microhabitats of Wadi Hoff. *J. Ecol.*, **53**, 139-45.

ABD EL RAHMAN, A. A., and EL HADIDY, M. N. 1958. Observations on the water output of the desert vegetation along the Suez Road. *Egypt. J. Bot.*, **1**, 19-38.

ABD EL RAHMAN, A. A., and EL HADIDY, M. N. 1959. Some observations on the effect of wind on the desert vegetation along Suez Road. *Bull. Soc. Géogr. Egypte*, **32**, 207-16.

ABD EL RAHMAN, A. A., and HAMMOUDA, M. A. 1960. Vegetation development in a fenced area at Ras El Hikma. *Unit. Arab. Rep. J. Bot.*, **3**, 1-11.

ARVIDSSON, I. 1958. Plants as dew collectors. *Asembl. Gener. Toronto 1957. C.r. Rapp.*, **2**, 481-84.

ARVIDSSON, I., and HELLSTRÖM, B. 1955. *Assn. Inst. Hydrol., Publ.*, **36**.

BATANOUNY, K. H. 1963. *Water economy of desert plants in Wadi Hoff*. Ph.D. Thesis, Cairo University.

BOULOS, L. 1960. *Flora of Gebel El-Maghara, North Sinai*. Agric. Ext. Dep. Herbar. Sect., Cairo.

BOYKO, H. 1954. A new plant-geographical subdivision of Israel (as an example for South-west Asia). *Vegetatio*, **5/6**, 309-18.

BUTZER, K. W. 1959. Studien zum vor- und frühgeschichtlichen Landschaftswandel der Sahara, III. Die Naturlandschaft Ägyptens während der

Vorgeschichte und der dynastichen Zeit. *Abh. Math. Naturw. Kl., Akad. Wiss. Mainz*, 2, 44-122.

Davis, P. H. 1953. The vegetation of the deserts near Cairo. *J. Ecol.*, **41,** 157-73.

Drar, M. 1952. A report on *Kochia indica* Wight in Egypt. *Bull. Inst. Désert Egypte*, 2, 54-58.

Draz, O. 1954. Some desert plants and their uses in animal feeding. *Publ. Inst. Desert*, 2, 1-95.

Dubuis, A., and Simonneau, P. 1954. Contribution à l'étude de la végétation de la region d'Ain Skrouna. *Serv. Et. Sci., Gouv. Gen. Algér.*

Evenari, M. 1938a. Root condition of certain plants in the wilderness of Judaea. *J. Linn. Soc.*, **51**, 382-88.

Evenari, M. 1938b. The physiological anatomy of the transpiratory organs and the conducting systems of certain plants typical of the wilderness of Judaea. *J. Linn. Soc.*, **51**, 389-407.

Evenari, M., and Koller, D. 1956. Desert agriculture problems and results in Israel. In *The future of arid lands*, pp. 390-413. Washington.

Evenari, M., and Richter, R. 1938. Physiological ecology investigations in the wilderness of Judaea. *J. Linn. Soc.*, **51**, 333-81.

Feinbrun, N., and Zohary, M. 1955. A geobotanical survey of Transjordan. *Bull. Res. Counc. Israel, Sect. Bot.*, **5**, 5-35.

Gradmann, R. 1934. Die Steppen des Morgenlandes. *Geogr. Abh., 3, Reihe*, **6**.

Karschon, R. 1953. The hamada in Wadi Araba, its properties and possibilities of afforestation. *Ilanoth* (*Israel*), **2**, 19-45.

Kassas, M. 1953. Landforms and plant cover in the Egyptian desert. *Bull. Soc. Géogr. Egypte*, **26**, 193-205.

Kassas, M., and El-Abyad, M. S. 1962. On the phytosociology of the desert vegetation of Egypt. *Ann. Arid Zone*, **1**, 54-83.

Kassas, M., and Girgis, W. A. 1964. Habitat and plant communities in the Egyptian desert. V. The limestone plateau. *J. Ecol.*, **52**, 107-19.

Kassas, M., and Imam, M. 1952, 1953, 1954, 1959. Habitat and plant communities in the Egyptian desert, I-IV. *J. Ecol.*, **40**, 342-51; **41**, 248-56; **42**, 424-41; **47**, 289-310.

Kassas, M., and Imam, M. 1957. Climate and microclimate in the Cairo desert. *Bull. Soc. Géogr. Egypte*, **30**, 25-52.

Kassas, M., and Zahran, M. A. 1962. Studies on the ecology of the Red Sea coastal land, I. *Bull. Soc. Géogr. Egypte*, **35**, 129-75.

Kokina, S. I. 1932. Water regime and internal factors of resistance in plants of the Karakum desert (in Russian). *Probl. Rasteniv i Osvoyen. Pustyn*, I, 99-195, Leningrad.

Kreeb, K. 1963. Untersuchungen zum Wasserhaushalt der Pflanzen unter extrem ariden Bedingungen. *Planta*, **59**, 442-458.

Litav, M. 1965. Mycorrhizal association in dwarf shrub species growing in soft Cretaceous rocks. *J. Ecol.*, **53**, 147-53.

Masson. 1948. *Bull. Inst. Fr. Afr. Noire*, **10**, 1-181.

Migahid, A. M., and Abd el Rahman, A. A. 1953a,b,c. Studies in the water economy of Egyptian desert plants. I Desert climate and its relation to vegetation. II Soil water conditions and their relation to vegetation. III Observations on the drought resistance of desert plants. *Bull. Inst. Désert Egypte*, **3**, 5-24: 25-57; 59-83.

MIGAHID, M., ABD EL RAHMAN, A. A., EL SHAFEI, M., and HAMMOUDA, M. A. 1955. Types of habitat and vegetation at Ras El Hikma. *Bull. Inst. Désert Egypte*, **5** (2).

MIGAHID, A. M., EL SHAFEI ALI, M., ABD EL RAHMAN, A. A., and HAMMOUDA, M. A. 1959. Ecological observations in western and southern Sinai. *Bull. Soc. Géogr. Egypte*, **32**, 165-206.

NEGBI, M., and EVENARI, M. 1961. The means of survival of some desert summer annuals. *Proc. Madrid Symp., Unesco Arid Zone Res.*, **16**, 249-59.

OPPENHEIMER, H. R. 1938. An account of the vegetation of the Huleh swamps. *Palestine J. Bot., Reh. Ser.*, **2**, 34-39.

OPPENHEIMER, H. R. 1949. The water turn-over of Valonea oak. *Palestine J. Bot., Reh. Ser.*, **7**, 177-79.

OPPENHEIMER, H. R. 1952. Geobotanical research in Palestine vegetation. *Vegetatio*, **3**, 301-20.

OPPENHEIMER, H. R. 1956. Pénétration active des racines de buissons mediterraneens dans les roches calcaires. *Bull. Res. Counc. Israel, D*, **5**, 219-22.

ORSHAN, G. 1953. Note on the application of Raunkiaer's system of life forms in arid regions. *Palestine J. Bot., Jerus. Ser.*, **6**, 120-22.

OZENDA, P. 1954. Observations sur la végétation d'une region semiaride. Les Hautes-Plâteaux du Sud-Algérois. *Bull. Soc. Hist. Nat. Afr. N.*, **45**, 134-69.

PAVER, G. L., and JORDAN, J. N. 1956. Report on reconnaissance of hydrological and geophysical observations in North Sinai coastal area of Egypt. *Publ. Inst. Désert Egypte*, **7**.

RANGE, P. 1921. Die Flora der Isthmuswüste. *Ges. Pal.-Forsch.*, **7**.

SHMUELI, E. 1948. Ecological investigations in Palestine, II. The water balance of some plants of the Dead Sea salty soils. *Palestine J. Bot., Jerus. Ser.*, **4**, 117-43.

STOCKER, O. 1926. Die ägyptisch-arabische Wüste. *Veg. Bilder*, **17**, Plates 25-36.

STOCKER, O. 1928. *Der Wasserhaushalt ägyptischer Wüsten- und Salzpflanzen.* Jena.

EL SHISHINY, E. D. H., and THODAY, D. 1953a. Inhibition and germination in *Kochia indica*. *J. Ecol.*, **4**, 10-22.

EL SHISHINY, E. D. H., and THODAY, D. 1953b. Effect of temperature and desiccation during storage on germination and keeping quality of *Kochia indica* seeds. *J. Ecol.*, **4**, 403-06.

TÄCKHOLM, V. and DRAR, M. 1949-54. *Flora of Egypt.* Vols. I-IV. Cairo.

TÄCKHOLM, V. Verbal communication.

TADROS, T. M. 1956. An ecological survey of the semi-arid coastal strip of the western desert of Egypt. *Bull. Inst. Désert Egypte*, **6**, 26-56.

TADROS, T. M. 1958. The plant communities of the barley fields and uncultivated desert areas of Mareotis (Egypt). *Vegetatio*, **8**, 161-75.

TADROS, T. M., and ATTA, B. A. M. 1958. Further contributions to the study of the sociology and ecology of the halophilous plant communities of Mareotis (Egypt). *Vegetatio*, **8**, 137-60.

THODAY, D., TADROS, T. M., and EL SHISHINY, E. D. G. 1956. *Kochia indica* Wight and its dispersal in Egypt. *Bull. Inst. Désert Egypte*, **6**, 57-66.

VASSILJEV, J. M. 1931. Über den Wasserhaushalt der Pflanzen der Sandwüste im südöstlichen Karakum. *Planta*, **14**, 225-309.

VOLKENS, G. 1887. *Die Flora der Ägyptisch-Arabischen Wüste*. Berlin.

WAISEL, Y. 1960, 1961. Ecological studies on *Tamarix aphylla* (L.) Karst. II The water economy. *Phyton*, **15**, 17-27. III The salt economy. *Plant and Soil*, **13**, 356-64.

WALTER, H. 1956. Das Problem der Zentralanatolischen Steppe. *Naturwissenschaften*, **43**, 97-102.

WRIGHT, R. L. 1964. *Arid zone*. Unesco.

ZOHARY, M. 1944. Vegetational transects through the desert of Sinai. *Palestine J. Bot., Jerus. Ser.*, **3**, 57-78.

ZOHARY, M. 1945. Outline of the vegetation of Wadi Araba. *J. Ecol.*, **32**, 204-13.

ZOHARY, M. 1947a. A vegetation map of western Palestine. *J. Ecol.*, **34**, 1-19.

ZOHARY, M. 1947b. A geobotanical soil map of western Palestine. *Palestine J. Bot., Jerus. Ser.*, **4**, 24-35.

ZOHARY, M. 1952. Ecological studies in the vegetation of Near Eastern deserts, I. Environment and vegetation classes. *Israel Expln J.*, **2**, 201-15.

ZOHARY, D. 1953. Ecological studies in the vegetation of the Near Eastern deserts. III Vegetation map of the central and southern Negev. *Palestine J. Bot., Jerus. Ser.*, **6**, 27-36.

ZOHARY, M. 1954. Hydro-economical types in the vegetation of Near East Deserts. In *Biology of Deserts*; *Proc. Symp. Biol. hot & cold Deserts*, 56-67. London.

ZOHARY, M. 1963. On the geobotanical structure of Iran. *Bull. Res. Counc. Israel, D* **11** (Suppl.) 1-113.

ZOHARY, D., and ORSHAN, C. 1954. The *Zygophylletum dumosi* and its hydroecology in the Negev of Israel. *Vegetatio*, **5/6**, 341-50.

ZOHARY, M., and ORSHAN, G. 1957. Ecological studies in the vegetation of Near Eastern deserts, II. Wadi Araba. *Vegetatio*, **7**, 15-37.

ZOHARY, M., and ORSHANSKY, G. 1947. The vegetation of the Huleh plain. *Palestine J. Bot., Jerus. Ser.*, **4**, 90-104.

ZOHARY, M., and ORSHANSKY, G. 1949. Structure and ecology of the vegetation in the Dead Sea region of Palestine. *Palestine J. Bot., Jerus. Ser.*, **4**, 177-206.

ZOHARY, M., and ORSHANSKY, G. 1951. Ecological studies on lithophytes. *Palestine J. Bot., Jerus. Ser.*, **5**, 119-28.

Index

*denotes an illustration in the text.
With a few exceptions, only the genus names of plants have been given.

Abies, 217, 308, 309
Abronia, 314
Abutilon, 1, 2, 292
Acacia, 32, 145, 191, 208, 220, 244, 248, 251, 258, 259, 308, 332, 388, 392, 408, 418, 436, 442, 458, 478, 501, 502, 506, 509
abyssinica, 283
albida, *259, 261, 283, 364, 445, 448, 510
aneura, 263, 410, 411, 415, 418, 420, 421, 422, 423, 424, 425, 426, 427, 429, 430; ecology, 424
arabica, 448
burkittii, 415, 418, 420, 421
cambregei, 414
campylacantha, 283
catechu, 447
constricta, 305
detinens, *249, *253, *255
drepanolobium, 232, 283
ehrenbergiana, 260
estrophiolata, 430
fistula, 283
flava, 260, 283
flute, 232
formicarum, 232
giraffae, *242, *253, 254, *256, 285, 364
greggii, 306
haematatoxylon, 285
harpophylla, 427
hebecladoides, 283
kempeana, 426
laeta, 458
latifolia, 122
macracantha, 382
malacocephala, 232
maras, *253
melanoxylon, 122
mellifera, 260, 261, 283
mummularia, 414
nilotica, 233
orfota, 283
raddiana, 260, 283, 448, 467, 501
saligna, 109, 490, 494
senegal, 261, 283
seyal, 232, 260, 261, 283, 448
sieberiana, 283
spirocarpa, 501, 502
stenocarpa, 458
sundra, 447
tetragonophylla, 414
tortilis, 245, 260, 467, 468, 508, 510
transpiration rates, *502
uniplumis, *255
xanthophloea, 146
Acalypha, 112, 446
Acanthosicyos, 364, *365
Acer, 209
Acheb, 448
Achillea, 482
Acrostichum, 153
Actiniopteris, 220, 359
Adansonia, 220, *221
Adaptation of plants, 36
Adenanthera, 447
Adenia, 126, 222, 353, *354, 357
Adenium, 222, *223
Adenocarpus, 197
Adesmia, 383
Adiantum, 121, 314, 431, 459, 466, 496
Adriana, 414
Aechmea, 135
Aegialitis, 157, 158
Aegiceras, 157, 158, 165
Aegopodium, 28
Aeluropus, 454, 455, 489
Aerva, 442
Affamomum, 121
Africa, East, map of, 221
South, comparison with Australia, *402
South-west, map of, 339
Agathis, 180
Agave, 30, 120, 170, 308
Ageratum, 173
Agropyrum, 488, 489
Agrostemma, 26
Agrostis, 196, 216
Agua, branca, 148
Agua preta, 148

Air currents at equinoxes, 46
Aizoon, 346, 350
Ajuga, 28
Albizzia, 105, 109, 214, 220, 261, 283, 285, 445
Alchemilla, 197, 198, 203
Aleurites, 446
Algae, 130, 137, 147, 289, 415
  filamentous, 37
  green, 35
  unicellular, 35
Alhagi, 435, 444, 501
Allanblackia, 105, 109
Allelopathic influences, 14
Alliona, 311, 312
Allium, 488
Almond, 445
Alnus, 32, 181
Aloe, 222, 258, 243, *356, 357, 358, *358, 390, 392, 393, 397
Alopecurus, 17
Alphitonia, 114
Alpinia, 446
Alsophilia, 122
Alstonia, 226, 446
Althaea, 443
Altitudinal belts, 13, 14, 184
  rainfall, 177
  zones, comparative scheme, 193
    in the tropics, 177
Altingia, 182
Aluminium, 95
Amani, 75, 85-7, 119, 201
  tropical virgin soil, 89-90
Amaranthos, 311, 312, 408
Amazon river, schematic profile, 149
Ammophila, 423, 489, 490
Amoona, 447
Amorphophallus, 118
Ampelocissus, 125, 126
Amygdalus, 115, 490, 512
Anabaena, 415
Anabasis, 291, 466, 488, 495, 501, 502, 506
  transpiration rates, *502
Anacampseros, 353, 390, 396, 397
Anacardium, 446
Anacystes, 345
Anagallis, 26
Anagyris, 504
Analysis, vegetation, 8
Ananas, 120, 122, 135
Anaphalis, 183, 186
Anaptychia, 382
Anarrhinum, 496
Anastatica, 448, 464, *465
Andes, cross-section, 384*
  vegetation profile of, 190
Andropogon, 186, 216, 455, 459
Androsace, 183
Anemone, 5, 196, 313
Angiopteris, 122
Angraecum, 134
Anisosperma, *125
Anogeissus, 261
Anthemis, 468
Anthocleista, 105, 106, 109
Anthriscus, 5
Anthrobryum, 385
Anthropogenic equilibrium, 7
Anthurium, 170
Antidesma, 446
Ants, 130, 131
Apicra, 390
Apple, 494
Apricots, 445
Arabidopsis, 496
Araceae, 123
Aralia, 446
Araucaria, 15, 180
Arbutus, 308, 516
Arctostaphylos, 308
Areca, 446
Areg, 434, 439
Arenga, 446
Argemone, 364, 393
Argyroderma, 390
Argyrolobium, 459
Arid areas, Australian, *405
    distribution of water in soil, 273
    ephemeral vegetation, 277
    plant types, 505
    rainfall, 271
    salt accumulation, 295-6
    sources of water, 270
  climate, types of, 268
  regions, concept of, 266
    plants and water factor, 285
    world, *267
Aristida, 242, 243, 244, 439, 510
  arenaria, 423, 428
  brevifolia, 412
  congesta, *389
  corvula, *389
  hochstetteriana, 352
  monosperma, 495
  namaquensis, 412
  nitidula, 430
  obtusa, 455
  plumosa, 483, 495
  pungens, 435, 441, 455, 483

scoparia, 495
uniplumis, 257
Aristolochia, 123
Arrhenatherum, 17
Arrowroot, 119, 120
Arroyos, *see* Washes
Artemisia, 473, 509, 515
fragrans, 512
fragrantissima, 481,
herba-alba, 291, 488, 490, 496, 506, 510, 512, 514
judaica, 459, 468, 496
monosperma, 291, 468, 483, 494, 505
tridentata, 309
Arthraerua, 350, 351, *351, 352, 365, 366, 371, 372
Arthrocarpus, 447
Arthrocnemum, 365, 368, 369, 419, 428, 429, 456, *457, 489, 493
Artocarpus, 106, 109, 216
Arundinaria, 217
Arundo, 217
Asparagus, 390, 396
Aspergillus, 371
Asperula, 5
Asphodelus, 443, 483, 484, 487, 488
Aspidistra, 122
Asplenium, 120, 121, 122, 131, 132-3, 136, 137, 170
Aster, 162
Asteriscus, 450, 464, 466, 482
Astragalus, 436, 459, 468, 495
Astrebla, 407, 408, 409, 414
Atalaya, 430
Atractylis, 498
Atriplex, 306, 385, 397, 415, 417, 449, 489
canescens, 309, 421, 422
cell sap and chloride concentration, 421
confertifolia, 421, 422
halimus, 391, 444, 450, 467, 506
inflata, 428
mauretanica, 515
mollis, *456
muralis, 368
nummularia, 398, 410, 416, 430
paludosa, 416, 428
portulacoides, 488
ragodioides, 410
spongiosa, 416
stipitata, 418
vesicaria, 409, 410, 414, 415, 416, 418 419, 420, 428, 429
Australia, arid areas, *405
central arid regions, 402ff
comparison with South Africa, *402
ecological investigations, 415
moist habitats, 427
rainfall patterns, 404
vegetation, *408, 409
West salt lakes, *413
Avena, 24
Avicennia, 155, 156, 157, 158, 159, 161, 162, 163, 165
Azolla, 147, 150
Azorella, 190, 385
Azotobacter, 138, 415

Babbagia, 421
Baccharis, 307
Backhausia, 15
Baeckea, 431
Balanites, 145, 232, 261, 442, 448, 510
Ballota, 459
Bambusa, 109, 117, 122
Banana, 119, 120, 172, 195, 445, 446
Banksia, 252
Baobab, 220
Barchans, 440
Barley, 445
Barringtonia, 167
Bassia, 414, 415, 418, 428, 429
Batatas, 114
Bauhinia, 208, 364, 371, 446
Bayada, 303
Beaumontia, 446
Begonia, 119, 122, 183, *195
Beijerinckia, 138
Beilschmiedia, *181
Bellis, 5
Berberis, 181
Bergeranthus, 395, 397
Berlinia, 105
Betula, 209
Bignonia, 446
Big-game animals, *see* Tropical grassland
Big-game, 254
Bilbergia, 136
Biotope, *see* Habitat
Birds of rain forest trees, 102
Blaeria, 197
Blennodia, 414
Boerhavia, 311, 408
Bogor, 75, 76, 78, 182
climate, 78
Bogs, 2
oligotrophic, 94
topogenous, 94
Bolpophyllum, 134, 226

Bombax, 170, 214, 446
Borassus, 145, 227, 233, 446
Boscia, 244, 257, 258, 260, 261, 458
Boswellia, 261
Bottle-trees, 225, 329
Bougainvillea, 447
Bouteloa, 311
Brachychiton, 208, 226
Brachystegia, 220, 232
Brassaia, 446
Brassica, 383
Breadfruit, 446
Brigalow, 427
Brillantaisia, 120
Brodiaea, 313
Broken Veld, 391
Bromnea, *104
Bromus, 17, 26, 459
Brugiuera, 153, *159, 164, 165, 459
Buddleia, 447
Buitenzorg, *see* Bogor
Bumelia, 447
Burkea, 218, 261
Bushmen's candle, 355
Butea, 447

Caatinga, 222, 225
Cacao, 110, 172
Cacti, drying-out process in, 321
  epiphytic, 135
  succulent, 191
Cadaba, 261
Caesalpinia, 382
Cairo, rainfall distribution, *462
  variation of distribution, *463
  water content of soil, 470
Calamagrostis, 186, 191
Calamus, 123, 148, 446
Calathea, 121
Caligonum, 441, 464, 502, 503, 505
Callistemon, 252, 447
Callitris, 430
Calophyllum, 112, 167
Calothrix, 383
Calotropis, 442, 445, 448, 510
Campo, 149
Campos cerrados, 222
Canavalia, 166
Cannistrum, 136
Canthum, 226
Caoutchouc, 136
Capparis, 260, 382, 430, 445, 446, 450, 452, 466
Caralluna, 222, *223, 356, *356, 390, 495
Carbohydrates, 29
Carbon and nitrogen content of solid, 90
Cardamine, 5
Cardamom, 119
Carduus, 205
Carex, 197, 198
Carica, 122, 382, 446
Carissa, 226, 446, 458
Carludovica, 120
Carnegiea, 304, *305, 316, 317, 318, 322, 327, *328
  osmotic value, 319
  water loss, 319
Carpirius, 4, 13,
Caryota, 446
Cassava, 119
Cassia, 112, 414, 415, 418, 424, 425, 446, 448, 459
Cassytha, 138, 168
Castanea, 32, 182
Castanopsis, 137, 215, 216
Casuarina, 109, 167, 382, 415, 419, 420, 421, 428, 429, 445
Catesbaea, 447
Catophractes, 244, 257, *258, 259, 401
Cauliflory, 112
Cecropia, 100
Cedrela, 112, 447
Cedrus, 515
Ceiba, 261
Celastrus, 226
Cell sap concentration, of *Asplenium*, 136
  of *Atriplex*, 421
  of epiphytic orchids, 136
  of ferns, 121
  of *Ficus*, 128
  of herbs, 121
  osmotic, 38
  of trees, 126
Celtis, 305, 306, 308
Cenchrus, 412, 459
Centaurea, *465, 468, 477, 482, 489, 496
Cephalotaxus, 446
Ceratonia, 112, 490, 516
Ceratophyllum, 147, 445
Ceratopteris, 150
Cercidium, 305, 308, 315, 327, 329, 335
  osmotic values, 330
Cercis, 112
Cereus, 222
Ceriops, 153
Ceterach, 392
Chamaegigas, 361
Chamaephytes, 31
Chaparales, 225
Charakterarten (characteristic species), 9

Chasmophytes, 513
Cheilanthes, 226, 314, 315, *315, 392, 398, 426, 459
Chenopodium, 382, 383, 414, 417, 487
Chilanthus, 392, 401
Chilean-Peruvian Coastal desert, 375
Chiropterochory, 113
Chlamydomonas, 35
Chlorella, 35
Chloride, 456, 483
Chlorophora, 214, 261
Chorisia, 447
Chrysanthemum, 5, 484
Chrysobalanus, 168
Chrysocoma, 388
Chrysophyllum, 447
Cinnamomum, 106, 447
Cissus, 222, 226, 359, *360
Cistus, 513
Citharexylum, 447
Citrullus, 364, 366, 430, 445, 448, 450, 452, 468, 471, 473
Citrus, 109, 122, 445, 494
Cladium, 148
Cladonia, 381
Cladophora, 37
Cladothrix, 311, 312
Classification, 8
  in alpine belt, 192
Clay pans, 434, 442
Clearing of tropical forests, *see* Tropical forests
Clematis, 306
Cleone, 364, 436, 459, 468
Clerodendron, 447
Climatic diagrams, 51ff, 53
  interpretation of, 55
  map, 57
    of Australia and Africa, 403
      Caatinga, 225
      Cairo, 463
      desert, regions, 269
      Huancayo, 190
      India, 210
      Karroo, 388
      Mersa Matruh, 462
      New Guinea and Fiji, 76
      Sahara, 435
      Sonoran desert, 300
      South-west Peru and North Chile, 377
      Swakopmund, 341
      Tonga, 155
  showing types, 58-60
Climatic zones, 47-8
Climatograms, 56-7
  of Ankara, 56
    Cairo, 474
    Hohenheim, 56
Climax, 12ff
Clinogyne, 121
Clostridium, 414
Clove-tree, 446
Clusia, 126
CO_2 assimilation, 112
Cobalt paper method, 372
Coccoloba, 168, *168, 169
Cochlospermum, 208
Cocoa, 446
Coconut, 446
  groves, 172
  tree, 168-9
Cocos, 112, 446
Codonanthe, 135
Coffea, 106, 107, 108, 110, *116, 122, 261, 447
  see also Coffee
Coffee, 87, 172, 195, 446
  climate of plantation, 88
Coix, 121
Coleogyen, 309
Colocasia, 119, 120, 122
Colophospermum, 220
Colutea, 495
Combretum, 232
Commiphora, 220, 358, *360
Communities, biotic, 2
  climax, 10
  dune, 4, 12
  halophytic, 4
  meadow, 4
  weed, 4
Competition, 14ff
  equilibrium, 16
  factors mediating, 14
  in desert plant communities, 477
  interspecific, 16, 25, 28
  intraspecific, 15, 16, 25
  pressure, 19
  root, 26, 27, 28
Competitive ability, 24ff
Condelia, 306
Convolvulus, 383, 468, 496
Copaifera, *259
Cordia, 446, 458, 510
Cornulaca, 435, 439
Corydalis, 5
Corypha, 117
Costa Rica, sunshine on, 81
Costus, 119, 121, 122

Cotoneaster, 496
Cotton, 445
Cotyledon, 343, 353, 357, *389, 390, 397
Coussapoa, 126
Crassula, 343, 390, 395, 397
Crataegus, 496, 512
Cratystylis, 428
Creosote bush, see *Larrea tridentata*
Crepis, 5
Crotolaria, 195, 414, 459, 468
Croton, 382
Crucianella, 489
Cryophytum, 365, 368, 369
Cryptandra, 431
Cryptostegia, 446
Cucurbita, 312
Culcasia, 126
Cupressus, 448, 496, 515
Cuscuta, 138, 459
Cussonia, 233, 392, 400, 401
Cyanidium, 383
Cyathea, 120, 121
Cycas, 120, 122
Cyclone activity, 81
  tropical, 7
Cyclosorus, 431
Cylicomorpha, 102, 106, 121
Cynanchum, 226
Cynodon, 146, 195, 455
Cyperus, 260, 436, 489
  conglomeratus, 441
  papyrus, 145, 148, 510
  rotundus, 146
Cyrtandra, 183

Dacrydium, 126, 181
Dacryodes, 99
Dactyloctenium, 408
Dactylis, 488
Daemia, 468
Dalbergia, *124, 261, 283, 445, 447
Dalechampia, *125
Daniella, 214
Danthonia, 186, 441
Darcydium, 180
Dasylirian, 308
Date palm, 445, 493, *494, 508
Datura, 311, 364, 368
Dayas, 442
De Saussure effect, 135
Deciduous species, in Australia, 208
  in New Zealand, 209
Deciduous tree life forms, xii
Dehydroangustione, 15
Delphinium, 313
Dendrobium, 117, 226
Deschampsia, 186, 197
Desert autochtonic, 438
  gravel, 434, 438
  pebble, 438
    vegetation, 467
  sand, 434, 439
  Simpson, 412
  stone, 434 ,437, *438
    vegetation, 464
  thorn-succulents, 189
  varnish, 438, 467
  with saline soils, 189
Dew measurements, 485ff
Dewfall, evaporation of, 488
  desert, 473
  at Ras El Hikma, *486
Dianthus, 496
Dichanthium, 409
Didymaotus, 390
Dierama, 196
Digitaria, 233, 409
Diospyros, 215, 446
Diplachne, 146, 366, 368
Diplopappus, 388
Diplotaxis, 459, 464, 481, 482
Dipterocarp forests, *see* Forests, dipterocarp
Dipterocarpus, 215
Disa, 196
Dischidia, 131
Distichlis, 385
Distribution of soils, 61
Distylium, 217
Dodonaea, 414, 423, 431, 447
Doline formation, 404
Dolines, 275
Dolomite, 513
Dombeya, 233
Drosomthemum, 346
Drought, 504
  -escaping plants, 506
  -resistant plants, 506
  -survival during, 504
  in Tucson, 302
Dry habitats, vegetation, 169
  matter production, 6, 21
  -valleys, 441
Drynaria, 131, 133, 136
Dunes, crescentic, 440
  gypsum, 412, 414, 428
  sand, 284, 415, 508
Dwarf shrubs, 31, 388

Echinocereus, 317, 327
Echinochloa, 145
Echinops, 459, 464, 468, 471, 473, 488
Echiochilon, 490
Echium, 489, 490
Ecological equilibrium, 7
niches, sequence of, 282
Ecology, experimental, xv
Ecosystem, components of, 33
Edaphic factors, in relation to vegetation, 95
Egyptian-Arabian Desert, 461ff
ecological investigations in Negev, 497
ecological research near Cairo, 468
habitats and plant communities, 461
rainfall, 462
transitional areas, 508
Egyptian clover, 445
Egyptian desert, ecological situations on Sinai peninsula, 493
vegetation along Mediterranean, 483
Ehretia, 226, 388, 401, 458
Eichhornia, 145, 146, 147, 150
Elaeis, 214, 446
Elatostema, 183
Elephants, 214-15
Elettaria, 122
Encelia, 304, *305, *330, 333, *334, 335
osmotic values, 333-4
transpiration, 333
Encephalartos, 120, 122, 169, 390
Endocarpon, 464
Energy cycles in ecosystems, 33ff
Engelhardtia, 183
Eperua, 93-4, 99
Ephedra, 305, 309, 441, 442, 448, 459, 464, 495, 496, 505
Ephemerals, changes in weight, 477
Ephemeropsis, 138
Epidendrum, 136
Epiphytes, 96, 99
decrease in fresh weights, 135
phanerogamic, 169
in tropical rain forest, 128
Episodic rains, distribution of, 440
Equisetum, 350, 496
Eragrostis, 216, 242, 243, 392, 412, 426, 436, 444, 509
Eremophila, 414, 415, 418, 421, 423, 425, 428, 429, 430
Erg, 434, 439
Eriachne, 414
Erianthus, 445
Erica, 196, 204
Eriocaulon, 148

Eriocephalus, 388, 393
Eriodendron, 446
Erodium, 311, 313, 382, 383, 450, 459, 464, 466, 468, 474, 476, 483
Erosion channel, salt deposition, 363
Eryobotrya, 447
Erythrina, 114, 218, 233
Erythroxylon, 226
Eschscholtzia, 314
Ethereal oils, 15
Eucalyptus, 172, 208, 210, 226, 227, 263, 271, 273, 382, 408, 409, 417, 418, 419, 421, 494
alba, 208
bigalerita, 208
camadulensis, 411, 430, 445
citriodora, 109
clavigera, 208
confertifolia, 208
coolabah, 414
deciduous, 208
dichromophoia, 430
diversicolor, 411
evergreen, 208
gamophylla, 424, 426
grandifolia, 208
lesouefii, 428
marginata, *227
melanophoia, 427
oleosa, 429
pachyphylla, 414
papuana, 430
populnea, 427
regnans, 31, 411
Euchylaena, 414
Euclea, *259, 364, 388, 392, 401
Eugenia, 149, 382, 447
Eupatorium, 173, 215
Euphorbia, 222, 311, 312, 353, 356, *356, 357, 358, *359, 390, 396, 397, 414, 436, 468, 488
Euryale, 147
Euryhydrous species, 292
Euryops, 198, 204, 388
Eustachys, 243
Evaporation in Sahara, 440
Evergreen forest, profile, on Trinidad, 99
Exotheca, 196

Fabiana, 383
Fagonia, 435, 450, 466, 468, 482
Fagopyrum, 25
Fagraea, 117
Fagus, 4, 13, 31, 115

Farsetia, 464, 468, 482, 483
Faurea, 233
Feijoa, 447
Ferocactus, 305, *305, 317, *317, 320, *321
osmotic value, 322
transpiration, 320
weight loss with water, 321
Ferula, 448
Festuca, 186, 191, 198, 205, 385
Ficus, 122, 126, 127-8, 130, 226, 261
carica var. *rupestris*, 496
elastica, 106
guerichiana, 359
macrophylla, 447
misorlusis, 447
nitida, 106, 447
pandurata, 447
platypoda, 431, 447
pseudosycomorus, 466, 496
repens, 447
salicifolia, 448, 459
sycomorus, 445, 447
vogelii, 447
Figs, 445, 494
Filago, 504
Filipendula, 22
Fire, 82, 239
Fish River Canyon, *309
Flacourtia, 446
Fleurya, 121
Fockea, *219, 390
Fog deserts, 268
ecological significance, 342
formation, 341
plants, 377, 378
precipitation, 341, 379, 380
vegetation, 343
Foliage, ericaceous, 204
Forest, alluvial, 150
beech, 2
border, *196
dipterocarp, 80, 84, 95, 100
flood-water, 150
grazing, 9
mangrove, 150
mist, 177
monsoon, 215
dry, 210
moist, 210
peat, 150
salt, 215
secondary, 171
spruce, 2
swamp, 150
teak, 172, 215
vegetation, relationship with rainfall, 212
Fouquieria, 44, 207, 291, *316, 327, *328
distribution, 328
osmotic value, 329
transpiration, 329
Fourcroya, 120
Frankenia, 428, 444, 449, 452, 455, 488, 489
Franseria, 305, 334
Fraximus, 307
Fredolia, 436, 449, 451, 452
root profile, *454
Freycinetia, 123, 180

Galenia, *389, *390
Galinsoga, 382
Galium, 496
Garcinia, 447
Garua, 375, 379
Gasteria, 390
Gaussia, 170
Gazania, 365
Geigeria, 459
Geijera, 226
Genista, 441
Geophytes, 448
Gezzu, 260
Gilgay soils, 427
Gleichenia, 172
Globularia, 459, 488, 495
Gomphocarpus, 393
Goodenia, 414
Grass-steppes, 510
Grasses, growth with water table, 18
Grasslands, ungrazed, 271
Grevillea, 414, 424, 246, 447
Griselinia, 126
Ground-water table, 7
Growth capacity, 24ff
Growth curves of plant species with and without competition, 20
Guavas, 445, 495
Gymnarrhena, 500
Gymnocarpus, 483, 488
Gymopteris, 315
Gypsophila, 468
Gypsum, 438
crystalline, 439
Gyrocarpus, 208, 226

Habitat, relative consistency with changing ecological niche, 281
principle of, 282
Hagenia, 195

Hakea, 252, 414, 424, 430
Halfa-grass, 510
Halocnemum, 444, 489, 490, 493
Halogeton, 468
Halophytes, 43, 167, 296, 447
  and salt factor, 293
  facultative, 168
Haloseres, 10, 11
Haloxylon, 439, 451, 467, 468, 469, 471 474, 476, 477, 494, 495, 498, 503
  articulatum, 291, 506
  persicum, 501, 502, 506
  salicornicum, 501
Hamada, 434, 437, 464, 494, 500
Hapoptelia, 215
Haworthia, 390, 397
Heat tolerance, 292
Hedera, 208
Height growth, 25
Helianthemum, 459, 464, 488, 489
Helianthus, 25, 112
Helichrysum, 196, 197, 198, 203, 204, 205, 365, 488
Heliophytes, 5, 23
Heliotropium, 368, 369, 468, 473, 482
Helipterum, 409, 414
Hemi-epiphytes in tropical rain forest, 126
Herb layer in tropical rain forest, 118
Hereroa, 353, 354, *355, 357
Hermannia, 399
Heterodendron, 415, 420, 421
Heteropogon, 305
Hevea, 110
Hibbertia, 431
Hibiscus, 167, 305, 431, 436
Hippomane, 168
Hoodia, 344, 353, 356, *356, 390
Hopea, 95, 117
Hordeum, 311, 313, 487
Human intervention, 7
Humiria, 93
Hura, 446
Hurricanes, 169
Hydnophytum, 130, 165, 166
Hydrature, 42ff, 285
  criterion for classification, 288
Hydrilla, 147
Hydrodea, 346, *347, 350
Hydroses, 10
Hygrohalophytes, 296, 428, 448
Hymenoclea, 307
Hymenophyllum, 134
Hymenosporum, 447
Hyoscyamus, 439, 468, 471, 473
Hyparrhenia, 145, 232
Hypericium, 195
Hyphaene, 145, 227, 233, 261, 283, 445, 448, 459
Hypochaeris, 19
Hypocyrta, 135

Idria, 329, 358
Ifloga, 468, 483
Igapo, 149
Ilex, 106, 195, 208, 446
Impatiens, 16, 119, 121
Imperata, 92, 174, 234
Indian monsoon area, phenological cycle, 216
Indigofera, 436, 447
Inula, 488
Iphiona, 464, 481, 482
Ipomoea, 148, 166, *166, 167, 168, 169, 442, 446
Iris, 495
Iron-oolites, 415
Irrigation, 110
Iseilema, 408
Isoberlinia, 220, 261
Ivy, 419
Ixora, 447

Jasminum, 447
Jatropha, 305, 329
Jelly-fish, 459
Juglans, 181
Juncellus, 146, 366, 368
Juncus, 162, 444, 489, 493, 496
Juniperus, 195, 309, 495, 512
Jurinaea, 6
Jussiaea, 147, 148, 150
Justicia, 365

Kallstroemia, 311, 312
Karroo, 387ff
  ecological investigations, 393
  general features, 387
  map of, *387
  succulents, osmotic values, 397
  vegetation of Upper, 391
Karst, 270, 404, 442
Kauri Pine, 180
  resin, 180
Kennarten (diagnostic species), 9
Kentia, 446
Keteleeria, 216
Khaya, 283, 447
Kibo-summit, *196
Kigelia, 446

Kilimanjaro, soil temperature, 201
vegetation profile, 194
see also Vegetation, altitudinal belts
Kingia, *227
Kleinia, 390, 395, 396, 397
Kniphofia, 196
Kochia, 415, 419, 429
georgeii, 420
glomerata, 428
indica, 484
lanosa, 420
planifolia, 418
pyramidata, 418, 428
sedifolia, 409, 410, 418, 420
tormentosa, 418
triptera, 418
vestita, 422
Koeleria, 6, 196, 198

Lacustrine deposits, 11
Lakes, silting up of, 11
sodium carbonates, 146
Landslide scars, 11
types, 240
Lantana, 173
Larrea, 305, 335
tridentata, 305, 315, *328, 330
distribution, 330
osmotic values, 331, 332
Larix, 32, 209
Lasiurus, 467
Latania, 446
Laterisation, 89
Lathraea, 5
Launaea, 448, 468, 489, 504
Laurus, 96, 515
Lavandula, 459, 481, 482, 495
Leaf-area index, 21ff
Leaf-fall, annual, ecological significance, 207
Leaf-shoots, nodding, 102
Leaf-size in tropical rain forest, 104
Lecanora, 464, 495
Lecidea, 464
Leitpflanzen (indicator species), 9
Lemaireocereus, *316, 327
Lemna, 147, 150, 445
Lepidium, 409
Leptadenia, 260
Lianas in tropical rain forest, 123
Lichens, 6, 130, 137, 166, 178, 288, 345, 381, 436, 495
Life forms and competitive ability, 28ff
Light requirement, 26
Lime-crusts, 452
Limestone, 513
Limnanthemum, 147
Limnophyton, 148
Limoniastrum, 449, 450, 451
Limonium, 466, 488, 493
Linasia, 468
Lindsaea, 431
Lippia, 305, 335
Liriodendron, 115
Lissochilus, 121
Lithocrapus, 192, 215
Lithophytes, 513
Lithops, 345, 353, *355, 357, 390, 393, *394, *395, 397
ecological investigation, 395
Liverworts, 137-8, 392
Livistona, 431, 446
Loasa, 382
Lobelia, 121, *185, 197, *197, 205
Loiseleuria, 204
Loma vegetation, 375, 381
Lonchitis, 121
Lonchocaprus, 215, 261
Lophira, 102
Lophocereus, 316, 327
Loranthus, 139, 156, 166, 425
Lotononis, 443
Lotus, 490
Loudetia, 232
Lower alpine belt, *197
Lumnitzera, 156
Lupinus, *186
Lycopodium, 137, 195, 196
Lychnis, 5
Lycium, 306, 308, 368, 385, 388, 392, 428, *435, 467, 468, 482, 488, 501, 502, 503
Lygeum, *441, 454, 455, 489, 514, 515
Lyperia, 365, 366

Maba, 226
Macaranga, 100, 106, 109
Machaerium, 447
Maclura, 447
Macrozamia, 430
Maerua, 260, 442, 445, 448, 510
Majanthemum, 27
Mallephora, 397
Malva, 311, 313, 459, 468, 471, 483
Mamba Station, 201
Mammillaria, 317, 327
Mangifera, 114, 446
Mango, 446
Mangrove, 150

competitive capacity, 161
distribution, 151
environmental factors, 151
forests, 150, 151
forest zonation, 162
leaf cross-section, 157
osmotic values, 163
pneumatophores, 164
river-mouth, 164ff
salt-excreting or non-salt-excreting, 157
trees, 151
water balance, 155
zonation, 153
Manihot, 119, 122, 446
Manioc, 446
Maranta, 119
Marattia, 85, 120, 121
Markhamia, 446
Marrubium, 313
Marsdenia, 431
Marsilia, 409
Martynia, 311
Maryuat lake, 484
Massonia, 395
Matthiola, 468, 483, 488
Maxillaria, 135, 136
Mbuga, 232
Medicago, 383, 445
Medinilla, 182
Mediterranean zone, *510, 515
Melaleuca, 172, 428, 430, 447
Melampyrum op., 5
Melastoma, 173
Melia, 226, 447
Melianthus, 393
Melilotus, 383
Melon holes, 427
Mentha, 496
Mercurialis, 5
Mersa Matruh, 485
Mesembryanthemun, 256, 278, 350, 388, 394, 448, 471, 476, 477
crystallinum, 489
ferox, 397
forskalei, 468, 483
hamatun, 395, 397
nodiflorum, 489
salicornioides, 346 347, *348, *349
saxicolum, 397
cf. spinosum, 393, 397
unidens, 395, 397
Metrodiseros, 126, 180, 181
Microcycas, 170
Microlepis, 121
Milium, 5
Mimusops, 261, 447
Miombo, 220
Miscanthidium, 145
Mold, 371
Mollugo, 311
Moltkia, 441
Monechma, 344
Monsoon areas, 210
Monstera, 120, 122, 126
Mopane-forest, *124
Mora, 99
Moringia, 359, 496, 510
Mortonia, 332
Moshi Station, 201
Moss, 6, 130, 133, 137, 147, 166, 436
epiphytic, 195
layer, 5
peat, 5
Mountain forest, dry type, *199
Muehlenbeckia, 414
Mulga, 263, 410, 411, 422, 423, 424, 426, 430
Mulka, amount of rainfall, *406
frequency curve of rainfall, *407
Murraya, 447
Musa, 112, 119, 122, 182
Musanga, 100
Mycetanthe, 138
Myoporum, 415, 418, 419, 420
Myrianthus, 105, 109
Myrica, 233
Myriophyllum, 147
Myrmecodia, 130, 131
Myrothamnus, 43, 288, 359, *362, *363
Myrtus, 459, 516

Najas, 431
Namib fog-desert, 338ff
plant habitats, *343
precipitation, 338
vegetation and ecology, 345
Nananthus, 393, *394, *395, 396, 397
Nanophanerophytes, 31
Naras gourd, 364
Near East, plant geographical subdivisions, *509
Needle trees, frost resistant, xii
Negev desert, 497
vegetation map, 497
Nelumbium, 147, 148
Neoreglia, 135
Neottia, 5
Nepenthes, 132, 148, 169
Nephelium, 447

Nephrolepis, 431
Neptunia, 150
Nerium, 450
Nesaea, 368
Nicotiana, 112, 364, 368, 382, 393, 430
Nidorella, 277
Nipa, 165
Nitraria, 414, 456, *457, 467, 482, 489, 493, 495
Nitrogen, 21, 27, 28, 129, 415
Noaea, 291, 488, 495, 506
Nolana, 382
Nolina, 308
Nomenclature, xii
Nostoc, 345, 381, 415
Notenia, 121
Nothofagus, 180, 181, 409
Notholaena, 314, 315, 359, 495
Nothopanax, 126
Nullarbor Plain, 404
Nymphaea, 147

Oak-forests, 220
Oases, 444
  plants at, 445
Ochradenus, 482, 501, 502
Odontopus, 371
Odontospermum, 468
Oil palms, 446
Olea, 195, 388, 391, 401, 448, 487, 490, 494
Oleandra, 137
Olearia, 423
Olives, 284, 445
Olneya, *305, *330
Onions, 448
Ononis, 489, 490
Opuntia, 191, 305, *305, 307, 309, 317, *328, 385
  osmotic values, 325
  slime content, 324
  temperature, 323
  transpiration, 323
  weight loss by drying, 324
Orange Free State, vegetation distribution, *392
Orchids, epiphytic, 209, 130-1, 136
  transpiration experiments, 134
Oreocereus, 385
Origanum, 496
Oroya, 191
Oryga, 150
Osmotic pressure, determination of relative, 286
Osmotic value, 285, 286, 287
  Acacia-Anabasis community, 503
  Carnegiea, 306
  cell sap, 168
  Cercidium, 330
  desert grasses, 455
  desert plants, 449
  Encelia, 333-4
  Ferocactus, 322
  Fouquieria, 329
  halophytes in Dead Sea area, 501
  Haloxylon persicum community, 504
  Larrea tridentata, 332
  mangroves, 156, 163
  Opuntia, 325, 326
  poikilohydrous species, 315
  rock-habitat plants, 357
  Solanum, 312
  species in Wadi Hoff, 482
  species in Ras El Hikma, 490
  succulents of Karroo, 397
  summer annuals, 312
  winter ephemerals, 313
Osyris, 388, 392, 400, 401
Othonna, 353, 357
Otostegia, 496
Overheating, 452
Oxytenanthera, 261

Pachira, 446
Pachycercus, 316, 322, 326, *326, 327
Pachypodium, 359, *360, 390
Palestine desert, transpiration rates of species, *507
  vegetation map of, 511
Pallaea, 392
Palm oil, 172
Pancratium, 448, 468
Pandonus, 148, 167, 177
Panicum, 225, 260, 262, 442, 458, 466, 467, 468, 474, 476, 477, 495
Pans, 442
Papaga Indians, 318
Papaya, 119, 445
Papyrus, 148
  -swamps, *144
Paramos, 184
  vegetation, *185
    profiles, 188
Parasites, phanerogamic, 138
Parietaria, 313
Parinarium, 105
Parkinsonia, *330, 364, 371
Parsonia, 226
Pasania, 215

Paspalum, 150
Pavement, stone, *see* Stone pavement
Paxiodendron, 105
Peas, 494
Pedilanthus, 316
Peganum, 450
Pelargonium, 353, 357, 390
Pellaea, 121, 220, 314, 315, *361
Penicereus, 317
Penkzia, *389
Pennisetum, 146, 233, 234, 442, 455, 466, 468, 480, 482
Penstemon, 313
Pentachistis, 198
Pentacme, 215
Pentzia, 388, 393
Peperomia, 130, 382
Perennial grasses, water relations of, 454
  plants, distribution, 478
Pergulasia, 482
Periodicity in tropical rain forest, 114
Perioptha1mus, 165
Periploca, 448
Peru, forest types, 213
  map of, 376
  orographic situation, 375
Peruvian coast, precipitation, 376
  profile, *379
Petalidium, 344, 365
Phacelia, 313
Phanerophytes, 101
Phaseolus, 25
Philippia, 94, 196, 197, 204
Phillyrea, 516
Philodendron, 126, 170
Phlomis, 291, 496
Phoenix, 120, 122, 145, 148, 445, 448, 450, 490
Photosynthesis, xv, 21, 22, 29
  in tropical rain forest, 113
Photosynthetic products, utilisation of (Assimilathaushalt), 21
Phragmites, 23, 24, 145, 148, 366, 368, 431, 444, 445, 452, 489
Phycopeltis, 138
Phyllanthus, 150
Phyllocactus, 130
Phyllocladus, 180
Physcia, 464
Phytocoenology, *see* Plant sociology
Picea, 309
Piceetum, 26
Pineapple, 120
Pinus, 5, 16, 32, 172, 183, 192, 216, 308, 309, 490, 516
Pipes, 121
Piptadenia, *103, 105, 109
Pistacia, 442, 448, 512, 516
Pistia, 145, 148, 150
Pittosporum, 409
Pituranthos, 442, 448, 468, 473, 481, 482, 488, 489, 490
  root system, *472
Plagianthus, 428, 429
Plagiosetum, 414
Plantago, 190, 313, 314, 436, 464, 468, 483, 488
Plant associations, 8
  community definition, 4
  general considerations, 1
  sociology, 1
Plants, autotrophic, 14
  ecological water relationships in, 34ff
Plasmolytic method, 449
Platanus, 308
Platycerium, 131
Plectrachne, 254, 263, 411, 422, 424
Pleigoynium, 226
Pleomele, 119, 120
Pneumatophores, 164
Podocarpus, 180, 181, 195, 233, 234
Podonosma, 513
Podzols, 93, 94
Poikilohydrous ferns, experiments with, 398
Polycarpaea, 487
Polygonum, 148, 150
Polylepis, 191, 192, 385
Polypodium, 130, 133, 137
Polyscias, 104, 106, 226
Pomegranates, 445
Populus, 102, 307, 333, 448
Porcupine grasses, 411, 422
  see also Spinifex grasses
Portulacaria, 390
Potamogeton, 147, 445
Poterium, 291, 513
Prairies, 6
Precipitation and soil moisture, *381
  distribution of, 49-50
  map, 50
  zones, oscillation, 47
Primula, 116, 183, 496
Production of dry matter, 21ff
Prosopis, 283, 305, 306, 307, 308, 333, 385, 445, 501
Protea, 196, 197, 233, 234
Prothalli, 137
Protoplasm, hydrature of, 285
Protosiphon, 345

Primus, 96, 181
Pseudo-oolites, 484
Pseudospondia, *127
Pseudostsuga, 308, 309
Psidium, 173, 494
Psila, 383
Psilocaulon, 393, 398
Psoralea, 408
Pteridium, 92, 120, 121, 172, 196
Pteris, 121
Pterocarpus, 220, 261
Pteronia, 388, *390, 393
Ptitotus, 414
Puna, 189, 383
  distribution, 189
Punica, 447
Pycnophyllum, 190, 385
Pyrenacantha, 222
Pyrethrum, 116, 496
Pyrophytes, 252
Pyrus, 126

Quartz, 289, 345, 415, 440
Queds, 434, 441
Quercus, 31, 192, 215, 216
  aegilops, 512
  coccifera var. *calliprinos*, 513, 516
  emoryi, 308
  hypoleuca, 308
  induta, 183
  infectoria, 512
  ithaburensis, 512
  macrolepis, 512
  oblongifolia, 308
  palustris, 19
  pedunculata, 115
  petraea, 4, 13
  pseudomoluccana, 183
  reticulata, 308
  robur, 4, 13
  rubra, 19
Quesnelia, 136
Quinine, 172

Radioactivity, 28
Rafflesia, 101, 138
Rain forests, tropical, 72ff
  of higher altitudes, 177ff
  subtropical, *see* Subtropical rain forests
Rain water, sodium chloride content, 295
Rainfall amount and distribution at Mulka, 406
  distribution in S.W. Africa, *272
  penetration, 280
Ramalina, 464
Randia, 447
Randonia, 439
Ranunculus, 5
Raphia, 148
Raphis, 446
Ras El Hikma, dewfall, *486
Rattans, 123
Ravenala, 119
Reaumuria, 464, 466, 495, 497, 498, 505, 506
Reed swamp, 489
Reg, 434, 438
Relative humidity, daily march, 77
  with height at Tjibodus, 86
Reseda, 436
Reserves, virgin forest, 171
Respiration, 21
  anaerobic, 162
Retama, 448, 452, 467, 468, 482, 485, 498, 502, 503, 505
Rhaetama, 441
Rhagodia, 414
Rhamnus, 512
Rhigozum, 244, 257, 258, *258, 401
Rhipsalis, 128, 132, 135, 136
Rhizophora, 153, *154, 159, *160, 161, 162, 163, *163, 164, 165
Rhododendron, 169, 182, 183, 192
Rhus, 364, 368, 392, 401, 458
Rhynchospora, 148
Ribes, 209
Riccia, 392
Ricciocarpus, 150
Rice, 177
Ricinus, 364, 431, 445, 494
  distribution, *494
Rivier-water, 361
Riviers, 361, 364, 366
  salt relations of, 366
Root excretions, growth-inhibiting, 14
  profile in Sahara, *453
Roots, buttressed, 102
  stilt, 102
Roplea, 224
Rosa, 209, 496
Rosette-plants, 6
Royena, 388, 392, 393
Roystonia, 446
Rubber, 172
Rubus, 27, 123
Rulingia, 431
Rumex, 313, 383, 459, 466
Run-off, 278

in deserts, 473
Ruschia, *390

Sabal, 446
Saccharum, 119, 122
see also Sugar cane
Sagittaria, 148
Sahara, 433ff
biology of plants, 447
eastern, map of, 461
evaporation, 440
map of, 433
rainfall, 434
relationships of vegetation in Tibesti mountains, 457
root profile, *453
soil and vegetation relations, 437
Saintpaulia, 122
Salares, *see* salt pans
Salicornia, 16, 162, 368, 385, 428, 456, 490, 493
Salinisation of soils at oases, 445
Salix, 209, 307, 333, 388, 393
Salsola, 256, 365, 391, 393, 397, 398, 436, 442, 444, 505
inermis, 498, 500
kali, 408, 409, 414, 418
tetrandra, 489, 501, 508
villosa, 506
zeyheri, 368, 369
Salt accumulation in Swalop valley, 294
of soils in arid areas, 295-6
Salt-bush, 263
Salt-crust, 294, 444, 456
Salt springs, 268
Salt deposition in erosion channel, 363
Salt dust, 294
Salt factor, 293
Salt lakes, 434, 444, 493
Salt pans, 275, 295, 364, 383, 410, 412, 419, 434, 443, 455
Salt sheets, 438, 443
Salvadora, 260, 364, 448, 510
Salvia, 399, 459, 477, 483, 488
Salvinia, 147, 150
Samolus, 162
Sambucus, 181, 209, 307
Sanchezia, 446
Sandstorms, 477
Sand-veld, 340
Sanguisorbia, 23, 24
Sanicula, 5
Sanseveria, *356, 357
Santalum, 447
Saprophytes, 5, 138
Saraca, 446
Sarcocaulon, 353, 354, 357, 390
Sarcostemma, 290, 431
Sasha-protection forest profile, 80
Savannah, 6, 238ff
anthropogenic, 334
competition equilibrium, 248, 262
definition, 238
dry, 234
edaphically-determined vegetation, 256
grasses, 242
moist, 212, 234
natural, 238, 241
palm-, 145
secondary, 214, 239
soils of, 240
termite, 228, 233
thorn-scrub, 239
woodland, *see* Tropical dry-woodland
woody plants, 244
Scaevola, 428
Schematic representation of transition from grassland, 250
Schima, 182
Schinus, 490
Schismus, 468, 474
Schizolobium, 100, 117
Schizotrix, 383
Schmidtia, *242
Schoenia, 414
Schotts, 434, 442, 443
Scirpus, 146, 445, 496
Sclerocarya, 261
Scopoletin, 14
Scrophularia, 482
Scrub, invasive, 252
Sea-water aerosol, 352
Sebchas, 434, 442, 443
Secondary forests, *see* Forests, secondary
Sedum, 156, 495
Sedumvivum, 136
Selaginella, 84, 97, 118, 121, 177, 195, 216, 288, 305, 314, 315, *315
Semi-desert, cheropodiaceous dwarf-shrub, 407
salt-bush, 415
Sempervivum, 136
Senecio, 197-8, *198, 203, 204, 205, 313, 353, *356, 357, 397, 436, 459
Sequoia, 31
Serir, 434, 438
Sesuvium, 166, 167, 353
Setaria, 145, 216
Shade-leaves, 23

Shifting cultivation, 173
Shorea, 117, 215
Shrub-grasses, 243
Shrub layer of tropical rain forest, 117
Sicyos, 313
Sida, 408, 414
Silene, 459, 489, 497
Simmondsia, 331
Simpson deserts, 412
Sinai peninsula, ecological situations, 493
Sinapis alba, 24-5, 26
Sisal-agaves, 229
Sisymbrium, 459
Site factors, 17
Sloanea, 99
Smilax, 126, 516
Sodium carbonate, 145, 366, 427
Sodium chloride, 293
  content of rain water, 295
Soil, chernozem, 256
  depth in rock outcrop areas, 11
  Gilgay, 427
  hard pan, 412
  mechanical composition, 90
  moisture, *380
  moisture and precipitation, *381
  non-saline, 285
  nutrient content of, 12
  saline, 266, 268
  salinity of, 11, 275
  solonetz, 295, 393
  temperature on Kilimanjaro, 201
  texture in arid regions, 278
  water content near Cairo, 469, 470
  zones of the world, 57ff
Solanum, 311, 312, *312, 382, 383, 510
Soncha, 382
Sonchus, 311, 313, 383
Sonneratia, *154, 161, 162, 164, 165
Sonoran desert, 299ff
  climate and habitat conditions, 299
  ecological plant types, 310
  map of, 300
  vegetation classification, 304
Sophora, 447
Sparmannia, 112
Spathelia, 170
Spathodea, 113, *113, 114, 446
Sphagnum, 5, 15, 94, 145, 148, 199
Sphaeralcea, 313
Sphenopus, 489
Spinifex, 167, 411
Spinifex, 410, 411, 430
  grasses, 263
  grassland, 422
Spirogyra, 37
Spondias, 446
Sporobolus, 146, 166-7, 366, 385, 392, 393, 487
Stachietum, 466
Stachys, 382, 383, 466, 481, 482, 513
Stapelia, 390
Statice, 449, 450
Stelechocarpus, 111, 112
Stellaria, 382
Stenocarpus, 447
Stenohydrous species, 292
Stenomesson, 382
Steppe, 238
  forest, 241
  zone, 12
Sterculia, 261, 283, 361, 447
Stereospermum, 261
Stipa, 186, 191, 412, 415, 455, 459, 488, 510, 514, 515
Stipagrostis, *352
Stomata, number on leaves, 122
Stomatium, 397
Stone pavement, 410, 413, 437, 464
Strand vegetation, 166ff
Stratification of vegetation layers, 6
Streptanthus, 313
Stretitzia, 446
Strobilanthes, 117, 183
Strychnos, 226, 261
Suaeda, 306, 365, 444, 493
  asphaltica, 505
  atramentifera, 393, 398
  fruticosa, 368, 369, 489, 490, 515
  palaestina, 505
  pruinosa, 489
  vermiculata, 450, 496
Subtropical rain forests, evergreen, 179
Succession, 10ff
  causes of, 10
  secondary, 11
Succulents, 251, 290, 388
  desert, xii
  halophilous, 346
  non-extreme, 136
Sudan, map of, 260
Sudd, 145
Sugar cane, 112, 172, 177
Sun-leaves, 23
Sunshine in Costa Rica, 81
Survival during drought, 504
Sweetia, *218
Swietania, 447
Symplocos, 216
Synadenium, 121

Synusia, 1
Syzgum, 233

Tabebuia, 446
Tabernaemontana, 109, 446
Taeniophyllum, 138
Tagetes, 393
Tamarindus, 261, 283, 446
Tamarix, 245, *259, 281, 364, 365, 371, 391, 444, 448, 467, 478, 493, 494, 506
  aphylla, 442, 508
  articulata, 281, 368, 441, 442, 467, 496
  gallica, *435, 445
  mannifera, 495
  nilotica, 445
  tetragyna, 496
Tanga Station, 201
Taraxacum, 5
Tarchonanthus, 400, 401
Taro, 119
Tavaresia, 356
Tea, 172
Teak forests, *see* Forests, teak
Tectona, 115, 172, 215, 447
Temperate plants, problems of, xvii
Temperature and relative humidity at Tjibodas, 83
Temperatures at different altitudes, 184
  of high-altitude stations, 187
Tephrocactus, 191
Tephrosia, 436
Terminalia, 167, 168, 214-5, 216, 285, 353, 446
Termites, 229
  see also Tropical rain forest
Termite savannah, *229
  see also Savannah
Terra firme, 149
Testudinaria, 390
Teucrium, 291, 431, 459, 495
Thailand, vegetation, 216
Thea, 109
Themeda, 217, 233, 234, 235, 392, 430
Theobroma, 110
Therophytes, 443
Thesium, 393
Thespesia, 168
Thevetia, 446
Thicket formation, 255
Thickets, thorn-scrub, *see* Tropical dry-woodland
Thorn scrub thickets, *see* Tropical dry-woodland
Thrinax, 170, 446
Thymus, 291, 495, 496
Thyrmelaea, 483, 484, 488, 490, 493, 494
Tibesti mountains, 457
  climate, 458
Tidal levels and mangrove distribution, 152
Tillandsia, *131, 137, 191, 377
  water balance, *378
Titanopsis, 390
Tjibodas, 75, 84, 86, 115, 182
Tournefortia, 167
Trachelospermum, 446
Tradescantia, 122
Traganum, 435, 444, 450
Tragopogon pratensis, 5
Transition in communities with altitude, 2-3
Transpiration in the tropics, 108; *see also* Tropical rain forests
  cuticular, 110
  of facultative shade plants, 119
  of obligatory shade plants, 119
  of tree species, 108
  of sclerophyllous woody plants, 400
  reductions, 455
Trapa, 147
Traveller's tree, 119
Tree layer, 5
Tree savannah, *241
Trees, mangrove, *see* Mangrove trees
Tree-seedling growth, 28
Trema, 226
Trentepohlia, 35, 137, 382
Trianthena, 311, 312
Tribulus, 311, 312, 364, 468
Trichocaulon, 353, *354, 356, 357, 390
Trichodesma, 414
Trientalis, 27
Trifolium, 22, 445
Triglochin, 162
Trigonella, 464, 468, 471, 483
Triodia, 254, 263, 407, 410, 411, 414, 422, 423, 424, 426, 430
  ecology of, 422
Tripteris, 365, 388, 393
Triticum, 24-5
Tropical alpine vegetation of Andes, 183
  temperature measurements, 184
Tropical dry-woodland and thorn-scrub thickets, 218
  age of trees, 220-2
  Australia, 226
  East Africa, 220
  precipitation, 222
  South America, 222

Tropical forests, clearing, 211-12
  herb layer, 215
  lianas and epiphytes, 215
  rain-green, 215
  rainfall, 214
  semi-evergreen and deciduous, 207ff, 213
  shrub layer, 215
  structure, 213
  tree strata, 215
Tropical grassland, 230
  big-game animals, 235
  ecological explanation, 233
  fire in, 234-6
  tree species, 233
Tropical parkland areas, 227
  in Africa, 228
  in Surinam, 230
Tropical rain forest, *129
  altitudinal levels, 79
  analysis of area, 182
  annual precipitation 72
  climate, 74
  distribution, 73, 74
  epiphytes, 128
  ecology of trees, 102
  evergreen, 171
  hemi-epiphytes, 126
  herb layer, 118
  leaf size, 104
  leaf-area index, 217
  lianas, 123
  microclimate, 82
  montane, 182
  periodicity, 114
  photosynthesis, 113
  regeneration, 100
  relative humidity, 85
  shrub layer, 117
  soil content, 93
  soil relations, 89
  stratification, 96
  structure, 95
  subalpine, 178
  termites, 101
  timber yields, 111
  transpiration, 108
  tree, *103
Tropics, cultivated landscape, 171
  humid, vegetation types, 144ff
Tucson, drought in, 300
  weather in, 300
Tundra, 63
Turgor pressure, 287
  negative, 424, 425
Turraca, 226
Tussilago, 204
'Tussock' grasses, 186
Typha, 148, 431, 444, 445
Typhoons, *see* hurricanes

Uapaca, 214, 233
Ulmus, 102
Urtica, 382
Usnea, 195, 215, 464
Utricularia, 137, 148, 445

Vegetation distribution in Orange Free State, *392
Vaccinium, 183, 192
  myrtillus, 26
Vanilla, 123
Varthenia, 513
Varzea, 149
Vaucheria, 37
Vegetation-free flat, *161
Vegetation, altitudinal belts on Kilimanjaro, 194
  azonal, 12
  distribution on cone-shaped mountain, 170
  distribution on Ruwenzori, 200
  of dry habitats, 169
  ecological investigations, 199
  evaporation, 203
  extrazonal, 12
  rainfall, 202
  swamp and aquatic, 144
  three-dimensional classification, 62
  temperature relations, 202
  zonal, 12
Vegetation profile of world, 65
  in S.W. Africa, 257
  of erosion channel, 246
  of Tsub Rivier, 247
Vegetation zones, asymmetrical relation, 64
Velamina, air-containing, 134
Veld, 388
Ventilago, 430
Verbena, 313
Veronica, 496
Verrucavia, 464
Veteviria, 145
Victoria, 147, 150
Vicia, 383, 445, 448
  sativa, 24
Vines, 445
Viniculture, 275